Civil and Environmental Engineering

A Series of Reference Books and Textbooks

Editor

Michael D. Meyer

Department of Civil and Environmental Engineering
Georgia Institute of Technology
Atlanta, Georgia

1. Preliminary Design of Bridges for Architects and Engineers, *Michele Melaragno*
2. Concrete Formwork Systems, *Awad S. Hanna*
. Multilayered Aquifer Systems: Fundamentals and Applications, *Alexander H.-D. Cheng*
Matrix Analysis of Structural Dynamics: Applications and Earthquake Engineering, *Franklin Y. Cheng*
Hazardous Gases Underground: Applications to Tunnel Engineering, *Barry R. Doyle*
old-Formed Steel Structures to the AISI Specification, *Gregory J. Hancock, Thomas M. Murray, and Duane S. Ellifritt*
ndamentals of Infrastructure Engineering: Civil Engineering Systems: ond Edition, Revised and Expanded, *Patrick H. McDonald*
dbook of Pollution Control and Waste Minimization, *Abbas Ghassemi*
duction to Approximate Solution Techniques, Numerical Modeling, inite Element Methods, *Victor N. Kaliakin*
chnical Engineering: Principles and Practices of Soil Mechanics undation Engineering, *V. N. S. Murthy*
ing Building Costs, *Calin M. Popescu, Kan Phaobunjong, tapong Ovararin*
l Grouting and Soil Stabilization: Third Edition, Revised and Expanded, *. Karol*
tional Cement-Based Materials, *Deborah D. L. Chung*
l Soil Engineering: Advances in Research and Practice, *Hoe I. Ling, insky, and Fumio Tatsuoka*
duling Handbook, *Jonathan F. Hutchings*
al Pollution Control Microbiology, *Ross E. McKinney*
Spillways and Energy Dissipators, *R. M. Khatsuria*
thquake Resistant Buildings: Structural Analysis and Design, anath
ater Treatment Systems, *Ronald W. Crites, Joe Middlebrooks, . Reed*

Natural
Wastewate
Treatment
System

4.

5.

6. C
 T

7. Fu
 Sec

8. Han

9. Intro
 and F

10. Geote
 and Fo

11. Estima
 and Nu

12. Chemica
 Reuben

13. Multifunc

14. Reinforce
 Dov Leshc

15. Project Sch

16. Environmen

17. Hydraulics o

18. Wind and Ea
 Bungale S. Ta

19. Natural Wastev
 and Sherwood

Natural Wastewater Treatment Systems

Ronald W. Crites
E. Joe Middlebrooks
Sherwood C. Reed

Taylor & Francis
Taylor & Francis Group
Boca Raton   London   New York

A CRC title, part of the Taylor & Francis imprint, a member of the
Taylor & Francis Group, the academic division of T&F Informa plc.

Published in 2006 by
CRC Press
Taylor & Francis Group
6000 Broken Sound Parkway NW, Suite 300
Boca Raton, FL 33487-2742

International Standard Book Number-10: 0-8493-3804-2 (Hardcover)
International Standard Book Number-13: 978-0-8493-3804-5 (Hardcover)
Library of Congress Card Number 2005041840

Library of Congress Cataloging-in-Publication Data

Reed, Sherwood C.
    Natural wastewater treatment systems / Sherwood C. Reed, Ronald W. Crites, E. Joe Middlebrooks.
        p. cm.
    Includes bibliographical references and index.
    ISBN 0-8493-3804-2 (alk. paper)
    1. Sewage--Purification--Biological treatment. 2. Sewage sludge--Management. I. Crites, Ronald W. II. Middlebrooks, E. Joe. III. Title.

TD755.R44 2005
628.3--dc22   2005041840

Taylor & Francis Group
is the Academic Division of T&F Informa plc.

Visit the Taylor & Francis Web site at
http://www.taylorandfrancis.com

and the CRC Press Web site at
http://www.crcpress.com

Dedication

We dedicate this book to the memory of Sherwood C. "Woody" Reed. Woody was the inspiration for this book and spent his wastewater engineering career planning, designing, evaluating, reviewing, teaching, and advancing the technology and understanding of natural wastewater treatment systems. Woody was the senior author of *Natural Systems for Waste Management and Treatment*, published in 1988, which introduced a rational basis for design of free water surface and subsurface flow constructed wetlands, reed beds for sludge treatment, and freezing for sludge dewatering. Woody passed away in 2003.

Preface

Natural systems for the treatment and management of municipal and industrial wastewaters and residuals feature processes that use minimal energy and minimal or no chemicals, and they produce relatively lower amounts of residual solids. This book is intended for the practicing engineers and scientists who are involved in the planning, design, construction, evaluation, and operation of wastewater management facilities.

The focus of the text is on wastewater management processes that provide passive treatment with a minimum of mechanical elements. Use of these natural systems often results in sustainable systems because of the low operating requirements and a minimum of biosolids production. Natural systems such as wetlands, sprinkler or drip irrigation, and groundwater recharge also result in water recycling and reuse.

The book is organized into ten chapters. The first three chapters introduce the planning procedures and treatment mechanisms responsible for treatment in ponds, wetlands, land applications, and soil absorption systems. Design criteria and methods of pond treatment and pond effluent upgrading are presented in Chapter 4 and Chapter 5. Constructed wetlands design procedures, process applications, and treatment performance data are described in Chapter 6 and Chapter 7. Land treatment concepts and design equations are described in Chapter 8. Residuals and biosolids management are presented in Chapter 9. A discussion of on-site wastewater management, including nitrogen removal pretreatment methods, is presented in Chapter 10. In all chapters, U.S. customary and metric units are used.

About the Authors

Ronald W. Crites is an Associate with Brown and Caldwell in Davis, California. As the Natural Systems Service Leader, he consults on land treatment, water recycling and reuse, constructed wetlands, biosolids land application, decentralized wastewater treatment, and industrial wastewater land application systems. He received his B.S. degree in Civil Engineering from California State University in Chico and his M.S. and Engineer's degree in Sanitary Engineering from Stanford University. He has 35 years of experience in wastewater treatment and reuse experience. He has authored or coauthored over 150 technical publications, including six textbooks. He is a registered civil engineer in California, Hawaii, and Oregon.

E. Joe Middlebrooks is a consulting environmental engineer in Lafayette, Colorado. He has been a college professor, a college administrator, researcher, and consultant. He received his B.S. and M.S. degrees in Civil Engineering from the University of Florida and his Ph.D. in Civil Engineering from Mississippi State University. He has authored or coauthored 12 books and over 240 articles. He has received numerous awards and is an internationally known expert in treatment pond systems.

Sherwood C. Reed (1932–2003) was an environmental engineer who was a leader in the planning and design of constructed wetlands and land treatment systems. He was the principal of Environmental Engineering Consultants (E.E.C.). He was a graduate of the University of Virginia (B.S.C.E., 1959) and the University of Alaska (M.S., 1968) and had a distinguished career with the U.S. Army Corps of Engineers, during which he spent most of his time at the Cold Regions Research and Engineering Laboratory (CRREL) in Hanover, New Hampshire, where he retired after an extended period of service from 1962 to 1989. His peers voted him into the CRREL Hall of Fame in 1991. After his retirement, he continued to teach, write, and accept both private and public sector consulting assignments. He was the author of four textbooks and over 100 technical articles.

Table of Contents

Chapter 1   Natural Waste Treatment Systems: An Overview 1

1.1   Natural Treatment Processes 1
    1.1.1   Background 1
    1.1.2   Wastewater Treatment Concepts and Performance Expectations 2
        1.1.2.1   Aquatic Treatment Units...2
        1.1.2.2   Wetland Treatment Units 5
        1.1.2.3   Terrestrial Treatment Methods...5
        1.1.2.4   Sludge Management Concepts 8
        1.1.2.5   Costs and Energy 8
1.2   Project Development...9
References 10

Chapter 2   Planning, Feasibility Assessment, and Site Selection 11

2.1   Concept Evaluation...11
    2.1.1   Information Needs and Sources..12
    2.1.2   Land Area Required 14
        2.1.2.1   Treatment Ponds 14
        2.1.2.2   Free Water Surface Constructed Wetlands 15
        2.1.2.3   Subsurface Flow Constructed Wetlands 16
        2.1.2.4   Overland Flow Systems 16
        2.1.2.5   Slow-Rate Systems 17
        2.1.2.6   Soil Aquifer Treatment Systems...18
        2.1.2.7   Land Area Comparison 18
        2.1.2.8   Biosolids Systems 19
2.2   Site Identification..19
    2.2.1   Site Screening Procedure 20
    2.2.2   Climate 26
    2.2.3   Flood Hazard...26
    2.2.4   Water Rights..27
2.3   Site Evaluation 28
    2.3.1   Soils Investigation 28
        2.3.1.1   Soil Texture and Structure 30
        2.3.1.2   Soil Chemistry 30
    2.3.2   Infiltration and Permeability 33
        2.3.2.1   Saturated Permeability 33
        2.3.2.2   Infiltration Capacity 35
        2.3.2.3   Porosity 35
        2.3.2.4   Specific Yield and Specific Retention 35
        2.3.2.5   Field Tests for Infiltration Rate 36

2.3.3   Subsurface Permeability and Groundwater Flow..............................39

       2.3.3.1   Buffer Zones 40

2.4   Site and Process Selection..41

References 41

Chapter 3   Basic Process Responses and Interactions 43

3.1   Water Management...43

    3.1.1   Fundamental Relationships...43

       3.1.1.1   Permeability 44

       3.1.1.2   Groundwater Flow Velocity...............................45

       3.1.1.3   Aquifer Transmissivity......................................45

       3.1.1.4   Dispersion 45

       3.1.1.5   Retardation..46

    3.1.2   Movement of Pollutants...47

    3.1.3   Groundwater Mounding...51

    3.1.4   Underdrainage 58

3.2   Biodegradable Organics...60

    3.2.1   Removal of BOD 60

    3.2.2   Removal of Suspended Solids 61

3.3   Organic Priority Pollutants..62

    3.3.1   Removal Methods 62

       3.3.1.1   Volatilization 62

       3.3.1.2   Adsorption..65

    3.3.2   Removal Performance...69

    3.3.3   Travel Time in Soils...70

3.4   Pathogens 71

    3.4.1   Aquatic Systems...71

       3.4.1.2   Bacteria and Virus Removal 71

    3.4.2   Wetland Systems 73

    3.4.3   Land Treatment Systems..75

       3.4.3.1   Ground Surface Aspects 75

       3.4.3.2   Groundwater Contamination..............................75

    3.4.4   Sludge Systems 76

    3.4.5   Aerosols 77

3.5   Metals..81

    3.5.1   Aquatic Systems...82

    3.5.2   Wetland Systems 84

    3.5.3   Land Treatment Systems..84

3.6   Nutrients 86

    3.6.1   Nitrogen..86

       3.6.1.1   Pond Systems 87

       3.6.1.2   Aquatic Systems...87

       3.6.1.3   Wetland Systems 88

       3.6.1.4   Land Treatment Systems....................................88

    3.6.2   Phosphorus 88

3.6.3   Potassium and Other Micronutrients 90
      3.6.3.1   Boron 91
      3.6.3.2   Sulfur 91
      3.6.3.3   Sodium 91
References 92

Chapter 4   Design of Wastewater Pond Systems 95

4.1   Introduction 95
    4.1.1   Trends 95
4.2   Facultative Ponds 96
    4.2.1   Areal Loading Rate Method 97
    4.2.2   Gloyna Method 99
    4.2.3   Complete-Mix Model 101
    4.2.4   Plug-Flow Model 102
    4.2.5   Wehner–Wilhelm Equation 103
    4.2.6   Comparison of Facultative Pond Design Models 107
4.3   Partial-Mix Aerated Ponds 109
    4.3.1   Partial-Mix Design Model 110
        4.3.1.1   Selection of Reaction Rate Constants 111
        4.3.1.2   Influence of Number of Cells 111
        4.3.1.3   Temperature Effects 112
    4.3.2   Pond Configuration 112
    4.3.3   Mixing and Aeration 113
4.4   Complete-Mix Aerated Pond Systems 123
    4.4.1   Design Equations 124
        4.4.1.1   Selection of Reaction Rate Constants 125
        4.4.1.2   Influence of Number of Cells 125
        4.4.1.3   Temperature Effects 126
    4.4.2   Pond Configuration 126
    4.4.3   Mixing and Aeration 127
4.5   Anaerobic Ponds 133
    4.5.1   Introduction 133
    4.5.2   Design 136
4.6   Controlled Discharge Pond System 140
4.7   Complete Retention Pond System 140
4.8   Hydrograph Controlled Release 140
4.9   High-Performance Aerated Pond Systems (Rich Design) 141
    4.9.1   Performance Data 142
4.10   Proprietary Systems 144
    4.10.1   Advanced Integrated Wastewater Pond Systems® 144
        4.10.1.1   Hotchkiss, Colorado 146
        4.10.1.2   Dove Creek, Colorado 147
    4.10.2   BIOLAC® Process (Activated Sludge in Earthen Ponds) 149
        4.10.2.1   BIOLAC® Processes 154
            4.10.2.1.1   BIOLAC-R System 155
            4.10.2.1.2   BIOLAC-L System 156

           4.10.2.1.3   Wave-Oxidation© Modification.....................157
           4.10.2.1.4   Other Applications 157
       4.10.2.2 Unit Operations...159
           4.10.2.2.1   Aeration Chains and Diffuser Assemblies 159
           4.10.2.2.2   Blowers and Air Manifold..............................159
           4.10.2.2.3   Clarification and Solids Handling 159
           4.10.2.2.4   BIOLAC-L Settling Basin 160
       4.10.2.3 Performance Data...160
       4.10.2.4 Operational Problems..164
    4.10.3 LEMNA Systems 164
       4.10.3.1 Lemna Duckweed System 164
       4.10.3.2 Performance Data...165
       4.10.3.3 LemTec™ Biological Treatment Process 165
    4.10.4 Las International, Ltd..171
    4.10.5 Praxair, Inc. 172
    4.10.6 Ultrafiltration Membrane Filtration 172
4.11   Nitrogen Removal in Lagoons 172
    4.11.1 Introduction 172
    4.11.2 Facultative Systems...173
       4.11.2.1 Theoretical Considerations 176
       4.11.2.2 Design Models 178
       4.11.2.3 Applications 181
       4.11.2.4 Summary 181
    4.11.3 Aerated Lagoons 182
       4.11.3.1 Comparison of Equations 182
       4.11.3.2 Summary 187
    4.11.4 Pump Systems, Inc., Batch Study..188
    4.11.5 Commercial Products 189
       4.11.5.1 Add Solids Recycle..189
       4.11.5.2 Convert to Sequencing Batch Reactor Operation 192
       4.11.5.3 Install Biomass Carrier Elements 192
       4.11.5.4 Commercial Lagoon Nitrification Systems 193
           4.11.5.4.1   ATLAS-IS™ 193
           4.11.5.4.2   CLEAR™ Process 193
           4.11.5.4.3   Ashbrook SBR 194
           4.11.5.4.4   AquaMat® Process 194
           4.11.5.4.5   MBBR™ Process...196
       4.11.5.5 Other Process Notes...196
       4.11.5.6 Ultrafiltration Membrane Filtration 198
       4.11.5.7 BIOLAC® Process (Parkson Corporation) 198
4.12   Modified High-Performance Aerated Pond Systems
    for Nitrification and Denitrification..199
4.13   Nitrogen Removal in Ponds Coupled with Wetlands
    and Gravel Bed Nitrification Filters 199
4.14   Control of Algae and Design of Settling Basins 200
4.15   Hydraulic Control of Ponds 200
4.16   Removal of Phosphorus...201

4.16.1 Batch Chemical Treatment 202
4.16.2 Continuous-Overflow Chemical Treatment 202
References 203

Chapter 5 Pond Modifications for Polishing Effluents 211

5.1 Solids Removal Methods 211
    5.1.1 Introduction 211
    5.1.2 Intermittent Sand Filtration 211
        5.1.2.1 Summary of Performance 214
        5.1.2.2 Operating Periods 215
        5.1.2.3 Maintenance Requirements 215
        5.1.2.4 Hydraulic Loading Rates 215
        5.1.3.5 Design of Intermittent Sand Filters 215
    5.1.3 Rock Filters 227
        5.1.3.1 Performance of Rock Filters 228
        5.1.3.2 Design of Rock Filters 230
    5.1.4 Normal Granular Media Filtration 230
    5.1.5 Coagulation–Flocculation 238
    5.1.6 Dissolved-Air Flotation 239
5.2 Modifications and Additions to Typical Designs 243
    5.2.1 Controlled Discharge 243
    5.2.2 Hydrograph Controlled Release 245
    5.2.3 Complete Retention Ponds 246
    5.2.4 Autoflocculation and Phase Isolation 247
    5.2.5 Baffles and Attached Growth 247
    5.2.6 Land Application 248
    5.2.7 Macrophyte and Animal Systems 248
        5.2.7.1 Floating Plants 248
        5.2.7.2 Submerged Plants 248
        5.2.7.3 *Daphnia* and Brine Shrimp 248
        5.2.7.4 Fish 249
    5.2.8 Control of Algae Growth by Shading and Barley Straw 249
        5.2.8.1 Dyes 249
        5.2.8.2 Fabric Structures 249
        5.2.8.3 Barley Straw 249
        5.2.8.4 Lemna Systems 250
5.3 Performance Comparisons with Other Removal Methods 250
References 252

Chapter 6 Free Water Surface Constructed Wetlands 259

6.1 Process Description 259
6.2 Wetland Components 261
    6.2.1 Types of Plants 261
    6.2.2 Emergent Species 262
        6.2.2.1 Cattail 262

          6.2.2.2 Bulrush 262
          6.2.2.3 Reeds 263
          6.2.2.4 Rushes 263
          6.2.2.5 Sedges 263
     6.2.3   Submerged Species 264
     6.2.4   Floating Species 264
     6.2.5   Evapotranspiration Losses..264
     6.2.6   Oxygen Transfer...265
     6.2.7   Plant Diversity 266
     6.2.8   Plant Functions 268
     6.2.9   Soils 267
     6.2.10 Organisms...268
6.3   Performance Expectations 268
     6.3.1   BOD Removal 269
     6.3.2   Suspended Solids Removal...269
     6.3.3   Nitrogen Removal 269
     6.3.4   Phosphorus Removal 272
     6.3.5   Metals Removal..273
     6.3.6   Temperature Reduction 274
     6.3.7   Trace Organics Removal 274
     6.3.8   Pathogen Removal..275
     6.3.9   Background Concentrations 277
6.4   Potential Applications 278
     6.4.1   Municipal Wastewaters 278
     6.4.2   Commercial and Industrial Wastewaters...281
     6.4.3   Stormwater Runoff...282
     6.4.4   Combined Sewer Overflow 283
     6.4.5   Agricultural Runoff..286
     6.4.6   Livestock Wastewaters 288
     6.4.7   Food Processing Wastewater..289
     6.4.8   Landfill Leachates 289
     6.4.9   Mine Drainage..291
6.5   Planning and Design..296
     6.5.1   Site Evaluation 297
     6.5.2   Preapplication Treatment 297
     6.5.3   General Design Procedures...297
6.6   Hydraulic Design Procedures 299
6.7   Thermal Aspects 302
     6.7.1   Case 1. Free Water Surface Wetland Prior to Ice Formation...........303
     6.7.2   Case 2. Flow Under an Ice Cover 304
     6.7.3   Case 3. Free Water Surface Wetland
           and Thickness of Ice Formation 305
     6.7.4   Summary 307
6.8   Design Models and Effluent Quality Prediction 308
     6.8.1   Volumetric Model..308
          6.8.1.1 Advantages 308
          6.8.1.2 Limitations 309

	6.8.2	Areal Loading Model	309
		6.8.2.1 Advantages	309
		6.8.2.2 Limitations	309
	6.8.3	Effluent Quality Prediction	309
	6.8.4	Design Criteria	314
6.9	Physical Design and Construction		314
	6.9.1	Earthwork	314
	6.9.2	Liners	316
	6.9.3	Inlet and Outlet Structures	316
	6.9.4	Vegetation	318
6.10	Operation and Maintenance		320
	6.10.1	Vegetation Establishment	320
	6.10.2	Nuisance Animals	323
	6.10.3	Mosquito Control	323
	6.10.4	Monitoring	324
6.11	Costs		324
	6.11.1	Geotechnical Investigations	325
	6.11.2	Clearing and Grubbing	326
	6.11.3	Earthwork	326
	6.11.4	Liners	327
	6.11.5	Vegetation Establishment	327
	6.11.6	Inlet and Outlet Structures	327
	6.11.7	Piping, Equipment, and Fencing	328
	6.11.8	Miscellaneous	328
6.12	Troubleshooting		328
References			329

Chapter 7 Subsurface and Vertical Flow Constructed Wetlands 335

7.1	Hydraulics of Subsurface Flow Wetlands		335
7.2	Thermal Aspects		339
7.3	Performance Expectations		343
	7.3.1	BOD Removal	344
	7.3.2	TSS Removal	344
	7.3.3	Nitrogen Removal	344
	7.3.4	Phosphorus Removal	345
	7.3.5	Metals Removal	345
	7.3.6	Pathogen Removal	345
7.4	Design of SSF Wetlands		346
	7.4.1	BOD Removal	346
	7.4.2	TSS Removal	347
	7.4.3	Nitrogen Removal	347
		7.4.3.1 Nitrification	349
		7.4.3.2 Denitrification	351
		7.4.3.3 Total Nitrogen	352
	7.4.4	Aspect Ratio	352

7.5 Design Elements of Subsurface Flow Wetlands 353
    7.5.1 Pretreatment..353
    7.5.2 Media..353
    7.5.3 Vegetation 353
    7.5.4 Inlet Distribution 354
    7.5.5 Outlet Collection 355
7.6 Alternative Application Strategies 355
    7.6.1 Batch Flow 355
    7.6.2 Reciprocating (Alternating) Dosing (TVA) 356
7.7 Potential Applications 356
    7.7.1 Domestic Wastewater 356
    7.7.2 Landfill Leachate 357
    7.7.3 Cheese Processing Wastewater 357
    7.7.4 Airport Deicing Fluids Treatment 357
7.8 Case Study: Minoa, New York 357
7.9 Nitrification Filter Bed 360
7.10 Design of On-Site Systems 364
7.11 Vertical-Flow Wetland Beds 366
    7.11.1 Municipal Systems 368
    7.11.2 Tidal Vertical-Flow Wetlands 369
    7.11.3 Winery Wastewater..369
7.12 Construction Considerations..370
    7.12.1 Vegetation Establishment 372
7.13 Operation and Maintenance...373
7.14 Costs...373
7.15 Troubleshooting 374
References 374

Chapter 8   Land Treatment Systems..379

8.1 Types of Land Treatment Systems 379
    8.1.1 Slow-Rate Systems...379
    8.1.2 Overland Flow Systems 379
    8.1.3 Soil Aquifer Treatment Systems 382
8.2 Slow Rate Land Treatment 384
    8.2.1 Design Objectives 384
        8.2.1.1 Management Alternatives...384
    8.2.2 Preapplication Treatment 384
        8.2.2.1 Distribution System Constraints 386
        8.2.2.2 Water Quality Considerations..386
        8.2.2.3 Groundwater Protection 388
    8.2.3 Design Procedure 388
    8.2.4 Crop Selection..388
        8.2.4.1 Type 1 System Crops 388
        8.2.4.2 Type 2 System Crops 390
    8.2.5 Hydraulic Loading Rates 392
        8.2.5.1 Hydraulic Loading for Type 1 Slow-Rate Systems 390
        8.2.5.2 Hydraulic Loading for Type 2 Slow-Rate Systems 391

	8.2.6	Design Considerations	392
		8.2.6.1	Nitrogen Loading Rate	392
		8.2.6.2	Organic Loading Rate	394
		8.2.6.3	Land Requirements	394
		8.2.6.4	Storage Requirements	396
		8.2.6.5	Distribution Techniques	400
		8.2.6.6	Application Cycles	401
		8.2.6.7	Surface Runoff Control	401
		8.2.6.8	Underdrainage	401
	8.2.7	Construction Considerations	401
	8.2.8	Operation and Maintenance	402
8.3	Overland Flow Systems		402
	8.3.1	Design Objectives	402
	8.3.2	Site Selection	403
	8.3.3	Treatment Performance	403
		8.3.3.1	BOD Loading and Removal	403
		8.3.3.2	Suspended Solids Removal	403
		8.3.3.3	Nitrogen Removal	405
		8.3.3.4	Phosphorus and Heavy Metal Removal	406
		8.3.3.5	Trace Organics	406
		8.3.3.6	Pathogens	407
	8.3.4	Preapplication Treatment	407
	8.3.5	Design Criteria	407
		8.3.5.1	Application Rate	408
		8.3.5.2	Slope Length	408
		8.3.5.3	Hydraulic Loading Rate	409
		8.3.5.4	Application Period	409
	8.3.6	Design Procedure	409
		8.3.6.1	Municipal Wastewater, Secondary Treatment	409
		8.3.6.2	Industrial Wastewater, Secondary Treatment	409
	8.3.7	Design Considerations	410
		8.3.7.1	Land Requirements	410
		8.3.7.2	Storage Requirements	411
		8.3.7.3	Vegetation Selection	412
		8.3.7.4	Distribution System	412
		8.3.7.5	Runoff Collection	412
	8.3.8	Construction Considerations	412
	8.3.9	Operation and Maintenance	412
8.4	Soil Aquifer Treatment Systems		413
	8.4.1	Design Objectives	413
	8.4.2	Site Selection	413
	8.4.3	Treatment Performance	413
		8.4.3.1	BOD and TSS Removal	413
		8.4.3.2	Nitrogen Removal	413
		8.4.3.3	Phosphorus Removal	415
		8.4.3.4	Heavy Metal Removal	415
		8.4.3.5	Trace Organics	415

        8.4.3.6   Endocrine Disruptors 419
        8.4.3.7   Pathogens 420
    8.4.4   Preapplication Treatment 420
    8.4.5   Design Procedure 420
    8.4.6   Design Considerations 421
        8.4.6.1   Hydraulic Loading Rates 422
        8.4.6.2   Nitrogen Loading Rates 422
        8.4.6.3   Organic Loading Rates 423
        8.4.6.4   Land Requirements 423
        8.4.6.5   Hydraulic Loading Cycle 423
        8.4.6.6   Infiltration System Design 424
        8.4.6.7   Groundwater Mounding 424
    8.4.7   Construction Considerations 425
    8.4.8   Operation and Maintenance 426
        8.4.8.1   Cold Climate Operation 426
        8.4.8.2   System Management 425
8.5   Phytoremediation 425
8.6   Industrial Wastewater Management 427
    8.6.1   Organic Loading Rates and Oxygen Balance 427
    8.6.2   Total Acidity Loading 429
    8.6.3   Salinity...430
References 431

Chapter 9   Sludge Management and Treatment 437

9.1   Sludge Quantity and Characteristics 437
    9.1.1   Sludges from Natural Treatment Systems 440
    9.1.2   Sludges from Drinking-Water Treatment 441
9.2   Stabilization and Dewatering..442
    9.2.1   Methods for Pathogen Reduction 442
9.3   Sludge Freezing 443
    9.3.1   Effects of Freezing 443
    9.3.2   Process Requirements 443
        9.3.2.1   General Equation 444
        9.3.2.2   Design Sludge Depth 445
    9.3.3   Design Procedures...445
        9.3.3.1   Calculation Methods 446
        9.3.3.2   Effect of Thawing 446
        9.3.3.3   Preliminary Designs...446
        9.3.3.4   Design Limits 446
        9.3.3.5   Thaw Period 448
    9.3.4   Sludge Freezing Facilities and Procedures....................................448
        9.3.4.1   Effect of Snow 449
        9.3.4.2   Combined Systems...449
        9.3.4.3   Sludge Removal 449
        9.3.4.4   Sludge Quality 450

9.4 Reed Beds 450
    9.4.1 Function of Vegetation 451
    9.4.2 Design Requirements 452
    9.4.3 Performance 453
    9.4.4 Benefits 454
    9.4.5 Sludge Quality 455
9.5 Vermistabilization 456
    9.5.1 Worm Species 456
    9.5.2 Loading Criteria 456
    9.5.3 Procedures and Performance 457
    9.5.4 Sludge Quality 458
9.6 Comparison of Bed-Type Operations 458
9.7 Composting 459
9.8 Land Application and Surface Disposal of Biosolids 464
    9.8.1 Concept and Site Selection 470
    9.8.2 Process Design, Land Application 471
        9.8.2.1 Metals 473
        9.8.2.2 Phosphorus 475
        9.8.2.3 Nitrogen 476
        9.8.2.4 Calculation of Land Area 478
    9.8.3 Design of Surface Disposal Systems 482
        9.8.3.1 Design Approach 482
        9.8.3.2 Data Requirements 483
        9.8.3.3 Half-Life Determination 483
        9.8.3.4 Loading Nomenclature 486
        9.8.3.5 Site Details for Surface Disposal Systems 487
References 488

Chapter 10 On-Site Wastewater Systems 493

10.1 Types of On-Site Systems 493
10.2 Effluent Disposal and Reuse Options 494
10.3 Site Evaluation and Assessment 494
    10.3.1 Preliminary Site Evaluation 497
    10.3.2 Applicable Regulations 497
    10.3.3 Detailed Site Assessment 498
    10.3.4 Hydraulic Assimilative Capacity 499
10.4 Cumulative Areal Nitrogen Loadings 499
    10.4.1 Nitrogen Loading from Conventional Effluent Leachfields 499
    10.4.2 Cumulative Nitrogen Loadings 500
10.5 Alternative Nutrient
Removal Processes 501
    10.5.1 Nitrogen Removal 501
        10.5.1.1 Intermittent Sand Filters 501
        10.5.1.2 Recirculating Gravel Filters 502
        10.5.1.3 Septic Tank with Attached Growth Reactor 505

              10.5.1.4 RSF2 Systems 507
              10.5.1.5 Other Nitrogen Removal Methods 509
      10.5.2 Phosphorus Removal..511
10.6 Disposal of Variously Treated Effluents in Soils 511
10.7 Design Criteria for On-Site Disposal Alternatives.........................512
      10.7.1 Gravity Leachfields 512
      10.7.2 Shallow Gravity Distribution 513
      10.7.3 Pressure-Dosed Distribution 515
      10.7.4 Imported Fill Systems 516
      10.7.5 At-Grade Systems 516
      10.7.6 Mound Systems..516
      10.7.7 Artificially Drained Systems...517
      10.7.8 Constructed Wetlands..517
      10.7.9 Evapotranspiration Systems 518
10.8 Design Criteria for On-Site Reuse Alternatives............................519
      10.8.1 Drip Irrigation 519
      10.8.2 Spray Irrigation 521
      10.8.3 Graywater Systems...521
10.9 Correction of Failed Systems 521
      10.9.1 Use of Effluent Screens 521
      10.9.2 Use of Hydrogen Peroxide 522
      10.9.3 Use of Upgraded Pretreatment 522
      10.9.4 Retrofitting Failed Systems...522
      10.9.5 Long-Term Effects of Sodium on Clay Soils................522
References 523

Appendices
Appendix 1. Metric Conversion Factors (SI to U.S. Customary Units) 529
Appendix 2. Conversion Factors for Commonly Used Design Parameters 533
Appendix 3. Physical Properties of Water 535
Appendix 4. Dissolved Oxygen Solubility in Freshwater 537

Index 539

1 Natural Waste Treatment Systems: An Overview

The waste treatment systems described in this book are specifically designed to utilize natural responses to the maximum possible degree when obtaining the intended waste treatment or management goal. In most cases, this approach will result in a system that costs less to build and operate and requires less energy than mechanical treatment alternatives.

1.1 NATURAL TREATMENT PROCESSES

All waste management processes depend on natural responses, such as gravity forces for sedimentation, or on natural components, such as biological organisms; however, in the typical case these natural components are supported by an often complex array of energy-intensive mechanical equipment. The term *natural system* as used in this text is intended to describe those processes that depend primarily on their natural components to achieve the intended purpose. A natural system might typically include pumps and piping for waste conveyance but would not depend on external energy sources exclusively to maintain the major treatment responses.

1.1.1 BACKGROUND

Serious interest in natural methods for waste treatment reemerged in the United States following passage of the Clean Water Act of 1972 (PL 92-500). The primary initial response was to assume that the "zero discharge" mandate of the law could be obtained via a combination of mechanical treatment units capable of advanced wastewater treatment (AWT). In theory, any specified level of water quality could be achieved via a combination of mechanical operations; however, the energy requirements and high costs of this approach soon became apparent, and a search for alternatives commenced.

Land application of wastewater was the first "natural" technology to be rediscovered. In the 19th century it was the only acceptable method for waste treatment, but it gradually slipped from use with the invention of modern devices. Studies and research quickly established that land treatment could realize all of the goals of PL 92-500 while at the same time obtaining significant benefit from the reuse of the nutrients, other minerals, and organic matter in the wastes. Land treatment of wastewater became recognized and accepted by the engineering profession as a viable treatment concept during the decade following passage of PL 92-500, and it is now considered routinely in project planning and design.

Other "natural" concepts that have never been dropped from use include lagoon systems and land application of sludges. Wastewater lagoons model the physical and biochemical interactions that occur in natural ponds, while land application of sludges model conventional farming practices with animal manures.

Aquatic and wetland concepts are essentially new developments in the United States with respect to utilization of wastewaters and sludges. Some of these concepts provide other cost-effective waste treatment options and are, therefore, included in this text. Several sludge management techniques, including conditioning, dewatering, disposal, and reuse methods, are also covered, as they also depend on natural components and processes. The sludge management (biosolids) procedures discussed in Chapter 9 of this book are compatible with current U.S. Environmental Protection Agency (EPA) regulations and guidelines for the use or disposal of sewage sludge (40 CFR Parts 257, 403, and 503).

1.1.2 WASTEWATER TREATMENT CONCEPTS AND PERFORMANCE EXPECTATIONS

Natural systems for effective wastewater treatment are available in three major categories: aquatic, terrestrial, and wetland. All depend on natural physical and chemical responses as well as the unique biological components in each process.

1.1.2.1 Aquatic Treatment Units

The design features and performance expectations for natural aquatic treatment units are summarized in Table 1.1. In all cases, the major treatment responses are due to the biological components. Aquatic systems are further subdivided in the process design chapters to distinguish between lagoon or pond systems. Chapter 4 discusses those that depend on microbial life and the lower forms of plants and animals, in contrast to the aquatic systems covered in Chapters 6 and 7 that also utilize the higher plants and animals. In most of the pond systems listed in Table 1.1, both performance and final water quality are dependent on the algae present in the system. Algae are functionally beneficial, providing oxygen to support other biological responses, and the algal–carbonate reactions discussed in Chapter 4 are the basis for effective nitrogen removal in ponds; however, algae can be difficult to remove. When stringent limits for suspended solids are required, alternatives to facultative ponds must be considered. For this purpose, controlled discharge systems were developed in which the treated wastewater is retained until the water quality in the pond and conditions in the receiving water are mutually compatible. The hyacinth ponds listed in Table 1.1 suppress algal growth in the pond because the plant leaves shade the surface and reduce the penetration of sunlight. The other forms of vegetation and animal life used in aquatic treatment units are described in Chapter 6 and Chapter 7.

TABLE 1.1
Design Features and Expected Performance for Aquatic Treatment Units

Concepts	Treatment Goals	Climate Needs	Typical Criteria			
			Detention Time (days)	Depth (ft; m)	Organic Loading (lb/ac-d; kg/ha-d)	Effluent Characteristics (mg/L)
Oxidation pond	Secondary	Warm	10–40	3–5; 1–1.5	36–107; 40–120	BOD, 20–40 TSS, 80–140
Facultative pond	Secondary	None	25–180	5–8; 1.5–2.5	20–60; 22–67	BOD, 30–40 TSS, 40–100
Partial-mix aerated pond	Secondary, polishing	None	7–20	6.5–20; 2–6	45–180; 50–200	BOD, 30–40 TSS, 30–60
Storage and controlled-discharge ponds	Secondary, storage, polishing	None	100–200	10–16; 3–5	—[a]	BOD, 10–30 TSS, 10–40
Hyacinth ponds	Secondary	Warm	30–50	<5; <1.5	<27; <30	BOD, <30 TSS, <30
Hyacinth ponds	AWT, with secondary input	Warm	>6	<3; <1	<45; <50	BOD, <10 TSS, <10 TP, <5 TN, <5

[a] First cell in system designed as a facultative or aerated treatment unit.

Note: AWT, advanced water treatment; BOD, biological oxygen demand; TSS, total suspended solids; TP, total phosphorus; TN, total nitrogen.

Source: Data from Banks and Davis (1983), Middlebrooks et al. (1981), and USEPA (1983).

1.1.2.2 Wetland Treatment Units

Wetlands are defined as land where the water table is at (or above) the ground surface long enough to maintain saturated soil conditions and the growth of related vegetation. The capability for wastewater renovation in wetlands has been verified in a number of studies in a variety of geographical settings. Wetlands used in this manner have included preexisting natural marshes, swamps, strands, bogs, peat lands, cypress domes, and systems specially constructed for wastewater treatment.

The design features and expected performance for the three basic wetland categories are summarized in Table 1.2. A major constraint on the use of many natural marshes is the fact that they are considered part of the receiving water by most regulatory authorities. As a result, the wastewater discharged to the wetland has to meet discharge standards prior to application to the wetland. In these cases, the renovative potential of the wetland is not fully utilized.

Constructed wetland units avoid the special requirements on influent quality and can also ensure much more reliable control over the hydraulic regime in the system; therefore, they perform more reliably than natural marshes. The two types of constructed wetlands in general use include the free water surface (FWS) wetland, which is similar to a natural marsh because the water surface is exposed to the atmosphere, and a subsurface flow (SSF) wetland, where a permeable medium is used and the water level is maintained below the top of the bed. Detailed descriptions of these concepts and variations can be found in Chapters 6 and 7. Another variation of the concept used for sludge drying is described in Chapter 9.

1.1.2.3 Terrestrial Treatment Methods

Typical design features and performance expectations for the three basic terrestrial concepts are presented in Table 1.3. All three are dependent on the physical, chemical, and biological reactions on and within the soil matrix. In addition, the slow rate (SR) and overland flow (OF) methods require the presence of vegetation as a major treatment component. The slow rate process can utilize a wide range of vegetation, from trees to pastures to row-crop vegetables. As described in Chapter 8, the overland flow process depends on perennial grasses to ensure a continuous vegetated cover. The hydraulic loading rates on rapid infiltration systems, with some exceptions, are typically too high to support beneficial vegetation. All three concepts can produce high-quality effluent. In the typical case, the slow rate process can be designed to produce drinking water quality in the percolate. Reuse of the treated water is possible with all three concepts. Recovery is easiest with overland flow because it is a surface system that discharges to ditches at the toe of the treatment slopes. Most slow rate and soil aquifer treatment systems require underdrains or wells for water recovery.

Another type of terrestrial concept is on-site systems that serve single-family dwellings, schools, public facilities, and commercial operations. These typically include a preliminary treatment step followed by in-ground disposal. Chapter 10 describes these on-site concepts. Small-scale constructed wetlands used for the preliminary treatment step are described in Chapters 6 and 7.

TABLE 1.2
Design Features and Expected Performance for Three Types of Wetlands

		Typical Criteria				
Concepts	Treatment Goals	Climate Needs	Detention Time (d)	Depth (ft; m)	Organic Loading (lb/ac-d; kg/ha-d)	Effluent Characteristics (mg/L)
Natural marshes	Polishing, AWT with secondary input	Warm	10	0.6–3; 0.2–1	90; 100	BOD, 5–10 TSS, 5–15 TN, 5–10
Constructed wetlands:						
Free water surface	Secondary to AWT	None	7–15	0.33–2; 0.1–0.6	180; 200	BOD, 5–10 TSS, 5–15 TN, 5–10
Subsurface flow	Secondary to AWT	None	3–14	1–2; 0.3–0.6	535; 600	BOD, 5–10 TSS, 5–20 TN, 5–10

Note: AWT, advanced water treatment; BOD, biological oxygen demand; TSS, total suspended solids; TN, total nitrogen.

Source: Data from Banks and Davis (1983), Middlebrooks et al. (1981), and Reed et al. (1984).

TABLE 1.3
Terrestrial Treatment Units, Design Features, and Performance

Concepts	Treatment Goals	Climate Needs	Vegetation	Typical Criteria		Effluent Characteristics (mg/L)
				Area (ac; ha)[a]	Hydraulic Loading (ft/yr; m/yr)	
Slow rate	Secondary or AWT	Warmer seasons	Yes	57–700; 23–280	1.6–20; 0.5–6	BOD, <2 TSS, <2 TN, <3[b] TP, <0.1 FC, 0[c]
Soil aquifer treatment	Secondary, AWT, or groundwater recharge	None	No	7.5–57; 3–23	20–410; 6–125	BOD, 5 TSS, 2 TN, 10 TP, <1[d] FC, 10
Overland flow	Secondary, nitrogen removal	Warmer seasons	Yes	15–100; 6–40	10–66; 3–20	BOD, 10 TSS, 10[e] TN, <10

| On-site | Secondary to tertiary | None | No | Not applicable for a flow of 1 mgd (3785 m^3/d). Size of bed and performance depend on the preliminary treatment level. See Chapter 10. |

[a] For design flow of 1 mgd (3785 m^3/d).
[b] Nitrogen removal depends on type of crop and management.
[c] Number/100 mL.
[d] Measured in immediate vicinity of basin; increased removal with longer travel distance.
[e] Total suspended solids depends in part on type of wastewater applied.

Note: AWT, advanced water treatment; BOD, biological oxygen demand; FC, fecal coliform; TSS, total suspended solids; TN, total nitrogen.

TABLE 1.4
Sludge Management with Natural Methods

Concept	Description	Limitations
Freezing	A method for conditioning and dewatering sludges in the winter months in cold climates; more effective and reliable than any of the available mechanical devices; can use existing sand beds	Must have freezing weather long enough to completely freeze the design sludge layer
Compost	A procedure to further stabilize and dewater sludges, with significant pathogen kill, so fewer restrictions are placed on end use of final product	Requires a bulking agent and mechanical equipment for mixing and sorting; winter operations can be difficult in cold climates
Reed beds	Narrow trenches or beds, with sand bottom and underdrained; planted with reeds; vegetation assists water removal	Best suited in warm to moderate climates; annual harvest and disposal of vegetation are required
Land apply	Application of liquid or partially dried sludge on agricultural, forested, or reclamation land	State and federal regulations limit the annual and cumulative loading of metals, etc.

1.1.2.4 Sludge Management Concepts

The freezing, composting, and reed bed concepts listed in Table 1.4 are intended to prepare the sludge for final disposal or reuse. The freeze/thaw approach described in Chapter 9 can easily increase sludge solids content to 35% or higher almost immediately upon thawing. Composting provides for further stabilization of the sludge and a significant reduction in pathogen content as well as a reduction in moisture content. The major benefits of the reed bed approach are the possibility for multiple-year sludge applications and drying before removal is required. Solids concentrations acceptable for landfill disposal can be obtained readily. Land application of sludge is designed to utilize the nutrient content in the sludge in agricultural, forest, and reclamation projects. Typically, the unit sludge loading is designed on the basis of the nutrient requirements for the vegetation of concern. The metal content of the sludge may then limit both the unit loading and the design application period for a particular site.

1.1.2.5 Costs and Energy

Interest in natural concepts was originally based on the environmental ethic of recycle and reuse of resources wherever possible. Many of the concepts described in the previous sections do incorporate such potential; however, as more and more systems were built and operational experience accumulated it was noticed that these natural systems, when site conditions were favorable, could usually be constructed and operated at less cost and with less energy than the more popular

TABLE 1.5
Guide to Project Development

Task	Description	See Chapter
Characterize waste	Define the volume and composition of the waste to be treated	Not covered in this text; see Metcalf & Eddy (1981, 2003)
Concept feasibility	Determine which, if any, of the natural systems are compatible for the particular waste and the site conditions and requirements	2, 3
Design limits	Determine the waste constituent that controls the design	3
Process design	Pond systems	4
	Aquatic systems	5
	Wetland systems	6, 7
	Terrestrial systems	8
	Sludge management	9
	On-site systems	10
Civil and mechanical details	Collection network in the community, pump stations, transmission piping, etc.	Not covered in this text; see Metcalf & Eddy, (1981, 2003)

and more conventional mechanical technologies. Numerous comparisons have documented these cost and energy advantages (Middlebrooks et al., 1982; Reed et al., 1979). It is likely that these advantages will remain and become even stronger over the long term. In the early 1970s, for example, about 400 municipal land treatment systems were using wastewater in the United States. That number had grown to at least 1400 by the mid-1980s and had passed 2000 by the year 2000. It is further estimated that a comparable number of private industrial and commercial systems also exist. These process selection decisions have been and will continue to be made on the basis of costs and energy requirements.

1.2 PROJECT DEVELOPMENT

The development of a waste treatment project, either municipal or industrial, involves consideration of institutional and social issues in addition to the technical factors. These issues influence and can often control decisions during the planning and preliminary design stages. The current regulatory requirements at the federal, state, and local level are particularly important. The engineer must determine these requirements at the earliest possible stage of project development to ensure that the concepts under consideration are institutionally feasible. Deese (1981), Forster and Southgate (1983), and USEPA (1981) provide useful guidance on the institutional and social aspects of project development. Table 1.5 provides summary guidance on the technical requirements for project development and indicates chapters in this book that describe the required criteria. Detailed information on waste characterization and the civil and mechanical engineering details of design are not unique to natural systems and are therefore not included in this text. Metcalf and Eddy (1981, 2003) are recommended for that purpose.

REFERENCES

Banks, L. and Davis, S. (1983). Wastewater and sludge treatment by rooted aquatic plants in sand and gravel basins, in *Proceedings of a Workshop on Low Cost Wastewater Treatment*, Clemson University, Clemson, SC, April 1983, pp. 205–218.

Bastian, R.K. and Reed, S.C., Eds. (1979). *Aquaculture Systems for Wastewater Treatment*, EPA 430/9-80-006, U.S. Environmental Protection Agency, Washington, D.C.

Deese, P.L. (1981). Institutional Constraints and public acceptance barriers to utilization of municipal wastewater and sludge for land reclamation and biomass production, in *Utilization of Municipal Wastewater and Sludge for Land Reclamation and Biomass Production*, EPA 430/9-81-013, U.S. Environmental Protection Agency, Washington, D.C.

Forster, D.L. and Southgate, D.D. (1983). Institutions constraining the utilization of municipal wastewaters and sludges on land, in *Proceedings of Workshop on Utilization of Municipal Wastewater and Sludge on Land*, University of California, Riverside, February 1983, pp. 29–45.

Metcalf & Eddy (1981). *Wastewater Engineering: Collection and Pumping of Wastewater*, McGraw-Hill, New York.

Metcalf & Eddy (2003). *Wastewater Engineering: Treatment and Reuse*, 4th ed., McGraw-Hill, New York.

Middlebrooks, E.J., Middlebrooks, C.H., and Reed. S.C. (1981). Energy requirements for small wastewater treatment systems, *J. Water Pollut. Control Fed.*, 53(7), 1172–1198.

Middlebrooks, E. J., Middlebrooks, C.H., Reynolds, J.H., Watters, G.Z., Reed, S.C., and George, D.B. (1982). *Wastewater Stabilization Lagoon Design, Performance and Upgrading*, Macmillan, New York.

Reed, S.C., Crites, R.W., Thomas, R.E., and Hais, A.B. (1979). Cost of land treatment systems, EPA 430/9-75-003, U.S. Environmental Protection Agency, Washington, D.C.

Reed, S.C., Bastian, R., Black, S., and Khettry, R.K. (1984). Wetlands for wastewater treatment in cold climates, in *Proceedings of the Water Reuse III Symposium*, American Water Works Association, August 1984, Denver, CO.

USEPA (1981). *Process Design Manual for Land Treatment of Municipal Wastewater*, EPA 625/1-81-013, U.S. Environmental Protection Agency, Washington, D.C.

USEPA (1983). *Process Design Manual for Municipal Wastewater Stabilization Ponds*, EPA 625/1-83-015, U.S. Environmental Protection Agency, Washington, D.C.

USEPA (1984). *Process Design Manual for Land Treatment of Municipal Wastewater: Supplement on Rapid Infiltration and Overland Flow*, EPA 625/1-81-013a, U.S. Environmental Protection Agency, Washington, D.C.

2 Planning, Feasibility Assessment, and Site Selection

When conducting a wastewater treatment and reuse/disposal planning study, it is important to evaluate as many alternatives as possible to ensure that the most cost-effective and appropriate system is selected. For new or unsewered communities, decentralized options should also be included in the mix of alternatives (Crites and Tchobanoglous, 1998). The feasibility of the natural treatment processes that are described in this book depends significantly on site conditions, climate, regulatory requirements, and related factors. It is neither practical nor economical, however, to conduct extensive field investigations for every process, at every potential site, during planning. This chapter provides a sequential approach that first determines potential feasibility and the necessary land requirements and site conditions of each alternative. The second step evaluates each site coupled with a natural treatment process based on technical and economic factors and selects one or more for detailed investigation. The final step involves detailed field investigations (as necessary), identification of the most cost-effective alternative, and development of the criteria necessary for the final design.

2.1 CONCEPT EVALUATION

One way of categorizing the natural systems is to divide them between discharging and nondischarging systems. Discharging systems would include those with a surface water discharge, such as treatment ponds, constructed wetlands, and overland flow land treatment. Underdrained slow rate or soil aquifer treatment (SAT) systems may also have a surface water discharge that would be permitted under the National Pollutant Discharge Elimination System (NPDES). Nondischarging systems would include slow rate land treatment and SAT, onsite methods, and biosolids treatment and reuse methods. Site topography, soils, geology, and groundwater conditions are important factors for the construction of discharging systems but are often critical components of the treatment process itself for nondischarging systems. Design features and performance expectations for both types of systems are presented in Table 2.1, Table 2.2, and Table 2.3. Special site requirements are summarized in Table 2.1 and Table 2.2 for each type of system for planning purposes. It is presumed that the percolate from a nondischarging system mingles with any groundwater that may be present. The typical regulatory

TABLE 2.1
Special Site Requirements for Discharge Systems

Concept	Requirement
Treatment ponds	Proximity to a surface water for discharge, impermeable soils or liner to minimize percolation, no steep slopes, out of flood plain, no bedrock or groundwater within excavation depth
Constructed wetlands	Proximity to a surface water for discharge, impermeable soils or liner to minimize percolation, slopes 0–6%, out of flood plain, no bedrock or groundwater within excavation depth
Overland flow (OF)	Relatively impermeable soils, clay and clay loams, slopes 0–12%, depth to groundwater and bedrock not critical but 0.5–1 m desirable, must have access to surface water for discharge or point of water reuse
Underdrained slow rate (SR) and soil aquifer treatment (SAT)	For SR, same as tables in Chapter 1 and Table 2.2 except for impermeable layer or high groundwater that requires the use of underdrains to remove percolating water; for SAT, wells or underdrains may remove percolating water for discharge

requirement for compliance is the quality measured in the percolate/groundwater as it reaches the project boundary.

As noted in Table 2.1, SR and SAT systems can include surface discharge from underdrains, recovery wells, or cutoff ditches. For example, the large SR system at Muskegon County, Michigan, has underdrains with a surface water discharge. For the forested SR system at Clayton County, Georgia, the subflow from the wastewater application leaves the site and enters the local streams. Although the subflow does emerge in surface streams, which are part of the community's drinking water supplies, the land treatment system is not considered to be a discharging system as defined by the U.S. Environmental Protection Agency (EPA) and the State of Georgia.

2.1.1 INFORMATION NEEDS AND SOURCES

A preliminary determination of process feasibility and identification of potential sites are based on the analysis of maps and other information. The requirements shown in Table 2.1 and Table 2.2, along with an estimate of the land area required for each of the methods, are considered during this procedure. The sources of information and type of information needed are summarized in Table 2.3.

TABLE 2.2
Special Site Requirements for Nondischarging System

Concept	Requirement
Wastewater Systems	
Slow rate (SR)	Sandy loams to clay loams: >0.15 to <15 cm/hr permeability preferred, bedrock and groundwater >1.5 m, slopes <20%, agricultural sites <12%
Soil aquifer treatment (SAT) or rapid infiltration (RI)	Sands to sandy loams: 5 to 50 cm/hr permeability, bedrock and groundwater >5 m preferred, >3 m necessary, slopes <10%; sites with slopes that require significant backfill for basin construction should be avoided; preferred sites are near surface waters where subsurface flow may discharge over non-drinking-water aquifers
Reuse wetlands	Slowly permeable soils, slopes 0 to 6%, out of flood plain, no bedrock or groundwater within excavation depth
Biosolids Systems	
Land application	Generally the same as for agricultural or forested SR systems
Composting, freezing, vermistabilization, or reed beds	Usually sited on the same site as the wastewater treatment plant; all three require impermeable barriers to protect groundwater; freezing and reed beds also require underdrains for the percolate

TABLE 2.3
Sources of Site Planning Information

Information Source	Information Items
Topographic maps	Elevations, slope, water and drainage features, building and road locations
Natural Resources Conservation Service soil surveys	Soil type, depth and permeability, depth to bedrock, slope
Federal Emergency Management Agency (FEMA) maps	Flood hazard
Community maps	Land use, water supply, sewerage systems
National Oceanic and Atmospheric Administration (NOAA)	Climatic data
U.S. Geological Survey (USGS) reports and maps	Geologic data, water quality data

2.1.2 LAND AREA REQUIRED

The land area estimates derived in this section are used with the information in Table 2.1 and Table 2.2 to determine, with a study of the maps, whether suitable sites exist for the process under consideration. These preliminary area estimates are very conservative and are intended only for this preliminary evaluation. These estimates should not be used for the final design.

2.1.2.1 Treatment Ponds

The types of treatment ponds (described in Chapter 4) include oxidation ponds, facultative ponds, controlled-discharge ponds, partial-mix aerated ponds, complete-mix ponds, proprietary approaches, and modifications to conventional approaches. The area estimate for pond systems will depend on the effluent quality required (as defined by biochemical oxygen demand [BOD] and total suspended solids [TSS]), on the type of pond system proposed, and on the climate in the particular geographic location. A facultative pond in the southern United States will require less area than the same process in Canada. The equations given below are for total project area and include an allowance for roads, levees, and unusable portions of the site.

Oxidation Ponds
The area for an aerobic pond assumes a depth of 3 ft (1 m), a warm climate, a 30-day detention time, an organic loading rate of 80 lb/ac·d (90 kg/ha·d), and an effluent quality of 30 mg/L BOD and >30 mg/L TSS. The planning area required is calculated using Equation 2.1:

$$A_{pm} = (k)(Q) \qquad (2.1)$$

where
  A_{pm} = Total project area, (ac; ha).
  k = Factor (3.0×10^{-5}, U.S. units; 3.2×10^{-3}, metric).
  Q = Design flow (gal/d; m³/d).

Facultative Ponds in Cold Climates
The area calculation in Equation 2.2 assumes an 80-day detention time, a pond 5 ft (1.5 m) deep, an organic loading of 15 lb/ac·d (16.8 kg/ha·d), an effluent BOD of 30 mg/L, and TSS > 30 mg/L. The area required is:

$$A_{fc} = (k)(Q) \qquad (2.2)$$

where
  A_{fc} = Facultative pond site area (ac; ha).
  k = Factor (1.6×10^{-4}, U.S. units; 1.68×10^{-2}, metric).
  Q = Design flow (gal/d; m³/d).

Facultative Ponds in Warm Climates

Assume more than 60 days of detention in a pond 5 ft (1.5 m) deep and an organic loading of 50 lb/ac·d (56 kg/ha·d); the expected effluent quality is BOD = 30 mg/L and TSS > 30 mg/L. The area required is:

$$A_{fw} = (k)(Q) \tag{2.3}$$

where

A_{fw} = Facultative pond site area, warm climate (ac; ha).

k = Factor (4.8×10^{-5}, U.S. units; 5.0×10^{-3}, metric).

Q = Design flow (gal/d; m³/d).

Controlled-Discharge Ponds

Controlled-discharge ponds are used in northern climates to avoid winter discharges and in warm climates to match effluent quality to acceptable stream flow conditions. The typical depth is 5 ft (1.5 m), maximum detention time is 180 days, and the expected effluent quality is BOD < 30 mg/L and TSS < 30 mg/L. The required site area is:

$$A_{cd} = (k)(Q) \tag{2.4}$$

where

A_{cd} = Controlled-discharge pond site area (ac; ha).

k = Factor (1.32×10^{-4}, U.S. units; 1.63×10^{-2}, metric).

Q = Design flow (gal/d; m³/d).

Partial-Mix Aerated Pond

The size of the partial-mix aerated pond site will vary with the climate; for example, shorter detention times are used in warm climates. For the purpose of this chapter, assume a 50-day detention time, a depth of 8 ft (2.5 m), and an organic loading of 89 lb/ac·d (100 kg/ha·d). Expected effluent quality is BOD = 30 mg/L and TSS > 30 mg/L. The site area can be calculated using Equation 2.5:

$$A_{pm} = (k)(Q) \tag{2.5}$$

where

A_{pm} = Aerated pond site area (ac; ha).

k = Factor (2.7×10^{-3}, U.S. units; 2.9×10^{-3}, metric).

Q = Design flow (gal/d; m³/d).

2.1.2.2 Free Water Surface Constructed Wetlands

Constructed wetlands are typically designed to receive primary or secondary effluent, to produce an advanced secondary effluent, and to operate year-round in moderately cold climates. The detention time is assumed to be 7 days, the depth is 1 ft (0.3 m), and the organic loading is <89 lb/ac·d (<100 kg/ha·d). The expected effluent quality is BOD = 10 mg/L, TSS = 10 mg/L, total N < 10 mg/L (during warm weather), and P > 5 mg/L. The estimated site area given in Equation 2.6 does not include the area required for a preliminary treatment system before the wetland:

$$A_{fws} = (k)(Q) \qquad (2.6)$$

where
A_{fws} = Site area for free water surface constructed wetland (ac; ha).
k = factor (4.03×10^{-5}, U.S. units; 4.31×10^{-3}, metric).
Q = Design flow (gal/d; m³/d)

2.1.2.3 Subsurface Flow Constructed Wetlands

Subsurface flow constructed wetlands generally require less site area for the same flow than do free water surface wetlands. The assumed detention time is 3 days, the water depth is 1 ft (0.3 m), with a media depth of 1.5 ft (0.45 m); the organic loading rate is <72 lb/ac·d (<80 kg/ha·d); and the expected effluent quality is similar to the free water surface wetlands above:

$$A_{ssf} = (k)(Q) \qquad (2.7)$$

where
A_{ssf} = Site area for subsurface flow constructed wetland (ac; ha).
k = Factor (1.73×10^{-5}, U.S. units; 1.85×10^{-3}, metric).
Q = Design flow (gal/d; m³/d).

2.1.2.4 Overland Flow Systems

The area required for an overland flow (OF) site depends on the length of the operating season. The recommended storage days for an overland flow system for planning purposes can be estimated from Figure 2.1. The effective flow to the OF site can then be estimated using Equation 2.8:

$$Q_m = q + (t_s)(q)/t_a \qquad (2.8)$$

where
Q_m = Average monthly design flow to the overland flow site (gal/mo; m³/mo).
q = Average monthly flow from pretreatment (gal/mo; m³/mo).
t_s = Number of months storage is required.
t_a = Number of months in the operating season.

The OF process can produce advanced secondary effluent from a primary effluent or equivalent. The expected effluent quality is BOD = 10 mg/L, TSS = 10 mg/L, total N < 10 mg/L, and total P < 6 mg/L. The site area given by Equation 2.9

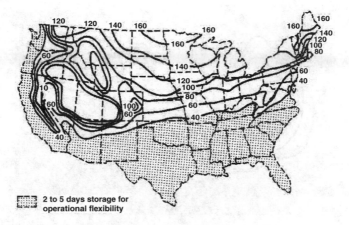

2 to 5 days storage for operational flexibility

FIGURE 2.1 Recommended storage days for overland flow systems.

includes an allowance for a 1-day aeration cell and for winter wastewater storage (if needed), as well as the actual treatment area, with an assumed hydraulic loading of 6 in./wk (15-cm/wk):

$$A_{of} = (3.9 \times 10^{-4})(Q_m + 0.05qt_s) \text{ (metric)} \tag{2.9a}$$

$$A_{of} = (3.7 \times 10^{-6})(Q_m + 0.04qt_s) \text{ (U.S.)} \tag{2.9b}$$

where

A_{of} = Overland flow project area (ac; ha).

Q_m = Average monthly design flow to the overland flow site, gal/mo (m³/mo).

q = Average monthly flow from pretreatment, gal/mo (m³/mo).

t_s = Number of months storage is required.

2.1.2.5 Slow-Rate Systems

Slow-rate (SR) systems are typically nondischarging systems. The size of the project site will depend on the operating season, the application rate, and the crop. The number of months of possible wastewater application is presented in Figure 2.2. The design flow to the SR system can be calculated from Equation 2.10. The land area will be based on either the hydraulic capacity of the soil or the nitrogen loading rate. The area estimate given in Equation 2.10 includes an allowance for preapplication treatment in an aerated pond as well as a winter storage allowance. The expected effluent (percolate) quality is BOD < 2 mg/L, TSS < 1 mg/L, total N < 10 mg/L (or lower if required), and total P < 0.1 mg/L:

$$A_{sr} = (6.0 \times 10^{-4})(Q_m + 0.03qt_s) \tag{2.10a}$$

$$A_{sr} = (5.5 \times 10^{-6})(Q_m + 0.04qt_s) \tag{2.10b}$$

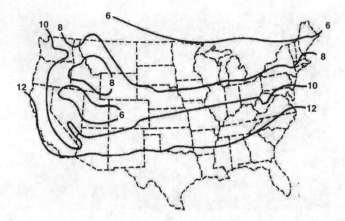

FIGURE 2.2 Approximate months per year that wastewater application is possible with slow rate land treatment systems.

where

A_{sr} = Slow rate land treatment project area (ac; ha).

Q_m = Average monthly design flow to the SR site (gal/mo; m³/mo).

q = Average monthly flow from pretreatment (gal/mo; m³/mo).

t_s = Number of months storage is required.

2.1.2.6 Soil Aquifer Treatment Systems

Typically a soil aquifer treatment (SAT) or rapid infiltration system is a nondischarging system. Year-round operation is possible in all parts of the United States so storage is not generally required. The hydraulic loading rate, which depends on the soil permeability and percolation capacity, controls the land area required. The expected percolate quality is BOD < 5 mg/L, TSS < 2 mg/L, total N > 10 mg/L, and total P < 1 mg/L:

$$A_{sat} = (k)(Q) \tag{2.11}$$

where

A_{sat} = SAT project site area (ac; ha).

k = Factor (4.8×10^{-7} U.S. units; 5.0×10^{-5}, metric).

Q_m = Average monthly design flow to the SAT site (gal/mo; m³/mo).

2.1.2.7 Land Area Comparison

The land area required for a community wastewater flow of 1 mgd (3785 m³/d) is estimated using the above equations for each of the processes and is summarized in Table 2.4. The three geographical locations in Table 2.4 reflect climate variations and the need for different amounts of storage: 5-month storage for SR and

TABLE 2.4
Planning Level Land Area Estimates for 1-mgd (3785-m³/d) Systems

Treatment System	North [ac (ha)]	Mid-Atlantic [ac (ha)]	South [ac (ha)]
Pond systems:			
Oxidation	NA	NA	30 (12.1)
Facultative	157 (63.6)	102 (41.3)	48 (19.3)
Controlled discharge	152 (61.7)	152 (61.7)	152 (61.7)
Partial-mix	48 (19.3)	36 (14.5)	27 (11.0)
Free water surface constructed wetlands	56 (22.7)	47 (19.1)	40 (16.3)
Subsurface flow constructed wetlands	24 (9.7)	20 (8.2)	17 (7.0)
Slow rate	311 (126)	240 (97)	168 (68)
Overland flow	215 (87)	160 (65)	111 (45)
Soil aquifer treatment	14 (5.7)	14 (5.7)	14 (5.7)

Note: 1 ac = 0.404 ha.

OF in the north, 3-month storage in the mid-Atlantic, and no storage in the warm climate (south). No storage is expected for constructed wetlands, but the temperature of the wastewater is reflected in the larger land area requirements in the colder north. Allowances are included in the area requirements for unusable land and preliminary treatment.

2.1.2.8   Biosolids Systems

The land area required for biosolids land application systems is summarized in Table 2.5. The actual rates depend on the climate and the biosolids characteristics, as discussed in Chapter 9.

2.2   SITE IDENTIFICATION

The information presented or developed in the previous sections is combined with maps of the community area to determine if feasible sites for wastewater treatment or biosolids land application exist within a reasonable distance. It is possible that a community or industry may not have suitable sites for all the natural system options listed in Table 2.1 and Table 2.2. All suitable sites should be located on the maps. Some options may be dropped from consideration because no suitable sites are located within a reasonable proximity from the wastewater source. In the next step, local knowledge regarding land use commitments, costs, and the technical ranking procedure (described in the next section) are considered

TABLE 2.5
Biosolids Loadings for Preliminary Site Area Determination

Option[a]	Application Schedule	Typical Loading Rate (Mg/ha)[b]
Agricultural	Annual	10
Forest	5-yr intervals	45
Reclamation	One time	100
Type B	Annual	340

[a] See Chapter 9 for a detailed description of options.
[b] Metric tons per hectare (Mg/ha) $\times$ 0.4461 = lb/ac.

to determine which processes and sites are technically feasible. A complex screening procedure is not usually required for pond and wetland systems, because close proximity and access to the point of discharge are usually most important in site selection for these systems. For land application systems for wastewater and biosolids, the economics of conveyance to the potential site may compete with the physical and land use factors described in the next section.

2.2.1 SITE SCREENING PROCEDURE

The screening procedure consists of assigning rating factors to each item for each site and then adding up the scores. Those sites with moderate to high scores are candidates for serious consideration, site investigation, and testing. Among the conditions included in the general procedure are site grades, depth of soil, depth to groundwater, and soil permeability (Table 2.6). Conditions for the wastewater treatment concepts include land use (current and future), pumping distance, and elevation (Table 2.7). The relative importance of the various conditions in Table 2.6 and Table 2.7 is reflected in the magnitude of the values assigned, so the largest value indicates the most important characteristic. The ranking for a specific site is obtained by summing the values from Table 2.6 and Table 2.7. The highest ranking site will be the most suitable. The suitability ranking can be determined according to the following ranges:

Low suitability   <18
Moderate suitability   18–34
High suitability   34–50

For land application of biosolids, a similar matrix of factors can be arrayed for each potential site using the rating factors shown in Table 2.8. The rating factors for forested sites are presented in Table 2.9 and Table 2.10. The restrictions on liquid biosolids (<7%) in Table 2.8 are intended to control runoff or erosion of

TABLE 2.6
Physical Rating Factors for Land Application of Wastewater

Condition	Slow Rate	Overland Flow	Soil Aquifer Treatment
Site grade (%)			
0–5	8	8	8
5–10	6	5	4
10–15	4	2	NS[a]
15–20	Forest only, 5	NS	NS
20–30	Forest only, 4	NS	NS
30–35	Forest only, 2	NS	NS
>35	Forest only, 0	NS	NS
Soil depth (m)[b]			
0.3–0.6	0	0	NS
0.6–1.5	3	5	NS
1.5–3.0	8	6	3
>3.0	9	7	8
Depth to groundwater (m)			
<1	0	4	NS
1 3	4	5	1
>3	6	6	6
Soil permeability of most restrictive soil layer (cm/hr)			
<0.15	1	10	NS
0.15–0.50	3	8	NS
0.50–1.50	5	6	1
1.50–5.10	8	1	6
>5.10	8	NS	9

[a] NS = not suitable.
[b] Soil depth to bedrock or impermeable barrier.

Source: Adapted from USEPA, *Onsite Wastewater Treatment Systems Manual*, EPA/625/R-00/008, CERI, Cincinnati, OH, 2002.

surface-applied biosolids. Injection of liquid biosolids is acceptable on 6 to 12% slopes but is not recommended on higher grades without effective runoff control.

The economical haul distance for a biosolids land application will depend on the solids concentration and other local factors and must be determined on a case-by-case basis. The values in Table 2.8 can be combined with the land use and land cost factors from Table 2.7 (if appropriate) to obtain an overall score for a

TABLE 2.7
Land Use and Economic Factors for Land Application of Wastewater

Condition	Rating Value
Distance from wastewater source (km)	
0–3	8
3–8	6
8–16	3
>16	1
Elevation difference from wastewater source (m)	
<0	6
0–15	5
15–60	3
>200	1
Land use, existing or planned	
Industrial	0
High density, residential or urban	0
Low density, residential or urban	1
Agricultural, or open space, for agricultural SR or OF	4
Forested:	
For forested sites	4
For agricultural SR or OF	1
Land cost and management	
No land cost, farmer or forest company management	5
Land purchased, farmer or forest company management	3
Land purchased, operated by industry or city	1

Note: SR, slow rate; OF, overland flow.

potential biosolids application site. These combinations produce the following ranges:

	Agricultural	Reclamation	Type B
Low suitability	<10	<10	<5
Moderate suitability	10–20	10–20	5–15
High suitability	20–35	20–35	15–25

TABLE 2.8
Physical Rating Factors for Land Application of Biosolids

Condition	Agricultural[a]	Reclamation
Site grade (%)		
0–3	8	8
3–6	6	7
6–12	4	6
12–15	3	5
>15	NS[b]	4
Soil depth (m)[c]		
<0.6	NS	2
0.6–1.2	3	5
>1.2	8	8
Soil permeability of the most restrictive soil layer (cm/hr)		
<0.08	1	3
0.08–0.24	3	4
0.24–0.8	5	5
0.8–2.4	3	4
>2.4	1	0
Depth to seasonal groundwater (m)		
<0.6	0	0
0.6–1.2	4	4
>1.2	6	6

[a] See Chapter 9 for a description of the processes.
[b] NS, not suitable.
[c] Soil depth to bedrock or impermeable barrier.

Source: Adapted from USEPA, *Process Design Manual: Land Application of Sewage Sludge and Domestic Septage*, EPA 625/R-95/001, CERI, Cincinnati, OH, 1995.

The transport distance is a critical factor and must be included in the final ranking. The rating values for distance given in Table 2.7 can also be used for agricultural biosolids operations. In general, it is economical to transport liquid biosolids (<7% solids) about 16 km (10 mi) from the source; for greater haul distances, it is usually more cost effective to dewater and haul the dewatered biosolids.

TABLE 2.9
Rating Factors for Biosolids or
Wastewater in Forests (Surface
Factors)

Condition	Rating Value[a]
Dominant vegetation	
Pine	2
Hardwood or mixed	3
Vegetation age (yr)	
Pine:	
>30	3
20–30	3
<20	4
Hardwood	
>50	1
30–50	2
<30	3
Mixed pine/hardwood	
>40	1
25–40	2
<25	3
Slope (%)	
>35	0
0–1	2
2–6	4
7–35	6
Distance to surface waters (m)	
15–30	1
30–60	2
>60	3
Adjacent land use	
High-density residential	1
Low-density residential	2
Industrial	2
Undeveloped	3

[a] Total rating: 3–4, not suitable; 5–6, poor; 9–14, good; >15, excellent.

Source: Adapted from Taylor, G.L., in *Proceedings of the Conference of Applied Research and Practice on Municipal and Industrial Waste*, Madison, WI, September 1980.

TABLE 2.10
Rating Factors for Biosolids or Wastewater in Forests (Subsurface Factors)

Condition	Rating Value[a]	Condition	Rating Value[a]
Depth to seasonal groundwater (m)		*NRCS shrink–swell potential for the soil*	
<1	0	High	1
1–3	4	Low	2
>10	6	Moderate	3
Depth to bedrock (m)		*Soil cation exchange capacity (mEq/100 g)*	
<1.5	0	<10	1
1.5–3.0	4	10–15	2
>3	6	>15	3
Type of bedrock		*Hydraulic conductivity of soil (cm/hr)*	
Shale	2	>15	2
Sandstone	4	<5	4
Granite–gneiss	6	5–10	6
Rock outcrops (% of total surface)		*Surface infiltration rate (cm/hr)*	
>33	0	<5	2
10–33	2	5–10	4
1–10	4	>15	6
None	6		
NRCS erosion classification			
Severely eroded	1		
Eroded	2		
Not eroded	3		

[a] Total rating: 5–10, not suitable; 15–25, poor; 25–30, good; 30–45, excellent.

Source: Adapted from Taylor, G.L., in *Proceedings of the Conference of Applied Research and Practice on Municipal and Industrial Waste*, Madison, WI, September 1980.

Forested sites for either wastewater or biosolids are presented as a separate category in Table 2.9 and Table 2.10. In the earlier cases, the type of vegetation to be used is a design decision to optimize treatment, and the appropriate vegetation is usually established during system construction. It is far more common for forested sites to depend on preexisting vegetation on the site, so the type and status of that growth become important selection factors (McKim et al., 1982). The total rating combines values from Table 2.9 and Table 2.10. The final ranking, as with other methods, must include the transport distance; the values in Table 2.7 can be used for wastewater systems.

TABLE 2.11
Climatic Influences on Land Application of Biosolids

Impact	Warm/Arid	Warm/Humid	Cold/Humid
Operating time	Year-round	Seasonal	Seasonal
Operating cost	Lower	Higher	Higher
Biosolids storage	Less	More	Most
Salt accumulation in the soil	High	Low	Moderate
Leaching potential	Low	High	Moderate
Runoff potential	Low	High	High

Source: Adapted from USEPA, *Process Design Manual: Land Application of Municipal Sludge*, EPA 625/1-83-016, CERI, Cincinnati, OH, 1983.

2.2.2 CLIMATE

The regional climate has a direct effect on the potential biosolids management options, as shown in Table 2.11. Climatic factors are not included in the rating procedure for wastewater systems, because seasonal constraints on operations are already included as a factor in the land area determinations. Seasonal constraints and the local climate are important factors in determining the design hydraulic loading rates and cycles for wastewater systems, as well as the length of the operating season and stormwater runoff conditions for all concepts. The pertinent climatic data required for the design of both wastewater and biosolids systems are listed in Table 2.12. At least a 10-year return period is recommended, although some agencies require a 100-year return period (see NOAA references).

2.2.3 FLOOD HAZARD

The location of wastewater and biosolids systems within a flood plain can be either an asset or a liability, depending on the approach used for planning and design. Flood-prone areas may be undesirable because of variable drainage characteristics and potential flood damage to the structural components of the system. On the other hand, flood plains and similar terrain may be the only deep soils in the area. If permitted by the regulatory authorities, utilization of such sites for wastewater or biosolids can be an integral part of a flood-plain management plan. Off-site storage of effluent or biosolids can be a design feature to allow the site to flood as needed.

Maps of flood-prone areas have been produced by the U.S. Geological Survey (USGS) in many areas of the United States as part of the Uniform National Program for Managing Flood Losses. The maps are based on the standard 7.5-minute USGS topographic sheets and identify areas with a potential of a 1-in-100

TABLE 2.12
Climatic Data Required for Land Application Designs

Condition	Required Data	Type of Analysis
Precipitation	As rain, as snow, annual averages, maxima, minima	Frequency, annual distribution
Storm events	Intensity, duration	Frequency
Temperature	Length of frost-free period	Frequency
Wind	Direction, velocity	Assess aerosol risk
Evapotranspiration	Annual and monthly averages	Annual distribution

Source: Adapted from Taylor, G.L., in *Proceedings of the Conference of Applied Research and Practice on Municipal and Industrial Waste*, Madison, WI, September 1980.

chance of flooding in a given year by means of a black-and-white overprint. Other detailed flood information is typically available from local offices of the U.S. Army Corps of Engineers and local flood-control districts. If the screening process identifies potential sites in flood-plain areas, local authorities should be consulted to identify regulatory requirements before beginning any detailed site investigation.

2.2.4 WATER RIGHTS

Riparian water laws, primarily in states east of the Mississippi River, protect the rights of landowners along a watercourse to use the water. Appropriative water rights laws in the western states protect the rights of prior users of the water. Adoption of any of the natural concepts for wastewater treatment can have a direct impact on water rights concerns:

- Site drainage, both quantity and quality, may be affected.
- A nondischarging system, or a new discharge location, will affect the quantity of flow in a body of water where the discharge previously existed.
- Operational considerations for land treatment systems may alter the pattern and the quality of discharges to a water body.

In addition to surface waters in well-defined channels or basins, many states also regulate or control other superficial waters and the groundwater beneath the surface. State and local discharge requirements for the appropriate case should be determined prior to initiation of design. If the project has any potential for legal entanglement, a water rights attorney should be consulted.

2.3 SITE EVALUATION

The next phase of the site and system selection process involves field surveys to confirm map data and then field testing (as needed) to confirm site information or to provide the data needed for design. This preliminary procedure includes an estimate of capital cost so the sites identified in previous steps can be evaluated economically. A concept and a site are then selected for final design. Each site evaluation must include the following information:

- Property ownership, physical dimensions of the site, and current and future land use
- Surface and groundwater conditions — location of any wells as well as use and quality of groundwater, surface waters, drainage and flooding problems, depth and fluctuations in depth to groundwater, groundwater flow direction
- Characterization of the soil profile to a depth of 5 ft (1.5 m) for SR and most biosolids systems and to a depth of 10 ft (3 m) for SAT and pond systems, as well as both physical and chemical properties
- Agricultural crops — history of cropping, yields, fertilizers used and amounts applied, tillage and irrigation methods and rates of irrigation, end use of crops
- Forest site — age and species of trees, commercial or recreational site, irrigation and fertilizer methods, vehicle access to and within the site
- Reclamation site — existing vegetation, historical causes for disturbance, previous reclamation efforts, and need for regrading or terrain modification

Investigation of SAT sites requires special consideration of the topography and of soil type and uniformity. Extensive cut-and-fill or related earthmoving operations not only are expensive but can also alter the necessary soil characteristics through compaction. Sites with significant and numerous changes in relief over a small area are not the best choice for SAT. Any soil with a significant clay fraction (>10%) would generally exclude SAT basin construction if fill is required by the design. Extremely nonuniform soils over the site do not absolutely preclude development of a SAT system, but they do significantly increase the cost and complexity of site investigation.

2.3.1 SOILS INVESTIGATION

A sequential approach to field testing to define the physical and chemical characteristics of the site soils is presented in Table 2.13. In addition to the site test pits and borings, examination of exposed soil profiles in road cuts, borrow pits, and plowed fields on or near the site should be part of the routine investigation. Backhoe test pits to a depth of 10 ft (3 m) are recommended, where site conditions permit, in each of the major soil types on the site. Soil samples should be obtained from critical layers, particularly from the layer being considered as the infiltration

TABLE 2.13
Sequence of Field Testing, Typical Order from Left to Right

Comments	Test Pits	Test Borings	Infiltration Tests[a]	Soil Chemistry[b]
Type of test	Backhoe pit; also inspect road cuts, drainage	Drilled or augered, also logs of local wells for soils data and water levels	Basin method, if possible	Review NRCS soil surveys
Data needed	Depth of profile, texture, structure, restricting layers	Depth to groundwater, depth to barrier	Infiltration rate	Nutrients, salts, pH, exchangeable sodium percentage
Then estimate	Need for hydraulic conductivity tests	Groundwater flow direction	Hydraulic capacity	Soil amendments, crop limitations
More tests for	Hydraulic conductivity, if necessary	Horizontal conductivity, if necessary	—	—
Also estimate	Loading rates	Groundwater mounding, need for drainage	—	Quality of percolate
Number of tests	Three to five minimum per site, more for larger sites or poor soil uniformity	Three-per-site minimum, more for soil aquifer treatment than slow rate, more for poor soil uniformity	Two-per-site minimum, more for large sites or poor soil uniformity	Depends on type of site, soil uniformity, wastewater characteristic

[a] Required only for land application of wastewater; some definition of subsurface permeability is necessary for pond and biosolids systems.
[b] Typically required only for land application of biosolids or wastewater.

Source: Adapted from USEPA, *Process Design Manual: Land Treatment of Municipal Wastewater*, EPA 6125/1-81-018, CERI, Cincinnati, OH, 1981.

surface for wastewater. These samples should be reserved for future testing. The walls of the backhoe pits should be examined carefully to define the characteristics listed in Table 2.14. Useful sources for more details are Crites et al. (2000) and USEPA (2002, 2005).

The test pit should be left open long enough to determine if groundwater seepage occurs, and then the highest level attained should be recorded. Equally important is any indication of seasonally high groundwater, most typically demonstrated by mottling of the soils. Soil borings should penetrate to below the groundwater table if the groundwater is within 30 to 50 ft (10 to 15 m) of the surface. At least one boring should be located in every major soil type on the

TABLE 2.14
Soil Characteristics in Field Investigations

Characteristic	Significance
Percent gravel, sand, fines (estimate)	Influences permeability, pollutant retention
Soil textural class	Influences permeability
Soil color	Indication of seasonal high groundwater, soil minerals
Plasticity of fines	Permeability, influence on cut or fill earthwork
Stratigraphy and structure	Ability to move water vertically and laterally
Wetness and consistency	Drainage characteristics

site. If generally uniform conditions prevail, medium to large systems might have one boring for every 5 to 10 ac (2 to 4 ha). Small systems (<12 ac, or 5 ha) should consider three borings as a minimum spaced over the entire site.

All of the parameters listed in Table 2.14 can be observed or estimated directly in the field by experienced personnel. This preliminary field identification serves to confirm or modify the published soils data obtained during the map survey phase. Laboratory tests with reserved samples are used to confirm field identification and provide criteria for design.

2.3.1.1 Soil Texture and Structure

Soil texture and structure are particularly important when infiltration of water is a design factor. The textural classes and the general terms used in soil descriptions are listed in Table 2.15. Soil structure refers to the aggregation of soil particles into clusters of particles referred to as *peds*. Well-structured soils with large voids between peds will transmit water more rapidly than structureless soils of the same texture. Even fine-textured soils that are well structured can transmit large quantities of water. Earthmoving and related construction activity can alter or destroy the *in situ* soil structure and significantly change the natural permeability. Soil structure can be observed in the side walls of a test pit; refer to Black, (1965), Crites, et al. (2000), and Richards (1954) for additional details.

2.3.1.2 Soil Chemistry

The chemical properties of a soil affect plant growth, control the removal of many waste constituents, and influence the hydraulic conductivity of the soil profile. Sodium can affect the permeability of fine-textured soils by dispersing clay particles and thereby changing a soil structure that initially allowed water movement. The problem is most severe in arid climates. Chapter 3 contains a discussion of the impact of sodium on clay soils. If the proposed concept involves land

TABLE 2.15
Soil Textural Classes and General Terminology Used in Soil Descriptions

Common Name	Texture	Class Name	USCS Symbol[a]
Sandy soils	Coarse	Sand; loamy sand	GW, GP, GM-d, SW
Loamy soils	Moderately coarse	Sandy loam; fine sandy loam	SP, SM-d
Clayey soils	Medium	Very fine sandy loam; loam; silt loam; silt	MH, ML
Clayey soils	Moderately fine	Clay loam; sandy clay loam; silty clay loam	SC
Clayey soils	Fine	Sandy clay; silty clay; clay	CH, CL

[a] USCS, Unified Soil Classification System.

Source: Adapted from USEPA, *Onsite Wastewater Treatment Systems Manual*, EPA/625/R-00/008, CERI, Cincinnati, OH, 2002.

application of biosolids or wastewater and in turn depends on surface vegetation as a treatment component, then soil chemistry is a very important factor in the development and future maintenance of the vegetation. The following tests are suggested for each of the major soil types on the site:

- pH, cation exchange capacity (CEC), exchangeable sodium percentage (ESP), and electrical conductivity (EC)
- Plant available nitrogen (N), phosphorus (P), potassium (K), and lime or gypsum requirements for pH adjustment and maintenance

A few standard test procedures are available for chemical analysis of soils (Black, 1965; Jackson, 1958). The interpretation of soil chemical test results can be aided by extension specialists and the use of Table 2.16.

The cation exchange capacity of a soil is a measure of the capacity of negatively charged soil colloids to adsorb cations from the soil solution. This adsorption is not necessarily permanent, because the cations can be replaced by others in the soil solution. These exchanges (except for excess sodium percentage in clay soils) do not significantly alter the structure of the soil colloids. The percentage of the CEC occupied by a particular cation is the *percent saturation* for that cation. The sum of the exchangeable hydrogen (H), sodium (Na), potassium (K), calcium (Ca), and magnesium (Mg), expressed as a percentage of the total CEC, is the *percent base saturation*. Optimum ranges for percent base saturation for various crop and soil combinations have been identified. It is important for Ca and Mg to be the dominant cations, rather than Na or K. The cation distribution in the natural soil can be changed easily by the use of soil amendments such as lime or gypsum.

TABLE 2.16
Interpretation of Soil Chemical Tests

Parameter and Test Result	Interpretation
pH of saturated soil paste	
<4.2	Too acid for most crops
5.2–5.5	Suitable for acid tolerant crops
5.5–8.4	Suitable for most crops
>8.4	Too alkaline for most crops
Cation exchange capacity (CEC) (mEq/100 g)	
1–10	Limited adsorption (sandy soils)
12–20	Moderate adsorption (silt loam)
>20	High adsorption (clay and organic soils)
Exchangeable cations	
	Desired range (as % of CEC):
Sodium	<5
Calcium	60–70
Potassium	5–10
Exchangeable sodium percentage (ESP) (as % of CEC)	
<5	Satisfactory
>10	Reduced permeability in fine-textured soils
>20	Reduced permeability in coarse-textured soils
Electrical conductivity (EC) (mmhos/cm)	
<2	No salinity problems
2–4	Restricts growth of very sensitive crops
4–8	Restricts growth of many crops
8–16	Only salt-tolerant crops will grow
>16	Only a very few salt-tolerant crops will grow

Source: Adapted from USEPA, *Process Design Manual: Land Treatment of Municipal Wastewater*, EPA 6125/1-81-018, CERI, Cincinnati, OH, 1981.

The nutrient status of the soil is important if vegetation is to become a component in the treatment system or if the soil system is otherwise to remove nitrogen or phosphorus. Potassium is also important for proper balance with the other nutrients. The N, P, and K ratios for wastewaters and biosolids are not always suitable for optimum crop growth, and in some cases it has been necessary to add supplemental potassium (see Chapter 3).

TABLE 2.17
Natural Resources Conservation Service (NRCS)
Permeability Classes for Saturated Soil

Permeability Class	Permeability (cm/hr)	Permeability (in./hr)
Very slow	<0.15	<0.06
Slow	0.15–0.5	0.06–0.2
Moderately slow	0.5–1.5	0.2–0.6
Moderate	1.5–5.1	0.6–2.0
Moderately rapid	5.1–15.2	2.0–6.0
Rapid	15.2–50	6.0–20
Very rapid	>50	>20

Source: Adapted from USEPA, *Process Design Manual: Land Treatment of Municipal Wastewater*, EPA 6125/1-81-018, CERI, Cincinnati, OH, 1981.

2.3.2 INFILTRATION AND PERMEABILITY

The ability of water to infiltrate the soil surface and then percolate vertically or laterally is a critical factor for most of the treatment concepts discussed in this book. On the one hand, excessive permeability can negate the design intentions for most ponds, wetlands, and OF systems. Insufficient permeability will limit the usefulness of SR and SAT systems and result in undesirable waterlogged conditions for land application of biosolids. The hydraulic properties of major concern are the ability of the soil surface to infiltrate water and the flow or retention of water within the soil profile. These factors are defined by the saturated permeability or hydraulic conductivity, the infiltration capacity, and the porosity, specific retention, and specific yield of the soil matrix.

2.3.2.1 Saturated Permeability

A material is considered permeable if it contains interconnected pores, cracks, or other passageways through which water or gas can flow. Hydraulic conductivity (synonymous with permeability as used in this text) is a measure of the ability of liquids and gases to pass through soil. A preliminary estimate of permeability can be found in most Natural Resources Conservation Service (NRCS) soil surveys. The final site and process selection and design should be based on appropriate field and laboratory tests to confirm the initial estimates. The permeability classes as defined by the NRCS are presented in Table 2.17. Natural soils at the low end of the permeability range are best suited for ponds, wetlands, OF, and treatment of industrial wastewaters and sludges that might have metals. Soils in the midrange are well suited for SR and for land application of biosolids; these soils can be rendered suitable for the former uses via amendments or special

TABLE 2.18
Typical Vertical Hydraulic Conductivity Values

Soil or Aquifer Material	K_v (m/d)	K_v (ft/d)
Clay soils (surface)	0.01–0.02	0.03–0.06
Deep clay beds	1×10^{-8}–0.01	3×10^{-8}–0.03
Clay, sand, gravel mixes (till)	0.001–0.1	0.003–0.3
Loam soils (surface)	0.1–1	0.3–3.0
Fine sand	1–5	3–16
Medium sand	5–20	16–66
Coarse sand	20–90	66–300
Sand and gravel mixes	5–100	16–330
Gravel	100–1000	330–3300

treatment. The soils at the upper end of the range are suited only for SAT systems in their natural state but can also be suitable for ponds, wetlands, or OF with construction of a proper liner.

The movement of water through soils can be defined using Darcy's equation:

$$q = Q/A = K(\Delta H/\Delta L) \tag{2.12}$$

where

q	=	Flux of water (the flow per unit cross-sectional area (in./hr; cm/hr).
Q	=	Volume of flow per unit time (in.^{3}/hr; cm^3/hr).
A	=	Unit cross-sectional area (in.^{2}; cm^2).
K	=	Permeability (hydraulic conductivity) (in./hr; cm/hr).
H	=	Total head (ft; m).
L	=	Hydraulic flow path (ft; m).
$\Delta H/\Delta L$	=	Hydraulic gradient (ft/ft; m/m).

The total head can be assumed to be the sum of the soil water pressure head (h) and the head due to gravity (Z); that is, $H = h + Z$. When the flow path is essentially vertical, the hydraulic gradient is equal to 1 and the vertical permeability (K_v) is used in Equation 2.12. Typical values of vertical permeability are presented in Table 2.18.

When the flow path is essentially horizontal, then the horizontal permeability (K_h) should be used. The permeability coefficient (K) is not a true constant but is a changing function of soil-water content. Even under saturated conditions, the K value may change due to swelling of clay particles and other factors, but for general engineering design purposes it can be considered a constant. The K_v will not necessarily be equal to the K_h for most soils. In general, the lateral K_h will be higher, because the interbedding of fine- and coarse-grained layers tends to restrict vertical flow. Typical values are given in Table 2.19.

TABLE 2.19
Horizontal Permeability

K_h (m/d)	K_h/K_v	Comments
42	2.0	Silty soil
75	2.0	—
56	4.4	—
100	7.0	Gravelly
72	20.0	Near terminal morain
72	10.0	Irregular succession of sand and gravel layers, from field measurements of K

Source: Adapted from Crites, R.W. et al., *Land Treatment Systems for Municipal and Industrial Wastes*, McGraw-Hill, New York, 2000.

2.3.2.2 Infiltration Capacity

The infiltration rate of a soil is defined as the rate at which water enters the soil from the surface. When the soil profile is saturated and there is negligible ponding at the surface, the infiltration rate is equal to the effective saturated permeability or conductivity of the immediate soil profile. Although the measured infiltration rate on a particular site may decrease with time due to surface clogging, the subsurface vertical permeability at saturation will generally remain constant. As a result, the short-term measurement of infiltration serves reasonably well as an estimate of the long-term saturated vertical permeability within the zone of influence for the test procedure being used.

2.3.2.3 Porosity

The ratio of voids to the total volume of the soil is referred to as the soil porosity. It is expressed either as a decimal fraction or as a percentage, as defined in Equation 2.13:

$$n = (V_t - V_s)/V_t = V_v/V_t \qquad (2.13)$$

where

n = Porosity (decimal).
V_t = Total unit volume of soil (ft^3; m^3).
V_s = Unit volume of soil fraction (ft^3; m^3).
V_v = Unit volume of voids (ft^3; m^3).

2.3.2.4 Specific Yield and Specific Retention

The porosity of a soil defines the maximum amount of water that a soil can contain when it is saturated. The specific yield is the portion of that water that

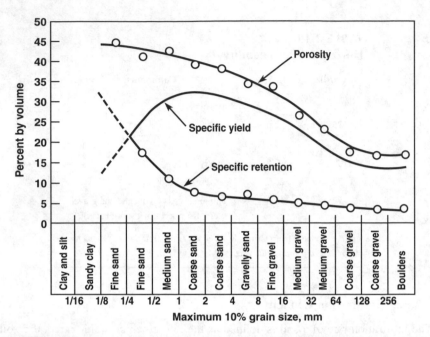

FIGURE 2.3 Porosity, specific retention, and specific yield variations with grain size (*in situ* consolidated soils, coastal basin, California).

will drain under the influence of gravity. The specific retention is the portion of the soil-water that will remain as a film and in very small voids. The porosity, therefore, is the sum of the specific yield and the specific retention. The relationship between the porosity, specific yield, and specific retention is illustrated for typical *in situ*, consolidated California soils in Figure 2.3. The specific yield is used when defining aquifer properties, particularly in calculating groundwater mounding beneath ponds and wastewater application sites. For relatively coarse-textured soils and deep water tables, it is acceptable to assume a constant value for the specific yield. Because the calculations are not especially sensitive to small changes in specific yield, it is usually satisfactory to estimate it from other properties, as shown in Figure 2.3 and Figure 2.4. Neither Figure 2.3 nor Figure 2.4 should be used to indicate the hydraulic properties of the medium in subsurface flow constructed wetlands. Groundwater mound analysis can be more complicated for finer-textured soils because of capillarity effects in the soil as the water table moves higher (Childs, 1969; Duke, 1972).

2.3.2.5   Field Tests for Infiltration Rate

In some cases it may be acceptable to use NRCS estimates of soil permeability after confirming the actual presence of the specific soil on the site during a field investigation. This should be sufficient for pond and OF systems on soils with

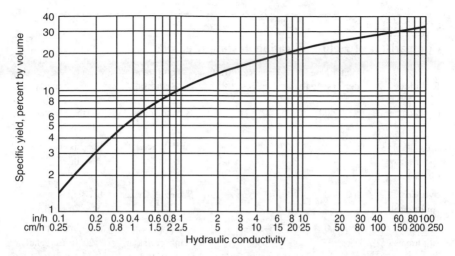

FIGURE 2.4 General relationship between specific yield and hydraulic conductivity for fine-textured soils.

naturally low permeability. Concepts where water flow in the soil is a major design consideration will require field and possibly laboratory testing. Infiltration testing in the field is recommended where infiltration rates are critical to the design. A variety of testing methods are available, as shown in Table 2.20, to measure surface infiltration rates. The reliability of test results is a function of the test area and the zone of subsurface material influenced. This relationship is shown in Table 2.20 by the volume of water required to conduct a single test. As indicated in Chapter 8, the increased confidence resulting from larger scale field tests allows a reduction in the safety factor for the design of some land treatment systems.

Flooding Basin Test

A basin test area of at least 75 ft² (7 m²) is suggested for all projects where infiltration and percolation of water are design expectations. The area can be surrounded by a low earthen berm with an impermeable plastic cover, or aluminum flashing can be partially set into the soil in a circular configuration to define the test area. The use of a bentonite seal around the aluminum flashing perimeter is recommended to prevent leakage of water. Tensiometers at a depth of 6 in. (15 cm) and 12 in. (30 cm) can be installed near the center of the circle to define saturated conditions at these depths as the test progresses (see Figure 2.5). The test basin should be flooded several times to ensure saturated conditions and to calibrate any instrumentation. The actual test run should be completed within 24 hr of the preliminary trials. This final test run may require 3 to 8 hr for coarse-textured soils.

The water level in the basin is observed and recorded with time. These values are plotted as intake rate vs. time. This intake rate will be relatively high, initially, and then will drop off with time. The test must continue until the intake rate

TABLE 2.20
Comparison of Field Infiltration Testing Methods

Technique	Water Needs per Test (gal)	Time Required per Test (hr)	Equipment Required	Comments
Flooding basin	600–3000	4–12	Backhoe or blade	See this chapter for details.
Air entry permeameter (AEP)	3	0.5–1	AEP device	See this chapter for details.
Cylinder infiltrometer	100–200	1–6	Standard device	See Crites et al. (2000) for details.
Sprinkler infiltrometer	250–300	1.5–3	Pump, pressure tank, sprinkler, collection cans	See Crites et al. (2000) for details.

approaches a "steady-state" condition. This steady-state rate can be taken as the limiting infiltration rate for the soil within the zone of influence of the test. A safety factor is then applied to that rate for system design, as described in Chapter 8.

Because it is the basic purpose of the test to define the hydraulic conductivity of the near-surface soil layers, the use of clean water (with about the same ionic composition as the expected wastewater) is acceptable in most cases. If, however, the wastewater is expected to have a high solids content that might clog the surface, then a similar liquid should be used for the field test.

The basin test is most critical for the SAT concept because large volumes of wastewater are applied to a relatively small surface area. Most SAT systems, as described in Chapter 8, are operated on an intermittent or cyclic pattern basis, alternating flooding and drying to maximize infiltration rates and allow alternating oxidation and reduction processes in the soil. If a particular project design calls for a continuously flooded seepage pond mode, then the initial field tests should be continued for a long enough period to simulate this condition.

If site conditions require construction of full-scale SAT basins on backfilled material (not recommended), a test fill should be constructed on the site with the equipment intended for full-scale use, and then the basin test described above should be conducted in that material. The test fill should be as deep as required by the site design or 5 ft (1.5 m), whichever is less. The top of the fill area should be at least 15 ft (5 m) wide and 15 ft (5 m) long to permit the installation of a flooding basin test near the center.

One flooding basin infiltration test should be conducted on each of the major soil types on the site. For large continuous areas, one test for up to 25 ac (10 ha) is typically sufficient. The test should be performed on the soil layer that will become the final infiltration surface in the constructed system.

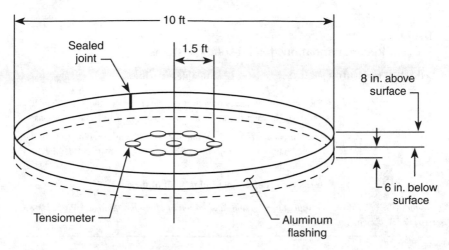

FIGURE 2.5 Basin test for infiltration.

Air Entry Permeameter

The air entry permeameter (AEP) was developed by the U.S. Department of Agriculture (USDA) to measure point hydraulic conductivity in the absence of a water table. The device is not commercially available, but specifications and fabrication details can be obtained from the USDA, Water Conservation Laboratory, 4332 East Broadway, Phoenix, AZ 85040. The unit defines conditions for a very small soil zone, but the small volume of water required and short time for a single test make it useful to verify site conditions between the larger scale flooding basin test locations. It can also be used in a test pit to define the *in situ* permeability with depth. The pit is excavated with one end inclined to the surface, benches are cut about 3 ft (1 m) wide by hand, and the AEP device is used on that surface.

2.3.3 SUBSURFACE PERMEABILITY AND GROUNDWATER FLOW

The permeability of deeper soils is usually measured via laboratory tests on undisturbed soil samples obtained during the field boring program. Such data are usually required only for the design of SAT systems or to ensure that subsoils are adequate to contain undesirable leachates. In many situations, it is desirable for the design of SAT systems to determine the horizontal permeability of the subsurface layers. This can be accomplished with a field test known as the *auger hole test*, which, in essence, requires pumping a slug of water out of a bore hole and then observing the time for the water level to recover via lateral flow. The U.S. Bureau of Reclamation has developed a standard procedure for this test (details can be found in Crites et al., 2000; USDOI, 1978).

Defining the groundwater position and flow direction is essential for most of the treatment concepts discussed in this book. Overland flow and wetland systems

TABLE 2.21
Setback Recommendations for Biosolids Systems

Setback Distance (ft)	Suitable Activities
50–200	Biosolids injection, and only near remote single dwellings, small ponds, 10-yr high water mark for streams, roads; no surface applications
300–1500	Injection or surface application near all of the above, plus springs and water supply wells; injection only near high-density residential developments
>1500	Injection or surface application at all of the above

Source: USEPA, *Process Design Manual: Land Application of Municipal Sludge*, EPA 625/1-83-016, CERI, Cincinnati, OH, 1983.

have little concern with deep groundwater tables but might still be affected by near-surface seasonally high groundwater. Evidence of seasonal groundwater may be observed in the test pits; water levels should be observed in any borings and in any existing wells on-site or on adjacent properties. These data can provide information on the general hydraulic gradient and flow direction for the area. These data are also necessary if groundwater mounding or underdrainage (as described in Chapter 3) are project concerns.

2.3.3.1 Buffer Zones

State and local requirements for buffer zones or setback distances from, for example, roads, wells, and property lines should be determined before the site investigation to be sure that the site is of an adequate area. Most requirements for buffer zones or separation distances are based on public health, aesthetics, or avoidance of odor complaints. The potential for aerosol transmission of pathogens has been a concern to some with regard to the operation of land application of wastewater and some kinds of sludge composting operations. A number of aerosol studies have been conducted at both conventional and land treatment facilities with no evidence of significant risk to adjacent populations (Crites et al., 2000). Extensive buffer zones for aerosol containment are not recommended. If the use of sprinklers is planned, a buffer zone to catch the droplets or mist on windy days should be considered. A strip 30 to 50 ft (10 to 15 m) wide planted with conifers should suffice. Odor potential is a concern for facultative-type pond systems, because the seasonal turnover may bring anaerobic materials to the liquid surface for a short period in the spring and fall. A typical requirement in these cases is to locate such ponds at least 0.25 mi (0.4 km) from habitations. Mosquito control for free water surface constructed wetlands may require setbacks from residences unless positive control measures are planned for the system. Recommended setback distances for land application of biosolids are listed in Table 2.21.

2.4 SITE AND PROCESS SELECTION

The evaluation procedure up to this point has resulted in potential sites being identified for a particular treatment alternative and then field investigations being conducted to obtain data for the feasibility determination. The evaluation of the field data will indicate whether or not the site requirements listed in Table 2.1 and Table 2.2 exist. If site conditions are favorable, it can be concluded that the site is apparently feasible for the intended concept. If only one site and related treatment concept result from this screening process, then the focus can shift to final design and possibly additional detailed field tests to support or refine the design. If more than one site for a particular concept or more than one concept remain technically feasible after the screening process, it will be necessary to conduct a preliminary cost analysis to identify the most cost-effective alternative. The criteria in Chapters 4 through 10 should be used for a preliminary design of the concept in question. The planning-level equations in this chapter should not be used for preliminary or final design. Cost-effectiveness should be determined from a detailed cost estimate of all the elements of the treatment system including pumping, preapplication treatment, storage, land, and final disposition of the treated water. In many cases, the final selection will also be influenced by non-financial factors, such as the social and institutional acceptability of the proposed site and concept to be developed on it.

REFERENCES

Black, A., Ed. (1965). *Methods of Soil Analysis*. Part 2. *Chemical and Microbiological Properties*, Agronomy 9, American Society of Agronomy, Madison, WI.

Bouwer, H. (1978). *Groundwater Hydrology*, McGraw-Hill, New York.

Childs, E.C. (1969). *An Introduction to the Physical Basis of Soil Water Phenomena*, John Wiley & Sons, London.

Crites, R.W. and Tchobanoglous, G. (1998). *Small and Decentralized Wastewater Management Systems*, McGraw-Hill, New York.

Crites, R.W., Reed, S.C., and Bastian R.K. (2000). *Land Treatment Systems for Municipal and Industrial Wastes*, McGraw-Hill, New York.

Duke, H.R. (1972). Capillary properties of soils: influence upon specific yields, *Trans. Am. Soc. Agric. Eng.*, 15, 688–691.

Jackson, M.L. (1958). *Soil Chemical Properties*, Prentice-Hall, Englewood Cliffs, NJ.

McKim, H.L., Cole, R., Sopper, W., and Nutter, W. (1982). *Wastewater Applications in Forest Ecosystems*, CRREL Report 82-19, U.S. Cold Regions Research and Engineering Laboratory (CRREL), Hanover, NH.

NOAA, *Climatic Summary of the United States* (a 10-year summary), National Oceanic and Atmospheric Administration, Rockville, MD.

NOAA, *Local Climatological Data*, annual summaries for selected locations, National Oceanic and Atmospheric Administration, Rockville, MD.

NOAA, *Monthly Summary of Climatic Data*, National Oceanic and Atmospheric Administration, Rockville, MD.

Richards, L.A., Ed. (1954). *Diagnosis and Improvement of Saline and Alkali Soils*, Agricultural Handbook No. 60, U.S. Department of Agriculture, Washington, D.C.

Taylor, G.L. (1980). A preliminary site evaluation method for treatment of municipal wastewater by spray irrigation of forest land, in *Proceedings of the Conference of Applied Research and Practice on Municipal and Industrial Waste*, Madison, WI, September 1980.

USDOI (1978). *Bureau of Reclamation: Drainage Manual*, U.S. Department of Interior, U.S. Government Printing Office, Washington, D.C.

USEPA (1981). *Process Design Manual: Land Treatment of Municipal Wastewater*, EPA 6125/1-81-018, Center for Environmental Research Information, U.S. Environmental Protection Agency, Cincinnati, OH.

USEPA (1983). *Process Design Manual: Land Application of Municipal Sludge*, EPA 625/1-83-016, Center for Environmental Research Information, U.S. Environmental Protection Agency, Cincinnati, OH.

USEPA (1995). *Process Design Manual: Land Application of Sewage Sludge and Domestic Septage*, EPA 625/R-95/001, Center for Environmental Research Information, U.S. Environmental Protection Agency, Cincinnati, OH.

USEPA (2002). *Onsite Wastewater Treatment Systems Manual*, EPA/625/R-00/008, Center for Environmental Research Information, U.S. Environmental Protection Agency, Cincinnati, OH.

USEPA (2005). *Process Design Manual: Land Treatment of Municipal Wastewater*, Center for Environmental Research Information, U.S. Environmental Protection Agency, Cincinnati, OH.

3 Basic Process Responses and Interactions

This chapter describes the basic responses and interactions among the waste constituents and process components of natural treatment systems. Many of these responses are common to more than one of the treatment concepts and are therefore discussed in this chapter. If a waste constituent is the limiting factor for design, it is also discussed in detail in the appropriate process design chapter. Water is the major constituent of all of the wastes of concern in this book, as even a "dried" sludge can contain more than 50% water. The presence of water is a volumetric concern for all treatment methods, but it has even greater significance for many of the natural treatment concepts because the flow path and the flow rate control the successful performance of the system. Other waste constituents of major concern include the simple carbonaceous organics (dissolved and suspended), toxic and hazardous organics, pathogens, trace metals, nutrients (nitrogen, phosphorus, potassium), and other micronutrients. The natural system components that provide the critical reactions and responses include bacteria, protozoa (e.g., algae), vegetation (aquatic and terrestrial), and the soil. The responses involved include a range of physical, chemical, and biological reactions.

3.1 WATER MANAGEMENT

Major concerns of water management include the potential for travel of contaminants with groundwater, the risk of leakage from ponds and other aquatic systems, the potential for groundwater mounding beneath a land treatment system, the need for drainage, and the maintenance of design flow conditions in ponds, wetlands, and other aquatic systems.

3.1.1 FUNDAMENTAL RELATIONSHIPS

Chapter 2 introduced some of the hydraulic parameters (e.g., permeability) that are important to natural systems and discussed methods for their determination in the field or laboratory. It is necessary to provide further details and definition before undertaking any flow analysis.

3.1.1.1   Permeability

The results from the field and laboratory test program described in the previous chapter may vary with respect to both depth and areal extent, even if the same basic soil type is known to exist over much of the site. The soil layer with the most restrictive permeability is taken as the design basis for those systems that depend on infiltration and percolation of water as a process requirement. In other cases, where there is considerable scatter to the data, it is necessary to determine a "mean" permeability for design.

If the soil is uniform, then the vertical permeability (K_v) should be constant with depth and area, and any differences in test results should be due to variations in the test procedure. In this case, K_v can be considered to be the arithmetic mean as defined by Equation 3.1:

$$K_{am} = \frac{K_1 + K_2 + K_3 + K_n}{n} \tag{3.1}$$

where K_{am} is the arithmetic mean vertical permeability, and K_1 through K_n are individual test results.

Where the soil profile consists of a layered series of uniform soils, each with a distinct K_v generally decreasing with depth, the average value can be represented as the harmonic mean:

$$K_{hm} = \frac{D}{\left(\dfrac{d_1}{K_1}\right) + \left(\dfrac{d_2}{K_2}\right) + \left(\dfrac{d_n}{K_n}\right)} \tag{3.2}$$

where
K_{hm} = Harmonic mean permeability.
D = Soil profile depth.
d_n = Depth of nth layer.

If no pattern or preference is indicated by a statistical analysis, then a random distribution of the K values for a layer must be assumed, and the geometric mean provides the most conservative estimate of the true K_v:

$$K_{gm} = \left[(K_1)(K_2)(K_3)(K_n)\right]^{1/n} \tag{3.3}$$

where K_{gm} is the geometric mean permeability (other terms are as defined previously).

Equation 3.1 or 3.3 can also be used with appropriate data to determine the lateral permeability, K_h. Table 2.17 presents typical values for the ratio K_h/K_v.

3.1.1.2 Groundwater Flow Velocity

The actual flow velocity in a groundwater system can be obtained by combining Darcy's law, the basic velocity equation from hydraulics, and the soil porosity, because flow can occur only in the pore spaces in the soil.

$$V = \frac{(K_h)(\Delta H)}{(n)(\Delta L)} \tag{3.4}$$

where

V = Groundwater *flow* velocity (ft/d; m/d).

K_h = Horizontal saturated permeability, mid (ft/d; m/d).

$\Delta H/\Delta L$ = Hydraulic gradient (ft/ft; m/m)

n = Porosity (as a decimal fraction; see Figure 2.4 for typical values for *in situ* soils).

Equation 3.4 can also be used to determine vertical flow velocity. In this case, the hydraulic gradient is equal to 1 and K_v should be used in the equation.

3.1.1.3 Aquifer Transmissivity

The transmissivity of an aquifer is the product of the permeability of the material and the saturated thickness of the aquifer. In effect, it represents the ability of a unit width of the aquifer to transmit water. The volume of water moving through this unit width can be calculated using Equation 3.5:

$$q = (K_h)(b)(w)\left(\frac{\Delta H}{\Delta L}\right) \tag{3.5}$$

where

q = Volume of water moving through aquifer (ft^3/d; m^3/d).

b = Depth of saturated thickness of aquifer (ft; m).

w = Width of aquifer, for unit width w = 1 ft (1 m).

$\Delta H/\Delta L$ = Hydraulic gradient (ft/ft; m/m).

In many situations, well pumping tests are used to define aquifer properties. The transmissivity of the aquifer can be estimated using pumping rate and draw-down data from well tests (Bouwer, 1978; USDOI, 1978).

3.1.1.4 Dispersion

The dispersion of contaminants in the groundwater is due to a combination of molecular diffusion and hydrodynamic mixing. The net result is that the concentration of the material is less, but the zone of contact is greater at downgradient locations. Dispersion occurs in a longitudinal direction (D_x) and transverse to the flow path (D_y). Dye studies in homogeneous and isotropic granular media have indicated that dispersion occurs in the shape of a cone about 6° from the application point (Danel, 1953). Stratification and other areal differences in the field will typically result in much greater lateral and longitudinal dispersion. For example, the divergence of the cone could be 20° or more in fractured rock (Bouwer, 1978). The dispersion coefficient is related to the seepage velocity as described by Equation 3.6:

$$D = (a)(v) \qquad\qquad (3.6)$$

where

D = Dispersion coefficient: D_x longitudinal, D_y transverse (ft²/d; m²/d).

a = Dispersivity: a_x longitudinal, a_y transverse (ft; m).

v = Seepage velocity of groundwater system (ft/d; m/d) = V/n, where V is the Darcy's velocity from Equation 3.5, and n is the porosity (see Figure 2.4 for typical values for *in situ* soils).

The dispersivity is difficult to measure in the field or to determine in the laboratory. Dispersivity is usually measured in the field by adding a tracer at the source and then observing the concentration in surrounding monitoring wells. An average value of 10 m²/d resulted from field experiments at the Fort Devens, Massachusetts, rapid infiltration system (Bedient et al., 1983), but predicted levels of contaminant transport changed very little after increasing the assumed dispersivity by 100% or more. Many of the values reported in the literature are site-specific, "fitted" values and cannot be used reliably for projects elsewhere.

3.1.1.5 Retardation

The hydrodynamic dispersion discussed in the previous section affects all the contaminant concentrations equally; however, adsorption, precipitation, and chemical reactions with other groundwater constituents retard the rate of advance of the affected contaminants. This effect is described by the retardation factor (R_d), which can range from a value of 1 to 50 for organics often encountered at field sites. The lowest values are for conservative substances, such as chlorides, which are not removed in the groundwater system. Chlorides move with the same velocity as the adjacent water in the system, and any change in observed chloride concentration is due to dispersion only, not retardation. Retardation is a function of soil and groundwater characteristics and is not necessarily constant for all locations. The R_d for some metals might be close to 1 if the aquifer is flowing through clean sandy soils with a low pH but close to 50 for clayey soils. The R_d for organic compounds depends on sorption of the compounds to soil organic matter plus volatilization and biodegradation. The sorptive reactions depend on the quantity of organic matter in the soil and on the solubility of the organic material in the groundwater. Insoluble compounds such as dichloro-diphenyl-trichloroethane (DDT), benzo[a]pyrenes, and some polychlorinated biphenyls

TABLE 3.1
Retardation Factors for Selected
Organic Compounds

Material	Retardation Factor (R_d)
Chloride	1
Chloroform	3
Tetrachloroethylene	9
Toluene	3
Dichlorobenzene	14
Styrene	31
Chlorobenzene	35

(PCBs) are effectively removed by most soils. Highly soluble compounds such as chloroform, benzene, and toluene are removed less efficiently by even highly organic soils. Because volatilization and biodegradation are not necessarily dependent on soil type, the removal of organic compounds via these methods tends to be more uniform from site to site. Table 3.1 presents retardation factors for a number of organic compounds, as estimated from several literature sources (Bedient et al., 1983; Daniel, 1953; Roberts et al., 1980).

3.1.2 Movement of Pollutants

The movement or migration of pollutants with the groundwater is controlled by the factors discussed in the previous section. This might be a concern for ponds and other aquatic systems as well as when utilizing the slow rate (SR) and rapid infiltration land treatment concepts. Figure 3.1 illustrates the subsurface zone of

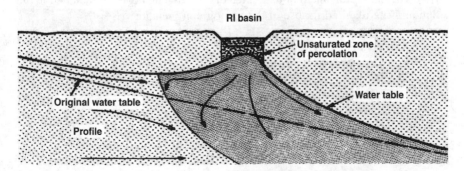

FIGURE 3.1 Subsurface zone of influence for SAT basin.

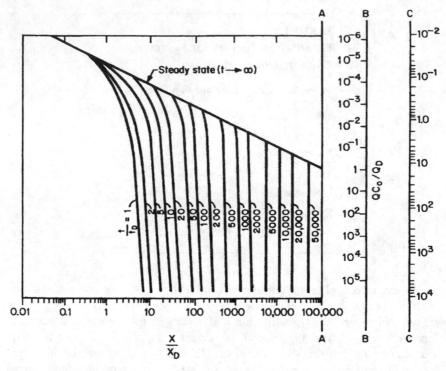

FIGURE 3.2 Nomograph for estimating pollutant travel.

influence for a rapid infiltration basin system or a treatment pond where significant seepage is allowed. It is frequently necessary to determine the concentration of a pollutant in the groundwater plume at a selected distance downgradient of the source. Alternatively, it may be desired to determine the distance at which a given concentration will occur at a given time or the time at which a given concentration will reach a particular point. Figure 3.2 is a nomograph that can be used to estimate these factors on the centerline of the downgradient plume (USEPA, 1985). The dispersion and retardation factors discussed above are included in the solution. Data required for use of the nomograph include:

- Aquifer thickness, z (m)
- Porosity, n (%, as a decimal)
- Seepage velocity, v (m/d)
- Dispersivity factors a_x and a_y (m)
- Retardation factor R_d for the contaminant of concern
- Volumetric water flow rate, Q (m³/d)
- Pollutant concentration at the source, C_0 (mg/L)
- Background concentration in groundwater, C_b (mg/L)
- Mass flow rate of contaminant QC_0 (kg/d)

Use of the nomograph requires calculation of three scale factors:

$$X_D = \frac{D_x}{v} \tag{3.7}$$

$$t_D = \frac{(R_d)(D_x)}{(v)^2} \tag{3.8}$$

$$Q_D = (16.02)(n)(z)\left[(D_x)(D_y)\right]^{1/2} \tag{3.9}$$

The procedure is best illustrated with an example.

Example 3.1

Determine the nitrate concentration in the centerline of the plume, 600 m down-gradient of a rapid infiltration system, 2 years after system startup. Data: aquifer thickness = 5 m; porosity = 0.35; seepage velocity = 0.45 m/d; dispersivity, a_x = 32 m, a_y = 6 m; volumetric flow rate = 90 m³/d; nitrate concentration in percolate = 20 mg/L; and nitrate concentration in background groundwater = 4 mg/L.

Solution

1. The downgradient volumetric flow rate combines the natural background flow plus the additional water introduced by the SAT system. To be conservative, assume for this calculation that the total nitrate at the origin of the plume is equal to the specified 20 mg/L. The residual concentration determined with the nomograph is then added to the 4-mg/L background concentration to determine the total downgradient concentration at the point of concern. Experience has shown that nitrate tends to be a conservative substance when the percolate has passed the active root zone in the soil, so for this case assume that the retardation factor R_d is equal to 1.
2. Determine the dispersion coefficients:

$$D_x = (a_x)(v) = (32)(0.45) = 14.4 \text{ m}^2/\text{d}$$

$$D_y = (a_y)(v) = (6)(0.45) = 2.7 \text{ m}^2/\text{d}$$

3. Calculate the scale factors:

$$X_D = D_x/v = 14.4/0.45 = 32 \text{ m}$$

$$t_D = R_d(D_x)/(v)^2 = 1(14.4)/(0.45)^2$$

$$Q_D = (16.02)(n)(z)[(D_x)(D_y)]^{1/2}$$
$$= (16.02)(0.35)(5)[(14.4)(2.7)]^{1/2} = 174.8 \text{ kg/d}$$

4. Determine the mass flow rate of the contaminant:

$$(Q)(C_0) = (90 \text{ m}^3/\text{d})(20 \text{ mg/L})/(1000 \text{ g/kg}) = 1.8 \text{ kg/d}$$

5. Determine the entry parameters for the nomograph:

$$\frac{x}{x_D} = \frac{600}{32} = 18.8$$

$$\frac{t}{t_D} = \frac{(2)(365)}{71} = 10.3 \quad \text{use} \quad \frac{t}{t_D} = 10 \text{ curve}$$

$$\frac{QC_0}{Q_D} = \frac{1.8}{174.8} = 0.01$$

6. Enter the nomograph on the x/x_D axis with the value of 18.8, draw a vertical line to intersect with the t/t_D curve = 10. From that point, project a line horizontally to the A–A axis. Locate the calculated value 0.01 on the B–B axis and connect this with the previously determined point on the A–A axis. Extend this line to the C–C axis and read the concentration of concern, which is about 0.4 mg/L.
7. After 2 years, the nitrate concentration at a point 600 m downgradient is the sum of the nomograph value and the background concentration, or 4.4 mg/L.

Calculations must be repeated for each contaminant using the appropriate retardation factor. The nomograph can also be used to estimate the distance at which a given concentration will occur in a given time. The upper line on the figure is the "steady-state" curve for very long time periods and, as shown in Example 3.2, can be used to evaluate conditions when equilibrium is reached.

Example 3.2
Using the data in Example 3.1, determine the distance downgradient where the groundwater in the plume will satisfy the U.S. Environmental Protection Agency (EPA) limits for nitrate in drinking-water supplies (10 mg/L).

Solution
1. Assuming a 4-mg/L background value, the plume concentration at the point of concern could be as much as 6 mg/L. Locate 6 mg/L on the C–C axis.
2. Connect the point on the C–C axis with the value 0.01 on the B–B axis (as determined in Example 3.1). Extend this line to the A–A axis. Project a horizontal line from this point to intersect the steady-state line. Project a vertical line downward to the x/x_D axis and read the value $x/x_D = 60$.
3. Calculate distance x using the previously determined value for x_D:

$$x = (x_D)(60) = (32)(60) = 1920 \text{ m}$$

3.1.3 GROUNDWATER MOUNDING

Groundwater mounding is illustrated schematically in Figure 3.1. The percolate flow in the unsaturated zone is essentially vertical and controlled by K_v. If a groundwater table, impeding layer, or barrier exists at depth, a horizontal component is introduced and flow is controlled by a combination of K_v and K_h within the groundwater mound. At the margins of the mound and beyond, the flow is typically lateral, and K_h controls.

The capability for lateral flow away from the source will determine the extent of mounding that will occur. The zone available for lateral flow includes the underground aquifer plus whatever additional elevation is considered acceptable for the particular project design. Excessive mounding will inhibit infiltration in a SAT system. As a result, the capillary fringe above the groundwater mound should never be closer than about 0.6 m (2 ft) to the infiltration surfaces in soil aquifer treatment (SAT) basins. This will correspond to a water table depth of about 1 to 2 m (3 to 7 ft), depending on the soil texture.

In many cases, the percolate or plume from a SAT system will emerge as base flow in adjacent surface waters, so it may be necessary to estimate the position of the groundwater table between the source and the point of emergence. Such an analysis will reveal if seeps or springs are likely to develop in the intervening terrain. In addition, some regulatory agencies require a specific residence time in the soil to protect adjacent surface waters, so it may be necessary to calculate the travel time from the source to the expected point of emergence. Equation 3.10 can be used to estimate the saturated thickness of the water table at any point downgradient of the source (USEPA, 1984). Typically, the calculation is repeated for a number of locations, and the results are converted to an elevation and plotted on maps and profiles to identify potential problem areas:

$$h = \left[(h_0)^2 - \left(\frac{(2)(Q_i)(d)}{K_h} \right) \right]^{1/2} \tag{3.10}$$

where

h = Saturated thickness of the unconfined aquifer at the point of concern (ft; m).

h_0 = Saturated thickness of the unconfined aquifer at the source (ft; m).

d = Lateral distance from the source to the point of concern (ft; m).

K_h = Effective horizontal permeability of the soil system, mid (ft/d).

Q_i = Lateral discharge from the unconfined aquifer system per unit width of the flow system (ft³/d·ft; m³/d·m):

$$Q_i = \frac{K_h}{2d_i} \left(h_0^2 - h_i^2 \right) \tag{3.11}$$

where
d_i = Distance to the seepage face or outlet point (ft; m).
h_i = Saturated thickness of the unconfined aquifer at the outlet point (ft; m).

The travel time for lateral flow is a function of the hydraulic gradient, the distance traveled, the K_h, and the porosity of the soil as defined by Equation 3.12:

$$t_D = \frac{(n)(d_i)^2}{(K_h)(h_0 - h_i)}$$

(3.12)

where
t_D = Travel time for lateral flow from source to the point of emergence in surface waters (ft; m).

K_h = Effective horizontal permeability of the soil system (ft/d; m/d).

h_0, h_i = Saturated thickness of the unconfined aquifer at the source and the outlet point, respectively (ft; m).

d_i = Distance to the seepage face or outlet point (ft; m).

n = Porosity, as a decimal fraction.

A simplified graphical method for determining groundwater mounding uses the procedure developed by Glover (1961) and summarized by Bianchi and Muckel (1970). The method is valid for square or rectangular basins that lie above level, fairly thick, homogeneous aquifers of assumed infinite extent; however, the behavior of circular basins can be adequately approximated by assuming a square of equal area. When groundwater mounding becomes a critical project issue, further analysis using the Hantush method (Bauman, 1965) is recommended. Further complications arise with sloped water tables or impeding subsurface layers that induce "perched" mounds or due to the presence of a nearby outlet point. References by Brock (1976), Kahn and Kirkham (1976), and USEPA (1981) are suggested for these conditions. The simplified method involves the graphical determination of several factors from Figure 3.3, Figure 3.4, Figure 3.5, or Figure 3.6, depending on whether the basin is square or rectangular.

It is necessary to calculate the values of $W/(4at)^{0.5}$ and R_t as defined in Equations 3.13 to 3.15:

$$\frac{W}{[(4)(\alpha)(t)]^{1/2}} = \text{dimensionless scale factor}$$

(3.13)

where W is the width of the recharge basin (ft; m), and

$$\alpha = \frac{(K_h)(h_0)}{Y_s}$$

(3.14)

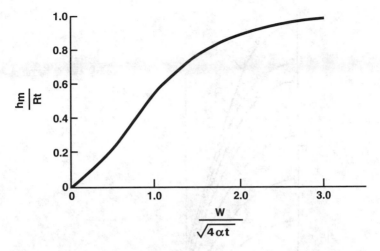

FIGURE 3.3 Groundwater mounding curve for center of a square recharge basin.

where
K_h = Effective horizontal permeability of the aquifer (ft/d; m/d).
h_0 = Original saturated thickness of the aquifer beneath the center of the recharge area (ft; m).
Y_s = Specific yield of the soil (use Figure 2.5 or 2.6 to determine) (ft³/ft³; m³/m³).

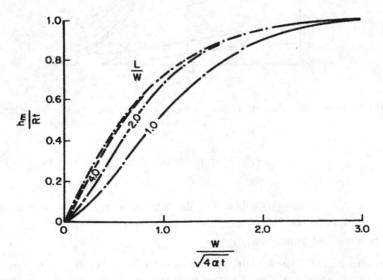

FIGURE 3.4 Groundwater mounding curves for center of a rectangle recharge area with different ratios of length (L) to width (W).

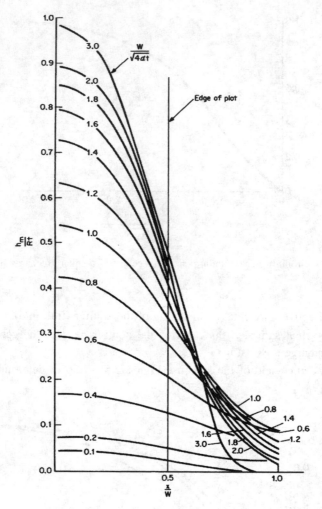

FIGURE 3.5 Rise and horizontal spread of a groundwater mound below a square recharge area.

$$(R)(t) = \text{scale factor (ft; m)} \qquad (3.15)$$

where

$R = (I)/(Y_s)$ (ft/d; m/d), where I is the infiltration rate or volume of water infiltrated per unit area of soil surface (ft³/ft²·d; m³/m²/d).

t = Period of infiltration, d.

Enter either Figure 3.3 or 3.4 with the calculated value of $W/(4(\alpha t)^{1/2}$ to determine the value for the ratio $h_m/(R)(t)$, where h_m is the rise at the center of the mound. Use the previously calculated value for $(R)(t)$ to solve for h_m. Figure 3.5 (for square areas) and Figure 3.6 (for rectangular areas, where $L = 2W$) can be used

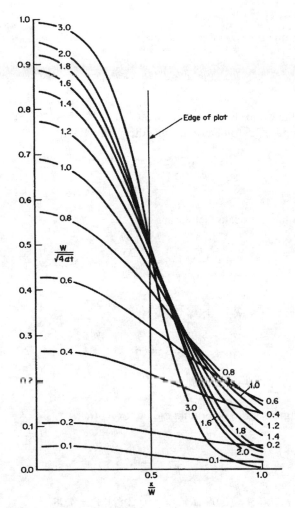

FIGURE 3.6 Rise and horizontal spread of a groudwater mound below a rectangular recharge area with a length equal to twice its width.

to estimate the depth of the mound at various distances from the center of the recharge area. The procedures involved are best illustrated with a design example.

Example 3.3

Determine the height and horizontal spread of a groundwater mound beneath a circular SAT basin 30 m in diameter. The original aquifer thickness is 4 m, and K_h as determined in the field is 1.25 m/d. The top of the original groundwater table is 6 m below the design infiltration surface of the constructed basin. The design infiltration rate will be 0.3 m/d and the wastewater application period will be 3 days in every cycle (3 days of flooding, 10 days for percolation and drying; see Chapter 8 for details).

Solution

1. Determine the size of an equivalent area square basin:

$$A = \frac{(3.14)(D)^2}{4} = 706.5 \text{ m}$$

Then the width (W) of an equivalent square basin is $(706.5)^{1/2} = 26.5$ m.

2. Use Figure 2.5 to determine specific yield (Y_s):

$$K_h = 1.25 \text{ m/d} = 5.21 \text{ cm/hr}$$

$$Y_s = 0.14$$

3. Determine the scale factors:

$$\alpha = \frac{(K_h)(h_0)}{Y_s} = \frac{(1.25)(4)}{0.14} = 35.7 \text{ m}^2/\text{d}$$

$$\frac{W}{(4\alpha t)^{1/2}} = \frac{26.5}{[(4)(35.7)(3)]^{1/2}} = 1.28$$

$$R = \frac{0.3}{0.14} = 2 \text{ m}/\text{d}$$

$$(R)(t) = (2)(3) = 6 \text{ m}$$

4. Use Figure 3.3 to determine the factor $h_m/(R)(t)$:

$$\frac{h_m}{(R)(t)} = 0.68$$

$$h_m = (0.68)(R)(t) = (0.68)(2)(3) = 4.08 \text{ m}$$

5. The original groundwater table is 6 m below the infiltration surface. The calculated rise of 4.08 m would bring the top of the mound within 2 m of the basin infiltration surface. As discussed previously, this is just adequate to maintain design infiltration rates. The design might consider a shorter (say, 2-day) flooding period, as discussed in Chapter 8, to reduce the potential for mounding somewhat.

6. Use Figure 3.5 to determine the lateral spread of the mound. Use the curve for $W/(4(\alpha t)^{1/2}$ with the previously calculated value of 1.28, enter the graph with selected values of x/W (where x is the lateral distance of concern), and read values of $h_m/(R)(t)$. Find the depth to the top of the mound 10 m from the centerline of basin:

$$x/W = 10/26.5 = 0.377$$

Enter the x/W axis with this value, project up to $W/(4\alpha t)^{1/2} = 1.28$, then read 0.58 on the $h_m/(R)(t)$ axis:

$$h_m = (0.58)(2)(3) = 3.48 \text{ m}$$

The depth to the mound at the 10-m point is 6 m – 3.48 m = 2.52 m. Similarly, at $x = 13$ m, the depth to the mound is 3.72 m, and at $x = 26$ m the depth to the mound is 5.6 m. This indicates that the water level is almost back to the normal groundwater level at a lateral distance about equal to two times the basin width. Changing the application schedule to 2 days instead of 3 would reduce the peak water level to about 3 m below the infiltration surface of the basin.

The procedure demonstrated in Example 3.3 is valid for a single basin; however, as described in Chapter 8, SAT systems typically include multiple basins that are loaded sequentially, and it is not appropriate to do the mounding calculation by assuming that the entire treatment area is uniformly loaded at the design hydraulic loading rate. In many situations, this will result in the erroneous conclusion that mounding will interfere with system operation.

It is necessary first to calculate the rise in the mound beneath a single basin during the flooding period. When hydraulic loading stops at time t, a uniform hypothetical discharge is assumed starting at t and continuing for the balance of the rest period. The algebraic sum of these two mound heights then approximates the mound shape just prior to the start of the next flooding period. Because adjacent basins may be flooded during this same period, it is also necessary to determine the lateral extent of their mounds and then add any increment from these sources to determine the total mound height beneath the basin of concern. The procedure is illustrated by Example 3.4.

Example 3.4

Determine the groundwater mound height beneath a SAT basin at the end of the operational cycle. Assume that the basin is square, 26.5 m on a side, and is one in a set of four arranged in a row (26.5 m wide by 106 m long). Assume the same site conditions as in Example 3.3. Also assume that flooding commences in one of the adjacent basins as soon as the rest period for the basin of concern begins. The operational cycle is 2 days flood, 12 days rest.

Solution

1. The maximum rise beneath the basin of concern would be the same as calculated in Example 3.3 with 2-day flooding: $h_m = 3.00$ m.
2. The influence from the next 2 days of flooding in the adjacent basin would be about equal to the mound rise at the 26-m point calculated in Example 3.3, or 0.4 m. All the other basins are beyond the zone of influence, so the maximum potential rise beneath the basin of concern is $(3.00) + (0.4) = 3.4$ m. The mound will actually not rise that high, because during the 2 days the adjacent basin is being flooded the first

basin is draining. However, for the purposes of this calculation, assume that the mound will rise the entire 3.4 m above the static groundwater table.

3. The R value for this "uniform" discharge will be the same as that calculated in Example 3.2, but t will now be 12 days: $(R)(t) = (2)(12) = 24$ m/d.

4. Calculate a new $W/(4\alpha t)^{1/2}$, as the "new" time is 12 days:

$$W/(4\alpha t)^{1/2} = 26.5/[(4)(35.7)(12)]^{1/2} = 0.62$$

5. Use Figure 3.3 to determine "h_m"$/(R)(t) = 0.30$: "h_m" = $(24)(0.3) = 7.2$ m. This is the hypothetical drop in the mound that could occur during the 10-day rest period; however, the water level cannot actually drop below the static groundwater table, so the maximum possible drop would be 3.4 m. This indicates that the mound would dissipate well before the start of the next flooding cycle. Assuming that the drop occurs at a uniform rate of 0.72 m/d, the 3.4-m mound will be gone in 4.7 days.

In cases where the groundwater mounding analysis indicates potential interference with system operation, several corrective options are available. As described in Chapter 8, the flooding and drying cycles can be adjusted or the layout of the basin sets rearranged into a configuration with less inter-basin interference. The final option is to underdrain the site to control mound development physically.

Underdrainage may also be required to control shallow or seasonal natural groundwater levels when they might interfere with the operation of either a land or aquatic treatment system. Underdrains are also sometimes used to recover the treated water beneath land treatment systems for beneficial use or discharge elsewhere.

3.1.4 UNDERDRAINAGE

In order to be effective, drainage or water recovery elements must either be at or within the natural groundwater table or just above some other flow barrier. When drains can be installed at depths of 5 m (16 ft) or less, underdrains are more effective and less costly than a series of wells. It is possible using modern techniques to install semiflexible plastic drain pipe enclosed in a geotextile membrane by means of a single machine that cuts and then closes the trench.

In some cases, underdrains are a project necessity to control a shallow groundwater table so the site can be developed for wastewater treatment. Such drains, if effective for groundwater control, will also collect the treated percolate from a land treatment operation. The collected water must be discharged, so the use of underdrains in this case converts the project to a surface-water discharge system unless the water is otherwise used or disposed of. In a few situations, drains have been installed to control a seasonally high water table. This type of system may

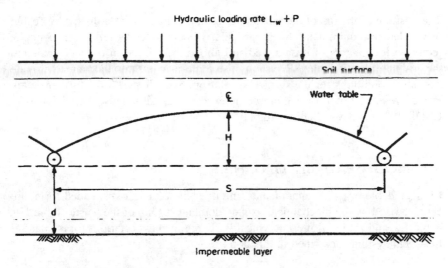

FIGURE 3.7 Definition sketch for calculation of drain spacing.

require a surface-water discharge permit during the period of high groundwater but will function as a nondischarging system for the balance of the year.

The drainage design consists of selecting the depth and spacing for placement of the drain pipes or tiles. In the typical case, drains may be at a depth of 1 to 3 m (3 to 10 ft) and spaced 60 m (200 ft) or more apart. In sandy soils, the spacing may approach 150 m (500 ft). The closer spacings provide better water control, but the costs increase significantly.

The Hooghoudt method (Luthin, 1973) is the most commonly used method for calculating drain spacing. The procedure assumes that the soil is homogeneous, that the drains are spaced evenly apart, that Darcy's law is applicable, that the hydraulic gradient at any point is equal to the slope of the water table above that point, and that a barrier of some type underlies the drain. Figure 3.7 defines the necessary parameters for drain design, and Equation 3.16 can be used for design:

$$S = \left[\left(\frac{(4)(K_h)(h_m)}{L_w + P} \right) (2d + h_m) \right]^{1/2} \tag{3.16}$$

where

S = Drain spacing (ft; m).
K_h = Horizontal permeability of the soil (ft/d; m/d).
h_m = Height of groundwater mound above the drains (ft; m).
L_w = Annual wastewater loading rate expressed as a daily rate (ft/d; m/d).
P = Average annual precipitation expressed as a daily rate (ft/d; m/d).
d = Distance from drain to barrier (ft; m).

The position of the top of the mound between the drains is established by design or regulatory requirements for a particular project. SAT systems, for example, require a few meters of unsaturated soil above the mound in order to maintain the design infiltration rate; SR systems also require an unsaturated zone to provide desirable conditions for the surface vegetation. See Chapter 8 for further detail. Procedures and criteria for more complex drainage situations can be found in USDI (1978) and Van Schifgaarde (1974).

3.2 BIODEGRADABLE ORGANICS

Biodegradable organic contaminants, in either dissolved or suspended form, are characterized by the biochemical oxygen demand (BOD) of the waste. Table 1.1, Table 1.2, and Table 1.3 present typical BOD removal expectations for the natural treatment systems described in this book.

3.2.1 Removal of BOD

As explained in Chapters 4 through 7, the biological oxygen demand (BOD) loading can be the limiting design factor for pond, aquatic, and wetland systems. The basis for these limits is the maintenance of aerobic conditions within the upper water column in the unit and the resulting control of odors. The natural sources of dissolved oxygen (DO) in these systems are surface reaeration and photosynthetic oxygenation. Surface reaeration can be significant under windy conditions or if surface turbulence is created by mechanical means. Observation has shown that the DO in unaerated wastewater ponds varies almost directly with the level of photosynthetic activity, being low at night and early morning, and rising to a peak in the early afternoon. The phytosynthetic responses of algae are controlled by the presence of light, the temperature of the liquid, and the availability of nutrients and other growth factors.

Because algae are difficult to remove and can represent an unacceptable level of suspended solids in the effluent, some pond and aquaculture processes utilize mechanical aeration as the oxygen source. In partially mixed aerated ponds, the increased depth of the pond and the partial mixing of the somewhat turbid contents limit the development of algae as compared to a facultative pond. Most wetland systems (Chapters 6 and 7) restrict algae growth, as the vegetation limits the penetration of light to the water column.

Emergent plant species used in wetlands treatment have the unique capability to transmit oxygen from the leaf to the plant root. These plants do not themselves remove the BOD directly; rather, they serve as hosts for a variety of attached growth organisms, and it is this microbial activity that is primarily responsible for the organic decomposition. The stems, stalks, roots, and rhizomes of the emergent varieties provide the necessary surfaces. This dependence requires a relatively shallow reactor and a relatively low flow velocity to ensure optimum contact opportunities between the wastewater and the attached microbial growth.

TABLE 3.2
Typical Organic Loading Rates for Natural
Treatment Systems

Process	Organic Loading (kg/ha/d)
Oxidation pond	40–120
Facultative pond	22–67
Aerated partial-mix pond	50–200
Hyacinth pond	20–50
Constructed wetland	100
Slow rate land treatment	45–450
Rapid infiltration land treatment	130–890
Overland flow land treatment	35–100
Land application of municipal sludge	27–930[a]

[a] These values were determined by dividing the annual rate by 365 days.

Wu et al. (2001) reported that little oxygen escaped from the roots of *Typha latifolia* in a constructed wetland, and in this system the major pathway of oxygen was atmospheric diffusion. These results were reported to be species specific, and other results for *Spartina pectinata* by Wu et al. (2000) indicate that the potential oxygen release could be 15 times that for *T. latifolia*. They also concluded that the amount of oxygen transferred to the wetlands through macrophyte roots and atmospheric diffusion were relatively small compared to the amount of oxygen required to oxidize ammonia.

The BOD of the wastewater or sludge is seldom the limiting design factor for the land treatment processes described in Chapter 8. Other factors, such as nitrogen, metals, toxics, or the hydraulic capacity of the soils, control the design so the system almost never approaches the upper limits for successful biodegradation of organics. Table 3.2 presents typical organic loadings for natural treatment systems.

3.2.2 REMOVAL OF SUSPENDED SOLIDS

The suspended solids content of wastewater is not usually a limiting factor for design, but the improper management of solids within the system can result in process failure. One critical concern for both aquatic and terrestrial systems is the attainment of proper distribution of solids within the treatment reactor. The use of inlet diffusers in ponds, step feed (multiple inlets) in wetland channels, and higher pressure sprinklers in industrial overland-flow systems is intended to achieve a more uniform distribution of solids and avoid anaerobic conditions at the head of the process. The removal of suspended solids in pond systems depends primarily on gravity sedimentation, and, as mentioned previously, algae can be a concern in some situations. Sedimentation and entrapment in the microbial growths are both contributing factors in wetland and overland-flow processes. Filtration in the soil matrix is the principal mechanism for SR and SAT systems. Removal expectations for the various processes are listed in Table 1.1, Table 1.2, and Table 1.3. Removal will typically exceed secondary treatment levels, except for some of the pond systems that contain algal solids in their effluents.

3.3 ORGANIC PRIORITY POLLUTANTS

Many organic priority pollutants are resistant to biological decomposition. Some are almost totally resistant and may persist in the environment for considerable periods of time; others are toxic or hazardous and require special management.

3.3.1 REMOVAL METHODS

Volatilization, adsorption, and then biodegradation are the principal methods for removing trace organics in natural treatment systems. Volatilization can occur at the water surface of ponds, wetlands, and SAT basins; in the water droplets from sprinklers used in land treatment; from the liquid films in overland-flow systems; and from the exposed surfaces of sludge. Adsorption occurs primarily on the organic matter in the treatment system that is in contact with the waste. In many cases, microbial activity then degrades the adsorbed materials.

3.3.1.1 Volatilization

The loss of volatile organics from a water surface can be described using first-order kinetics, because it is assumed that the concentration in the atmosphere above the water surface is essentially zero. Equation 3.17 is the basic kinetic equation, and Equation 3.18 can be used to determine the "half-life" of the contaminant of concern (see Chapter 9 for further discussion of the half-life concept and its application to sludge organics):

$$\frac{C_t}{C_0} = e^{-(k_{vol})(t)(y)}$$

(3.17)

where
C_t = Concentration at time t (mg/L or g/L).
C_0 = Initial concentration at $t = 0$ (mg/L or g/L).
k_{vol} = Volatilization mass transfer coefficient (cm/hr) = $(k)(y)$.
k = Overall rate coefficient (hr^{-1}).
y = Depth of liquid (cm).

$$t_{1/2} = \frac{(0.693)(y)}{k_{vol}} \qquad (3.18)$$

where $t_{1/2}$ is the time when concentration $C_t = (1/2)(C_0)$ (hr), and the other terms are as defined previously.

The volatilization mass transfer coefficient is a function of the molecular weight of the contaminant and the air/water partition coefficient as defined by the Henry's law constant, as shown by Equation 3.19:

$$k_{vol} = \left(\frac{B_1}{y}\right)\left(\frac{H}{\left(B_2 + H\right)(M)^{1/2}}\right) \qquad (3.19)$$

where
k_{vol} = Volatilization coefficient (hr^{-1}).
H = Henry's law constant (10^5 atm·m^3·mol^{-1}).
M = Molecular weight of contaminant of concern (g/mol).

The coefficients B_1 and B_2 are specific to the physical system of concern. Dilling (1977) determined values for a variety of volatile chlorinated hydrocarbons at a well-mixed water surface:

$$B_1 = 2.211, B_2 = 0.01042$$

Jenkins et al. (1985) experimentally determined values for a number of volatile organics on an overland flow slope:

$$B_1 = 0.2563, B_2 = (5.86)(10^{-4})$$

The coefficients for the overland-flow case are much lower because the flow of liquid down the slope is nonturbulent and may be considered almost laminar flow (Reynolds number = 100 – 400). The average depth of flowing liquid on this slope was about 1.2 cm (Jenkins et al., 1985).

Using a variation of Equation 3.19, Parker and Jenkins (1986) determined volatilization losses from the droplets at a low-pressure, large-droplet wastewater sprinkler. In this case, the y term in the equation is equal to the average droplet radius; as a result, their coefficients are valid only for the particular sprinkler system used. The approach is valid, however, and can be used for other sprinklers and operating pressures. Equation 3.20 was developed by Parker and Jenkins for the organic compounds listed in Table 3.3:

$$\ln\frac{C_t}{C_0} = (4.535)\left[k'_{vol} + 11.02(10^{-4})\right] \qquad (3.20)$$

TABLE 3.3
Volatile Organic Removal
by Wastewater Sprinkling

Substance	Calculated k'_{vol} for Equation 3.20 (cm/min)
Chloroform	0.188
Benzene	0.236
Toluene	0.220
Chlorobenzene	0.190
Bromoform	0.0987
m-Dichlorobenzene	0.175
Pentane	0.260
Hexane	0.239
Nitrobenzene	0.0136
m-Nitrotoluene	0.0322
PCB 1242	0.0734
Naphthalene	0.114
Phenanthrene	0.0218

Source: Parker, L.V. and Jenkins, T.F., *Water Res.*,
20(11), 1417–1426, 1986. With permission.

Volatile organics can also be removed by aeration in pond systems. Clark et al. (1984a) developed Equation 3.21 to determine the amount of air required to strip a given quantity of volatile organics from water via aeration:

$$\left(\frac{A}{W}\right) = (76.4)\left(1 - \frac{C_t}{C_0}\right)^{12.44}(S)^{0.37}(V)^{-0.45}(M)^{-0.18}(0.33)^s \qquad (3.21)$$

where
(A/W) = Air-to-water ratio.
S = Saturated condition of the compound of concern equal to 0, for unsaturated organics; 1, for saturated compounds).
V = Vapor pressure (mmHg).
M = Molecular weight (g/mol).
s = Solubility of organic compound (mg/L).

The values in Table 3.4 can be used in Equation 3.21 to calculate the air-to-water ratio required for some typical volatile organics.

TABLE 3.4
Properties of Selected Volatile Organics
for Equation 3.21

Chemical	M	S	s
Trichloroethylene	132	1000	0
1,1,1-Trichloroethane	133	5000	1
Tetrachloroethlyene	166	145	0
Carbon tetrachloride	154	800	1
cis-1,2-Dichloroethylene	97	3500	0
1,2-Dichloroethane	99	8700	1
1,1-Dichloroethylene	97	40	0

Source: Love, O.T. et al., *Treatment of Volatile Organic Chemicals in Drinking Water*, EPA 600/8-83-019, U.S. Environmental Protection Agency, Municipal Engineering Research Laboratory, Cincinnati, OH, 1983.

3.3.1.2 Adsorption

Sorption of trace organics to the organic matter present in the treatment system is thought to be the primary physicochemical mechanism of removal (USEPA, 1982a). The concentration of the trace organic that is sorbed relative to that in solution is defined by a partition coefficient K_p, which is related to the solubility of the chemical. This value can be estimated if the octanol–water partition coefficient (K_{ow}) and the percentage of organic carbon in the system are defined, as shown by Equation 3.22:

$$\log K_{oc} = (1.00)(\log K_{ow}) - 0.21 \qquad (3.22)$$

where

K_{oc} = Sorption coefficient expressed on an organic carbon basis equal to K_{sorb}/O_c.

K_{sorb} = Sorption mass transfer coefficient (cm/hr).

O_c = Percentage of organic carbon present in the system.

K_{ow} = Octanol–water partition coefficient.

Hutchins et al. (1985) presented other correlations and a detailed discussion of sorption in soil systems.

   Jenkins et al. (1985) determined that sorption of trace organics on an overland-flow slope could be described with first-order kinetics with the rate constant defined by Equation 3.23:

$$k_{sorb} = \left(\frac{B_3}{y}\right)\left(\frac{K_{ow}}{(B_4 + K_{ow})(M)^{1/2}}\right) \tag{3.23}$$

where

k_{sorb} = Sorption coefficient (hr^{-1}).

B_3 = Coefficient specific to the treatment system, equal to 0.7309 for the overland-flow system studied.

y = Depth of water on the overland-flow slope (1.2 cm).

K_{ow} = Octanol–water partition coefficient.

B_4 = Coefficient specific to the treatment system = 170.8 for the overland-flow system studied.

M = Molecular weight of the organic chemical (g/mol).

In many cases, the removal of trace organics is due to a combination of sorption and volatilization. The overall process rate constant (k_{sv}) is then the sum of the coefficients defined with Equations 3.19 and 3.23, and the combined removal is described by Equation 3.24:

$$\frac{C_t}{C_0} = e^{(k_{sv})(t)} \tag{3.24}$$

where

C_t = Concentration at time t (mg/L or µg/L).

C_0 = Initial concentration at t equal to 0 (mg/L or µg/L).

k_{sv} = Overall rate constant for combined volatilization and sorption equal to $k_{vol} + k_{sorb}$.

Table 3.5 presents the physical characteristics of a number of volatile organics for use in the equations presented above for volatilization and sorption.

Example 3.5

Determine the removal of toluene in an overland-flow system. Assume a 30-m-long terrace; hydraulic loading of 0.4 m^3·hr·m (see Chapter 8 for discussion); mean residence time on slope of 90 min; wastewater application with a low-pressure, large-droplet sprinkler; physical characteristics for toluene (Table 3.5) of K_w = 490, H = 515, M = 92; depth of flowing water on the terrace = 1.5 cm; concentration of toluene in applied wastewater = 70 µg/L.

Solution

1. Use Equation 3.20 to estimate volatilization losses during sprinkling:

$$\ln\frac{C_t}{C_0} = 4.535\left[k'_{vol} + 11.02(10)^{-4}\right]$$

$$C_t = 70\left[e^{-(4.535)(0.220)+0.00112)}\right] = 25.6\,\mu\text{g}/\text{L}$$

TABLE 3.5
Physical Characteristics for Selected Organic Chemicals

Substance	K_{ow}[a]	H[b]	Vapor Pressure[c]	M[d]
Chloroform	93.3	314	194	119
Benzene	135	435	95.2	78
Toluene	490	515	28.4	92
Chlorobenzene	692	267	12	113
Bromoform	189	63	5.68	253
m-Dichlorobenzene	2.4×10^3	360	2.33	147
Pentane	1.7×10^3	125,000	520	72
Hexane	7.1×10^3	170,000	154	86
Nitrobenzene	70.8	1.9	0.23	123
m-Nitrotoluene	282	5.3	0.23	137
Diethylphthalate	162	0.056	7×10^{-4}	222
PCB 1242	3.8×10^5	30	4×10^{-4}	26
Naphthalene	2.3×10^3	36	8.28×10^{-2}	128
Phenanthrene	2.2×10^4	3.9	2.03×10^{-4}	178
2,4-Dinitrophenol	34.7	0.001	—	184

[a] Octanol-water partition coefficient.
[b] Henry's law constant, 10^5 atm-m³/mol at 20°C and 1 atm.
[c] At 25°C.
[d] Molecular weight (g/mol).

2. Use Equation 3.19 to determine the volatilization coefficient during flow on the overland-flow terrace:

$$k_{vol} = \left(\frac{B_1}{y}\right)\left(\frac{H}{(B_2 + H)(M)^{1/2}}\right)$$

$$k_{vol} = \left(\frac{0.2563}{1.5}\right)\left(\frac{515}{(5.86)(10)^{-4} + (515)(92)^{1/2}}\right)$$

$$= (0.17087)(0.1042)$$

$$= 0.0178$$

3. Use Equation 3.23 to determine the sorption coefficient during flow on the overland-flow terrace:

TABLE 3.6
Removal of Organic Chemicals in Land Treatment Systems

	Slow Rate[a]		Overland Flow[b] (%)	Rapid Infiltration[c] (%)
Substance	Sandy Soil (%)	Silty Soil (%)		
Chloroform	98.57	99.23	96.5	>99.99
Toluene	>99.99	>99.99	99.00	99.99
Benzene	>99.99	>99.99	98.09	>99.99
Chlorobenzene	99.97	99.98	98.99	>99.99
Bromoform	99.93	99.96	97.43	>99.99
Dibromochloromethane	99.72	99.72	98.78	>99.99
m-Nitrotoluene	>99.99	>99.99	94.03	—[d]
PCB 1242	>99.99	>99.99	96.46	>99.99
Naphthalene	99.98	99.98	98.49	96.15
Phenanthrene	>99.99	>99.99	99.19	—
Pentachlorophenol	>99.99	>99.99	98.06	—
2,4-Dinitrophenol	—	—	93.44	—
Nitrobenzene	>99.99	>99.99	88.73	—
m-Dichlorobenzene	>99.99	>99.99	—	82.27
Pentane	>99.99	>99.99	—	—
Hexane	99.96	99.96	—	—
Diethylphthalate	—	—	—	90.75

[a] Parker and Jenkins (1986).
[b] Jenkins et al. (1985).
[c] Love et al. (1983).
[d] Not reported.

$$k_{sorb} = \left(\frac{B_3}{y}\right)\left(\frac{K_{ow}}{(B_4 + K_{ow})(M)^{1/2}}\right)$$

$$k_{sorb} = \left(\frac{0.7309}{1.5}\right)\left(\frac{490}{(170.8 + 490)(92)^{1/2}}\right)$$

$$k_{sorb} = (0.4873)(0.0774)$$

$$k_{sorb} = 0.0377$$

4. The overall rate constant is the sum of k_{vol} and k_{sorb}:

$$k_t = 0.0178 + 0.0377 = 0.055$$

5. Use Equation 3.24 to determine the toluene concentration in the overland-flow runoff:

$$\frac{C_t}{C_0} = e^{-(k_t)(t)}$$

$$C_t = (25.68)\left[e^{-(0.0555)(90)}\right] = 0.17\,\mu g\,/\,L$$

This represents about 99.8% removal.

3.3.2 REMOVAL PERFORMANCE

The land treatment systems are the only natural treatment systems that have been studied extensively to determine the removal of priority-pollutant organic chemicals. This is probably due to the greater concern about groundwater contamination with these systems. Results from these studies have been generally positive. As indicated previously, the more soluble compounds such as chloroform tend to move through the soil system more rapidly than the less soluble materials such as some PCBs. In all cases, the amount escaping the treatment system with percolate or effluent is very small. Table 3.6 presents removal performance for the three major land treatment concepts. The removals observed in the SR system were after 1.5 m of vertical travel in the soils indicated, and a low-pressure, large-droplet sprinkler was used for the application. The removals in the OF system were measured after flow on a terrace about 30 m long, with application via gated pipe at the top of the slope at a hydraulic loading of 0.12 $m^3 \cdot m \cdot hr$. The SAT data were obtained from wells about 200 m downgradient of the application basins.

The removals reported in Table 3.6 for SR systems represent concentrations in the applied wastewater ranging from 2 to 111 µg/L and percolate concentrations ranging from 0 to 0.4 µg/L. The applied concentrations in the overland-flow system ranged from 25 to 315 µg/L and the effluent from 0.3 to 16 µg/L. Concentrations of the reported substances applied to the SAT system ranged from 3 to 89 µg/L, and the percolate ranged from 0.1 to 0.9.

The results in Table 3.6 indicate that the SR system was more consistent and gave higher removals than the other two concepts. This is probably due to the use of the sprinkler and the enhanced opportunity for sorption on the organic matter in these finer-textured soils. Chloroform was the only compound to appear consistently in the percolate, and that was at very low concentrations. Although they were slightly less effective than SR, the other two concepts still produced very high removals. If sprinklers had been used in the OF system, it is likely that the removals would have been even higher. Based on these data, it appears that all three concepts are more effective for trace organic removal than activated sludge and other conventional mechanical treatment systems.

Quantitative relationships have not yet been developed for trace organic removal from natural aquatic systems. The removal due to volatilization in pond and free water surface wetland systems can at least be estimated with Equations 3.19 and 3.24. The liquid depth in these systems is much greater than on an OF slope, but the detention time is measured in terms of many days instead of minutes, so the removal can still be very significant. Organic removal in subsurface flow wetlands may be comparable to the SAT values in Table 3.6, depending on the media used in the wetland. See Chapters 6 and 7 for data on removal of priority pollutants in constructed wetlands.

In a modification of land treatment, Wang et al. (1999) have demonstrated the successful removal by hybrid poplar trees (H11-11) of carbon tetrachloride (15 mg/L in solution). The plant degrades and dechlorinates the carbon tetrachloride and releases the chloride ions to the soil and carbon dioxide to the atmosphere. Indian mustard and maize have been studied for the removal of metals from contaminated soils (Lombi et al., 2001). Alfalfa has been used to remediate a fertilizer spill (Russelle et al., 2001).

In microcosm studies, Bankston et al. (2002) concluded that trichloroethylene (TCE) could be attenuated in natural wetlands which would imply that similar results would be expected in constructed wetlands. The presence of broad-leaved cattails increased the rate of mineralization of TCE above that observed by the indigenous soil microorganisms.

3.3.3 Travel Time in Soils

The rate of movement of organic compounds in soils is a function of the velocity of the carrier water, the organic content of the soil, the octanol–water partition coefficient for the organic compound, and other physical properties of the soil system. Equation 3.25 can be used to estimate the movement velocity of an organic compound during saturated flow in the soil system:

$$V_c = \frac{(K)(G)}{n - (0.63)(p)(O_c)(K_{ow})} \qquad (3.25)$$

where
V_c = Velocity of organic compound (ft/d; m/d).
K = Saturated permeability of soil (ft/d; m/d), in vertical or horizontal direction.
K_v = Saturated vertical permeability (ft/d; m/d).
K_h = Saturated horizontal permeability (ft/d; m/d).
G = Hydraulic gradient of flow system (ft/ft; m/m), equal to 1 for vertical flow.
   = $\Delta H/\Delta L$ for horizontal flow (ft/ft; m/m); see Equation 3.4 for definition.
n = Porosity of the soil (%, as a decimal); see Figure 2.4.

p = Bulk density of soil (lb/in.^{3}; g/cm^3).

O_c = Organic content of soil (%, as a decimal).

K_{ow} = Octanol–water partition coefficient.

3.4 PATHOGENS

Pathogenic organisms may be present in both wastewaters and sludges, and their control is one of the fundamental reasons for waste management. Many regulatory agencies specify bacterial limits on discharges to surface waters. Other potential risks are impacts on groundwaters from both aquatic and land treatment systems, the contamination of crops or infection of grazing animals on land treatment sites, and the off-site loss of aerosolized organisms from pond aerators or land treatment sprinklers. Investigations have shown that the natural aquatic, wetland, and land treatment concepts provide very effective control of pathogens (Reed et al., 1979).

3.4.1 AQUATIC SYSTEMS

The removal of pathogens in pond-type systems is due to natural die-off, predation, sedimentation, and adsorption. Helminths, Ascaris, and other parasitic cysts and eggs settle to the bottom in the quiescent zone of ponds. Facultative ponds with three cells and about 20 days' detention time and aerated ponds with a separate settling cell prior to discharge provide more than adequate helminth and protozoa removal. As a result, there is little risk of parasitic infection from pond effluents or from use of such effluents in agriculture. Some risk may arise when sludges are removed for disposal. These sludges can be treated, or temporary restrictions on public access and agricultural use can be placed on the disposal site.

3.4.1.2 Bacteria and Virus Removal

The removal of both bacteria and viruses in multiple-cell pond systems is very effective for both the aerated and unaerated types, as shown in Table 3.7 and Table 3.8. The effluent in all three of the cases in Table 3.8 was undisinfected. The viruses measured were the naturally occurring enteric types and not seeded viruses or bacteriophage. Table 3.8 presents seasonal averages; see Bausum (1983) for full details. The viral concentrations in the effluent were consistently low at all times, although, as shown in the table, the removal efficiency did drop slightly in the winter at all three locations.

Numerous studies have shown that the removal of fecal coliforms in ponds depends on detention time and temperature. Equation 3.25 can be used to estimate the removal of fecal coliforms in pond systems. The detention time used in the equation is the actual detention time in the system as measured by dye studies. The actual detention time in a pond can be as little as 45% of the theoretical design detention time due to short-circuiting of flow. If dye studies are not practical or possible, it would be conservative to assume for Equation 3.26 that the "actual" detention time is 50% of the design residence time:

TABLE 3.7
Fecal Coliform Removal in Pond Systems

Location	Number of Cells	Detention Time (d)	Fecal Coliforms (No./100 mL)	
			Influent	Effluent
Facultative ponds				
Peterborough, New Hampshire	3	57	4.3×10^6	3.6×10^5
Eudora, Kansas	3	47	2.4×10^6	2.0×10^2
Kilmichael, Mississippi	3	79	12.8×10^6	2.3×10^4
Corinne, Utah	7	180	1.0×10^6	7.0×10^0
Partial-mix aerated ponds				
Windber, Pennsylvania	3	30	1×10^6	3.0×10^2
Edgerton, Wisconsin	3	30	1×10^6	3.0×10^1
Pawnee, Illinois	3	60	1×10^6	3.3×10^1
Gulfport, Mississippi	2	26	1×10^6	1.0×10^5

Source: USEPA, *Design Manual Municipal Wastewater Stabilization Ponds*, EPA 625/1-83-015, U.S. Environmental Protection Agency, Center for Environmental Research Information, Cincinnati, OH, 1983.

$$\frac{C_f}{C_i} = \frac{1}{\left(1 + t(k_T)\right)^n} \qquad (3.26)$$

where

C_f = Effluent fecal coli concentration (number/100 mL).

C_i = Influent fecal coli concentration (number/100 mL).

t = Actual detention time in the cell (d).

k_T = Temperature-dependent rate constant (d^{-1}), equal to $(2.6)(1.19)^{(T_w-20)}$.

T_w = Mean water temperature in pond (°C).

n = Number of cells in series.

See Chapter 4 for a method of determining the temperature in the pond; for the general case, it is safe to assume that the water temperature will be about equal to the mean monthly air temperature, down to a minimum of 2°C.

   Equation 3.26 in the form presented assumes that all cells in the system are the same size. See Chapter 4 for the general form of the equation when the cells are different sizes. The equation can be rearranged and solved to determine the

TABLE 3.8
Enteric Virus Removal in Facultative Ponds

Location	Enteric Virus (PFU/L)[a]	
	Influent	Effluent
Shelby, Mississippi (3 cells, 72 d)		
Summer	791	0.8
Winter	52	0.7
Spring	53	0.2
El Paso, Texas (3 cells, 35 d)		
Summer	348	0.6
Winter	87	1
Spring	74	1.1
Beresford, South Dakota (2 cells, 62 d)		
Summer	94	0.5
Winter	44	2.2
Spring	50	0.4

[a] PFU/L, plaque-forming units per liter.

Source. Dausum, H.I., *Enteric Virus Removal in Wastewater Treatment Lagoon Systems*, PB83-234914, National Technical Information Service, Springfield, VA, 1983.

optimum number of cells needed for a particular level of pathogen removal. In general, a three- or four-cell (in series) system with an actual detention time of about 20 days will remove fecal coliforms to desired levels. Model studies with polio and coxsackie viruses indicated that the removal of viruses proceeds similarly to the first-order reaction described by Equation 3.26. Hyacinth ponds and similar aquatic units should also perform in accordance with Equation 3.26.

3.4.2 WETLAND SYSTEMS

Pathogen removal in many wetland systems is due to essentially the same factors described above for pond systems. Equation 3.26 can also be used to estimate the removal of bacteria or virus in wetland systems where the water flow path is above the surface. The detention time will be less in most constructed wetlands as compared to ponds, but the opportunities for adsorption and filtration will be greater. The subsurface-flow wetland systems described in Chapter 7 remove pathogens in essentially the same ways as land treatment systems. Table 3.9 summarizes pathogen removal information for selected wetlands. A study of over

TABLE 3.9
Pathogen Removal in Constructed Wetland Systems

	System Performance	
Location	Influent	Effluent[a]
Santee, California (bullrush wetland)[b]		
Winter season (October–March)		
Total coliforms (number/100 mL)	5×10^7	1×10^5
Bacteriophage (PFU/mL)	1900	15
Summer season (April–September)		
Total coliforms (number/100 mL)	6.5×10^7	3×10^5
Bacteriophage (PFU/mL)	2300	26
Iselin, Pennsylvania (cattails and grasses)[c]		
Winter season (November–April)		
Fecal coliforms (number/100 mL)	1.7×10^6	6200
Summer season (May–October)		
Fecal coliforms (number/100 mL)	1.0×10^6	723
Arcata, California (bullrush wetland)[d]		
Winter season		
Fecal coliforms (number/100 mL)	4300	900
Summer season		
Fecal coliforms (number/100 mL)	1800	80
Listowel, Ontario (cattails)[d]		
Winter season		
Fecal coliforms (number/100 mL)	556,000	1400
Summer season		
Fecal coliforms (number/100 mL)	198,000	400

[a] Undisinfected.
[b] Gravel bed, subsurface flow.
[c] Sand bed, subsurface flow.
[d] Free water surface.

Source: Reed, S.C. et al., in *Proceedings AWWA Water Reuse III*, American Water Works Association, Denver, CO, 1985, 962–972.

40 constructed wetlands in Colorado was funded by the Colorado Governor's Office of Energy Management and Conservation (2000) and performed by HDR Engineering, Inc., and ERO Resources. The performance of and deficiencies in

the various systems were evaluated, and a comprehensive report is available by contacting the office listed in the references.

3.4.3 LAND TREATMENT SYSTEMS

Because land treatment systems in the United States are typically preceded by some form of preliminary treatment or a storage pond, parasites should be of little concern. The evidence in the literature of infection of grazing animals (Reed, 1979) is due to the direct ingestion of, or irrigation with, essentially raw waste-water. The removal of bacteria and viruses in land treatment systems is due to a combination of filtration, dessication, adsorption, radiation, and predation.

3.4.3.1 Ground Surface Aspects

The major concerns relate to the potential for the contamination of surface vegetation or off-site runoff, as the persistence of bacteria or viruses on plant surfaces could then infect people or animals if the plants were consumed raw. To eliminate these risks, it is generally recommended in the United States that agricultural land treatment sites not be used to grow vegetables that may be eaten raw. The major risk is then to grazing animals on a pasture irrigated with waste-water. Typical criteria specify a period ranging from 1 to 3 weeks after sprinkling undisinfected effluent before allowing animals to graze. Systems of this type are divided into relatively small paddocks, and the animals are moved in rotation around the site. Control of runoff is a design requirement of SR and SAT land treatment systems (as described in Chapter 8), so these sources should present no pathogenic hazard. Runoff of the treated effluent is the design intention of overland flow systems, which typically can achieve about 90% removal of applied fecal coliforms. It is a site-specific decision by the regulatory agency regarding the need for final disinfection of treated OF runoff. Overland flow slopes also collect precipitation of any intensity that may happen to occur. The runoff from these rainfall events can be more intense than the design treatment rate, but the additional dilution provided results in equal or better water quality than the normal runoff.

3.4.3.2 Groundwater Contamination

Because percolate from SR and SAT land treatment can reach groundwater aquifers, the risk of pathogenic contamination must be considered. The removal of bacteria and viruses from the finer-textured agricultural soils used in SR systems is quite effective. A 5-year study in Hanover, New Hampshire, demon-strated almost complete removal of fecal coliforms within the top 5 ft of the soil profile (Reed, 1979). Similar studies in Canada (Bell and Bole, 1978) indicated that fecal coliforms were retained in the top 8 cm (3 in.) of the soil. About 90% of the bacteria died within the first 48 hr, and the remainder was eliminated over the next 2 weeks. Virus removal, which depends initially on adsorption, is also very effective in these soils.

TABLE 3.10
Typical Pathogen Levels in Wastewater Sludges

Pathogen	Untreated (No./100 mL)	Anaerobically Digested (No./100 mL)
Viruses	2500–70,000	100–1000
Fecal coliforms	1.0×10^6	30,000–6×10^5
Salmonella	8000	3–62
Ascaris lumbricoides	200–1000	0–1000

The coarse-textured soils and high hydraulic loading rates used in SAT systems increase the risk of bacteria and virus transmission to groundwater aquifers. A considerable research effort, both in the laboratory and at operational systems, has focused on viral movement in SAT systems (Reed, 1979). The results of this work indicate minimal risk for the general case; movement can occur with very high viral concentrations if the wastewater is applied at very high loading rates on very coarse-textured soils. It is unlikely that all three factors will be present in the majority of cases. Chlorine disinfection prior to wastewater application in a SAT system is not recommended, as the chlorinated organic compounds formed represent a greater threat to the groundwater than does the potential transmission of a few bacteria or viruses.

3.4.4 SLUDGE SYSTEMS

As shown by the values in Table 3.10, the pathogen levels in raw and digested sludge can be quite high. The pathogen content of sludge is especially critical when the sludge is to be used in agricultural operations or when public exposure is a concern. The sludge utilization guidelines developed by the U.S. EPA are discussed in detail in Chapter 9. Sludge stabilization with earthworms (vermistabilization) is also described in Chapter 9, and some evidence suggests that a reduction in pathogenic bacteria occurs during the process. The freeze–dewatering process will not kill pathogens but can reduce the concentration in the remaining sludge due to enhanced drainage upon thawing. The reed-bed drying concept can achieve significant pathogen reduction due to desiccation and the long detention time in the system. Pathogens are further reduced after sludge is land applied, by the same mechanisms discussed previously for land application of wastewater. There is little risk of transmission of sludge pathogens to groundwater or in runoff to surface waters if the criteria in Chapter 9 are used in system design.

TABLE 3.11
Organism Concentration in Wastewater
and Downwind Aerosol

Organism	Wastewater Concentration [(No./100 mL) × 10⁶]	Aerosol Concentration at Edge of Sprinkler Impact Circle (No./m³ of air sampled)
Standard plate count	69.9	2578
Total coliforms	7.5	5.6
Fecal coliforms	0.8	1.1
Coliphage	0.22	0.4
Fecal streptococci	0.007	11.3
Pseudomonas	1.1	71.7
Klebsiella	0.39	<1.0
Clostridium perfringes	0.005	1.4

Source: Sorber, C.A. and Sagik, B.P., *in Wastewater Aerosols and Disease*, EPA 600/9-80-078, U.S. Environmental Protection Agency, Health Effects Research Laboratory, Cincinnati, OH, 1980, 23–35.

3.4.5 AEROSOLS

Aerosol particles may be up to 20 μm in diameter, which is large enough to transport bacteria or virus. Aerosols will be produced any time that liquid droplets are sprayed into the air, or at the boundary layer above agitated water surfaces, or when sludges are moved about or aerated. Aerosol particles can travel significant distances, and the contained pathogens remain viable until inactivated by desiccation or ultraviolet light. The downwind travel distance for aerosol particles depends on the wind speed, turbulence, temperature, humidity, and presence of any barrier that might entrap the particle. With the impact sprinklers commonly used in land application of wastewater, the volume of aerosols produced amounts to about 0.3% of the water leaving the nozzle (Sorber et al., 1976). If no barrier is present, the greatest travel distance will occur with steady, nonturbulent winds under cool, humid conditions, which are generally most likely to happen at night. The concentration of organisms entering a sprinkler nozzle should be no different than the concentration in the bulk liquid or sludge. Immediately after aerosolization, temperature, sunlight, and humidity have an immediate and significant effect on organism concentration. This *aerosol shock* is demonstrated in Table 3.11.

As the aerosol particle travels downwind, the microorganisms continue to die off at a slower, first-order rate due to desiccation, ultraviolet radiation, and possibly trace compounds in the air or in the aerosol. This die-off can be very significant for bacteria, but the rates for viruses are very slow so it is prudent to assume no further downwind inactivation of viruses by these factors. Equations 3.27 and 3.28 form a predictive model that can be used to estimate the downwind concentration of aerosol organisms:

$$C_d = \left(C_n\right)\left(D_d\right)(e)^{(xa)} + B \qquad (3.27)$$

where

C_d = Concentration at distance d (number/ft^3; number/m^3).
C_n = Concentration released at source (number/s).
D_d = Atmospheric diffusion factor (s/ft^3; s/m^3).
x = Decay or die-off rate (s^{-1}).
       = −0.023 for bacteria (derived for fecal coliforms).
       = 0.00 for viruses (assumed).
a = Downwind distance d/wind velocity (ft·ft·s; m·m·s),
B = Background concentration in upwind air (number/ft^3; number/m^3).

The initial concentration C_n leaving the nozzle area is a function of the original concentration in the bulk wastewater (W), the wastewater flow rate (F), the aerosolization efficiency (E), and a survival factor (I), all as described by Equation 3.28:

$$C_n = (W)(F)(E)(I) \qquad (3.28)$$

where

C_n = Organisms released at source (number/ft^3; number/m^3).
W = Concentration in bulk wastewater (number/100 mL).
F = Flow rate (0.631 gal/min; L/s).
E = Aerosolization efficiency.
       = 0.003 for wastewater.
       = 0.0004 for sludge spray guns.
       = 0.000007 for sludge applied with tank truck sprinklers.
I = Survival factor.
       = 0.34 for total coliforms.
       = 0.27 for fecal coliforms.
       = 0.71 for coliphage.
       = 3.6 for fecal streptococci.
       = 80.0 for enteroviruses.

The atmospheric dispersion factor (D_d) in Equation 3.27 depends on a number of related meteorological conditions. Typical values for a range of expected conditions are given below; USEPA (1982b) should be consulted for a more exact determination:

Field Condition	D_d (s/m^3)
Wind < 6 km/hr, strong sunlight	176×10^{-6}
Wind <6 km/hr, cloudy daylight	388×10^{-6}
Wind 6-16 km/hr, strong sunlight	141×10^{-6}
Wind 6-16 km/hr, cloudy daylight	318×10^{-6}
Wind > 16 km/hr, strong sunlight	282×10^{-6}
Wind > 16 km/hr, cloudy daylight	600×10^{-6}
Wind > 11 km/hr, night	600×10^{-6}

The following example illustrates the use of this predictive model.

Example 3.6

Find the fecal coliform concentration in aerosols 8 m downwind of a sprinkler impact zone. The sprinkler has a 23-m impact circle and is discharging at 30 L/s, fecal coliforms in the bulk wastewater are 1×10^5, the sprinkler is operating on a cloudy day with a wind speed of about 8 km/hr, and the background concentration of fecal coliforms in the upwind air is zero.

Solution

1. The distance of concern is 31 m downwind of the nozzle source, and the wind velocity is 2.22 m/s, so we can calculate the a factor:

$$a = \frac{\text{Downwind distance}}{\text{Wind velocity}} = \frac{31}{2.22} = 13.96 \text{ s}^{-1}$$

2. Calculate the concentration leaving the nozzle area using Equation 3.28:

$$C_n = (W)(F)(E)(I)$$
$$= (1 \times 10^5)(30 \text{ L/s})(0.003)(0.27)$$
$$= 2430 \text{ fecal coliforms released per second at the nozzle}$$

3. Calculate the concentration at the downwind point of concern using Equation 3.27:

$$D_d = 318 \times 10^{-6}$$
$$C_d = (C_n)(D_d)(e)^{(xa)} + B$$
$$= (2340)(318 \times 10^{-6})(e)^{(-0.023)(13..96)} + 0.0$$
$$= 0.54 \text{ fecal coliforms per m}^3 \text{ of air,}$$
$$8 \text{ m downwind of the wetted zone of the sprinkler}$$

This is an insignificant level of risk.

The very low concentration predicted in Example 3.6 is typical of the very low concentrations actually measured at a number of operational land treatment sites. Table 3.12 provides a summary of data collected at an intensively studied

TABLE 3.12
Aerosol Bacteria and Viruses at Pleasanton, California, Land Treatment System Using Undisinfected Effluent

Location	Fecal Coliform	Fecal Streptococcus	Coliphage	Pseudomonas	Enteroviruses
Wastewater (number/100 mL)	1×10^5	8.8×10^3	2.6×10^5	2.6×10^5	2.8
Upwind (number/m^3)	0.02	0.23	0.01	0.03	ND[a]
Downwind (number/m^3):					
10–30 m	0.99	1.45	0.34	81	0.01
31–80 m	0.46	0.6	0.39	46	ND
81–200 m	0.23	0.42	0.21	25	ND

[a] ND, none detected

system where undisinfected effluent was applied to the land. It seems clear that the very low aerosolization efficiencies (E) as defined in Equation 3.27 for sludge spray guns and truck-mounted sprinklers indicate very little risk of aerosol transport of pathogens from these sources, and this has been confirmed by field investigations (Sorber et al., 1984)

Composting is a very effective process for inactivating most microorganisms, including viruses, due to the high temperatures generated during the treatment (see Chapter 9 for details); however, the heat produced in the process also stimulates the growth of thermophilic fungi and actinomycetes, and concerns have been expressed regarding their aerosol transport. The aerosols in this case are dust particles released when the compost materials are aerated, mixed, screened, or otherwise moved about the site.

A study was conducted at four composting operations involving 400 on-site and off-site workers (Clark et al., 1984b). The most significant finding was a higher concentration of the fungus *Aspergillus fumigatus* in the throat and nasal cultures of the actively involved on-site workers, but this finding was not correlated with an increased incidence of infection or disease. The fungus was rarely detected in on-site workers involved only occasionally or in the off-site control group.

The presence of *Aspergillus fumigatus* is due to the composting process itself and not because wastewater sludges are involved. The study results suggest that workers who are directly and frequently involved with composting operations have a greater risk of exposure, but the impact on those who are exposed only occasionally or on the downwind off-site population is negligible. It should be possible to protect all concerned with respirators for the exposed workers and a boundary screen of vegetation around the site.

3.5 METALS

Metals at trace-level concentrations are found in all wastewaters and sludges. Industrial and commercial activities are the major sources, but wastewater from private residences can also have significant metal concentrations. The metals of greatest concern are copper, nickel, lead, zinc, and cadmium, and the reason for the concern is the risk of their entry into the food chain or water supply. A large percentage of the metals present in wastewater will accumulate in the sludges produced during the wastewater treatment process. As a result, metals are often the controlling design parameter for land application of sludge, as described in detail in Chapter 9. Metals are not usually the critical design parameter for wastewater treatment or reuse, with the possible exception of certain industries. Table 3.13 compares the metal concentrations in untreated municipal wastewaters and the requirements for irrigation and drinking-water supplies.

TABLE 3.13
Metal Concentrations in Wastewater and Requirements for
Irrigation and Drinking Water Supplies

Metal	Untreated Wastewater[a] (mg/L)	Drinking Water (mg/L)	Irrigation (mg/L)	
			Continuous[b]	Short-Term[c]
Cadmium	<0.005	0.01	0.01	0.05
Lead	0.008	0.05	5.0	10.0
Zinc	0.04	0.05	2.0	10.0
Copper	0.18	1.0	0.2	5.0
Nickel	0.04	—	0.2	2.0

[a] Median values for typical municipal wastewater.
[b] For waters used for an infinite time period on any kind of soil.
[c] For waters used for up to 20 years on fine-textured soils when sensitive crops
   are to be grown.

Source: USEPA, *Process Design Manual Land Treatment of Municipal Wastewater*, EPA 625/1-81-013, U.S. Environmental Protection Agency, Center for Environmental Research Information, Cincinnati, OH, 1981.

3.5.1 AQUATIC SYSTEMS

Trace metals are not usually a concern for the design or performance of pond systems that treat typical municipal wastewaters. The major pathways for removal are adsorption on organic matter and precipitation. Because the opportunity for both is somewhat limited, the removal of metals in most pond systems will be less effective than with activated sludge — for example, where more than 50% of the metals present in the untreated wastewater can be transferred to the sludge in a relatively short time period. Sludges from pond systems can, however, contain relatively high concentrations of metals due to the long retention times and infrequent sludge removal. The metal concentrations found in lagoon sludges at several locations are summarized in Table 3.14. The concentrations shown in Table 3.14 are within the range normally found in unstabilized primary sludges and therefore would not inhibit further digestion or land application as described in Chapter 9. Table 9.4 and Table 9.5 in Chapter 9 list other characteristics of pond sludges. The data in Table 3.14 are from lagoons in cold climates. It is likely that sludge metal concentrations may be higher than these values in lagoons in warm climates that receive a significant industrial wastewater input. In these cases, the benthic sludge will undergo further digestion, which reduces the organic content and sludge mass but not the metals content so their concentrations should increase with time.

TABLE 3.14
Metal Concentrations in Sludges
from Treatment Lagoons

Metal	Facultative Lagoons[a]	Partial-Mix Aerated Lagoons[b]
Copper		
Wet sludge (mg/L)	3.8	10.1
Dry solids (mg/kg)	53.8	809.2
Iron		
Wet sludge (mg/L)	0.1	1.2
Dry solids (mg/kg)	9.0	9.2
Lead		
Wet sludge (mg/L)	8.9	21.1
Dry solids (mg/kg)	144	394
Mercury		
Wet sludge (mg/L)	0.1	0.2
Dry solids (mg/kg)	2.4	4.7
Zinc		
Wet sludge (mg/L)	54.6	85.2
Dry solids (mg/kg)	840	2729

[a] Average of values from two facultative lagoons in Utah.
[b] Average of values from two partial-mix aerated lagoons in Alaska.

Source: Schneiter, R.W. and Middlebrooks, E.J., *Cold Region Wastewater Lagoon Sludge: Accumulation, Characterization, and Digestion,* Contract Report DACA89-79-C0011, U.S. Cold Regions Research and Engineering Laboratory, Hanover, NH, 1981.

If metal removal is a process requirement and the local climate is close to subtropical, the use of water hyacinths in shallow ponds may be considered. Tests with full-scale systems in both Louisiana and Florida (Kamber, 1982) have documented excellent removal, with uptake by the plant itself being a major factor. The plant tissue concentrations may range from hundreds to thousands of times that of the water or sediment concentrations, indicating that bioaccumulation of trace elements by the plant occurs. Metal removals in a pilot hyacinth system in central Florida are presented in Table 3.15. Hyacinths have also been shown to be particularly effective in extracting metals from photoprocessing wastewater at a system in Louisiana (Kamber, 1982)

TABLE 3.15
Metal Removal in Hyacinth Ponds

Metal	Influent Concentration	Percent Removal[a]
Boron	0.14 mg/L	37
Copper	27.6 g/L	20
Iron	457.8 g/L	34
Manganese	18.2 g/l	37
Lead	12.8 g/L	68
Cadmium	0.4 g/L	46
Chromium	0.8 g/L	22
Arsenic	0.9 g/L	18

[a] Average of three parallel channels, with a detention time about 5 days.

Source: Kamber, D.M., *Benefits and Implementation Potential of Wastewater Aquaculture*, EPA Contract Report 68-01-6232, U.S. Environmental Protection Agency, Office of Water Regulations and Standards, Washington, D.C., 1982.

3.5.2 WETLAND SYSTEMS

Excellent metal removals have been demonstrated in the type of constructed wetlands described in Chapter 6 and Chapter 7. Tests at pilot wetlands in southern California, with about 5.5 days' hydraulic residence time, indicated 99, 97, and 99% removal for copper, zinc, and cadmium, respectively (Gersberg et al., 1983); however, plant uptake by the vegetation accounted for less than 1% of the metals involved. The major mechanisms responsible for metal removal were precipitation and adsorption interactions with the organic benthic layer.

3.5.3 LAND TREATMENT SYSTEMS

Removal of metals in land treatment systems can involve both uptake by any vegetation and adsorption, ion exchange, precipitation, and complexation in or on the soil. As explained in Chapter 9, zinc, copper, and nickel are toxic to vegetation long before they reach a concentration in the plant tissue that would represent a risk to human or animal food chains. Cadmium, however, can accumulate in many plants without toxic effects and may represent some health risk. As a result, cadmium is the major limiting factor for application of sludge on agricultural land.

The near-surface soil layer in land treatment systems is very effective for removal, and most retained metals are found in this zone. Investigations at a rapid-infiltration system that had operated for 33 years on Cape Cod, Massachusetts,

TABLE 3.16
Metal Content of Grasses at Land Treatment Sites

	Locations ¯Concentrations (mg/kg)					
	Melbourne, Australia Started 1896 Sampled 1972		Fresno, California Started 1907 Sampled 1973	Manteca, California Started 1961 Sampled 1973	Livermore, California Started 1964 Sampled 1973
Metal	Control Site	Measurement			
Cadmium	0.77	0.89	0.9	1.6	0.3
Copper	6.5	12.0	16.0	13.0	10.0
Nickel	2.7	4.9	5.0	45.0	2.0
Lead	2.5	2.5	13.0	15.0	10.0
Zinc	50.0	63.0	93.0	161.0	103.0

indicated that essentially all of the metals applied could be accounted for in the top 50 cm (20 in.) of the sandy soil, and over 95% were contained within the top 15 cm (6 in.) (Reed, 1979).

Although the metal concentrations in typical wastewaters is low, concerns have been expressed regarding long-term accumulation in the soil that might then affect the future agricultural potential of the site. Work by Hinesly and others as reported by Reed (1979) seems to indicate that most of the metals retained over a long period in the soil are in forms that are not readily available to most vegetation. The plants will respond to the metals applied during the current growing season but are not significantly affected by previous accumulations in the soil. The data in Table 3.16 demonstrate the same relationship. At Melbourne, Australia, after 76 years of application of raw sewage, the cadmium concentration in the grass was just slightly higher than in the grass on the control site, which received no wastewater. The other locations are newer systems in California, where the cadmium content is the same order of magnitude as measured at Melbourne, suggesting that the vegetation in all these locations is responding to the metals applied during the current growing season and not to prior soil accumulation. The significantly higher lead in the three California sites as compared to Melbourne is believed to be due to motor vehicle exhaust from adjacent highways.

Metals do not pose a threat to groundwater aquifers, even at the very high hydraulic loadings used in rapid-infiltration systems. Experience at Hollister, California, demonstrates that the concentration of cadmium in the shallow groundwater beneath the site is not significantly different than normal offsite groundwater quality (Pound and Crites, 1979). After 33 years of operation at this site, the accumulation of metals in the soil was still below or near the low end of the range normally expected for agricultural soils. Had the site been operated in the slow rate mode, it would have taken over 150 years to apply the same volume of wastewater and contained metals.

3.6   NUTRIENTS

A dual concern with respect to nutrients is that their control is necessary to avoid adverse health or environmental effects but the same nutrients are essential for the performance of the natural biological treatment systems discussed in this book. The nutrients of major importance for both purposes are nitrogen, phosphorus, and potassium. Nitrogen is the controlling parameter for the design of many land treatment and sludge application systems, and those aspects are discussed in detail in Chapter 6, Chapter 7, and Chapter 9. This section covers the potential for nutrient removal using the other treatment concepts and the nutrient requirements of the various system components.

3.6.1   NITROGEN

Nitrogen is limited in drinking water to protect the health of infants and may be limited in surface waters to protect fish life or to avoid eutrophication. As described in Chapter 8, land treatment systems are typically designed to meet the

10-mg/L nitrate drinking-water standard for any percolate or groundwater leaving the project boundary. In some cases, nitrogen removal may also be necessary prior to discharge to surface waters. More often, it is necessary to oxidize or otherwise remove the ammonia form of nitrogen, as this is toxic to many fish and can also represent a significant oxygen demand on the stream.

Nitrogen is present in wastewaters in a variety of forms because of the various oxidation states represented, and it can readily change from one state to another depending on the physical and biochemical conditions present. The total nitrogen concentration in typical municipal wastewaters ranges from about 15 to over 50 mg/L. About 60% of this is in ammonia form, and the remainder is in organic form.

Ammonia can be present as molecular ammonia (NH_3) or as ammonium ions (NH_4^+). The equilibrium between these two forms in water is strongly dependent on pH and temperature. At pH 7 essentially only ammonium ions are present, while at pH 12 only dissolved ammonia gas is present. This relationship is the basis for air-stripping operations in advanced wastewater treatment plants and for a significant portion of the nitrogen removal that occurs in wastewater treatment ponds.

3.6.1.1 Pond Systems

Nitrogen can be removed in pond systems by plant or algal uptake, nitrification and denitrification, adsorption, sludge deposition, and loss of ammonia gas to the atmosphere (volatilization). In facultative wastewater treatment ponds, the dominant mechanism is believed to be volatilization, and under favorable conditions up to 80% of the total nitrogen present can be lost. The rate of removal depends on pH, temperature, and detention time. The amount of gaseous ammonia present at near-neutral pH levels is relatively low, but when some of this gas is lost to the atmosphere additional ammonium ions shift to the ammonia form to maintain equilibrium. Although the unit rate of conversion and loss may be very low, the long detention time in these ponds compensates, resulting in very effective removal over the long term. Chapter 4 presents equations describing this nitrogen removal in ponds that can be used for design. Because nitrogen is often the controlling design parameter for land treatment, a reduction in pond effluent nitrogen can often permit a very significant reduction in the land area needed for wastewater application, with a comparable savings in project costs.

3.6.1.2 Aquatic Systems

Nitrogen removal in hyacinth ponds, due primarily to nitrification/denitrification and plant uptake, can be very effective. The plant uptake will not represent permanent removal, however, unless the plants are routinely harvested. A complete harvest is not typically possible, as another function of the hyacinth plant is to shade the water surface so restricted light penetration will limit algal growth. Because harvest might remove only 20 to 30% of the plants in the basin at any one time, the full nitrogen-removal potential of the plants is never realized.

Nitrification and denitrification are possible in shallow hyacinth ponds even if mechanical aeration is used, due to the presence of aerobic and anaerobic micro-sites within the dense root zone of the floating plant and the presence of the carbon sources needed for denitrification. Nitrogen removals observed in hyacinth ponds range from less than 10 to over 50 kg/ha/d (9 to 45 lb/ac/d), depending on the season and frequency of harvest. Some of these were carefully managed pilot-scale or research facilities.

3.6.1.3 Wetland Systems

Volatilization of ammonia, denitrification, and plant uptake (if the vegetation is harvested) are the potential methods of nitrogen removal in wetland systems (Gersberg et al., 1983). Studies in Canada (Wile et al., 1985) demonstrated that a regular harvest of cattails still accounted for only about 10% of the nitrogen removed by the system. These findings have been confirmed elsewhere, which indicates that the major pathway for nitrogen removal is nitrification followed by denitrification.

3.6.1.4 Land Treatment Systems

Nitrogen is usually the limiting design parameter for slow-rate land treatment of wastewater, and the criteria and procedures for nitrogen are presented in Chapter 8. Nitrogen can also limit the annual application rate for many sludge systems, as described in Chapter 9. The removal pathways for both types of systems are similar, and include plant uptake, ammonia volatilization, and nitrification/deni-trification. Ammonium ions can be adsorbed onto soil particles, thus providing a temporary control; soil microorganisms then nitrify this ammonium, restoring the original adsorptive capacity. Nitrate, on the other hand, will not be chemically retained by the soil system. Nitrate removal by plant uptake or denitrification can occur only during the hydraulic residence time of the carrier water in the soil profile. The overall capability for nitrogen removal will be improved if the applied nitrogen is ammonia or other less well-oxidized forms. Nitrification and denitri-fication are the major factors for nitrogen removal in rapid-infiltration systems, and crop uptake is a major method for both slow rate and overland flow systems. Volatilization and denitrification also occur with the latter two types of system and may account for from 10 to over 50% of the applied nitrogen, depending on waste characteristics and application methods, as described in Chapter 8. Design procedures based on nitrogen uptake of agricultural and forest vegetation can be found in Chapter 8.

3.6.2 Phosphorus

Phosphorus has no known health significance but is the wastewater constituent that is most often associated with eutrophication of surface waters. Phosphorus in wastewater can occur as polyphosphates, orthophosphates (which can originate from a number of sources), and organic phosphorus, which is more commonly found in industrial discharges. The potential removal pathways in natural treatment systems include vegetation uptake, other biological processes, adsorption, and precipitation. The vegetative uptake can be significant in the slow-rate and overland flow land treatment processes when harvest and removal are routinely practiced. In these cases, the harvested vegetation might account for 20 to 30% of the applied phosphorus. The vegetation typically used in wetland systems is not considered a significant factor for phosphorus removal, even if harvesting is practiced. If the plants are not harvested, their decomposition releases phosphorus back to the water in the system. Phosphorus removal by water hyacinths and other aquatic plants is limited to plant needs and will not exceed 50 to 70% of the phosphorus present in the wastewater, even with careful management and regular harvests.

Adsorption and precipitation reactions are the major pathways for phosphorus removal when wastewater has the opportunity for contact with a significant volume of soil. This is always the case with slow rate and rapid infiltration systems, as well as some wetland systems where infiltration and lateral flow through the subsoil are possible. The possibilities for contact between the wastewater and the soil are more limited with the overland flow process, as relatively impermeable soils are used.

The soil reactions involve clay, oxides of iron and aluminum, calcium compounds present, and the soil pH. Finer-textured soils tend to have the greatest potential for phosphorus sorption due to the higher clay content but also to the increased hydraulic residence time. Coarse-textured, acidic, or organic soils have the lowest capacity for phosphorus. Peat soils are both acidic and organic, but some have a significant sorption potential due to the presence of iron and aluminum.

A laboratory-scale adsorption test can estimate the amount of phosphorus that a soil can remove during short application periods. Actual phosphorus retention in the field will be at least two to five times the value obtained during a typical 5-day adsorption test. The sorption potential of a given soil layer will eventually be exhausted, but until that occurs the removal of phosphorus will be almost complete. It has been estimated that a 30-cm depth of soil in a typical slow-rate system might become saturated with phosphorus every 10 years. The phosphorus concentrations in the percolate from slow-rate systems usually approach background levels for the native groundwater within 2 m of travel in the soil. The coarser textured soils utilized for rapid infiltration might require an order-of-magnitude greater travel distance.

Phosphorus is not usually a critical issue for groundwater quality; however, when the groundwater emerges in a nearby surface stream or pond, eutrophication concerns may arise. Equation 3.29 can be used to estimate the phosphorus concentration at any point on the infiltration/percolation, groundwater flow path. The equation was originally developed from rapid infiltration system responses, so it provides a very conservative basis for all soil systems (USEPA, 1981):

$$P_x = P_0 e^{k_p t} \qquad\qquad (3.29)$$

where

P_x = Total P at a distance x on the flow path (mg/L).

P_0 = Total P in applied wastewater (mg/L).

k_p = 0.048 at pH 7 (d^{-1}) (pH 7 gives the lowest value).

t = Detention time (d) = $(x)(W)/(K_x)(G)$, where:

x = Distance along flow path (ft; m).

W = Saturated soil water content; assume 0.4.

K_x = Hydraulic conductivity of soil in direction x (ft/d; m/d); thus, K_v = vertical and K_h = horizontal.

G = Hydraulic gradient for flow system:

= 1 for vertical flow.

= $\Delta H/\Delta L$ for lateral flow.

The equation is solved in two steps: first for the vertical flow component, from the soil surface to the subsurface flow barrier (if one exists), and then for the lateral flow to the adjacent surface water. The calculations are based on assumed saturated conditions, so the lowest possible detention time will result. The actual vertical flow in most cases will be unsaturated, so the actual detention time will be much longer than is calculated with this procedure. If the equation predicts acceptable removal, we have some assurance that the site should perform reliably and detailed tests should not be necessary for preliminary work. Detailed tests should be conducted for final design of large-scale projects.

3.6.3 POTASSIUM AND OTHER MICRONUTRIENTS

As a wastewater constituent, potassium usually has no health or environmental effects. It is, however, an essential nutrient for vegetative growth, and it is not typically present in wastewaters in the optimum combination with nitrogen and phosphorus. If a land or aquatic treatment system depends on vegetation for nitrogen removal, it may be necessary to add supplemental potassium to maintain plant uptake of nitrogen at the optimum level. Equation 3.30 can be used to estimate the supplemental potassium that may be required for aquatic systems and for land systems where the soils have a low level of natural potassium:

$$K = (0.9)(U) - K_{ww} \qquad (3.30)$$

where

K = Annual supplemental potassium needed (kg/ha).

U = Estimated annual nitrogen uptake of vegetation (kg/ha).

K_{ww} = Amount of potassium in the applied wastewater (kg/ha).

Most plants also require magnesium, calcium, and sulfur and, depending on soil characteristics, there may be deficiencies in some locations. Iron, manganese, zinc, boron, copper, molybdenum, and sodium are other micronutrients that are important for vegetative growth. Generally, wastewater contains a sufficient amount of these elements, and in some cases the excess can lead to phytotoxicity problems. Some high-rate hyacinth systems may require supplemental iron to maintain vigorous plant growth.

3.6.3.1 Boron

Boron is at the same time essential for plant growth and toxic to sensitive plants at low concentrations. Experience has shown that soil systems have very limited capacity for boron adsorption, so it is conservative to assume a zero removal potential for land treatment systems. Industrial wastewaters may have a higher boron content than typical municipal effluents; the boron content may influence the type of crop selected but will not control the feasibility of land treatment. Tolerant crops such as alfalfa, cotton, sugar beets, and sweet clover might accept up to 2 to 4 mg/L boron in the wastewater; semi-tolerant crops such as corn, barley, milo, oats, and wheat might accept 1 to 2 mg/L; and sensitive crops such as fruits and nuts should receive less than 1 mg/L.

3.6.3.2 Sulfur

Wastewaters contain sulfur in either the sulfite or the sulfate form. Municipal wastewaters do not usually contain enough sulfur to be a design problem, but industrial wastewaters from petroleum refining and Kraft paper mills can be a concern. Sulfate is limited to 250 mg/L in drinking waters and 200 to 600 mg/L for irrigation, depending on the type of vegetation. Sulfur is weakly adsorbed on soils, so the major pathway for removal is by plant uptake. The grasses typically used in land treatment can remove 2 to 3 kg of sulfur per 1000 kg (4 to 7 lb per 2200 lb) of material harvested (Overcash and Pal, 1979). The presence of sulfites or sulfates in wastewater can lead to serious odor problems if anaerobic conditions develop. This has occurred with some hyacinth systems, and supplemental aeration is then needed to maintain aerobic conditions in the basin.

3.6.3.3 Sodium

Sodium is not limited by primary drinking-water standards, and the sodium content of typical municipal wastewaters is not a significant water-quality concern. A sudden change to high sodium content will adversely affect the biota in an aquatic system, but most systems can acclimate to gradual changes. Sodium and also calcium influence soil alkalinity and salinity, which in turn can affect the vegetation in land treatment systems. The growth of the plant and its ability to absorb moisture from the soil are influenced by salinity. The structure of clay soils can be damaged when there is an excess of sodium with respect to calcium and magnesium in the wastewater. The resulting swelling of some clay particles changes the hydraulic capacity of the soil profile. The *sodium adsorption ratio* (SAR) as shown by Equation 3.31 defines the relationship among these three elements:

$$SAR = \frac{[Na]}{\left\{\frac{([Ca]+[Mg])}{2}\right\}^{1/2}} \tag{3.31}$$

where

    SAR = Sodium adsorption ratio.

    [Na] = Sodium concentration (mEq/L) = (mg/L in wastewater)/22.99.

    [Ca] = Calcium concentration (mEq/L) = (mg/L in wastewater)(2)/40.08.

    [Mg] = Magnesium concentration (mEq/L) = (mg/L in wastewater)(2)/24.32.

The SAR for typical municipal effluents seldom exceeds a value of 5 to 8, so it should not be a problem with most soils in any climate. Soils with up to 15% clay can tolerate a SAR of 10 or less, while soils with little clay or with nonswelling clays can accept SARs up to about 20. Industrial wastewaters can have a high SAR, and periodic soil treatment with gypsum or some other inexpensive source of calcium may be necessary to reduce clay swelling. Soil salinity is managed by adding an excess of water above that required for crop growth to leach the salts from the soil profile. A "rule of thumb" for total water required to prevent salt buildup in arid climates is to apply the crop needs plus about 10% (Pettygrove and Asano, 1985). A report by the USEPA (1984) provides further details.

REFERENCES

Bankston, J.L., Sola, D.L., Komor, A.T., and Dwyer, D.F. (2002). Degradation of trichloroethylene in wetland microcosms containing broad-leaved cattail and Eastern cottonwood, *Water Res.*, 36, 153–1546.

Bauman, P. (1965). Technical development in ground water recharge, in Chow, V.T., Ed., *Advances in Hydroscience*, Vol. 2, Academic Press, New York, 209–279.

Bausum, H.T. (1983). *Enteric Virus Removal in Wastewater Treatment Lagoon Systems*, PB83-234914, National Technical Information Service, Springfield, VA.

Bedient, P.B., Springer, N.K., Baca, E., Bouvette, T.C., Hutchins, S.R., and Tomson, M.B. (1983). Ground-water transport from wastewater infiltration, *ASCE EED Div. J.*, 109(2), 485–501.

Bell, R.G. and Bole, J.B. (1978). Elimination of fecal coliform bacteria from soil irrigated with municipal sewage lagoon effluent, *J. Environ. Qual.*, 7, 193–196.

Bianchi, W.C. and Muckel, C. (1970). *Ground Water Recharge Hydrology*, ARS 41-161, U.S. Department of Agriculture, Agricultural Research Service, Beltsville, MD.

Bouwer, H. (1978). *Groundwater Hydrology*, McGraw-Hill, New York.

Brock, R.P. (1976). Dupuit–Forchheimer and potential theories for recharge from basins, *Water Resources Res.*, 12, 909–911.

Clark, C.S., Bjornson, H.S., Schwartz-Fulton, J., Holland, J.W., and Gartside, P.S. (1984a). Biological health risks associated with the composting of wastewater treatment plant sludge, *J. Water Pollut. Control Fed.*, 56(12), 1269–1276.

Clark, R.M., Eilers, R.C., and Goodrich, J.A. (1984b). VOCs in drinking water: cost of removal, *ASCE EED Div. J.*, 110(6), 1146–1162.

Colorado Governor's Office of Energy Management and Conservation. (2000). *Colorado Constructed Treatment Wetlands Inventory*, 225 E. 16th Ave., Suite 650, Denver, CO 80203.

Danel, P. (1953). The measurement of ground-water flow, in *Proceedings of the Ankara Symposium on Arid Zone Hydrology*, UNESCO, Paris, 99–107.

Dilling, W.L. (1977). Interphase transfer processes. II. Evaporation of chloromethanes, ethanes, ethylenes, propanes, and propylenes from dilute aqueous solutions: comparisons with theoretical predictions, *Environ. Sci. Technol.*, 11, 405–409.

Gersberg, R.M., Elkins, B.V., and Goldman, C.R. (1983). Nitrogen removal in artificial wetlands, *Water Res.*, 17(9), 1009–1014.

Gersberg, R.M., Lyon, S.R., Elkins, B.V., and Goldman, C.R. (1985). The removal of heavy metals by artificial wetlands, in *Proceedings AWWA Water Reuse III*, American Water Works Association, Denver, CO, 639–645.

Glover, R.E. (1961). *Mathematical Derivations as Pertaining to Groundwater Recharge*, U.S. Department of Agriculture, Agricultural Research Service, Beltsville, MD.

Hutchins, S.R., Tomsom, M.B., Bedient, P.B., and Ward, C.H. (1985). Fate of trace organics during land application of municipal wastewater, *CRC Crit. Rev. Environ. Control*, 15(4), 355–416.

Jenkins, T.F., Leggett, D.C., Parker, L.V., and Oliphant, J.L. (1985). Toxic organics removal kinetics in overland flow land treatment, *Water Res.*, 19(6), 707–718.

Kahn, M.Y. and Kirkham, D. (1976). Shapes of steady state perched groundwater mounds, *Water Resources Res.*, 12, 429–436.

Kamber, D.M. (1982). *Benefits and Implementation Potential of Wastewater Aquaculture*, EPA Contract Report 68-01-6232, U.S. Environmental Protection Agency, Office of Water Regulations and Standards, Washington, D.C.

Lombi, E., Zhao, F.J., Dunham, S.J., and McGrath, S.P. (2001). Phytoremediation of heavy metal-contaminated soils: natural hyperaccumulation versus chemically enhanced phytoextraction, *J. Environ. Qual.*, 30, 1919–1926.

Love, O.T., Miltner, R., Eilers, R.G., and Fronk-Leist, C.A. (1983). *Treatment of Volatile Organic Chemicals in Drinking Water*, EPA 600/8-83-019, U.S. Environmental Protection Agency, Municipal Engineering Research Laboratory, Cincinnati, OH.

Luthin, J.N. (1973). *Drainage Engineering*, Kreiger, Huntington, NY.

Overcash, M.R. and Pal, D. (1979). *Design of Land Treatment Systems for Industrial Wastes: Theory and Practice*, Ann Arbor Science, Ann Arbor, MI.

Parker, L.V. and Jenkins, T.F. (1986). Removal of trace-level organics by slow-rate land treatment, *Water Res.*, 20(11), 1417–1426.

Pettygrove, G.S. and Asano, T., Eds. (1985). *Irrigation with Reclaimed Municipal Wastewater: A Guidance Manual*, prepared for California State Water Resources Control Board, reprinted by Lewis Publishers, Chelsea, MI.

Pound, C.E. and Crites, R.W. (1979). *Long Term Effects of Land Application of Domestic Wastewater: Hollister, California*, EPA 600/2-78-084, U.S. Environmental Protection Agency, Office of Research and Development, Washington, D.C.

Reed, S.C. (1979). *Health Aspects of Land Treatment*, GPO 1979-657-093/7086, U.S. Environmental Protection Agency, Center for Environmental Research Information, Cincinnati, OH.

Reed, S.C., Bastian, R., Black, S., and Khettry, R. (1985). Wetlands for wastewater treatment in cold climates, in *Proceedings AWWA Water Reuse III*, American Water Works Association, Denver, CO, 962–972.

Roberts, P.V., McCarty, P.L., Reinhard, M., and Schriner, J. (1980). Organic contaminant behavior during groundwater recharge, *J. Water Pollut. Control Fed.*, 52(l), 161–172.

Russelle, M.P., Lamb, J.F.S., Montgomery, B.R., Elsenheimer, D.W., Miller, B.S., and Vance, C.P. (2001). Alfalfa rapidly remediates excess inorganic nitrogen at a fertilizer spill site, *J. Environ. Qual.*, 30, 30–36.

Schneiter, R.W. and Middlebrooks, E.J. (1981). *Cold Region Wastewater Lagoon Sludge: Accumulation, Characterization, and Digestion*, Contract Report DACA89-79-C-0011, U.S. Cold Regions Research and Engineering Laboratory, Hanover, NH.

Sorber, C.A. and Sagik, B.P. (1980). Indicators and pathogens in wastewater aerosols and factors affecting survivability, in *Wastewater Aerosols and Disease*, EPA 600/9-80-078, U.S. Environmental Protection Agency, Health Effects Research Laboratory, Cincinnati, OH, 23–35.

Sorber, C.A., Bausum, H.T., Schaub, S.A., and Small, M.J. (1976). A study of bacterial aerosols at a wastewater irrigation site, *J. Water Pollut. Control Fed.*, 48(10), 2367–2379.

Sorber, C.A., Moore, B.E., Johnson, D.E., Hardy, H.J., and Thomas R.E. (1984). Microbiological aerosols from the application of liquid sludge to land, *J. Water Pollut. Control Fed.*, 56(7), 830–836.

USDOI. (1978). *Drainage Manual*, U.S. Department of the Interior, Bureau of Reclamation, U.S. Government Printing Office, Washington, D.C.

USEPA. (1981). *Process Design Manual Land Treatment of Municipal Wastewater*, EPA 625/1-81-013, U.S. Environmental Protection Agency, Center for Environmental Research Information, Cincinnati, OH.

USEPA. (1982a). *Fate of Priority Pollutants in Publicly Owned Treatment Works*, EPA 440/1-82-303, U.S. Environmental Protection Agency, Washington, D.C.

USEPA. (1982b). *Estimating Microorganism Densities in Aerosols from Spray Irrigation of Wastewater*, EPA 600/9-82-003, U.S. Environmental Protection Agency, Center for Environmental Research Information, Cincinnati, OH.

USEPA. (1983). *Design Manual Municipal Wastewater Stabilization Ponds*, EPA 625/1-83-015, U.S. Environmental Protection Agency, Center for Environmental Research Information, Cincinnati, OH.

USEPA. (1984). *Process Design Manual Land Treatment of Municipal Wastewater Supplement on Rapid Infiltration and Overland Flow*, EPA 625/1-81-013a, U.S. Environmental Protection Agency, Center for Environmental Research Information, Cincinnati, OH.

USEPA. (1985). *Protection of Public Water Supplies from Ground-Water Contamination*, EPA 625/4-85-016, U.S. Environmental Protection Agency, Center for Environmental Research Information, Cincinnati, OH.

Van Schifgaarde, J. (1974). Drainage for agriculture, *Am. Soc. Agron. Ser. Agron.*, No. 17.

Wang, X., Newman, L.E., and Gordon M.P. (1999). Biodegradation of carbon tetrachloride by poplar trees: results from cell culture and field experiments, in *Phytoremediation and Innovative Strategies for Specialized Remedial Applications*, Battelle Press, Columbus, OH.

Wile, I., Miller, G., and Black, S. (1985). Design and use of artificial wetlands, in *Ecological Considerations in Wetlands Treatment of Municipal Wastewaters*, Godfrey, P.J. et al., Eds., Van Nostrand Reinhold, New York, 26–37.

Wu, M., Franz, E.H., and Chen, S. (2001). Oxygen fluxes and ammonia removal efficiencies in constructed treatment wetlands, *Water Environ. Res.*, 73(6), 661–666.

4 Design of Wastewater Pond Systems

4.1 INTRODUCTION

4.1.1 TRENDS

Changes in the basic approach to the design of pond systems in the past 20 years have been limited primarily to the introduction of floating plastic partitions to improve the hydraulic characteristics of the pond system, modifications to basic designs, and the development of a wider selection of more efficient aeration equipment. Several modifications of complete mix and partial mix combinations have been developed and evaluated, but the basic design concepts are unchanged. Examples of these modifications are the BIOLAC® processes, the Rich aerated lagoon design procedures, and the LemTec® biological treatment process. These among other modifications are described in this chapter.

The importance of hydraulic characteristics was emphasized in the 1983 EPA lagoon design manual (USEPA, 1983) and has been restated numerous times in many publications; however, based on the number of pond systems constructed over the past 20 years with poor hydraulic characteristics, one would assume that many designers have not read the literature or have ignored what they read. More recent designs have improved hydraulic designs considerably, and it is hoped that this trend will continue.

The trend toward omitting redundancy in the design of lagoon systems has been alarming. It appears that little thought is given to the need for maintenance in the future. Operating costs associated with aerated lagoon systems frequently have been ignored or overlooked in comparing options available to a community. The initial cost of systems without redundancy obviously is lower than that obtained with systems that include flexibility in operation, but the cost to the environment and the owner will be far greater when maintenance is required.

Several design procedures have been proposed and implemented since the 1983 design manual was written. Several have been applied, some with great success and others with moderate success. Some of these modifications have been developed and operated successfully in warm climates, and the application of these systems has expanded to cold climates. The degree of success with these systems has varied, and much of the lack of success can be attributed to a lack of valid design information and considerable experimentation on the part of designers. An extensive description of these systems along with a comparison with more conventional design methods are sorely needed. The major procedures, processes, and design methods, old and new, are listed below and described in the following sections:

- Facultative ponds
- Partial-mix ponds
- Complete-mix ponds
- Anaerobic ponds
- Controlled discharge pond
- Complete retention pond
- Hydrograph controlled release
- High-performance aerated pond systems (Rich design)
- Proprietary systems
  - Advanced Integrated Wastewater Pond Systems® (AIWPS®) (Oswald Design)
  - BIOLAC® process (activated sludge in earthen ponds)
  - Lemna systems
  - LAS International, Ltd.
  - Praxair, Inc.
- Nitrogen removal in pond systems (including proprietary systems)
- Modified high-performance aerated pond systems for nitrification and denitrification
- Nitrogen removal in ponds coupled with wetlands and gravel bed nitrification filters
- Control of algae and design of settling basins
- Hydraulic control of ponds
- Phosphorus removal

4.2 FACULTATIVE PONDS

Facultative pond design is based on biological oxygen demand (BOD) removal; however, the majority of the suspended solids will be removed in the primary cell of a facultative pond system. Sludge fermentation feedback of organic compounds to the water in a pond system is significant and has an effect on the performance. During the spring and fall, the thermal overturn of the pond contents can result in significant quantities of benthic solids being resuspended. The rate of sludge accumulation is affected by the liquid temperature, and additional volume is added for sludge accumulation in cold climates. Although total suspended solids (TSS) have a profound influence on the performance of pond systems, most design equations simplify the incorporation of the influence of TSS by using an overall reaction rate constant. Effluent TSS generally consist of suspended organism biomass and do not include suspended waste organic matter.

Several empirical and rational models for the design of these ponds have been developed. These include the ideal plug flow and complete mix models, as well as models proposed by Fritz et al. (1979), Gloyna (1971), Larson (1974), Marais

(1970), McGarry and Pescod (1970), Oswald et al. (1970), and Thirumurthi (1974). Middlebrooks (1987) presented a summary of many models, including the ones referenced in the preceding sentence, that have been developed to evaluate and design facultative pond systems (Table 4.1). This is not an exhaustive list, and most of these models are variations of the ones in the references listed above. Several produce satisfactory results, but the use of some may be limited because of the difficulty in evaluating coefficients or by the complexity of the model. The methods and equations used most are discussed in the following paragraphs.

4.2.1 AREAL LOADING RATE METHOD

A cursory review of state design standards since Canter and Englande (1970) reported that most states have design criteria for organic loading and hydraulic detention times for facultative ponds shows that little has changed since 1970; however, individual states should be contacted to obtain the latest information. These criteria are assumed to ensure satisfactory performance; however, repeated violations of effluent standards by pond systems that meet state design criteria indicate the inadequacy of the criteria. A summary of the state design criteria for each location and actual design values for organic loading and hydraulic detention time for four facultative pond systems evaluated by the U.S. Environmental Protection Agency (EPA) (Middlebrooks et al., 1982; USEPA, 1983) are shown in Table 4.2. Also included is a list of the months the federal effluent standards for BOD_5 were exceeded. The actual organic loading for the four systems is nearly equal, but the system in Corinne, Utah, consistently satisfied the federal effluent standard. This may be a function of the larger number of cells in the Corinne system — seven as compared to three for the others. More hydraulic short-circuiting is likely to occur in the three-cell systems, resulting in actual detention times shorter than those for the Corinne system. The detention time may also be affected by the location of the pond cell inlet and outlet structures. Many of the design faults in the systems referenced in Table 4.2 have been corrected since 1983.

Based on many years of experience, the following loading rates for various climatic conditions are recommended for use in designing facultative pond systems. For average winter air temperatures above 59°F (15°C), a BOD_5 loading rate range of 40 to 80 lb/ac·d (45 to 90 kg/ha·d) is recommended. When the average winter air temperature ranges between 32 and 59°F (0 and 15°C), the organic loading rate should range between 20 and 40 lb/ac·d (22 and 45 kg/ha·d). For average winter temperatures below 32°F (0°C), the organic loading rates should range from 10 to 20 lb/ac·d (11 to 22 kg/ha·d).

The BOD loading rate in the first cell is usually limited to 35 lb/ac·d (40 kg/ha·d) or less, and the total hydraulic detention time in the system is 120 to 180 days in climates where the average air temperature is below 32°F (0°C). In mild climates where the air temperature is greater than 59°F (15°C), loadings on the primary cell can be 89 lb/ac·d (100 kg/ha·d).

TABLE 4.1
Design Equations Developed for Facultative Ponds

1. $C_0 - C = kt$

2. $\ln\left(\dfrac{C_0}{C_e}\right) = kt$

3. $\left[\dfrac{1}{C_e} - \dfrac{1}{C_0}\right] = kt$

4. $t = \left(\dfrac{K_S}{\mu}\right)\ln\left(\dfrac{C_0}{C_e}\right) + \left(\dfrac{C_0 - C_e}{\mu}\right)$

5. $t = \left(\dfrac{K_S}{\mu}\right)\left(\dfrac{1}{C_e} - \dfrac{1}{C_0}\right) + \left(\dfrac{1}{\mu}\right)\ln\left(\dfrac{C_0}{C_e}\right)$

6. $t = \left(\dfrac{K_S}{2\mu}\right)\left(\dfrac{1}{(C_e)^2} - \dfrac{1}{(C_0)^2}\right) + \left(\dfrac{1}{\mu}\right)\left(\dfrac{1}{C_e} - \dfrac{1}{C_0}\right)$

7. $t = \left(\dfrac{C_0}{C_e} - 1\right)\bigg/ k$

8. $C_e = C_0\left(\dfrac{4ae^{1/2D}}{(1+a)^2 e^{a/2D} - (1-a)^2 e^{a/2D}}\right)$

9. $\left(\dfrac{C_0 - C_e}{X_1}\right) = \left(\dfrac{k_d}{Y}\right)t + \dfrac{1}{Y}$

10. $\dfrac{t}{(1 + tk_d)} = \left(\dfrac{K_S}{\mu}\right)\left(\dfrac{1}{C_e}\right) + \dfrac{1}{\mu}$

11. $AR = 10.37 + 0.725(AL)$

12. $t = 3.5 \times 10^{-5}(C_{0u})\left(\theta^{(35-T)}\right)ff'$

13. $t = 3.5 \times 10^{-5}(C_{0u})(1.099)^{L(35-T)/250}\,ff'$

14. $\dfrac{C_e}{C_0} = k(t)^a(C_0)^b(L)^c(T)^d$

15. $\dfrac{C_e}{C_0} = \dfrac{1}{1 + (a + bT)te^{(C-dT)(\text{pH}-6.6)}}$

16. $\dfrac{C_e}{C_0} = k(t)^a(C_0)^b(L)^c(T)^d(\text{pH})^3$

17. $\dfrac{C_e}{C_0} = k(t)^a(C_0)^b(L)^c(T)^d(\text{pH})^3 + Z$

18. $\ln\left(\dfrac{C_e}{C_0}\right) = k(t-a)(T-b)(\text{pH}-c)(L)$

19. $\ln\left(\dfrac{C_e}{C_0}\right) = -kte^{aT - b(\text{pH}) - cL}$

20. $\ln\left(\dfrac{C_e}{C_0}\right) = ktT(\text{pH})L$

21. $\dfrac{(C_0 - C_e)}{t} = kt^a$

22. $\dfrac{(C_0 - C_e)}{t} = kt^a(C_o)^b(L)^c(T)^d$

23. $\dfrac{k_2}{k_1} = \theta^{(T_2 - T_1)}$

TABLE 4.1 (cont.)
Design Equations Developed for Facultative Ponds

Note: t, hydraulic residence time (days); C_0, influent BOD_5 concentration (mg/L); C_e, effluent BOD_5 concentration (mg/L); k, reaction rate constant (units vary); μ, maximum reaction rate for Monod-type kinetics (units vary); K_S, substrate concentration at 0.5 μm; $a = \sqrt{1 + 4ktD}$; D, dimensionless dispersion number; e, base of natural logarithms (2.7183); k_d, decay rate (d^{-1}); Y, yield coefficient (mass of TSS or VSS formed per mass of BOD_5 removed); AR, areal BOD_5 removal (kg/ha·d); AL, areal BOD_5 loading (kg/ha·d); C_{0u}, ultimate influent BOD or COD (mg/L); θ, temperature coefficient (dimensionless); T, pond water temperature (°C); f, algal toxicity factor (dimensionless); f', sulfide oxygen demand (dimensionless); L, light intensity (langleys); pH, pH value; a, b, c, d, e, reaction orders; Z, constant; k_2, k_1, T_2, T_1, reaction rate constants and temperatures.

Source: Middlebrooks, E.J., *Water Sci. Technol.*, 19, 12, 1987. With permission.

4.2.2 GLOYNA METHOD

Gloyna (1976) proposed the following empirical equation for the design of facultative wastewater stabilization ponds:

$$V = (3.5 \times 10^{-5})(Q)(La)[\theta^{(35-T)}](f)(f') \tag{4.1}$$

where

V	=	Pond volume (m³).
Q	=	Influent flow rate (L/d).
La	=	Ultimate influent BOD or chemical oxygen demand (COD) (mg/L).
θ	=	Temperature correction coefficient = 1.085.
T	=	Pond temperature (°C).
f	=	Algal toxicity factor.
f'	=	Sulfide oxygen demand.

The BOD_5 removal efficiency is projected to be 80 to 90% based on unfiltered influent samples and filtered effluent samples. A pond depth of 5 ft (1.5 m) is suggested for systems with significant seasonal variations in temperature and major fluctuations in daily flow. The surface area design using Equation 4.1 should always be based on a depth of 3 ft (1 m). The algal toxicity factor (f) is assumed to be equal to 1.0 for domestic wastes and many industrial wastes. The sulfide oxygen demand (f') is also equal to 1.0 for sulfate equivalent ion concentration of less than 500 mg/L. The design temperature is usually selected as the average pond temperature in the coldest month. Sunlight is not considered to be critical in pond design but can be incorporated into Equation 4.1 by multiplying the pond volume by the ratio of sunlight at the design location to the average found in the southwestern United States.

TABLE 4.2
Design and Performance Data from EPA Pond Studies

Location	Organic Loading (kg BOD ha⁻¹ d⁻¹)			Theoretical Detention Time			Months Effluent BOD Exceeded 30 mg/L
	State Design Standard	Design	Actual (1974–1975)	State Design Standard	Design	Actual (1974–1975)	
Peterborough, New Hampshire	39.3	19.6	16.2	None	57	107	October, February, March, April
Kilmichael, Mississippi	56.2	43	17.5	None	79	214	November, July
Eudora, Kansas	38.1	38.1	18.8	None	47	231	March, April, August
Corinne, Utah	45.0[a]	36.2[a]	29.7[a]/14.6[b]	180	180	70/88[c]	None

[a] Primary cell.
[b] Entire system.
[c] Estimated from dye study.

Source: Data from Middlebrooks et al. (1982) and USEPA (1983).

Design of Wastewater Pond Systems

The Gloyna method was evaluated using th
The equation giving the best fit of the data is
despite the considerable scatter to the data, the re
icant:

$$V = 0.035Q(\text{BOD})(1.099)^{LI}$$

where
BOD = BOD$_5$ in the system influent (mg/l
$LIGHT$ = Solar radiation (langleys).
V = Pond volume (m^3).
Q = Influent flow rate (m^3/day).
T = Pond temperature (°C).

4.2.3 COMPLETE-MIX MODEL

The Marais and Shaw (1961) equation is based on a complete-mix model and
first-order kinetics. The basic relationship is shown in Equation 4.3:

$$\frac{C_n}{C_0} = \left(\frac{1}{1 + k_c t_n}\right)^n \tag{4.3}$$

where
C_n = Effluent BOD$_5$ concentration (mg/L).
C_0 = Influent BOD$_5$ concentration (mg/L).
k_c = Complete-mix first-order reaction rate (d^{-1}).
t_n = Hydraulic residence time in each cell (d).
n = Number of equal-sized pond cells in series.

The proposed upper limit for the BOD$_5$ concentration $(C_e)_{max}$ in the primary cells
is 55 mg/L to avoid anaerobic conditions and odors. The permissible depth of
the pond, d (in meters), is related to $(C_e)_{max}$ as follows:

$$(C_e)_{max} = \frac{700}{1.9d + 8} \tag{4.4}$$

where $(C_e)_{max}$ is the maximum effluent BOD (55 mg/L), and d is the design depth
of the pond (in meters).

The influence of water temperature on the reaction rate is estimated using
Equation 4.5:

$$k_{cT} = k_{c35}(1.085)^{T-35} \tag{4.5}$$

**TABLE 4.3
Variation of the Plug-Flow
Reaction Rate Constant with
Organic Loading Rate**

Organic Loading Rate	
(kg/ha·d)	k_p (d^{-1})[a]
22	0.045
45	0.071
67	0.083
90	0.096
112	0.129

[a] Reaction rate constant at 20°C.

Source: Neel, J.K. et al., *J. Water Pollut. Control Fed.*, 33, 6, 603–641, 1961. With permission.

---

where

k_{cT} = Reaction rate at water temperature T (d^{-1}).

k_{c35} = Reaction rate at 35°C = 1.2 (d^{-1}).

T = Operating water temperature (°C).

4.2.4 PLUG-FLOW MODEL

The basic equation for the plug-flow model is:

$$\frac{C_e}{C_0} = \exp\left[-k_p t\right]$$ (4.6)

where

C_e = Effluent BOD$_5$ concentration (mg/L).

C_0 = Influent BOD$_5$ concentration (mg/L).

k_p = Plug-flow first-order reaction rate (d^{-1}).

t = Hydraulic residence time (d).

The reaction rate (k_p) was reported to vary with the BOD loading rate as shown in Table 4.3 (Neel et al., 1961). Theoretically, the reaction rate should not vary with loading rate; however, that is what was reported.

The influence of water temperature on the reaction rate constant can be determined with Equation 4.7:

$$k_{pT} = k_{p20}(1.09)^{T-20} \qquad (4.7)$$

where

k_{pT} = Reaction rate at temperature T (d^{-1}).

k_{p20} = Reaction rate at 20°C (d^{-1}).

T = Operating water temperature (°C).

4.2.5 WEHNER–WILHELM EQUATION

Thirumurthi (1974) found that the flow pattern in facultative ponds is somewhere between ideal plug flow and complete mix, and he recommended the use of the following chemical reactor equation developed by Wehner and Wilhelm (1956) for chemical reactor design:

$$\frac{C_e}{C_0} = \frac{4ae^{1/(2D)}}{(1+a)^2\left(e^{a/(2D)}\right) - (1-a)^2\left(e^{-a/(2D)}\right)} \qquad (4.8)$$

where

C_e = Effluent BOD concentration (mg/L).

C_0 = Influent BOD concentration (mg/L).

a = $(1 + 4ktD)^{0.5}$, where k is a first-order reaction rate constant (d^{-1}), t is the hydraulic residence time (d), and D is a dimensionless dispersion number = $H/vL = Ht/L^2$, where H is the axial dispersion coefficient (area per unit time), v is the fluid velocity (length per unit time), and L is the length of travel path of a typical particle.

e = Base of natural logarithms (2.7183).

A modified form of the chart prepared by Thirumurthi (1974) is shown in Figure 4.1 to facilitate the use of Equation 4.8. The dimensionless term kt is plotted vs. the percentage of BOD remaining for dispersion numbers ranging from zero for an ideal plug flow unit to infinity for a completely mixed unit. Dispersion numbers measured in wastewater ponds range from 0.1 to 2.0, with most values being less than 1.0. The selection of a value for D can dramatically affect the detention time required to produce a given quality effluent. The selection of a design value for k can have an equal effect. If the chart in Figure 4.1 is not used, Equation 4.8 can be solved on a trial-and-error basis as shown in Example 4.1 (see below). Middlebrooks (2000) has developed a spreadsheet that calculates the dimensions for a pond system using the Wehner–Wilhelm equation with options to use a wide range of variables. The spreadsheet takes the tedium out of the design procedure and eliminates the need to read an imprecise table. A copy can be obtained by contacting the author.

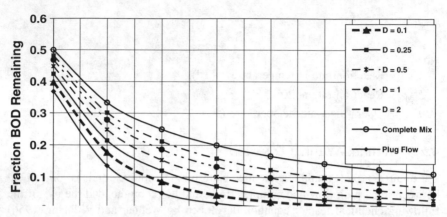

FIGURE 4.1 Wehner–Wilhelm equation. (Adapted from Thirumurthi, D., *J. Water Pollut. Control Fed.*, 46, 2094–2106, 1974.)

To improve on the selection of a D value for use in Equation 4.8, Polprasert and Bhattarai (1985) developed Equation 4.9 based on data from pilot and full-scale pond systems:

$$D = \frac{0.184[tv(W + 2d)]^{0.489}(W)^{1.511}}{(Ld)^{1.489}} \qquad (4.9)$$

where
D = Dimensionless dispersion number.
t = Hydraulic residence time (d).
v = Kinematic viscosity (m^2/d).
W = Width of pond (m).
L = Length of pond (m).
d = Liquid depth of pond (m).

The hydraulic residence times used to derive Equation 4.9 were determined by tracer studies; therefore, it is still difficult to estimate the value of D to use in Equation 4.8. A good approximation is to assume that the actual hydraulic residence time is half that of the theoretical hydraulic residence time.

The variation of the reaction rate constant k in Equation 4.8 with the water temperature is determined with Equation 4.10:

$$k_T = k_{20}(1.09)^{T-20} \qquad (4.10)$$

where
k_T = Reaction rate at water temperature T (d^{-1}).
k_{20} = Reaction rate at 20°C = 0.15 d^{-1}.
T = Operating water temperature (°C).

TABLE 4.5
Results From Facultative Pond Design Methods

Method	Detention Time (d) Primary Cell	Detention Time (d) Total System	Volume (m³) Primary Cell	Volume (m³) Total System	Surface Area (ha) Primary Cell	Surface Area (ha) Total System	Primary Cell Depth (m)	Number of Cells in Series	Organic Loading (kg BOD ha⁻¹ d⁻¹) Primary	Organic Loading (kg BOD ha⁻¹ d⁻¹) Total
Areal loading rate	53[a]	71	82,900[a]	135,300	6.3	11.5	1.7 (1.4)[c]	4	60	33
Gloyna	—	65	82,900[a]	123,000	—	12.3	1.5 (1.0)[c]	—	—	31
Marais and Shaw	17[b]	34	32,000[b]	64,000	1.3	2.6	2.4	2[d]	290	145
Plug flow	53[a]	53	82,900[a]	123,000	6.3	6.3	1.7 (1.4)[c]	1[d]	60	60
Wehner and Wilhelm	53[a]	36–58	82,900[a]	68,100–109,80]	6.3	4.8–7.8	1.7 (1.4)[c]	4	—	80–50

[a] Controlled by state standards and equal to value calculated for an areal loading rate of 60 kg/ha·d and an effective depth of 1.4 m.
[b] Also would be controlled by state standards for areal loading rate; however, the method includes a provision for calculating a value, and this calculated value is shown.
[c] Effective depth.
[d] Baffling recommended to improve hydraulic characteristics.

Source: Reed, S.C. et al., Natural Systems for Waste Management and Treatment, 2nd ed., McGraw-Hill, New York, 1995. With permission.

Example 4.1

Determine the design detention time in a facultative pond by solving Equation 4.8 on a trial-and-error basis. Assume $C_e = 30$ mg/L, $C_0 = 200$ mg/L, $k_{20} = 0.15$, $D = 0.1$, $d = 1.5$ m, $Q = 3785$ m³/day, and the water temperature is 0.5°C.

Solution

1. Calculate k_T using Equation 4.10:

$$k_T = k_{20}(1.09)^{T-20}$$

$$k_T = 0.15(1.09)^{0.5-20}$$

$$k_T = 0.028$$

2. Assume for the first iteration that $t = 50$ d, and solve for a:

$$a = (1 + 4\, k_T Dt)0.5$$
$$a = (1 + 4 \times 0.028 \times 0.1 \times 50)0.5$$
$$a = 1.25$$

3. Solve Equation 4.8 and see if the two sides are equal:

$$\frac{C_e}{C_0} = \frac{30}{200} = \frac{(4)(1.25)e^{1/(2\times0.1)}}{(1+1.25)^2\, e^{1.25/(2\times0.1)} - (1-1.25)^2\, e^{-1.25/(2\times0.1)}}$$

$$= 0.15 = \frac{742.07}{(5.0625)(518.01) - (0.0625)(0.00193)} = 0.283$$

Because 0.15 does not equal 0.283, repeat the calculation.

4. The final iteration assumes that $t = 80$ d:

$$0.15 = \frac{817.46}{(5.65)(977.50) - (0.142)(0.00102)} = 0.148$$

The agreement is adequate, so use a design time of 80 days.

5. Using Equation 4.9, determine the length-to-width ratio that will yield a D of approximately the assumed value of 0.1:

   $v = 0.1521$ m²/d.

   Volume = 80 d × 3785 m³/day = 302,800 m³.

   Divide the flow into two streams.

   Volume in one half of the system = 151,400 m³.

   Divide the half system into four equal-volume ponds.

   Volume in one pond = 37,850 m³.

   Surface area of one pond = 37,850/1.5 = 25,233 m².

   Theoretical hydraulic detention time in each pond = 80/4 = 20 d.

   Assume length-to-width ratio is 4:1.

Surface area = 4W × W = 25,233 m².
W = 79.4 m.
L = 317.7 m.

Equation 4.9 was developed using the hydraulic detention time determined by dye studies; therefore, it is reasonable to assume that the theoretical hydraulic detention time is not the correct value to use in Equation 4.9. A good approximation of the measured hydraulic detention time is to use a value of one half that of the theoretical value:

$$D_{10} = \frac{0.184[10 \times 0.1521(79.4 + 2 \times 1.5)]^{0.489}(79.4)^{1.511}}{(317.7 \times 1.5)^{1.489}}$$

$$D_{10} = \frac{1450.1}{9720.8} = 0.149$$

$$D_{20} = 0.209$$

To illustrate the effect of using the theoretical hydraulic detention time, a D value is calculated using the theoretical value, and both values of D are used in Equation 4.8 to calculate the effluent BOD_5 concentration. The theoretical hydraulic detention time is used in Equation 4.8 because it was developed based on the theoretical value. The total detention time is used because the equation represents the entire system and not a component of the system:

$$\frac{C_e}{C_0} = \frac{4ae^{1/(2D)}}{(1+a)^2 \left(e^{a/(2D)}\right) - (1-a)^2 \left(e^{-a/(2D)}\right)}$$

$$a = (1 + 4ktD)^{0.5}$$

Let $D = 0.1$ and $t = 80$:

$$a = [1 + 4(0.028 \times 0.1 \times 80)]^{0.5} = 1.377$$

$$\frac{C_e}{C_0} = \frac{4(1.377)e^{\frac{1}{(2 \times 0.1)}}}{(1+1.377)^2 e^{\frac{1.377}{(2 \times 0.1)}}} = \frac{817.5}{5523.0} = 0.148$$

$$C_e = 200 \times 0.148 = 29.6 \text{ mg/L}$$

The latter part of the denominator in Equation 4.8 was omitted because it is insignificant in this and most situations. For $D = 0.149$, $C_e = 32.5$ mg/L, and for $D = 0.209$, $C_e = 35.4$ mg/L. As shown by these calculations, small changes in D can have a significant influence on the effluent quality.

TABLE 4.4
Conditions for Facultative Design Comparisons

Q = design flow rate = 1893 m³/day (0.5 mgd)

C_o = influent BOD = 200 mg/L

C_e = required effluent BOD = 30 mg/L

T = water temperature at critical part of year = 10°C

T_a = average winter air temperature = 5°C

Adequate light intensity

Suspended solids (SS) = 250 mg/L

Sulfate = <500 mg/L

4.2.6 COMPARISON OF FACULTATIVE POND DESIGN MODELS

Because of the many approaches to the design of facultative ponds and the lack of adequate performance data for the latest designs, it is not possible to recommend the "best" procedure. An evaluation of the design methods presented above, with operational data referenced in Table 4.2, failed to show that any of the models are superior to the others in terms of predicting the performance of facultative pond systems (USEPA, 1983; Middlebrooks, 1987). Many other studies of facultative pond systems with limited data have been conducted and reached much the same conclusions (Pearson and Green, 1995). Each of the design models presented above in detail was used to design a facultative pond for the conditions presented in Table 4.4, and the results are summarized in Table 4.5.

The limitations on the various design methods make it difficult to make direct comparisons; however, an examination of the hydraulic detention times and total volume requirements calculated by all of the methods show considerable consistency if the Marais–Shaw method is excluded and a value of 1.0 is selected for the dispersion factor in the Wehner–Wilhelm method. The major limitation of all these methods is the selection of a reaction rate constant or other factors in the equations. Even with this limitation, if the pond hydraulic system is designed and constructed so the theoretical hydraulic detention time is approached, reasonable success can be assured with all of the design methods. Short-circuiting is the greatest deterrent to successful pond performance, barring any toxic effects. The importance of the hydraulic design of a pond system cannot be overemphasized.

The surface loading rate approach to design requires a minimum of input data and is based on operational experiences in various geographical areas of the United States. It is probably the most conservative of the design methods, but the hydraulic design still cannot be neglected.

The Gloyna method is applicable only for 80 to 90% BOD removal efficiency, and it assumes that solar energy for photosynthesis is above the saturation level.

Provisions for removals outside this range are not made; however, an adjustment for other solar conditions can be made as described previously. Mara (1975) should be consulted if a detailed critique of the Gloyna method is needed.

The Marais–Shaw method is based on complete-mix hydraulics, which is not approached in facultative ponds, but the greatest weakness in the approach may lie in the requirement that the primary cell must not turn anaerobic. Mara (1975, 1976) provides a detailed discussion of this model.

Plug flow hydraulics and first-order reaction kinetics have been found to adequately describe the performance of many facultative pond systems (Neel et al., 1961; Thirumurthi, 1974; Middlebrooks et al., 1982; Middlebrooks, 1987; Pearson and Green, 1995). A plug-flow model was found to best describe the performance of the four pond systems evaluated in an EPA study as well as several others (Middlebrooks et al., 1982; USEPA, 1983). Because of the arrangement of most facultative ponds into a series of three or more cells, logically it would be expected that the hydraulic regime could be approximated by a plug-flow model. Reaction rates calculated from the USEPA (1983) data are very low primarily because of the long hydraulic detention times in the pond systems (70 to 231 days) and will yield designs that are too conservative.

Use of the Wehner–Wilhelm equation requires knowledge of both the reaction rate and the dispersion factor, further complicating the design procedure. If knowledge of the hydraulic characteristics of a proposed pond configuration are known or can be determined (Equation 4.9), the Wehner–Wilhelm equation will yield satisfactory results; however, because of the difficulty of selecting both parameters, a design using one of the simpler equations is likely to be as good as one using this model. The Wehner–Wilhelm equation is used in many countries around the world to design facultative ponds and apparently has been used successfully. In summary, all of the design methods discussed can provide a valid design if the proper parameters are selected and the hydraulic characteristics of the system are controlled.

4.3 PARTIAL-MIX AERATED PONDS

Changes in the basic approach to the design of partial-mix aerated ponds since the publication of USEPA's 1983 design manual have been limited primarily to the introduction of floating plastic partitions to improve the hydraulic characteristics of the pond system and the development of a wider selection of more efficient aeration equipment (WEF/ASCE, 1991). The importance of hydraulic characteristics was emphasized in the 1983 design manual and has been restated numerous times in many publications; however, based on the number of pond systems constructed over the past 20 years with poor hydraulic characteristics, one would assume that many designers have not read the literature or have ignored what they read.

The trend toward omitting redundancy in the design of aerated lagoon systems has been alarming. It appears that little thought is given to the need for maintenance in the future. Operating costs associated with aerated lagoon systems frequently have been ignored or overlooked when comparing options available to a community. The initial costs of systems without redundancy are obviously lower than those for systems that include flexibility in operation, but the cost to the environment and the owner will be far greater when maintenance is required.

In the partial-mix aerated pond system, the aeration serves only to provide an adequate oxygen supply, and no attempt is made to keep all of the solids in suspension in the pond as is done with complete-mix and activated sludge systems. Some mixing obviously occurs and keeps portions of the solids suspended; however, an anaerobic degradation of the organic matter that settles does occur. The system is sometimes referred to as a *facultative aerated pond system*.

Even though the pond is only partially mixed, it is conventional to estimate the BOD removal using a complete mix model and first-order reaction kinetics. Studies by Middlebrooks et al. (1982) have shown that a plug-flow model and first-order kinetics more closely predict the performance of these ponds when either surface or diffused aeration is used. However, most of the ponds evaluated in this study were lightly loaded, and the reaction rates calculated are very conservative because it appears that the rate decreases as the organic loading decreases (Neel et al., 1961). Because of the lack of better design reaction rates, it is still necessary to design partial-mix ponds using complete-mix kinetics.

4.3.1 PARTIAL-MIX DESIGN MODEL

The design model using first-order kinetics and operating n number of equal-sized cells in series is given by Equation 4.11 (Middlebrooks et al., 1982; Great Lakes–Upper Mississippi River Board of State Sanitary Engineers, 1990; WEF/ASCE, 1991):

$$\frac{C_n}{C_0} = \frac{1}{[1 + kt/n]^n} \tag{4.11}$$

where
- C_n = Effluent BOD concentration in cell n (mg/L).
- C_0 = Influent BOD concentration (mg/L).
- k = First-order reaction rate constant (d^{-1}) = 0.276 d^{-1} at 20°C (assumed to be constant in all cells).
- t = Total hydraulic residence time in pond system (d).
- n = Number of cells in the series.

If other than a series of equal volume ponds are to be employed and it is desired to use varying reaction rates, it is necessary to use the following general equation:

$$\frac{C_n}{C_0} = \left(\frac{1}{1 + k_1 t_1}\right)\left(\frac{1}{1 + k_2 t_2}\right)\cdots\left(\frac{1}{1 + k_n t_n}\right) \tag{4.12}$$

where $k_1, k_2, ..., k_n$ are the reaction rates in cells 1 through n (all usually assumed equal for lack of better information) and $t_1, t_2, ..., t_n$ are the hydraulic residence times in the respective cells.

Mara (1975) has shown that a number of equal volume reactors in series is more efficient than unequal volumes; however, due to site topography or other factors, in some cases it may be necessary to construct cells of unequal volume.

4.3.1.1 Selection of Reaction Rate Constants

The selection of the k value is the critical decision in the design of any pond system. The Ten-States Standards (Great Lakes–Upper Mississippi River Board of State Sanitary Engineers, 1990) recommended a design value of 0.276 d^{-1} at 20°C and 0.138 d^{-1} at 1°C. Using these values to calculate the temperature coefficient yields a value of 1.036. Boulier and Atchinson (1975) recommended values of k of 0.2 to 0.3 at 20°C and 0.1 to 0.15 at 0.5°C. A temperature coefficient of 1.036 results when the two lower or higher values of k are used in the calculation. Reid (1970) suggested a k value of 0.28 at 20°C and 0.14 at 0.5°C based on research with partial-mix ponds aerated with perforated tubing in central Alaska. These values are essentially identical to the recommendations of the Ten-States Standards.

4.3.1.2 Influence of Number of Cells

When using the partial-mix design model, the number of cells in series has a pronounced effect on the size of the pond system required to achieve the specified degree of treatment. The effect can be demonstrated by rearranging Equation 4.11 and solving for t:

$$t = \frac{n}{k}\left[\left(\frac{C_0}{C_n}\right)^{1/n} - 1\right] \qquad (4.13)$$

All terms in this equation have been defined previously.

Example 4.2

Compare detention times for the same BOD removal levels in partial-mix aerated ponds having one to five cells. Assume $C_0 = 200$ mg/L, $k = 0.28$ d^{-1}, and $T_w = 20$°C.

Solution

1. Solve Equation 4.13 for a single-cell system:

$$t = \frac{n}{k}\left[\left(\frac{C_0}{C_n}\right)^{1/n} - 1\right]$$

$$t = \frac{1}{0.28}\left[\left(\frac{200}{30}\right)^{1/1} - 1\right]$$

$$t = 20.2 \text{ d}$$

2. Similarly, when:
    $n = 2$, then $t = 11$ d.
    $n = 3$, then $t = 9.4$ d.
    $n = 4$, then $t = 8.7$ d.
    $n = 5$, then $t = 8.2$ d.

3. Continuing to increase n will result in the detention time being equal to the detention time in a plug-flow reactor. It can be seen from the tabulation above that the advantages diminish after the third or fourth cell.

4.3.1.3 Temperature Effects

The influence of temperature on the reaction rate is defined by Equation 4.14:

$$k_T = k_{20}\theta^{T_w-20} \tag{4.14}$$

where
k_T = Reaction rate at temperature T (d^{-1}).
k_{20} = Reaction rate at 20°C (d^{-1}).
θ = Temperature coefficient = 1.036.
T_w = Temperature of pond water (°C).

The pond water temperature (T_w) can be estimated using the following equation developed by Mancini and Barnhart (1976):

$$T_w = \frac{AfT_a + QT_i}{Af + Q} \tag{4.15}$$

where
T_w = Pond water temperature (°C).
A = Surface area of pond (m^2).
f = Proportionality factor = 0.5.
T_a = Ambient air temperature (°C).
Q = Wastewater flow rate (m^3/day).

An estimate of the surface area is made based on Equation 4.13, corrected for temperature, and then the temperature is calculated using Equation 4.15. After several iterations, when the water temperature used to correct the reaction rate coefficient agrees with the value calculated with Equation 4.15, the selection of the detention time in the system is completed.

4.3.2 POND CONFIGURATION

The ideal configuration of a pond designed on the basis of complete-mix hydraulics is a circular or a square pond; however, even though partial-mix ponds are designed using the complete mix model, it is recommended that the cells be configured with a length-to-width ratio of 3:1 or 4:1. This is done because it is recognized that the hydraulic flow pattern in partial-mix systems more closely resembles the plug-flow condition. The dimensions of the cells can be calculated using Equation 4.16:

$$V = \left[LW + (L - 2sd)(W - 2sd) + 4(L - sd)(W - sd)\right]\frac{d}{6} \qquad (4.16)$$

where
  V = Volume of pond or cell (m³).
  L = Length of pond or cell at water surface (m).
  W = Width of pond or cell at water surface (m).
  s = Slope factor (e.g., for 3:1 slope, $s = 3$).
  d = Depth of pond (m).

4.3.3 MIXING AND AERATION

The oxygen requirements control the power input required for partial-mix pond systems. Several rational equations are available to estimate the oxygen requirements for pond systems (Benefield and Randall, 1980; Gloyna, 1971, 1976; Metcalf & Eddy, 1991, 2003). In most cases, partial-mix system design is based on the BOD entering the system to estimate the biological oxygen requirements. After the required rate of oxygen transfer has been calculated, equipment manufacturers' catalogs should be used to determine the zone of complete oxygen dispersion by surface, helical, or air gun aerators or by the proper spacing of perforated tubing. Schematic sketches of several of the various types of aerators used in pond systems are shown in Figure 4.2. Photographs of some of the aeration equipment illustrated in Figure 4.2 plus additional photographs of installed aeration equipment are shown in Figure 4.3. Equation 4.17 is used to estimate oxygen transfer rates:

$$N = \frac{N_a}{\alpha\left[\dfrac{(C_{sw} - C_L)}{C_s}\right](1.025)^{(T_w - 20)}} \qquad (4.17)$$

where
  N = Equivalent oxygen transfer to tapwater at standard conditions (kg/hr).
  N_a = Oxygen required to treat the wastewater (kg/hr) (usually taken as 1.5× the organic loading entering the cell).
  α = (Oxygen transfer in wastewater)/(oxygen transfer in tapwater) = 0.9.
  C_{sw} = $\beta(C_{ss})P$ = oxygen saturation value of the waste (mg/L), where β = (wastewater saturation value)/tapwater oxygen saturation value) = 0.9; C_{ss} is the tapwater oxygen saturation value at temperature T_w; and P is the ratio of barometric pressure at the pond site to barometric pressure at sea level (assume 1.0 for an elevation of 100 m).
  C_L = Minimum DO concentration to be maintained in the wastewater (assume 2 mg/L).

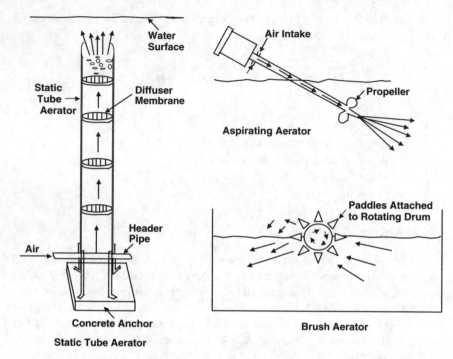

FIGURE 4.2 Schematics of aeration equipment used in wastewater ponds.

C_s = Oxygen saturation value of tapwater at 20°C and one atmosphere pressure = 9.17 mg/L.

T_w = Wastewater temperature (°C).

Equation 4.15 can be used to estimate the water temperature in the pond during the summer months that will be the critical period for design. The partial-mix design procedure is illustrated in Example 4.3. The four-cell system can be obtained by using floating plastic partitions such as those shown in Figure 4.4.

Example 4.3

Design a four-cell partial-mix aerated pond with two trains to remove BOD$_5$ for the following environmental conditions and wastewater characteristics: $Q = 1136$ m³/d (0.3 mgd), $C_0 = 220$ mg/L, C_e from the fourth cell is 30 mg/L, $k_{20} = 2.5$ d⁻¹, winter air temperature = 8°C, summer air temperature = 25°C, elevation = 50 m (164 ft), pond depth = 4 m (13.1 ft).

Solution

        Flow rate = Q = 568.00 m³/d
        Influent BOD = 220.00 mg/L
        Influent TSS = 200.00 mg/L
        Total nitrogen = 30.00 mg/L

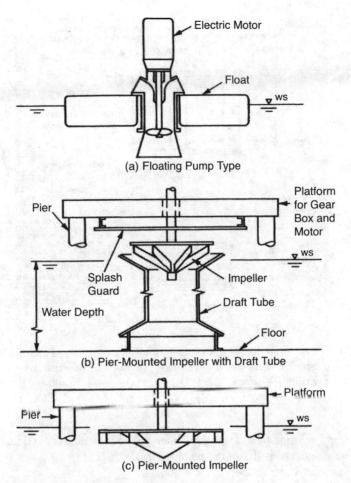

(a) Floating Pump Type

(b) Pier-Mounted Impeller with Draft Tube

(c) Pier-Mounted Impeller

FIGURE 4.2 (cont.) (From Reynolds, T.D. and Richards, P.A., *Unit Operations and Processes in Environmental Engineering*, 2nd ed., PWS Publishing, New York, 1996. With permission.)

Total phosphorus = 10.00 mg/L
Reaction rate at 20°C = 0.276 d^{-1}
Influent temperature = 15.00°C
Winter air temperature = T_a = 8.00°C
Summer air temperature = T_a = 25.00°C
f = units conversion factor = 0.50
Temperature correction coefficient = 1.09
Surface elevation = 50.00 m
Minimum DO concentration = 2.00 mg/L
Depth = 4.00 m
Length-to-width ratio = 2.00
Side slope = 3.00

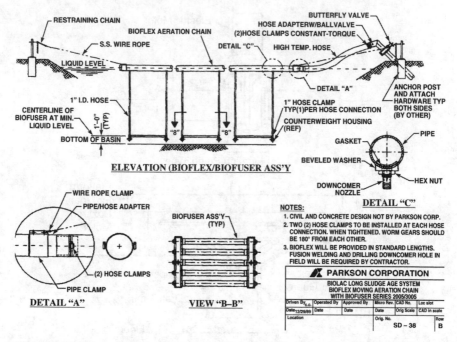

FIGURE 4.3 Diagram of the BIOLAC® aeration chain (courtesy of Parkson Corp.) and photographs of aeration equipment used in pond systems (R.H. Bowman, West Slope Unit Leader, Water Quality Division, Colorado Department of Public Health and Environment, personal communication, 2000).

1. Begin solution by assuming a winter pond temperature and determine the volume of cell 1 in the pond system.
   Assumed water temperature = 12.06°C.
   Correct reaction rate for temperature: $k_T = k_{20}(1.036)^{(T-20)} = 0.210$ d^{-1}.
   Hydraulic residence time in cell 1 is 3.60 d.
   Effluent BOD in cell 1 = $C_0/(1 + kt) = 125.69$ mg/L.
   Volume in cell 1 = 2044.80 m^3.

2. Calculate the dimensions of cell 1 at the water surface and the surface area:
       Depth = 4.00 m.
       Width = 24.51 m.
       Length = 49.02 m.
       Surface area in cell 1 = 1201.61 m^2 = 0.134 ac.

3. Check pond temperature using cell area calculated above and equation shown below:

$$\text{Cell 1 } T_w = (AfT_a + QT_i)/(Af + Q) = 11.40°C$$

Aspirating Aerator Up Close and Mixing Effects of Aerator

Floating Aerator in Operation and Out of Service

Floating Aerators During Winter Operation

FIGURE 4.3 (cont.).

If the calculated T_w differs from the assumed water temperature, another iteration is necessary.

Add a freeboard = 0.90 m.

Dimensions at top of dike in cell 1:

   W top of dike = 29.91 m.

   L top of dike = 54.42 m.

FIGURE 4.3 (cont.).

4. For cell 2:
   Entering water temperature = 11.40°C
   Correct reaction rate for temperature: $k_T = k_{20}(1.09)^{(T-20)} = 0.20$ d^{-1}.
   Influent BOD in cell 2 = 125.69 mg/L.
   Hydraulic residence time in cell 2 = 3.50 d.
   Effluent BOD cell 2 = 73.39 mg/L

Inca Grid Aeration System in Operation and in Dry Pond

Covered Aerated Pond

**Erosion Pad in Shallow Pond
with Surface Aerators**

Mixers

FIGURE 4.3 (cont.).

Volume in cell 2 = 1988.00 m³.
Calculate dimensions of cell 2 at water surface and the surface area:
Depth = 4.00 m.
Width = 24.28 m.
Length = 48.56 m.
Area = 1179.11 m² = 0.134 ac.

Cell 2 $T_w = (AfT_a + QT_i)/(Af + Q) = 9.67°C$

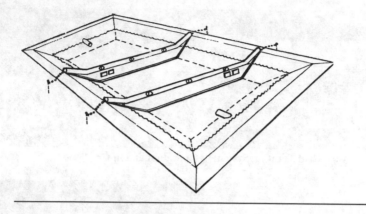

TYPICAL ANCHORING DETAILS
ENVIRONETICS FLOATING BAFFLE

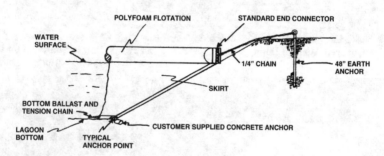

FIGURE 4.4 Floating baffle. (Courtesy of Environetics, Inc.; Lockport, IL.)

Add a freeboard = 0.90 m.
Dimensions at top of dike in cell 2:
    W top of dike = 29.68 m.
    L top of dike = 53.96 m.
5. For cell 3:
Entering water temperature = 9.67°C.
$k_T = 0.19$ d^{-1}.
Influent BOD to cell 3 = 73.39 mg/L.
Hydraulic residence time in cell 3 = 3.00 d.
Effluent BOD in cell 3 = 46.61 mg/L.
Volume in cell 3 = 1704.00 m^3.
Calculate dimensions of cell 3 at the water surface and the surface area:
    Depth = 4.00 m.
    Width = 23.07 m.
    Length = 46.14 m.
    Area = 1064.56 m^2 = 0.134 ac.

$$\text{Cell 3 } T_w = (AfT_a + QT_i)/(Af + Q) = 8.86°C$$

Add a freeboard = 0.90 m.
Dimensions at top of dike in cell 3:
    W top of dike = 28.47 m.
    L top of dike = 51.54 m.

6. For cell 4:
   Entering water temperature = 8.86°C.
   $k_T = 0.19$ d^{-1}.
   Influent BOD to cell 4 = 46.61 mg/L.
   Hydraulic residence time in cell 4 = 3.00 d.
   Effluent BOD in cell 4 = 29.91 mg/L.
   Volume in cell 4 = 1704.00 m^3.
   Calculate dimensions of cell 4 at the water surface and the surface area:
       Depth = 4.00 m.
       Width = 23.07 m.
       Length = 46.14 m.
       Area = 1064.56 m^2 = 0.263 ac.

$$\text{Cell 4 } T_w = (AfT_a + QT_i)/(Af + Q) = 8.44°C$$

Add a freeboard = 0.90 m.
Dimensions at top of dike:
    W top of dike = 28.47 m.
    L top of dike = 51.54 m.

7. Determine the oxygen requirements for pond system based on organic loading and water temperature. Maximum oxygen requirements will occur during the summer months:
   T_w for summer cell 1 = $(AfT_a + QT_i)/(Af + Q)$ = 20.14°C.
   T_w for summer cell 2 = $(AfT_a + QT_i)/(Af + Q)$ = 22.62°C.
   T_w for summer cell 3 = $(AfT_a + QT_i)/(Af + Q)$ = 23.77°C.
   T_w for summer cell 4 = $(AfT_a + QT_i)/(Af + Q)$ = 24.36°C.
   Organic load (OL) in the influent wastewater:
   OL on cell 1 = $C_0 \times Q$ = 5.21 kg/hr.
       Calculate effluent BOD from first cell using equations below at T_w
       for summer:
   $k_{Tw} = k_{20} \times (\text{temperature coefficient})^{(T_w-20)}$ = 0.28 d^{-1}.
   $C_1 = C_0/[(kt) + 1]$ = 110.08 mg/L.
   Winter = 125.69 mg/L.
   OL on cell 2 = $C_1 \times Q$ = 2.61 kg/hr.
   $k_{Tw} = k_{20} \times (\text{temperature coefficient})^{(T_w-20)}$ = 0.30 d^{-1}.
   $C_2 = C_1/[(kt) + 1]$ = 53.45 mg/L.
   Winter = 73.39 mg/L.
   OL on cell 3 = $C_2 \times Q$ = 1.26 kg/hr.
   $k_{Tw} = k_{20} \times (\text{temperature coefficient})^{(T_w-20)}$ = 0.32 d^{-1}.
   $C_3 = C_2/[(kt) + 1]$ = 27.46 mg/L.
   Winter = 46.61 mg/L.

OL on cell 4 = $C_3 \times Q$ = 0.65 kg/hr.

$k_{Tw} = k_{20} \times$ (temperature coefficient)$^{(T_w-20)}$ = 0.32 d^{-1}.

$C_4 = C_3/[(kt) + 1]$ = 13.97 mg/L.

Winter = 29.91 mg/L.

Oxygen demand (OD) is assumed to be a multiple of organic loading (OL) (with a multiplying factor of 1.50):

OD in cell 1 = OL1 × multiplying factor = 7.81 kg/hr.

OD in cell 2 = OL2 × multiplying factor = 3.91 kg/hr.

OD in cell 3 = OL3 × multiplying factor = 1.90 kg/hr.

OD in cell 4 = OL4 × multiplying factor = 0.97 kg/hr.

8. Use the following equation to calculate equivalent oxygen transfer:

$$N = N_{OD}/(a[(C_{sw} - C_L)/C_s](\text{temperature factor})^{(T_w-20)})$$

where N_{OD} = oxygen demand in various cells; $C_{sw} = b \times C_{ss} \times P$; b = 0.90; P = ratio of barometric pressure at pond site to pressure at sea level = 0.80.

Cell 1 tapwater oxygen saturation value C_{ss} = 9.15 mg/L.

Cell 2 tapwater oxygen saturation value C_{ss} = 8.74 mg/L.

Cell 3 tapwater oxygen saturation value C_{ss} = 8.56 mg/L.

Cell 4 tapwater oxygen saturation value C_{ss} = 8.46 mg/L.

Cell 1 C_{sw} = 6.59 mg/L.

Cell 2 C_{sw} = 6.29 mg/L.

Cell 3 C_{sw} = 6.16 mg/L.

Cell 4 C_{sw} = 6.09 mg/L.

a = (oxygen transfer in wastewater)/(oxygen transfer in tapwater) = 0.90.

C_L = minimum oxygen concentration to be maintained in wastewater (usually assumed to be 2 mg/L) = 2.00 mg/L

C_s = oxygen saturation value of tapwater at 20°C and 1 atm = 9.17 mg/L.

Temperature factor (normally 1.025) = 1.025.

N1 = 17.29 kg/hr.   N3 = 4.23 kg/hr.

N2 = 8.70 kg/hr.   N4 = 2.18 kg/hr.

9. Evaluate surface and diffused air aeration equipment to satisfy oxygen requirement only:

Power requirement for surface aerators is approximately 1.9 kg O_2 per kWh, or 1.40 kg O_2 per hp per hr.

Power requirement for diffused air is approximately 2.70 kg O_2 per kWh, or 2.00 kg O_2 per hp per hr.

Total power for surface aeration in one train:

Cell 1: 9.10 kW or 12.35 hp.

Cell 2: 4.58 kW or 6.21 hp.

Cell 3: 2.23 kW or 2.99 hp.

Cell 4: 1.15 kW or 1.54 hp.

Total power for diffused aeration in one train:
   Cell 1: 6.40 kW or 8.64 hp.
   Cell 2: 3.22 kW or 4.35 hp.
   Cell 3: 1.57 kW or 2.12 hp.
   Cell 4: 0.81 kW or 1.09 hp.
These surface and diffused aerator power requirements must be corrected for gearing and blower efficiency:
   Gearing efficiency = 0.90.
   Blower efficiency = 0.90.
Total power requirements corrected for efficiency for one train are:
   Cell 1 surface aerators: 10.11 kW or 3.56 hp.
   Cell 2 surface aerators: 5.09 kW or 6.83 hp.
   Cell 3 surface aerators: 2.48 kW or 3.33 hp.
   Cell 4 surface aerators :1.27 kW or 1.70 hp.
   Total power for surface aerators: 18.95 kW or 25.41 hp.
   Power cost per kilowatt-hour = $0.06/kWh.
Total power costs for surface aerators per year for one train = $9958.02/yr.
   Cell 1 diffused aeration: 7.11 kW or 9.53 hp.
   Cell 2 diffused aeration: 3.58 kW or 4.80 hp.
   Cell 3 diffused aeration: 1.74 kW or 2.33 hp.
   Cell 4 diffused aeration: 0.90 kW or 1.21 hp.
   Total power diffused aeration: 13.33 kW or 17.87 hp.
   Power cost per kilowatt-hour = $0.06/kWh.
Total power costs for diffused aerators per year for one train = $7007.49/yr.

These power requirements are approximate values and are used for the preliminary selection of equipment. These values are used in conjunction with equipment manufacturers' catalogs to select the proper equipment.

Surface aeration equipment is subject to potential icing problems in cold climates, but many options are available to avoid this problem (see Figure 4.2 and Figure 4.3). Improvements have been made in fine bubble perforated tubing, but a diligent maintenance program is still a good policy. In the past, a number of communities experienced clogging of the perforations, particularly in hardwater areas, and corrective action required purging with HCl gas. The final element recommended in this partial-mix aerated pond system is a settling cell with a 2-day detention time.

4.4 COMPLETE-MIX AERATED POND SYSTEMS

There are many configurations of complete-mix pond systems, but most are similar in design. Examples of several types that utilize the complete-mix concept are discussed in the following sections: high-performance aerated pond systems, nitrogen removal in pond systems, modified high-performance aerated lagoon systems for nitrification and denitrification, the BIOLAC® process, and Lemna systems. Most complete-mix systems are designed using the equations that are presented in the following paragraphs with minor modifications. As noted previously, the trend toward omitting redundancy in design of aerated lagoon systems has been alarming. It appears that little thought is given to the need for maintenance in the future, as operating costs associated with aerated lagoon systems frequently have been ignored or overlooked when comparing options available to a community. The initial costs of systems without redundancy obviously are lower than those for systems that include flexibility in operation, but the cost to the environment and the owner will be far greater when maintenance is required.

An examination of Example 4.5 reveals the similarity between the design for the high-performance aerated pond system and the complete-mix design presented below when the final three cells of the complete-mix design (Example 4.5) are supplied only enough dissolved oxygen to satisfy the BOD. This is not to imply that both are the same but only demonstrates the similarity between the two design methods.

4.4.1 DESIGN EQUATIONS

The design model using first-order kinetics and operating n number of equal-sized cells in series is given by Equation 4.18 (Middlebrooks et al. 1982; Great Lakes–Upper Mississippi River Board of State Sanitary Engineers, 1990; WEF/ASCE, 1991).

$$\frac{C_n}{C_0} = \frac{1}{\left[(1 + kt/n)\right]^n} \tag{4.18}$$

where
C_n = Effluent BOD concentration in cell n (mg/L).
C_0 = Influent BOD concentration (mg/L).
k = First-order reaction rate constant (d^{-1}) = 2.5 d^{-1} at 20°C (assumed to be constant in all cells).
t = Total hydraulic residence time in pond system (d).
n = Number of cells in the series.

If other than a series of equal volume ponds are to be employed or it is desired to use varying reaction rates, it is necessary to use the following general equation:

$$\frac{C_n}{C_0} = \left(\frac{1}{1 + k_1 t_1}\right)\left(\frac{1}{1 + k_2 t_2}\right)\cdots\left(\frac{1}{1 + k_n t_n}\right) \tag{4.19}$$

where $k_1, k_2, \ldots, k_n$ are the reaction rates in cells 1 through n (all usually assumed equal for lack of better information) and $t_1, t_2, \ldots, t_n$ are the hydraulic residence times in the respective cells.

Mara (1975) has shown that a number of equal-volume reactors in series is more efficient than unequal volumes; however, due to site topography or other factors in some cases it may be necessary to construct cells of unequal volume.

4.4.1.1 Selection of Reaction Rate Constants

Selection of the k value is one of the critical decisions in the design of any pond system. A design value of 2.5 d^{-1} at 20°C (68°F) is recommended by Reynolds and Middlebrooks (1990) based on a study of a complete-mix aerated lagoon system located in Colorado. Higher values are recommended by others, but designs based on this value of k_C have worked well at full design load, whereas designs with higher k_C values have not functioned as well as design flow. In most cases, the designer will be constrained by state design standards, and in many states the prescribed reaction rate will exceed the value of 2.5 d^{-1}.

4.4.1.2 Influence of Number of Cells

When using the complete-mix design model, the number of cells in series has a pronounced effect on the size of the pond system required to achieve the specified degree of treatment. Rearranging Equation 4.18 and solving for t can demonstrate the effect:

$$t - \frac{n}{k}\left[\left(\frac{C_0}{C_n}\right)^{1/n} - 1\right] \tag{4.20}$$

All terms in this equation have been defined previously.

Example 4.4

Compare detention times for the same BOD removal levels in complete-mix aerated ponds having one to five cells. Assume $C_0 = 200$ mg/L, $k = 2.5$ d^{-1}, $T_w = 20$°C.

Solution

1. Solve Equation 4.20 for a single-cell system:

$$t = \frac{n}{k}\left[\left(\frac{C_0}{C_n}\right)^{1/n} - 1\right]$$

$$t = \frac{1}{2.5}\left[\left(\frac{200}{30}\right)^{1/1} - 1\right]$$

$$t = 2.27 \text{ d}$$

2. Similarly, when:

    $n = 2$, then $t = 1.03$ d.
    $n = 3$, then $t = 0.75$ d.
    $n = 4$, then $t = 0.64$ d.
    $n = 5$, then $t = 0.58$ d.

3. Continuing to increase n will result in the detention time being equal to the detention time in a plug-flow reactor. It can be seen from the tabulation above that the advantages diminish after the third or fourth cell.

4.4.1.3 Temperature Effects

The influence of temperature on the reaction rate is defined by Equation 4.21:

$$k_T = k_{20}\theta^{T_w - 20} \tag{4.21}$$

where

k_T = Reaction rate at temperature T (d^{-1}).
k_{20} = Reaction rate at 20°C (d^{-1}).
θ = Temperature coefficient = 1.036.
T_w = Temperature of pond water (°C).

The pond water temperature (T_w) can be estimated using the following equation developed by Mancini and Barnhart (1976):

$$T_w = \frac{AfT_a + QT_i}{Af + Q} \tag{4.22}$$

where

T_w = Pond water temperature (°C).
A = Surface area of pond (m^2).
f = Proportionality factor = 0.5.
T_a = Ambient air temperature (°C).
Q = Wastewater flow rate (m^3/d).

An estimate of the surface area is made based on Equation 4.20 and corrected for temperature, then the temperature is calculated using Equation 4.22. After several iterations, when the water temperature used to correct the reaction rate coefficient agrees with the value calculated with Equation 4.22, the selection of the detention time in the system is completed.

4.4.2 POND CONFIGURATION

The ideal configuration of a pond designed on the basis of complete-mix hydraulics is a circular or a square pond; however, even though complete-mix ponds are designed using the complete-mix model, it is recommended that the cells be configured with a length-to-width ratio of 3:1 or 4:1. This is done because it is recognized that the hydraulic flow pattern in complete-mix designed systems more closely resembles the plug-flow condition. The dimensions of the cells can be calculated using Equation 4.23:

$$V = \left[LW + (L - 2sd)(W - 2sd) + 4(L - sd)(W - sd) \right] \frac{d}{6} \qquad (4.23)$$

where
- V = Volume of pond or cell (m^3).
- L = Length of pond or cell at water surface (m).
- W = Width of pond or cell at water surface (m).
- s = Slope factor (e.g., for a 3:1 slope, $s = 3$).
- d = Depth of pond (m).

4.4.3 MIXING AND AERATION

The mixing requirements usually control the power input required for complete-mix pond systems. Several rational equations are available to estimate the oxygen requirements for pond systems (Benefield and Randall, 1980; Gloyna, 1971, 1976; Metcalf & Eddy, 1991, 2003). Complete-mix systems are designed by estimating the BOD entering the system to estimate the biological oxygen requirements and then checked to ensure that adequate power is available to provide complete mixing. After calculating the required rate of oxygen transfer, equipment manufacturers' catalogs should be used to determine the zone of complete mixing and oxygen dispersion by surface, helical, aeration chain, or air gun aerators or by the proper spacing of perforated tubing. Schematic sketches of several of the various types of aerators used in pond systems are shown in Figure 4.2. Photographs and drawings of some of the aeration equipment illustrated in Figure 4.2 plus additional photographs of installed aeration equipment are shown in Figure 4.3.

Equation 4.24 is used to estimate the oxygen transfer rates:

$$N = \frac{N_a}{\alpha \left[\dfrac{(C_{sw} - C_L)}{C_s} \right] (1.025)^{(T_w - 20)}} \qquad (4.24)$$

where
- N = Equivalent oxygen transfer to tapwater at standard conditions (kg/hr).
- N_a = Oxygen required to treat the wastewater (kg/hr) (usually taken as 1.5 × the organic loading entering the cell).
- α = (Oxygen transfer in wastewater)/(oxygen transfer in tapwater) = 0.9.

C_{sw} = $\beta(C_{ss})P$ = oxygen saturation value of the waste (mg/L), where β = (wastewater saturation value)/(tapwater oxygen saturation value); C_{ss} is the tapwater oxygen saturation value at temperature T_w; and P is the ratio of barometric pressure at the pond site to barometric pressure at sea level (assume 1.0 for an elevation of 100 m).

C_L = Minimum DO concentration to be maintained in the wastewater, assume 2 mg/L.

C_s = Oxygen saturation value of tapwater at 20°C and one atmosphere pressure = 9.17 mg/L.

T_w = Wastewater temperature (°C).

Equation 4.22 can be used to estimate the water temperature in the pond during the summer months that will be the critical period for design for biological activity; however, with power to provide complete mixing, adequate dissolved oxygen is normally readily available. The complete mix design procedure is illustrated in Example 4.5. The four-cell system can be obtained by using floating plastic partitions such as those shown in Figure 4.4, and the aeration equipment can be selected from the types shown in Figures 4.2 and 4.3.

Example 4.5

Design a four-cell complete-mix aerated pond with two trains to remove BOD_5 for the following environmental conditions and wastewater characteristics: Q = 1136 m³/d (0.3 mgd), C_0 = 220 mg/L, C_e from the fourth cell is 30 mg/L, k_{20} = 2.5 d⁻¹, winter air temperature = 8°C, summer air temperature = 25°C, elevation = 50 m (164 ft), pond depth = 4 m (13.1 ft); maintain a minimum DO concentration of 2 mg/L in all cells.

Solution

       Flow rate = Q = 568.00 m³/d
       Influent BOD = 220.00 mg/L
       Influent TSS = 200.00 mg/L
       Total nitrogen = 30.00 mg/L
       Total phosphorus = 10.00 mg/L
       Reaction rate at 20°C = 2.500 d⁻¹
       Influent temperature = 15.00°C
       Winter air temperature = T_a = 8.00°C
       Summer air temperature = T_a = 25.00°C
       f = units conversion factor = 0.50
       Temperature correction coefficient = 1.09
       Surface elevation = 50.00 m
       Minimum DO concentration = 2.00 mg/L
       Depth = 4.00 m
       Length-to-width ratio = 2.00
       Side slope = 3.00

1. Begin solution by assuming a winter pond temperature and determine the volume of cell 1 in the pond system.
   Assumed water temperature = 12.74°C.
   Correct reaction rate for temperature: $k_T = k_{20}(1.09)^{(T-20)} = 1.34$ d^{-1}.
   Hydraulic residence time in cell 1 is 1.00 d.
   Effluent BOD in cell 1 = 94.13 mg/L.
   Volume in cell 1 = 568.00 m^3.

2. Calculate the dimensions of cell 1 at the water surface and the surface area:
   Depth = 4.00 m.
   Width = 16.48 m.
   Length = 32.97 m.
   Surface area in cell 1 = 543.40 m^2 = 0.134 ac.

3. Check pond temperature using cell area calculated above and equation shown below:

$$\text{Cell 1 } T_w = (AfT_a + QT_i)/(Af + Q) = 12.74°C$$

   If the calculated T_w differs from the assumed water temperature, another iteration is necessary.
   Add a freeboard = 0.90 m.
   Dimensions at top of dike in cell 1:
   W top of dike = 21.88 m.
   L top of dike = 38.37 m.

4. For cell 2:
   Entering water temperature = 12.74°C
   Correct reaction rate for temperature: $k_T = k_{20}(1.09)^{(T-20)} = 1.34$ d^{-1}.
   Influent BOD in cell 2 = 94.13 mg/L.
   Hydraulic residence time in cell 2 = 1.00 d.
   Effluent BOD cell 2 = 40.28 mg/L
   Volume in cell 2 = 568.00 m^3.
   Calculate dimensions of cell 2 at water surface and the surface area:
   Depth = 4.00 m.
   Width = 16.48 m.
   Length = 32.97 m.
   Area = 543.40 m^2 = 0.134 ac.

$$\text{Cell 2 } T_w = (AfT_a + QT_i)/(Af + Q) = 11.20°C.$$

   Add a freeboard = 0.90 m.
   Dimensions at top of dike in cell 2:
   W top of dike = 21.88 m.
   L top of dike = 38.37 m.

5. For cell 3:
   Entering water temperature = 11.20°C.
   $k_T = 1.17$ d^{-1}.

Influent BOD to cell 3 = 40.28 mg/L.
Hydraulic residence time in cell 3 = 1.00 d.
Effluent BOD in cell 3 = 18.55 mg/L.
Volume in cell 3 = 568.00 m^3.
Calculate dimensions of cell 3 at the water surface and the surface area:
  Depth = 4.00 m.
  Width = 16.48 m.
  Length = 32.97 m.
  Area = 543.40 m^2 = 0.134 ac.

$$\text{Cell 3 } T_w = (AfT_a + QT_i)/(Af + Q) = 10.17°C$$

Add a freeboard = 0.90 m.
Dimensions at top of dike in cell 3:
  W top of dike = 21.88 m.
  L top of dike = 38.37 m.
6. For cell 4:
Entering water temperature = 10.17°C.
$k_T = 1.07$ d^{-1}.
Influent BOD to cell 4 = 18.55 mg/L.
Hydraulic residence time in cell 4 = 1.00 d.
Effluent BOD in cell 4 = 8.96 mg/L.
Volume in cell 4 = 568.00 m^3.
Calculate dimensions of cell 4 at the water surface and the surface area:
  Depth = 4.00 m.
  Width = 16.48 m.
  Length = 32.97 m.
  Area = 543.40 m^2 = 0.134 ac.

$$\text{Cell 4 } T_w = (AfT_a + QT_i)/(Af + Q) = 9.47°C$$

Add a freeboard = 0.90 m.
Dimensions at top of dike:
  W top of dike = 21.88 m.
  L top of dike = 38.37 m.
7. Determine the oxygen requirements for pond system based on organic loading and water temperature. Maximum oxygen requirements will occur during the summer months:
  T_w for summer cell 1 = $(AfT_a + QT_i)/(Af + Q)$ = 18.24°C.
  T_w for summer cell 2 = $(AfT_a + QT_i)/(Af + Q)$ = 20.42°C.
  T_w for summer cell 3 = $(AfT_a + QT_i)/(Af + Q)$ = 21.90°C.
  T_w for summer cell 4 = $(AfT_a + QT_i)/(Af + Q)$ = 22.91°C.
Organic load (OL) in the influent wastewater:
OL on cell 1 = $C_0 \times Q$ = 5.21 kg/hr.
  Calculate effluent BOD from first cell using equations below at T_w
  for summer:
  $k_{Tw} = k_{20} \times$ (temperature coefficient)$^{(T_w-20)}$ = 2.15 d^{-1}.

$C_1 = C_0/[(kt) + 1] = 69.90$ mg/L.
Winter = 94.13 mg/L.
OL on cell 2 = $C_1 \times Q = 1.65$ kg/hr.
$k_{Tw} = k_{20} \times$ (temperature coefficient)$^{(T_w-20)} = 2.59$ d^{-1}.
$C_2 = C_1/[(kt) + 1] = 19.45$ mg/L.
Winter = 40.28 mg/L.
OL on cell 3 = $C_2 \times Q = 0.46$ kg/hr.
$k_{Tw} = k_{20} \times$ (temperature coefficient)$^{(T_w-20)} = 2.95$ d^{-1}.
$C_3 = C_2/[(kt) + 1] = 4.93$ mg/L.
Winter = 18.55 mg/L.
OL on cell 4 = $C_3 \times Q = 0.12$ kg/hr.
$k_{Tw} = k_{20} \times$ (temperature coefficient)$^{(T_w-20)} = 3.21$ d^{-1}.
$C_4 = C_3/[(kt) + 1] = 1.17$ mg/L.
Winter = 8.96 mg/L.
Oxygen demand (OD) is assumed to be a multiple of organic loading (OL) (with a multiplying factor of 1.50):
OD in cell 1 = OL1 × multiplying factor = 7.81 kg/hr.
OD in cell 2 = OL2 × multiplying factor = 2.48 kg/hr.
OD in cell 3 = OL3 × multiplying factor = 0.69 kg/hr.
OD in cell 4 = OL4 × multiplying factor = 0.18 kg/hr.
8. Use the following equation to calculate equivalent oxygen transfer:

$$N = N_{OD}/(a[(C_{rw} - C_L)/C_s](\text{temperature factor})^{(T_w-20)})$$

where N_{OD} = oxygen demand in various cells; $C_{sw} = b \times C_{ss} \times P$; $b = 0.90$; P = ratio of barometric pressure at pond site to pressure at sea level = 0.80.
Cell 1 tapwater oxygen saturation value $C_{ss} = 9.49$ mg/L.
Cell 2 tapwater oxygen saturation value $C_{ss}' = 9.10$ mg/L.
Cell 3 tapwater oxygen saturation value $C_{ss} = 8.85$ mg/L.
Cell 4 tapwater oxygen saturation value $C_{ss} = 8.69$ mg/L.
Cell 1 $C_{sw} = 6.84$ mg/L.
Cell 2 $C_{sw} = 6.55$ mg/L.
Cell 3 $C_{sw} = 6.37$ mg/L.
Cell 4 $C_{sw} = 6.26$ mg/L.
a = (oxygen transfer in wastewater)/(oxygen transfer in tapwater) = 0.90.
C_L = minimum oxygen concentration to be maintained in wastewater (usually assumed to be 2 mg/L) = 2.00 mg/L.
C_s = oxygen saturation value of tapwater at 20°C and 1 atm = 9.17 mg/L.
Temperature factor (normally 1.025) = 1.025.
$N_1 = 17.19$ kg/hr.
$N_2 = 5.50$ kg/hr.
$N_3 = 1.54$ kg/hr.
$N_4 = 0.39$ kg/hr.

9. Evaluate surface and diffused air aeration equipment to satisfy oxygen requirement only:

Power requirement for surface aerators is approximately 1.9 kg O_2 per kWh, or 1.40 kg O_2 per hp per hr.

Power requirement for diffused air is approximately 2.70 kg O_2 per kWh, or 2.00 kg O_2 per hp per hr.

Total power for surface aeration:

Cell 1: 9.05 kW or 12.28 hp.

Cell 2: 2.89 kW or 3.93 hp.

Cell 3: 0.81 kW or 1.10 hp.

Cell 4: 0.21 kW or 0.28 hp.

Total power for diffused aeration:

Cell 1: 6.37 kW or 8.60 hp.

Cell 2: 2.04 kW or 2.75 hp.

Cell 3: 0.57 kW or 0.77 hp.

Cell 4: 0.14 kW or 0.19 hp.

These surface and diffused aerator power requirements must be corrected for gearing and blower efficiency:

Gearing efficiency = 0.90.

Blower efficiency = 0.90.

Total power requirements corrected for efficiency are:

Cell 1 surface aerators: 10.05 kW or 13.48 hp.

Cell 2 surface aerators: 3.21 kW or 4.31 hp.

Cell 3 surface aerators: 0.90 kW or 1.20 hp.

Cell 4 surface aerators :0.23 kW or 0.31 hp.

Total power for surface aerators: 14.39 kW or 19.30 hp.

Power cost per kilowatt-hour = $0.06/kWh.

Total power costs for surface aerators per year = $7564.74/yr.

Cell 1 diffused aeration: 7.07 kW or 9.49 hp.

Cell 2 diffused aeration: 2.26 kW or 3.03 hp.

Cell 3 diffused aeration: 0.63 kW or 0.85 hp.

Cell 4 diffused aeration: 0.16 kW or 0.22 hp.

Total power diffused aeration: 10.13 kW or 13.58 hp.

Power cost per kilowatt-hour = $0.06/kWh.

Total power costs for diffused aerators per year are $5323.33/yr.

These power requirements are approximate values and are used for the preliminary selection of equipment. These values are used in conjunction with equipment manufacturers' catalogs to select the proper equipment.

10. Evaluation of power requirements for maintaining a complete-mix reactor:

Power required to maintain solids suspension = 6.00 kW/1000 m^3, or 30.48 hp/MG.

Total power required in cell 1 = 3.41 kW or 4.57 hp.

Total power required in cell 2 = 3.41 kW or 4.57 hp.

Total power required in cell 3 = 3.41 kW or 4.57 hp.
Total power required in cell 4 = 3.41 kW or 4.57 hp.

11. Total power required in the system will be the sum of the maximum power required in each cell as measured above. Assuming that complete mixing is to occur in all cells, use the first set shown below. Another alternative is to use the power calculated for each cell to satisfy oxygen demand or a mixture of complete-mix and oxygen requirements.

Power required for complete mix in all cells in one train:
Cell 1 = 3.41 kW.
Cell 2 = 3.41 kW.
Cell 3 = 3.41 kW.
Cell 4 = 3.41 kW.
Total = 13.63 kW.
Power costs = $7164.98/yr.

Power requirements for each cell based on BOD removal in one train:
Cell 1 = 10.05 kW or 13.48 hp.
Cell 2 = 3.21 kW or 4.31 hp.
Cell 3 = 0.90 kW or 1.20 hp.
Cell 4 = 0.23 kW or 0.31 hp.
Total = 14.39 kW or 19.30 hp.
Power costs = $7564.74/yr.

The system is over-designed so it is necessary to make another iteration and change some of the reactors. Another possibility is to reduce the number of cells in series. Many combinations will yield a satisfactory solution. It is not advisable to reduce the hydraulic residence time below 1 day, and 1.5 days is preferable. Because of the small size of the reactors, more aeration horsepower is required for BOD reduction than is required to maintain complete-mix conditions; normally, the opposite would be true.

4.5 ANAEROBIC PONDS

4.5.1 INTRODUCTION

Anaerobic lagoons or ponds have been used for treatment of municipal, agricultural, and industrial wastewaters. The primary function of anaerobic lagoons is to stabilize large concentrations of organic solids contained in wastewater and not necessarily to produce a high-quality effluent. Most often anaerobic lagoons are operated in series with aerated or facultative lagoons. A three-cell lagoon system can produce a stable, high-quality effluent throughout its design life. Proper design and operation of an anaerobic lagoon should consider the biological reactions that stabilize organic waste material.

In the absence of oxygen, insoluble organics are hydrolyzed by extracellular enzymes to form soluble organics (i.e., carbohydrates such as glucose, cellobiose, xylose). The soluble carbohydrates are biologically converted to volatile acids. These organic (volatile) acids are predominantly acetic, proprionic, and butyric. The group of facultative organisms that transforms soluble organic molecules to short-chain organic acids is known as *acid formers* or *acid producers*. The next sequential biochemical reaction that occurs is the conversion of the organic acid to methane and carbon dioxide by a group of strict, anaerobic bacteria know as *methane formers* or *methane producers*.

Anaerobic decomposition of carbohydrate to bacterial cells with the formation of organic acids can be illustrated as:

$$5(CH_2O)_x \rightarrow (CH_2O)_x + 2CH_3COOH + Energy$$

Bicarbonate buffer present in solution neutralizes the acid formed in the above reaction:

$$2CH_3COOH + 2NH_4HCO_3 \rightarrow 2CH_3COONH_4 + 2H_2O + 2CO_2$$

During the growth of methane bacteria, ammonia acetate (CH_3COONH_4) is decomposed to methane and regeneration of the bicarbonate buffer, NH_4HCO_3:

$$2CH_3COONH_4 + 2H_2O \rightarrow 2CH_4 + 2NH_4HCO_3$$

If sufficient buffer is not available, the pH will decrease, which will inhibit the third reaction.

The facultative acid formers are not as sensitive to ambient environmental factors such as pH value, heavy metals, and sulfides. Acid formers are normally very plentiful in the system and are not the rate-limiting step. The rate-limiting step in anaerobic digestion is the methane fermentation process. Methane-producing bacteria are quite sensitive to such factors as pH changes, heavy metals, detergents, alterations in alkalinity, ammonia nitrogen concentration, temperature, and sulfides. Furthermore, methane-fermenting bacteria have a slow growth rate.

Environmental factors that affect methane fermentation are shown in Table 4.6. In addition, work by Kotze et al. (1968), Chan and Pearson (1970), Hobson et al. (1974), Ghosh et al. (1974), and Ghosh and Klass (1974) provides some evidence that the hydrolysis step may become rate limiting in the digestion of particulates and cellulosic feeds. Design and operation of anaerobic lagoons should be founded on the fundamental biochemical and kinetic principles that govern the process. Most anaerobic lagoons, however, have been empirically designed.

A major problem associated with anaerobic lagoons is the production of odors. Odors can be controlled by providing an aerobic zone at the surface to oxidize the volatile organic compounds that cause odors. Recirculation from an aerobic pond to the primary anaerobic pond can alleviate odors by providing dissolved oxygen from the aerobic pond effluent that overlays the anaerobic pond

TABLE 4.6
Environmental Factors Influencing Methane Fermentation

Variable	Optimal	Extreme
Temperature (°C)	30–35	25–40
pH	6.8–7.4	6.2–7.8
Oxidation/reduction	−520 to −530	−490 to −550
Potential (MV):		
Volatile acids (mg/L as acetic)	50–500	2000
Alkalinity (mg/L as CaCO$_3$)	2000–3000	1000–5000

and oxidizes sulfide odors (Oswald, 1968). To avoid contact of anaerobic processes with oxygen, influent wastewater can be introduced to the anaerobic pond at the center into a chamber in which the sludge accumulates to some depth as shown in Figure 4.5 (Oswald, 1968). Mixing of the influent with the active anaerobic sludge will enhance BOD removal efficiency and reduce odors (Parker et al., 1968).

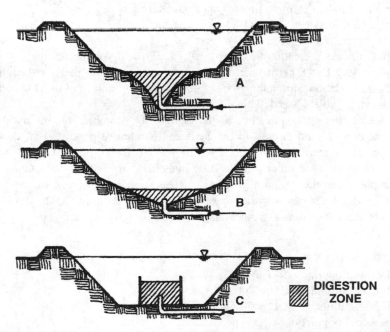

FIGURE 4.5 Method of creating a digestion chamber in the bottom of an anaerobic lagoon. (From Oswald, W.J., in *Advances in Water Quality Improvement*, Gloyna, E.F. and Eckenfelder, W.W., Jr., Eds., University of Texas Press, Austin, TX, 1968. With permission.)

As stated earlier, the purpose of anaerobic lagoons is the decomposition and stabilization of organic matter. Water purification is not the primary function of anaerobic lagoons. Anaerobic lagoons are used as sedimentation basins to reduce organic loads on subsequent treatment units. A general compilation of information about the design of municipal anaerobic lagoons is presented in the following text.

4.5.2 DESIGN

There is no agreement on the best approach to the design of anaerobic stabilization ponds. Systems are designed on the basis of surface loading rate, volumetric loading rate, and hydraulic detention time. Although done frequently, design on the basis of surface loading rate probably is inaccurate. Proper design should be based on the volumetric loading rate, temperature of the liquid, and hydraulic detention time. Areal loading rates that have been used around the world are shown in Table 4.7. It is possible to approximate the volumetric loading rates by dividing by the average depth of the ponds and converting to the proper set of units. Based on these loading rates, it is obvious that there has been little consistency in the design loading rates for anaerobic ponds. In climates where the temperature exceeds 22°C, the following design criteria should yield a BOD_5 removal of 50% or better (WHO, 1987):

- Volumetric loading up to 300 g BOD_5 per m^3 per d
- Hydraulic detention time of approximately 5 d
- Depth between 2.5 and 5 m

In cold climates, detention times as great as 50 d and volumetric loading rates as low as 40 g BOD_5 per m^3 per d may be required to achieve 50% reduction in BOD_5. The relationships among temperature, detention time, and BOD reduction are shown inTable 4.8 and Table 4.9.

One of the best approaches to the design of anaerobic lagoons has been presented by Oswald (1996). In his advanced facultative pond design, Oswald incorporates a deep anaerobic pond within the facultative pond. The anaerobic pond design is based on organic loading rates that vary with water temperature in the pond, and the design is checked by determining the volume of anaerobic pond provided per capita, which is one of the methods used for the design of separate anaerobic digesters. An example of this design approach is presented in Example 4.6.

Example 4.6

Design flow rate = 947 m^3/d.
Influent ultimate BOD = 400 mg/L.
Effluent ultimate BOD = 50 mg/L.
Sewered population = 6000 people.
Maximum bottom water temperature in local bodies of water = 20°C.
Temperature of pond water at bottom of pond = 10°C.

TABLE 4.7
Design and Operational Parameters for Anaerobic Lagoons Treating Municipal Wastewater

Areal BOD$_5$ Mass Loading Rates (lb/ac-d)		Estimated Volumetric Loading Rates (lb/1000 ft^3-d)		BOD$_5$ Removal (%)		Depth (ft)	Hydration Detention Time (d)	Ref.
Summer	Winter	Summer	Winter	Summer	Winter			
360	—	2.34	—	75	—	3–4	—	Parker (1970)
280	—	1.84	—	65	—	3–4	—	Parker (1970)
100	—	0.66	—	86	—	3–4	—	Parker (1970)
170	—	1.11	—	52	—	3–4	—	Parker (1970)
560	400	3.67	2.62	89	60	3–4	—	Parker (1970)
400	100	—	—	70	—	—	—	Oswald (1968)
900–1200	675	5.17–6.89	3.88	60–70	—	3–5	2–5	Parker et al. (1959)
—	—	—	—	—	—	8–10	30–50	Eckenfelder (1961)
220–600	—	—	0.51–1.38	—	—	—	15–160	Cooper (1968)
500	—	—	1.15	70	—	8–12	5	Oswald et al. (1967)
—	—	—	—	—	—	8–12	2 (summer)	Malina and Rios (1976)
—	—	—	—	—	—	—	5 (winter)	Malina and Rios (1976)

TABLE 4.8
Five-Day BOD$_5$ Reduction as a
Function of Detention Time for
Temperatures Greater Than 20°C

Detention Time (d)	BOD$_5$ Reduction (%)
1	50
2.5	60
5	70

Source: WHO, *Wastewater Stabilization Ponds: Principles of Planning and Practice*, WHO Tech. Publ. 10, Regional Office for the Eastern Mediterranean, World Health Organization, Alexandria, 1987.

Solution

1. Calculate the BOD loading:

    BOD loading = Influent BOD × flow rate/1000 = 378.8 kg/d

2. Design the anaerobic pond (fermentation pits). Except for systems with flows less than 200 m^3/day, always use two ponds, so one will be available for desludging when the pond is filled. The surface area of the anaerobic pond should be limited to 1000 m^2, and it should be made as deep as possible to avoid turnover with oxygen intrusion. Minimum pit depth should be 4 m.

    Number of anaerobic ponds in parallel = minimum of two ponds = 2.
    BOD loading on single pond = 189.4 kg/d.

    First, size pond on basis of load per unit volume:

    Load per unit volume (varies with temperature of water) = 0.189 kg/m^3/d.
    Volume in one pond = 1002.7 m^3.
    Hydraulic residence time in ponds = 2.12 d.
    Pond depth = minimum of 4 m = 4 m.
    Pond surface area (assuming vertical walls) = 250.7 m^2.
    Maximum pond surface area = 1000 m^2; number of ponds = 0.25 (round to next largest number of ponds = 1.00).
    Overflow rate in ponds = (total surface area)/(total flow rate) = 1.89 m/d.
    Overflow rates of less than 1.5 m/d should retain parasite eggs and other particles as small as 20 μm, which includes all but the smallest parasite eggs (ova). The size of the pond should be increased to reduce the overflow rate to 1.5 m/d.

TABLE 4.9
**Five-Day BOD_5 Reduction as a Function
of Detention Time and Temperature**

Temperature (°C)	Detention Time (d)	BOD Reduction (%)
10	5	0–10
10–15	4–5	30–40
15–20	2–3	40–50
20–25	1–2	40–60
25–30	1–2	60–80

Source: WHO, *Wastewater Stabilization Ponds: Principles of Planning and Practice*, WHO Tech. Publ. 10, Regional Office for the Eastern Mediterranean, World Health Organization, Alexandria, 1987.

Check pond volume per capita:

Total volume in ponds = (total BOD loading)/(loading rate) = 2005 m^3.

Pond volume/capita = (total volume)/(population) = 0.33 m^3/capita. Pond volume/capita should be greater than 0.0566 m^3/person as used in conventional separate digesters. When pit volume/capita exceeds 0.0566 m^3/person, fermentation can go to completion with only grit and refractory organics left to accumulate

Designs of anaerobic ponds based on information in Table 4.8 and Table 4.9 are presented in Example 4.7 (Reed et al., 1995).

Example 4.7

Temperature = 10°C, detention time (d) = 5, BOD reduction (%) = 0–10.

Temperature = 10–15°C, detention time (d) = 4–5, BOD reduction (%) = 30–40.

Temperature = 15–20°C, detention time (d) = 2–3, BOD reduction (%) = 40–50.

Temperature = 20–25°C, detention time (d) = 1–2, BOD reduction (%) = 40–60.

Temperature = 25–30°C, detention time (d) = 1–2, BOD reduction (%) = 60–80.

Climates with temperatures exceeding 22°C:

Volumetric loading — up to 300 g BOD_5 per m^3 per d

Hydraulic detention time — approximately 5 d

Depth — 2.5 to 5 m

Cold climates (50% estimated reduction in BOD_5):

Volumetric loading — as low as 40 g BOD_5 per m^3 per d

Hydraulic detention time — approximately 50 d

Design input:

Flow = 18,925 m^3/d.
Influent BOD_5 = 250 mg/L.
Temperature = 10°C.
Depth = 3 m.
Length-to-width ratio = 1
Volumetric loading = 60 g BOD_5 per m^3 per d.
Detention time = 5 d.
Slope (e.g., for a 3:1 slope, s = 3).

Output (volumetric loading):

Volume = 78854 m^3
Length = 71 mL.
Width = 171 m.

Output (detention time):

Volume = 94,625 m^3.
Length = 187 m.
Width = 187 m.
Detention time = 5 d.

Oswald's design procedure is semirational, whereas the other approaches are empirical. It is possible that some of the newer approaches to anaerobic reactor design may be applicable to the design of anaerobic ponds; however, it is likely that the controls required in the newer approaches will be impractical for pond design and operation.

4.6   CONTROLLED DISCHARGE POND SYSTEM

See Chapter 5 for details.

4.7   COMPLETE RETENTION POND SYSTEM

See Chapter 5 for details.

4.8   HYDROGRAPH CONTROLLED RELEASE

See Chapter 5 for details.

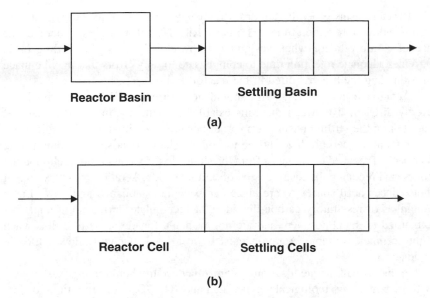

FIGURE 4.6 Flow diagram of dual-power, multicellular (DPMC) aerated lagoon system: (a) two basins in series utilizing floating baffles in the settling cells; (b) a single basin using floating baffles to divide various unit processes. (From Rich, L.G., *High-Performance Aerated Lagoon Systems*, American Academy of Environmental Engineers, Annapolis, MD, 1999. With permission.)

4.9 HIGH-PERFORMANCE AERATED POND SYSTEMS (RICH DESIGN)

The high-performance aerated pond system (HPAPS) described by Rich (1999) has frequently been referred to in the literature as a dual-power, multicellular (DPMC) system. The system consists of two aerated basins in series. Screens to remove large solids precede the system. A reactor basin for bioconversion and flocculation is followed by a settling basin dedicated to sedimentation, solids stabilization, and sludge storage. Algae growth is controlled by limited hydraulic retention time and dividing the settling basin into cells in series. Disinfection facilities follow the settling basin (Figure 4.6).

Aeration is provided in both the reactor portion and the settling basin. Aeration in the reactor is provided at a level of approximately 6 W/m³ to keep the solids suspended, and a minimum hydraulic detention time of 1.5 days is required. In small systems, the reactor and the settling basin can be placed in the same earthen basin; however, in large systems, it is best to put the reactor in a separate basin. Using a separate basin makes it easier to modify the system for upgrading to include nitrification and denitrification. (Nitrification and denitrification will be discussed in another section.)

Reactor basins generally are designed using Monod kinetics but with a minimum hydraulic retention time of 1.5 days. Rich (1999) strongly discourages the use of a safety factor when designing the reactor, because the settling basin provides adequate retention time to compensate for any errors that may be made in estimating the time required in the reactor basin.

Aeration in the settling basin should not exceed 1.8 W/m^3 and should be evenly distributed between the cells established with floating plastic dividers. Aeration in the settling basin is important because it maintains an aerobic water column and an aerobic layer at the top of the sludge deposit, thus minimizing feedback of reduced compounds from the sludge to the water column, eliminating odors, and reducing the resuspension of bottom solids. Aeration provides mixing that reduces dead spaces where algae can become established and grow. Large quantities of respiratory carbon dioxide that accumulate during night hours are exhausted to the atmosphere and are not available for the algae to utilize when light becomes available. Aeration must be at a level that will allow settleable solids to settle.

Problems with aerated lagoon systems may occur when treating wastewaters with carbonaceous biological oxygen demand ($CBOD_5$) concentrations of less than 100 mg/L because few settleable solids may be produced. This is particularly a problem when the wastewater has been presettled. Application of HPAPS at schools and seasonal recreational areas should be avoided. At these operations, lagoon volumes are often too small to provide adequate depth; with side slopes of 3:1, commercially available aerators are too large to be used in the settling basin, and flow is intermittent, leading to long hydraulic retention times and excessive algae growth. Design procedures are available for the HPAPS system (Rich, 1999).

4.9.1 PERFORMANCE DATA

Several sets of performance data for the HPAPS systems are available, but all are for locations in mild climates such as South Carolina and Georgia. It is likely that the process has been introduced in areas with more severe climates, and these data should used to design in more severe climates. Performance data for the DPMC system in Berkeley County, South Carolina, are presented in Figure 4.7. Data in the figure are for 6 years of operation, but Rich (2000) presented an additional 3 years of data on the Internet showing similar results. The system has functioned as designed for over 9 years. The performance has been exceptional for several years, but sludge removal data are not available.

Continuous operation of the aeration system is essential to obtain maximum efficiency, as illustrated by Figure 4.7 and Figure 4.8. The performance data for Berkeley County shown in Figure 4.7 were obtained with continuous aeration, while performance data for a similar system also located in South Carolina were obtained under conditions of intermittent aeration (operation 50% of the time). Results with continuous aeration were improved by about 50%.

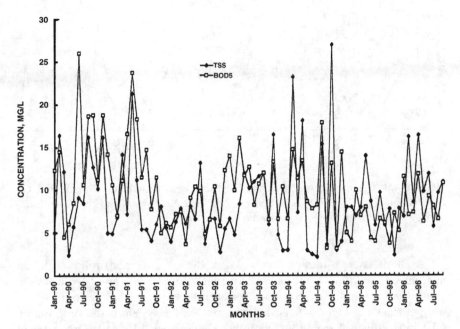

FIGURE 4.7 Performance of dual-power, multicellular (DPMC) aerated lagoon system in Berkley County, South Carolina, with aerators operating continuously. (From Rich, L.G., *High-Performance Aerated Lagoon Systems*, American Academy of Environmental Engineers, Annapolis, MD, 1999. With permission.)

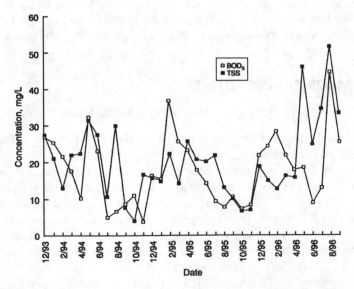

FIGURE 4.8 Effluent TSS and BOD_5 from a dual-power, multicellular (DPMC) aerated lagoon system with aerators operating intermittently. (From Rich, L.G., *High-Performance Aerated Lagoon Systems*, American Academy of Environmental Engineers, Annapolis, MD, 1999. With permission.)

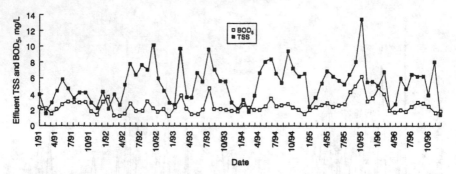

FIGURE 4.9 Monthly average BOD$_5$ and TSS from Ocean Drive plant. (From Rich, L.G., *High-Performance Aerated Lagoon Systems*, American Academy of Environmental Engineers, Annapolis, MD, 1999. With permission.)

A DPMC system (design flow = 3.4 mgd or 12,870 m^3/d) followed by an intermittent sand filter at the Ocean Drive plant located in North Myrtle Beach, South Carolina, has been in service for over 12 years and has performed very well, as shown in Figure 4.9. A flow diagram for the system is shown in Figure 4.10. Final effluent TSS concentrations have not exceeded 15 mg/L. Only effluent data are available; however, in October 1997 the USEPA, Region 4, collected two 24-hr composite samples from the DPMC aerated lagoons. The data from this evaluation are presented in Table 4.10 (Rich, 1999), and the sampling locations are shown on Figure 4.10. A similar plant, the Crescent Beach at Myrtle Beach, South Carolina, also performed well, as shown in Figure 4.11. When designed and operated properly, the DPMC systems perform admirably.

4.10 PROPRIETARY SYSTEMS

4.10.1 ADVANCED INTEGRATED WASTEWATER POND SYSTEMS®

The Advanced Integrated Wastewater Pond Systems® (AIWPS®) has evolved over a 50-year period of research by Dr. William J. Oswald at the University of California, Berkeley, and other locations. The majority of the research and operational experience has been obtained in areas with moderate climates. The greatest advantage to AIWPS® appears to be the elimination of or great reduction in the need for sludge disposal. The facility at St. Helena, California, has not had to dispose of primary sludge in over 30 years. Other facilities in moderate climates have had similar experiences with sludge disposal. Early indications are that the use of deep sludge pits in cold climates will provide significant reductions in solids. Examples of systems located in cold climates with deep sludge pits are presented below. Another advantage of the system is the ability of the sludge blanket to adsorb toxic materials. Toxicity tests have shown that AIWPS® is capable of producing an effluent from municipal and industrial wastewaters that

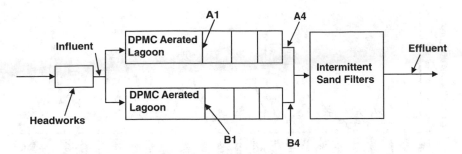

FIGURE 4.10 Sketch of a dual-power, multicellular (DPMC) aerated lagoon–intermittent sand filter system at North Myrtle Beach, South Carolina (unpublished paper by Rich, Bowden, and Henry, 1998).

will satisfy most regulations. Costs to construct and operate AIWPS® are much less than those for conventional wastewater treatment processes. Oswald (1996) has reported that the Hollister, California, system cost about one third as much to construct and only about one fifth as much to operate as a comparable mechanical plant located nearby. The system has been in operation for over 25 years. Information about the process and operational data are available in the following

TABLE 4.10
Performance of Dual-Power, Multicellular (DPMC) Aerated Lagoon at North Myrtle Beach, South Carolina

Characteristic	Influent	Effluent Aerated Reactor A1	Effluent Aerated Reactor B1	Effluent Settling Pond A4	Effluent Settling Pond B4	Effluent Intermittent Sand Filter
BOD_5	160	21	23	10	12	2
$CBOD_5$	165	16	20	8	6	1
$SCBOD_5$	62	5	5	4	4	1
TSS	185	79	77	8	4	4
Alkalinity	195	190	190	210	220	17
NH_3–N	25	25	28	31	30	1
NO_3–N	0.07	0.05	0.05	0.09	0.44	32
TKN	37	35	40	34	33	2
TP	5.9	2.8	3.3	0.6	1.2	0.8
Chlorophyl-*a*	—	—	—	0.056	0.043	—

Source: Unpublished paper by Rich, Bowden and Henry, 1998.

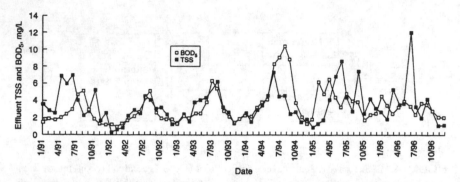

FIGURE 4.11 Monthly average effluent BOD₅ and TSS at Crescent Beach Plant, North Myrtle Beach, South Carolina. (From Rich, L.G., *High-Performance Aerated Lagoon Systems*, American Academy of Environmental Engineers, Annapolis, MD, 1999. With permission.)

references: Oswald (1990a,b, 1995, 1996, 2003), Oswald et al. (1994), Green et al. (1995), Nurdogan and Oswald (1995), Green et al. (1995, 1996, 2003), USEPA (2000), and Downing et al. (2002).

4.10.1.1 Hotchkiss, Colorado

The wastewater treatment facility at Hotchkiss, Colorado, is similar to an AIWPS® system without the high-rate raceway system. Although discussed in this section, it is not a proprietary system. This facility is located about 60 miles (97 km) east of Grand Junction, Colorado, at an elevation of 5300 feet (1616 m) above sea level. It serves approximately 800 people. The facility went online in October 1997. Annual mean temperature is about 50°F (10°C); winter temperatures are as low as –20°F (–29°C), and summer temperatures are between 90 and 100°F (32 and 38°C). Annual precipitation averages 13 in. (33 cm) of water, most of which is in the form of spring snows, late-evening rains in the late summer, and fall storms. A summary of the characteristics of the treatment facility was extracted from the Colorado Discharge Permit and is presented in Table 4.11, and effluent limits required by the State of Colorado are summarized in Table 4.12. A flow diagram of the treatment facility is shown in Figure 4.12. The system differs from conventional AIWPS® in that the anaerobic pond is a separate pond preceding the aerated ponds rather than being located within the first facultative pond.

Performance data from October 1997 through 2000 are summarized in Table 4.13. For over 3 years, the system has functioned well, as shown in Figure 4.13 through Figure 4.17; however, the flow rate entering the plant is approximately 35% of the design flow of 0.494 mgd (1870 m³/d) (Figure 4.16). The maximum influent flow does on occasion exceed 0.3 mgd (1136 m³/d). During the over 3 years of operation, BOD₅ removal has averaged 92.9%, TSS removal 88.7%, and NH₃–N removal 79.8%. NH₃–N removal has improved materially as the plant

TABLE 4.11
Description of Hotchkiss, Colorado, Wastewater Treatment Facility

Unit Process	Unit Process Features/Description	Capacity (Hydraulic/Organic)
Lagoon #1		
Anaerobic portion	Volume = 1.75 MG, depth = 18.5–21.5 ft, t = 3.5 d	315 lb BOD$_5$ d^{-1}
Aerobic portion	Volume = 2.9 MG, depth = 13 ft, t = 5.9 d	
Aeration	2- to 5-hp and 1- to 10-hp surface aerators, FTR = 1.40 lb O$_2$ hp^{-1} hr^{-1}	
Lagoon #2	Volume = 5.0 MG, depth = 13 ft, t = 10.0 d	
Aeration	1- to 5-hp and 1- to 10-hp surface aerators, 1- to 5-hp aspirating aerator, FTR = 1.46 lb O$_2$ hp^{-1} hr^{-1}	
Polishing pond	Volume = 1.74 MG, depth = 12 ft, t = 3.5 d	0.494 mgd
Recirculation	0.5-hp pump rated at 100 gpm	
Chlorination	Two 150-lb gas cylinders, 0 to 4 lb/d and 0 to 10 lb/d; regulators, 2.5 mg/L maximum dosage	
Chlorine contact chamber	Serpentine basin; length = 190 ft, width = 3.5 ft, 54:1 length-to-width ratio; volume = 34,800 gal; t = 30 min	1.67 mgd
Effluent flow measuring	450 V-notch weir, height = 15 in.	
Irrigation pumping	A pump (of undetermined size) will pump a portion of the effluent to an irrigation ditch supplying 70 acres of farm land	
Dechlorination	SO$_2$ gas, same equipment as gas chlorination equipment	0.494 mgd

has matured, with an average effluent concentration of 1.78 mg/L for 1999 (Figure 4.14). In 1999, the effluent NH$_3$–N ranged from 5.8 mg/L in February to 0.43 mg/L in June. As shown in Figure 4.15, NH$_3$–N removal is closely correlated with the effluent water temperature.

4.10.1.2 Dove Creek, Colorado

Dove Creek is located approximately 30 miles (48 km) north of Cortez, Colorado, at an elevation of 6750 feet (2060 m) above sea level. Although listed in this section, it is not a proprietary system. Air temperatures range from 0°F (–18°C) to greater than 90°F (>32°C). The wastewater treatment plant serves approximately 700 people with an average design flow rate of 60,000 gpd (227 m^3/d).

TABLE 4.12
Hotchkiss, Colorado, Effluent Limits

Parameters	Limit	Rationale
Outfall 001A		
Flow (mgd)	0.494[a]	Design capacity
BOD$_5$ (mg/L)	30/45[b]	State effluent regulations
TSS (mg/L)	75/110[b]	State effluent regulations
Fecal coliform (number/100 mL)	6000/12,000[c]	State fecal coliform policy
pH (minimum–maximum)	6.0–9.0	State effluent regulations
Oil and grease (mg/L)	10[d]	State effluent regulations
Salinity	Report[a]	Discharge permit regulations
Outfall 002A		
Flow (mgd)	0.494[a]	Design capacity
Total residual chlorine (mg/L)	0.5[a]	State effluent regulations
Total ammonia (mg/L as N):		Water quality standards
December–February	30[a]	
March–April	25[a]	
May and November	15[a]	
June–August:		
0.25 > Flow < 0.494 mgd	7.0[a]	
0.20 > Flow < 0.25 mgd	12[a]	
September:		
0.34 > Flow < 0.494 mgd	8.5[a]	
0.27 > Flow < 0.34 mgd	12[a]	
Flow < 0.27 mgd	15[a]	
October:		
0.45 > Flow < 0.494 mgd	11[a]	
0.36 > Flow < 0.45 mgd	12[a]	
Flow < 0.36 mgd	15[a]	

[a] 30-day average.
[b] 30-day average/7-day average.
[c] 30-day geometric mean/7-day geometric mean.
[d] Daily maximum.
[e] 30-day average/daily maximum.

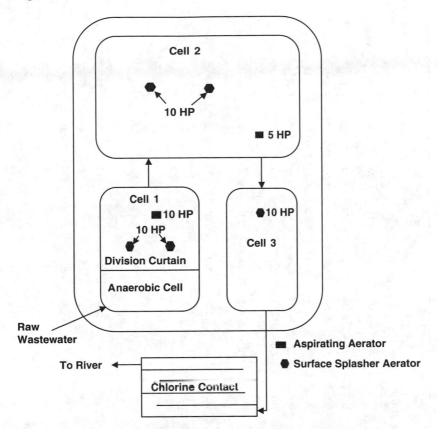

FIGURE 4.12 Plan view of Hotchkiss, Colorado, wastewater plant. (Joanne Fagan, Consolidated Consulting Services, Delta, CO, personal communication, 2000.)

The system is permitted for a design flow of 0.115 mgd (435 m³/d) and 288 lb (131 kg/d) of BOD_5 per day. The system is similar to the Hotchkiss, Colorado, facility in that it also has an anaerobic pond preceding the aerated cells, which are followed by a free water surface wetland. Plan view and cross-sectional views of the fermentation pit are shown in Figure 4.18 and Figure 4.19, respectively. The fermentation pit has a total volume of 31,947 ft³, or 239,123 gallons (905 m³).

4.10.2 BIOLAC® PROCESS (ACTIVATED SLUDGE IN EARTHEN PONDS)

The U.S. Environmental Protection Agency published an excellent summary of the status of BIOLAC® processes in the United States as of 1990 (USEPA, 1990). Pertinent information has been extracted from that report and is presented in the following text, figures, and tables. Additional information was provided by the Parkson Corporation. Since the report was published, over 600 BIOLAC® systems have been installed in the United States and throughout the world. Much of the

TABLE 4.13
Hotchkiss, Colorado, Performance Data for 1997 to 2000

Parameter	O	N	D	J	F	M	A	M	J	J	A	S	O	Mean
Average influent Q	0.238	0.169	0.143	0.127	0.113	0.112	0.11	0.139	0.189	0.227	0.241	0.27	0.274	—
Maximum	0.3245	0.199	0.153	0.162	0.12	0.133	0.136	0.182	0.208	0.276	0.265	0.339	0.309	—
Average effluent Q	0.2205	0.159	0.168	0.142	0.127	0.131	0.109	0.143	0.203	0.262	0.273	0.319	0.331	—
Maximum	0.255	0.176	0.194	0.155	0.136	0.131	0.208	0.201	0.249	0.361	0.308	0.396	0.377	—
Influent temperature (°C)	—	—	—	11.4	11.2	10.4	12.6	15.1	16.9	18.8	19.6	19.5	18	—
Effluent temperature (°C)	—	—	—	6.3	8	8	12	16.5	21	24	25	25	18	—
Influent BOD	138	168	180	282	258	198	186	216	174	210	126	126	102	—
Effluent BOD	10	10	13	7	8	10	11	28	30	23	18	14	7	—
% BOD removal	—	—	—	—	—	—	—	—	—	—	—	—	—	—
BOD loading	—	—	—	—	—	—	—	—	—	—	—	—	—	—
cBOD (effluent)	5	8	7	5	6	5	5	18	17	14	16	10	3	—
Filtered BOD	3	—	—	—	—	—	—	—	—	—	—	—	—	—
Influent TSS	84	238	172	306	296	218	224	238	156	84	110	128	126	—
Effluent TSS	8	19	40	46	24	26	12	30	28	26	44	4	9	—
Fecal	100	130	3200	<30	<30	170	130	<30	<30	<30	<30	<30	130	—
TR Cl_2 average	0.11	0.21	0.18	—	—	—	—	—	—	—	—	—	—	—
Maximum	0.3	0.35	0.35	—	—	—	—	—	—	—	—	—	—	—
pH minimum	7.24	8.18	7.59	7.8	7.75	7.19	7.14	7.8	7.52	7.27	7.72	7.81	7.57	—

Parameter	N	D	J	F	M	A	M	J	J	A	S	O	N	Mean
pH maximum	8.6	8.54	8.46	8.61	8.82	8.97	8.92	8.91	8.35	8.98	8.98	8.1	8.05	—
TDS raw	109	112	140	142	140	136	148	.146	84	84	102	104	—	—
TDS influent	1282	1058	968	750	826	632	586	666	1040	1348	1478	1362	1388	—
TDS effluent	1456	1304	1174	1068	1048	932	904	870	784	994	1162	1488	1430	—
Influent ammonia	13	12.8	19.9	13.8	29.5	34.9	22.2	27.6	21.6	11.1	15.7	13.7	8.7	—
Effluent ammonia	6	2.9	7.6	7.9	13.5	13.7	10.3	9.4	4.3	0.86	1.28	1.28	0.54	—
Well TDS	1318	1718	1790	1750	1958	1884	1892	1812	936	1680	1968	2752	1280	—
Pinion TDS	—	—	—	—	—	—	—	—	—	2090	—	2014	—	—
Average influent Q	0.179	0.132	0.124	0.111	0.101	0.097	0.167	0.221	0.24	0.238	0.242	0.205	0.149	—
Max	0.237	0.145	0.135	0.124	0.106	0.175	0.205	0.268	0.292	0.277	0.404	0.242	0.169	—
Average effluent Q	0.216	0.149	0.142	0.121	0.112	0.124	0.171	0.228	0.258	0.269	0.267	0.234	0.169	—
Maximum	0.294	0.179	0.162	0.135	0.13	0.173	0.203	0.3	0.364	0.323	0.318	0.273	0.197	—
Influent T (°C)	15.6	13.6	10	10.1	12.2	11.9	13.3	16	19.8	19.8	19.8	17.8	15.9	—
Effluent T (°C)	13	9	3	5	8	9	13	19	25	24.5	23	16.5	11	—
Influent BOD	120	174	240	188	186	270	156	138	168	114	114	122	156	—
Effluent BOD	2	4	4	7	12	25	17	25	10	6	4	5	4	—
% BOD removal	—	—	98.3	96.3	93	90.7	89.1	81.9	94	94.7	96.5	99.6	97.4	—
BOD loading	—	—	248	174	157	218	217	254	336	226	230	2086	194	—
cBOD (effluent)	1	4	2	5	9	9	9	8	7	4	1	4	2	—
Filtered BOD	—	—	—	—	—	—	—	—	—	—	—	—	—	—
Influent TSS	214	232	358	244	260	192	162	166	248	152	110	216	148	—
Effluent TSS	3	7	16	21	35	32	45	36	26	6	4	3	10	—

TABLE 4.13 (cont.)
Hotchkiss, Colorado, Performance Data for 1997 to 2000

Parameter	N	D	J	F	M	A	M	J	J	A	S	O	N	Mean
Fecal	30	<30	<30	200	<30	<30	<30	30	<30	<30	<30	<30	<30	—
TR Cl₂ average	—	—	0.22	2	0.15	21	0.24	0.28	0.21	0.32	0.23	0.24	0.3	—
Maximum	—	—	0.45	0.35	0.2	0.35	0.35	0.35	0.47	0.47	0.4	0.42	0.45	—
pH minimum	7.61	7.68	8.03	8.02	7.95	7.11	7.61	7.46	7.43	7.49	7.7	6.55	6.17	—
pH maximum	8.13	8.34	8.47	8.48	8.39	8.34	8.97	8.8	8.99	8.3	8.19	8.1	8.65	—
TDS raw	—	—	138	198	166	144	124	100	1.06	102	102	130	110	—
TDS influent	—	—	854	912	650	656	816	1036	1442	1532	596	912	1660	—
TDS effluent	—	—	1056	1112	1012	992	886	906	1476	1416	1318	1192	1430	—
Influent ammonia	15	17.6	23.6	19.1	28.6	26.9	21	14.4	14.2	10.6	13.6	14.1	13.9	—
Effluent ammonia	0.65	0.27	2.7	5.8	4.8	0.94	0.59	0.43	0.57	0.78	0.95	1.38	1.35	—
Well TDS	—	—	2064	1960	2154	2416	764	896	3128	2256	2586	2302	2566	—
Pinon TDS	—	—	1550	1784	1766	1440	1800	1596	2242	1522	1796	1734	1726	—

Parameter	D	J	F	M	A	M	J	J	A	S	O	N	D	Mean
Average influent Q	0.126	0.113	0.106	0.097	0.096	0.143	0.153	0.205	0.276	0.293	0.229	0.158	0.098	0.17054
Maximum	0.134	0.126	0.122	0.112	0.117	0.301	0.264	0.222	0.36	0.371	0.308	0.202	0.114	0.21381
Average effluent Q	0.145	0.139	0.115	0.11	0.102	0.15	0.17	0.207	0.269	0.271	0.188	0.163	0.122	0.18458
Maximum	0.166	0.156	0.127	0.12	0.12	0.17	0.272	0.243	0.365	0.324	0.246	0.258	0.131	0.2281
Influent T (°C)	12.9	10	9	11	12.5	15	18	20	19	19	17	14	10	14.9083

Effluent T (°C)	8.5	3	5	9	13	19	23	25	26	23	19	12	5.5	14.8278
Influent BOD	114	216	198	378	246	390	162	162	126	60	120	138	174	179.333
Effluent BOD	14	10	7	14	19	12	14	18	6	14	9	1	3	11.6667
% BOD removal	87.7	95.4	96.5	96.3	92.3	96.9	91.4	88.9	95.2	76.7	92.5	99.3	98.3	43.4822
BOD loading	120	204	175	306	197	465	207	277	290	147	229	182	142	303.336
cBOD (effluent)	10	10	7	13	12	10	8	8	3	4	5	1	1	7.07692
Filtered BOD	–	–	–	–	–	–	–	–	–	–	–	–	–	3
Influent TSS	142	218	214	250	328	330	174	180	90	59	88	144	166	191.41
Effluent TSS	26	18	24	37	47	25	13	18	5	36	13	2	2	21.1795
Fecal	<30	<30	<30	<30	<30	<30	<30	<30	<30	<30	<30	<30	<30	457.778
TR Cl$_2$ average	0.35	0.28	0.22	0.23	0.21	0.27	0.13	0.17	0.34	0.4	0.38	0.35	0.4	1.08963
Maximum	0.5	0.47	0.35	0.4	0.33	0.35	0.45	0.48	0.5	0.48	0.47	0.46	0.48	0.40667
pH minimum	7.76	8.09	7.24	7.78	7.65	7.59	7.48	7.16	7.58	8.01	8.01	8.18	8.1	7.59026
pH maximum	8.82	8.72	8.67	8.63	8.19	7.72	8.1	8.07	7.97	8.7	8.36	8.6	8.56	8.50128
TDS raw	136	244	256	186	170	96	80	98	84	62	126	142	162	127.373
TDS influent	818	748	762	682	704	700	922	1046	1156	1162	1204	978	742	974.973
TDS effluent	1346	1222	1070	898	992	902	880	960	1050	1150	1238	1158	1148	1119.57
Influent ammonia	26.6	26.7	33.7	29.1	29.3	23.8	16.8	11.6	12.4	9.6	12.9	16	24.2	19.2256
Effluent ammonia	1.02	6.5	10.2	9.6	0.24	3.09	4.01	3.36	8.14	3.39	1.45	1.46	0.96	3.95359
Well TDS	2640	2504	630	1990	2408	2576	2528	2676	2498	1788	2398	1558	2404	2012.78
Pinon TDS	1692	1556	1782	1808	1918	1842	1850	1782	1640	2318	1680	1704	1664	1787.56

Source: Joanne Fagan, Consolidated Consulting Services, Delta, CO, personal communication, 2000.

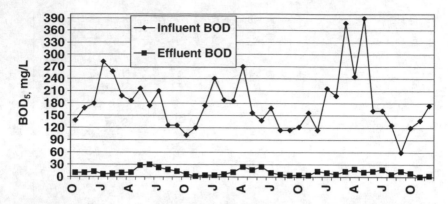

FIGURE 4.13 Performance data for BOD at Hotchkiss, Colorado (October 1997 to December 2000). (Joanne Fagan, Consolidated Consulting Services, Delta, CO, personal communication, 2000.)

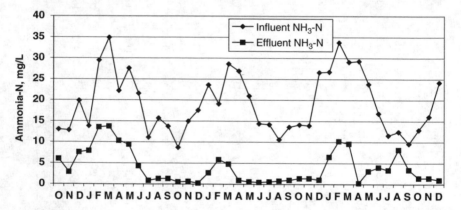

FIGURE 4.14 Performance data for ammonia nitrogen at Hotchkiss, Colorado (October 1997 to December 2000). (Joanne Fagan, Consolidated Consulting Services, Delta, CO, personal communication, 2000.)

information presented by the USEPA (1990) was obtained after a relatively short operating period for most of the plants; therefore, it is important to evaluate the database for the previously sampled installations and include data from facilities constructed since the 1990 report.

4.10.2.1 BIOLAC® Processes

The BIOLAC® processes have several variations. The basic processes are extended aeration activated sludge with and without recirculation of solids. The three basic systems are BIOLAC-R, which is an extended aeration process with recycle of solids; the BIOLAC-L system, which is an aerated lagoon system without recycle of solids; and BIOLAC Wave-Oxidation© modification, which is

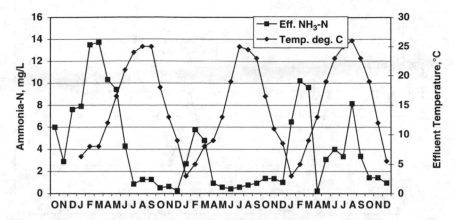

FIGURE 4.15 Effluent NH_3–N variation with effluent temperature at Hotchkiss, Colorado (October 1997 to December 2000). (Joanne Fagan, Consolidated Consulting Services, Delta, CO, personal communication, 2000.)

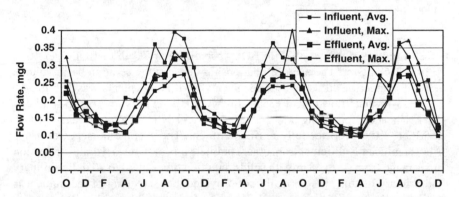

FIGURE 4.16 Flow data for Hotchkiss, Colorado (October 1997 to December 2000). (Joanne Fagan, Consolidated Consulting Services, Delta, CO, personal communication, 2000.)

used to nitrify and denitrify wastewater. In addition to these systems, floating aeration chains used in the above processes have been installed in existing lagoon systems as an upgrade.

4.10.2.1.1 BIOLAC-R System

The BIOLAC-R system, shown in Figure 4.20, is an extended aeration process operating within earthen embankments or other types of structures. Recommended design criteria are shown in Table 4.14. Conservative design parameters are used, and loadings typically are 7 to 12 lb BOD_5 per day per 1000 ft^3 (0.11 to 0.16 kg/m^3) of aeration pond, with food-to-microorganism ratios of 0.03 to 0.1 lb BOD_5 per lb mixed liquor volatile suspended solids (MLVSS) (0.014 to 0.045 kg/kg). The average loading rate for 25 BIOLAC-R plants reported in the

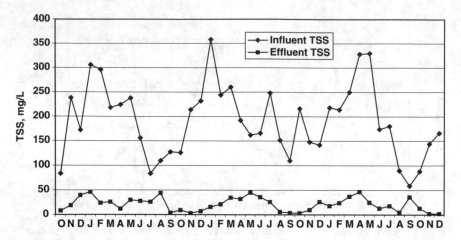

FIGURE 4.17 TSS performance data for Hotchkiss, Colorado (October 1997 to December 2000). (Joanne Fagan, Consolidated Consulting Services, Delta, CO, personal communication, 2000.)

USEPA report (1990) was 975 lb BOD_5 per d·MG or 7.3 lb of BOD_5 per 1000 ft^3 or 0.1168 kg/m^3. The average relationship between the aeration basin volume and the number of diffusers used for the 25 BIOLAC-R plants was 385 diffusers/MG with an airflow rate of 1350 scfm/MG (0.01 scm/min/m^3) or 3.5 scfm/diffuser. The actual operating horsepower at the 25 BIOLAC-R plants averaged 45 hp/MG (34 kW/MG) for fully nitrified effluent. The average horsepower usage is not significantly different from other complete mix systems. Hydraulic retention times range from 24 to 48 hr with solids retention times of 30 to 70 d. Preliminary and primary treatment is normally not provided, but screening of the influent is desirable. Depths in the aeration ponds range from 8 to 20 ft (2.5 to 6.1 m) with the lower depths being found in retrofits or where deep construction is impractical. Integral clarifiers are the most common form of solids separation for return to the aeration tanks; however, some systems have conventional clarifiers (Bowman, 2000). A relatively small waste sludge tank is provided because of the low sludge production. A polishing basin is not recommended.

4.10.2.1.2   BIOLAC-L System

The BIOLAC-L system is a typical flow-through aerated lagoon without recycle of solids and a waste sludge pond. The flow diagram is the same as that shown in Figure 4.20 without the clarifier and sludge pond. Design of the BIOLAC-L system is normally based on hydraulic retention time, and values range from 6 to 20 days. Equivalent loadings of 0.5 to 1.8 lb BOD_5 per d per 1000 ft^3 (0.008 to 0.029 kg/m^3·d) are used. The polishing pond required for the BIOLAC-L system has a hydraulic retention time of 2 to 4 days. Sludge storage and decomposition occur in the polishing pond.

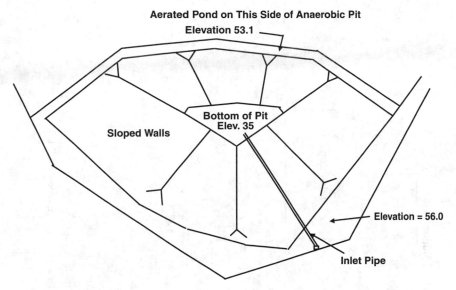

FIGURE 4.18 Plan view of fermentation pit at Dove Creek, Colorado. (Joanne Fagan, Consolidated Consulting Services, Delta, CO, personal communication, 2000.)

4.10.2.1.3 Wave-Oxidation© Modification

Carbon oxidation and nitrification/denitrification occur in the Wave Oxidation® modification (Figure 4.21). This BIOLAC-R system operates at low dissolved oxygen concentrations and provides automatic control of the airflow rate in each aeration chain. Airflow is alternated such that several moving oxic and anoxic zones are created in the aeration basin. This modification has been used successfully for nitrogen removal.

4.10.2.1.4 Other Applications

The BIOLAC® floating aeration chains are used as retrofits for existing lagoons and are installed as original aeration equipment. Several operations around the country are using BIOLAC® aeration equipment.

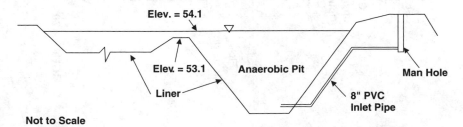

FIGURE 4.19 Cross-sectional view of anaerobic pit at Dove Creek, Colorado. (Joanne Fagan, Consolidated Consulting Services, Delta, CO, personal communication, 2000.)

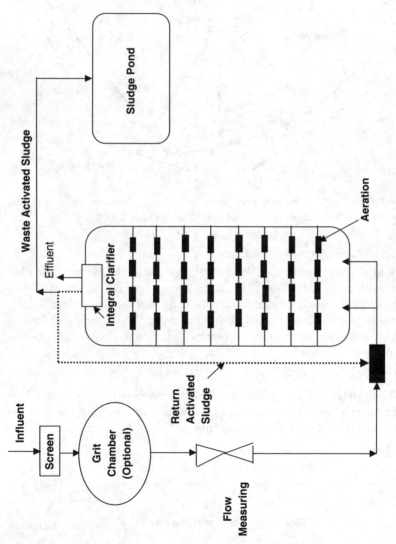

FIGURE 4.20 Flow diagram of BIOLAC-R system. (Courtesy of Parkson Corp., Ft. Lauderdale, FL.)

TABLE 4.14
Manufacturer's Typical Design Criteria for BIOLAC-R Systems vs. Conventional Extended Aeration Systems

Parameter	BIOLAC-R	Extended Aeration
Hydraulic residence time (hr)	24–48	18–36
Solids retention time (d)	30–70	20–30
Ratio of food to microorganisms (lb BOD_5 per d per lb MLVSS)	0.03–0.1	0.05–0.15
Volumetric loading (lb BOD_5 per d per 1000 ft^3)	7–12	10–25
MLSS (mg/L)	1500–5000	3000–6000
Basin mixing (hp/MG)	12–15	80–150

Source: Courtesy of Parkson Corp., Ft. Lauderdale, FL.

4.10.2.2 Unit Operations

Major components of the BIOLAC® systems are the aeration equipment and the clarifier and solids handling equipment.

4.10.2.2.1 Aeration Chains and Diffuser Assemblies

The unique feature of BIOLAC® systems is the floating aeration chain system (Figure 4.2). Fine bubble diffusers are suspended from a floating aeration chain that carries air to the diffusers. The floating aeration chain is attached to an anchor on the embankment and is allowed to move in a controlled way to create the oxic and anoxic zones discussed above. Each diffuser assembly can support two, three, four, or five diffusers. Each diffuser is rated at 2 to 10 scfm and normally operates at an airflow rate of 6 scfm. Diffuser membranes are expected to last about 5 to 8 years before replacement is required.

4.10.2.2.2 Blowers and Air Manifold

Continuous-service positive displacement rotary blowers are generally used. In larger systems, multistage centrifugal blowers may be more economical. Most systems use three blowers, each capable of providing 50% of the required airflow and one unit serving as a spare.

4.10.2.2.3 Clarification and Solids Handling

An integral clarifier is used with the BIOLAC-R system, although conventional clarifiers are used on occasion. BIOLAC-L systems require installation of a polishing basin for solids separation and storage. A cross-sectional view of the integral clarifier is shown in Figure 4.22. The integral clarifier is constructed in the aeration basin but is separated from the aeration zone by a partition wall. Flow enters the clarifier along the bottom over the entire length of the partition wall to minimize short-circuiting. A flocculating rake moves the length of the clarifier sludge trough to concentrate and distribute the sludge. Sludge return and waste are removed with an air-lift pump.

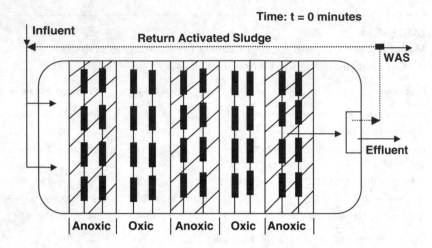

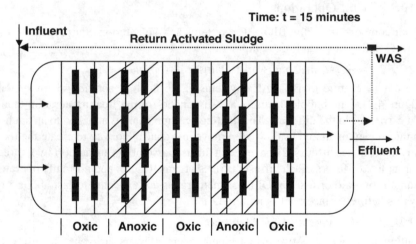

FIGURE 4.21 Wave-Oxidation© modification of the BIOLAC-R system. (Courtesy of Parkson Corp., Ft. Lauderdale, FL.)

4.10.2.2.4 BIOLAC-L Settling Basin

A minimum hydraulic retention of one day is normally provided in the unaerated section of the polishing basin. Sludge storage of up to 1 to 2 decades is provided in the quiescent zone of the polishing or settling basin. Further sludge degradation of 40 to 60% occurs under anaerobic conditions in the settling basin.

4.10.2.3 Performance Data

Mean performance data for 13 BIOLAC® systems are shown in Table 4.15, and monthly performance data are available in the USEPA (1990) report. All but the

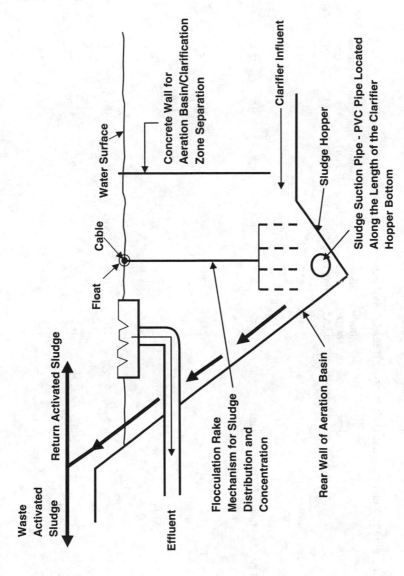

FIGURE 4.22 Cross-sectional view of integral BIOLAC-R clarifier. (Courtesy of Parkson Corp., Ft. Lauderdale, FL.)

TABLE 4.15
Summary of Average Performance Data from BIOLAC® Systems

Location	Period	Type	Flow (mgd)	% Design	Influent BOD (mg/L)	Effluent BOD (mg/L)	% BOD Removal	Loading (lb BOD d^{-1})	Influent TSS (mg/L)	Effluent TSS (mg/L)	% TSS Removal	Effluent NH_3–N (mg/L)
Morgantown WWTP (Morgantown, Kentucky)	4/89–9/89	R	0.29	58	243	12.7	92.3	575	188	11.7	95.7	0.1
Greenville WWTP (Greenville, Kentucky)	5/88–8/89	R	0.40	55.3	178	6.2	96.5	528	213	12.4	94.7	0.5
New Brockton WWTP (New Brockton, Alabama)	6/89–8/89	R	0.05	27.8	233	8.7	95.5	111.5	257	10.7	94.4	1.9
Edmonton WWTP (Edmonton, Kentucky)	7/89–11/89	R	0.2	39.2	203	11.6	91.1	185	266	18.4	89.5	3.2
Fincastle WWTP (Fincastle, Virginia)	9/88–8/89	L	0.05	62.5	218	18.6	91.2	86.9	190	21.5	89.7	ND
Lowell WWTP (Lowell, Ohio)	7/89–9/89	R	0.11	204	186	13.3	91.8	167	172	26	85.3	6.7
Hanceville WWTP (Hanceville, Alabama)	6/89–9/89	R	0.5	87.8	134	9.7	92.0	514	97.8	9.0	92.0	0.8

Facility	Period											
Livinston Manor WWTP (Rockland, New York)	6/86–8/89	R	0.5	62.5	260	5.1	97.9	1062	217	8.7	95.3	1.9
Blytheville West WWTP (Blytheville, Arkansas)	7/89–10/89	R	0.39	26.0	ND	7.6	—	—	ND	14.9	—	2.2
Blytheville North WWTP (Blytheville, Arkansas)	4/89–10/89	R	0.39	48.8	ND	13.8	—	—	ND	26.3	—	26.0
Blytheville South WWTP (Blytheville, Arkansas)	4/89–10/89	R	0.60	42.8	ND	15.1	—	—	ND	18.1	—	30.9
Bay WWTP (Bay, Arkansas)	6/89–9/89	R	0.27	180	ND	10.4	—	—	ND	6.7	—	11.3
Piggot WWTP (Piggot, Arkansas)	6/89–9/89	R	0.35	58.0	ND	20.8	—	—	ND	34.8	—	ND

Note: ND, no data.

Source: Data reported by USEPA (1990); additional data available from Parkson Corp. (Ft. Lauderdale, FL).

facility located in Fincastle, Virginia, are BIOLAC-R systems. All of the systems, with the exception being the plant in Piggot, Arkansas, produced an effluent that satisfied secondary standards of 30 mg/L of BOD_5 and TSS. Most of these plants had been operating for only a few months, so the data may or may not be indicative of long-term performance. Richard H. Bowman (2000), with the Colorado Department of Health and Environment, has reported that the BIOLAC® systems in Colorado have satisfied secondary standards and ammonia nitrogen removal requirements where required for many years. Parkson Corporation (2004) reported that the Nevada, Ohio, BIOLAC® system (100,000 gpd) produced a 2-year average effluent containing 4.1 mg/L BOD, 6.9 mg/L TSS, and 0.7 mg/L NH_3–N. Additional data available from the Parkson Corporation show BOD and TSS concentrations of less than 10 mg/L, ammonia nitrogen concentrations of less than 1.0 mg/L, and total nitrogen concentrations of less than 8 mg/L.

4.10.2.4 Operational Problems

The U.S. Environmental Protection Agency (USEPA, 1990) presented a summary of the problems encountered at various BIOLAC® plants. The difficulties appear to be typical mechanical failures and excessive debris and floating sludges with excessive oil and grease in the clarifier. Most of the problems appear to be correctable with routine maintenance.

4.10.3 LEMNA Systems

Numerous references to the use of duckweed in lagoon wastewater treatment systems date back to the early 1970s, but this discussion is limited to the application of proprietary processes produced by Lemna Technologies, Inc. (Culley and Epps, 1973; Reed et al., 1995; Wolverton and McDonald, 1979; Zirschky and Reed, 1988). Lemna Technologies offers two basic systems for wastewater treatment: the Lemna duckweed system, in which floating partitions keep the plants evenly distributed over the surface of the pond, and the LemTec™ Biological Treatment Process. In addition to these basic units, the company produces the LemTec™ Modular Cover System, Lemna Polishing Reactor™, LemTec™ C-4 Chlorine Contact Chamber-Cleaner, LemTec™ Anaerobic Lagoon System, and LemTec™ Gas Collection Cover. Lemna Technologies reported in a recent press release that over 150 municipal and industrial installations exist worldwide; it is assumed that the 150 installations include regular lemna and biological treatment process systems as well as the other systems produced by the company (Lemna Technologies, Inc., 1999a,b). The descriptions and discussions of processes in this chapter are limited to the Lemna duckweed system with floating partitions and the LemTec™ Biological Treatment Process.

4.10.3.1 Lemna Duckweed System

The duckweed system can be used in retrofitting an existing facultative or aerated lagoon system or can be an original design. An original design consists of a regular facultative or aerated lagoon followed in series by Lemna system components, including a floating barrier grid to prevent clustering of the duckweed and baffles to improve the hydraulics of the system. These basic components are followed by disinfection, if required, and reaeration of the effluent that is anaerobic beneath the duckweed cover. A diagram and flow scheme for a typical Lemna system design are shown in Figure 4.23 (Lemna Technologies, Inc., 2000). The Lemna system has been installed in several locations, ranging from Georgia to North Dakota in the United States and in Poland in Europe. Flow diagrams of several of these systems are shown in Figure 4.24

For the Lemna system to function properly, it is necessary to harvest the duckweed on a regular basis. LemTec™ harvesters are available for use in ponds utilizing the floating barrier grid to ensure even distribution of the duckweed (Figure 4.25). The harvesters operate by depressing the floating barrier and removing the duckweed from the water surface. Biomass harvested from the Lemna system can be managed via land application of the duckweed, composting the duckweed, or the production of pelletized feedstuff. Other than land application, these management methods can be expensive, and additional data are required to evaluate the economic feasibility of these two options.

4.10.3.2 Performance Data

A typical performance data summary reported by Lemna is shown in Table 4.16. Similar effluent quality is reported for the systems shown in Figure 4.24. Buddhavarapu and Hancock (1989) reported on the performance of two pilot-scale Lemna systems located in Devils Lake, North Dakota, and DeRidder, Louisiana. The DeRidder system was operated from October 1988 to December 1989, but the period of operation for the Devils Lake facility was only 3 months. The pilot-scale systems produced a good-quality effluent, with average BOD_5 concentrations of less than 10 mg/L at both facilities. TSS concentrations were less than 20 mg/L at both sites. Total Kjeldahl nitrogen (TKN) was less than 5 mg/L at both locations. The Devils Lake pilot plant reported TP concentrations of less than 1 mg/L; however, the system was operated for only 3 months during warmer months of the year.

4.10.3.3 LemTec™ Biological Treatment Process

The LemTec™ Biological Treatment Process uses the LemTec™ Modular Cover to completely cover the system rather than a mat to retain duckweed (Figure 4.26). The process is still a lagoon-based treatment process composed of a series of aerobic cells followed by an anaerobic settling pond. Cells in series consist of a complete-mix aerated reactor, a partial-mix aerated reactor, a covered anaerobic settling pond, and a Lemna polishing reactor. The polishing reactor is aerated and has submerged, attached-growth media modules to supplement BOD and NH_3–N reduction. Sludge removal from the settling pond is expected to be required about every 5 to 12 years. Frequency of cleaning will vary with climate and strength of the wastewater.

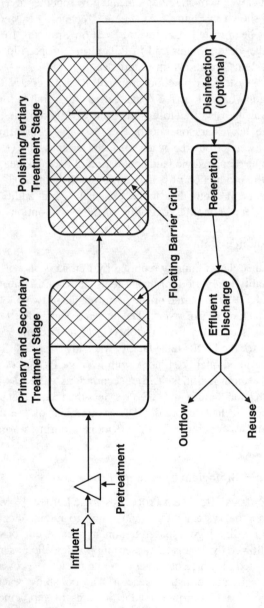

FIGURE 4.23 Flow diagram for a typical Lemna system. (Courtesy of Lemna Technologies, Inc., Minneapolis, MN.)

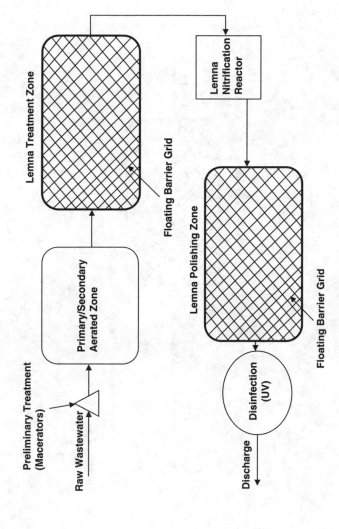

FIGURE 4.24 Flow diagram for Lemna system application. (Courtesy of Lemna Technologies, Inc., Minneapolis, MN.)

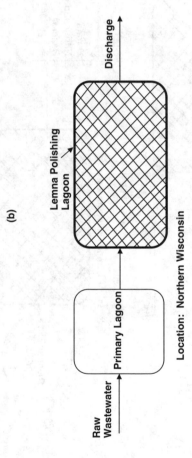

(b)

Location: Northern Wisconsin

FIGURE 4.24 (cont.)

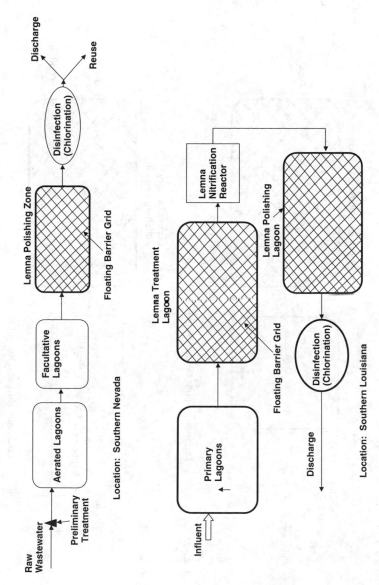

FIGURE 4.24 (cont.)

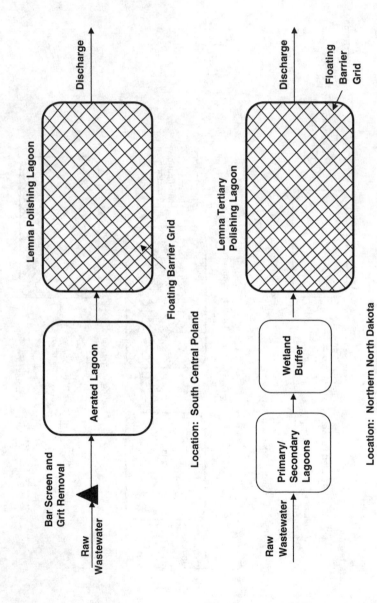

(d)

Location: South Central Poland

Location: Northern North Dakota

FIGURE 4.24 (cont.)

FIGURE 4.25 Photograph of Lemna harvesting equipment and floating barrier grid. (Courtesy of Lemna Technologies, Inc., Minneapolis, MN.)

4.10.4 Las International, Ltd.

Accel-o-Fac™ and Aero-Fac™ systems are offered as upgrades and original installations. Accel-o-Fac™ is a facultative pond with wind-driven aerators, and Aero-Fac™ is a partial-mix aerated lagoon utilizing an Aero-Fac™ diffused air bridge and LAS Mark 3 wind and electric aerators. Systems have been installed in several countries including Great Britain, Canada, and the United States. Performance data are limited, as with most lagoon systems, and data presented in the company literature are primarily limited to operation during warm months

TABLE 4.16
Typical Effluent Qualities Expected from Lemna Systems

Parameter	Influent	Effluent
BOD (mg/L)	250–200	<30–10
TSS (mg/L)	300–250	<30–10
TN (mg/L)	80–40	<20–5
NH_3–N, (mg/L)	50–10	<10–2
TP (mg/L)	20–10	<1

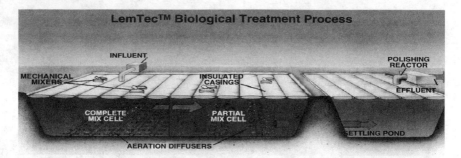

FIGURE 4.26 LemTec™ Biological Treatment Process. (Courtesy of Lemna Technologies, Inc., Minneapolis, MN.)

of the year. Winter performance data are limited but are necessary to evaluate the processes; however, it is expected that the systems will perform essentially as other partial-mix lagoon systems with equivalent aeration. The advantage of the processes is a savings in power costs if adequate wind velocity is available. A disadvantage of the Accel-o-Fac™ is the lack of control of the aeration process.

4.10.5 PRAXAIR, INC.

The Praxair® I-SO™ systems have been installed in over 100 locations throughout the world (Figure 4.27). Each unit is capable of transferring 240 lb oxygen per hr. Praxair has reported that the total power required to operate the Praxair® I-SO™ System, including the generation of oxygen, is as much as 60% less than the air systems replaced. Plants located near an oxygen pipeline supply can decrease power costs up to 90%.

4.10.6 ULTRAFILTRATION MEMBRANE FILTRATION

In 2001, John Thompson Engineering of New South Wales, Australia, designed and installed a 0.264-mgd (1000 m³/d) Zenon membrane facility to polish an effluent from a lagoon facility. Operating data are not available, but all indications are that the facility is functioning well. Very little operating experience is available with membranes in lagoons, but it is an option that should be evaluated.

4.11 NITROGEN REMOVAL IN LAGOONS

4.11.1 INTRODUCTION

The BOD and suspended solids removal capability of lagoon systems has been reasonably well-documented, and reliable designs are possible; however, the nitrogen removal capability of wastewater lagoons has been given little consideration in system designs until recently. Nitrogen removal can be critical in many situations because ammonia nitrogen in low concentrations can adversely affect some young fish in receiving waters, and the addition of nitrogen to surface waters

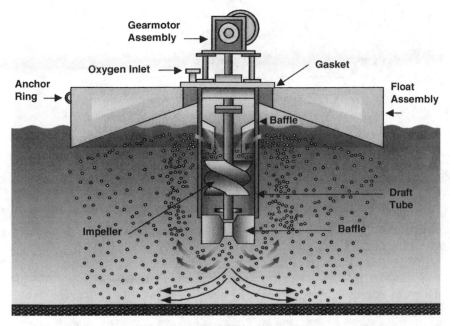

FIGURE 4.27 Praxair® In-Situ Oxygenation (I-SO™) system. (Courtesy of Praxair Technology, Inc., Danbury, CT.)

can cause eutrophication. In addition, nitrogen is often the controlling parameter for the design of land treatment systems. Any nitrogen removal in the preliminary lagoon units can result in very significant savings in land and costs for the final land treatment site. The following sections describe several conventional and commercial products that have been developed for nitrogen removal.

4.11.2 FACULTATIVE SYSTEMS

Nitrogen loss from streams, lakes, impoundments, and wastewater lagoons has been observed for many years. Extensive data on nitrogen losses in lagoon systems were insufficient for a comprehensive analysis of this issue until the early 1980s, and no agreement was reached on the removal mechanisms. Various investigators have suggested algae uptake, sludge deposition, adsorption by bottom soils, nitrification, denitrification, and loss of ammonia as a gas to the atmosphere (volatilization). Evaluations by Pano and Middlebrooks (1982), USEPA (1983), Reed (1984), and Reed et al. (1995) suggest that a combination of factors may be responsible, with the dominant mechanism under favorable conditions being volatilization losses to the atmosphere, as shown by the relative size of the arrows in Figure 4.28.

The U.S. Environmental Protection Agency sponsored comprehensive studies of facultative wastewater lagoon systems in the late 1970s (Bowen, 1977; Hill and Shindala, 1977; McKinney, 1977; Reynolds et al., 1977). These results

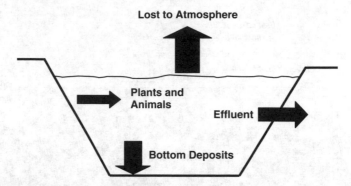

FIGURE 4.28 Nitrogen pathways in wastewater lagoons under favorable conditions.

provided verification that significant nitrogen removal does occur in lagoon systems. Key findings from those studies are summarized in Table 4.17. These results verify the consensus of previous investigators that nitrogen removal was in some way related to pH, detention time, and temperature in the lagoon system. The pH fluctuates as a result of algae–carbonate interactions in the lagoon, so wastewater alkalinity is important. Under ideal conditions, up to 95% nitrogen removal can be achieved from facultative wastewater stabilization lagoons.

Several recent studies of nitrogen removal have been completed, but the quantity of data is limited. A study of 178 facultative lagoons in France showed an average nitrogen removal of 60 to 70%; however, only a limited quantity of data was available from each lagoon system (Racault et al., 1995). Wrigley and Toerien (1990) studied four small-scale facultative lagoons in series for 21 months and observed an 82% reduction in ammonia nitrogen, but an extensive sampling program similar to those conducted by the USEPA in the late 1970s was not carried out.

Shilton (1995) quantified the removal of ammonia nitrogen from a facultative lagoon treating piggery wastewater and found that the rate of volatilization varied from 0.07 to 0.314 lb/1000 ft^2·d (355 to 1534 mg/m^2·day). The rate of volatilization increased at higher concentrations of ammonia nitrogen and TKN.

Soares et al. (1995) monitored ammonia nitrogen removal in a wastewater stabilization lagoon complex of varying geometries and depths in Brazil. The ammonia nitrogen concentrations were lowered to 5 mg/L in the maturation lagoons, thus making the effluent satisfactory for discharge to surface waters. It was found that the ammonia removal in the facultative and maturation lagoons could be modeled by the equations based on the volatilization mechanism proposed by Pano and Middlebrooks (1982).

Commercial products, as mentioned in the introduction to this section, appear to offer improvements that may remove significant amounts of ammonia nitrogen and some total nitrogen. Some of the options are described below.

TABLE 4.17
Data Summary from EPA Facultative Wastewater Pond Studies (Annual Values)

Location	Detention Time (d)	Water Temperature (°C)	pH (median)	Alkalinity (mg/L)	Influent Nitrogen (mg/L)	Removal (%)
Peterborough, New Hampshire (three cells)	107	11	7.1	85	17.8	43
Kilmichael, Mississippi (three cells)	214	18.4	8.2	116	35.9	80
Eudora, Kansas (three cells)	231	14.7	8.4	284	50.8	82
Corinne, Utah (first three cells)	42	10	9.4	555	14.0	46

4.11.2.1 Theoretical Considerations

Ammonia nitrogen removal in facultative wastewater stabilization lagoons can occur through the following three processes:

- Gaseous ammonia stripping to the atmosphere
- Ammonia assimilation in algal biomass
- Biological nitrification

The low concentrations of nitrates and nitrites in lagoon effluents indicate that nitrification generally does not account for a significant portion of ammonia nitrogen removal. Ammonia nitrogen assimilation in algal biomass depends on the biological activity in the system and is affected by temperature, organic load, detention time, and wastewater characteristics. The rate of gaseous ammonia losses to the atmosphere depends mainly on the pH value, temperature, and the mixing conditions in the lagoon. Alkaline pH shifts the equilibrium equation $NH_3 + H_2O \leftrightarrow NH_4^+ + OH^-$ toward gaseous ammonia, whereas the mixing conditions affect the magnitude of the mass-transfer coefficient. Temperature affects both the equilibrium constant and mass-transfer coefficient.

At low temperatures, when biological activity decreases and the lagoon contents are generally well mixed because of wind effects, ammonia stripping will be the major process for ammonia nitrogen removal in facultative wastewater stabilization lagoons. The ammonia stripping lagoons may be expressed by assuming a first-order reaction (Stratton, 1968, 1969). The mass balance equation will be:

$$VdC / dt = Q(C_0 - C_e) - kA(NH_3) \tag{4.25}$$

where

V = Volume of the pond (m^3).
C = Average lagoon contents concentration of ($NH_4^+ + NH_3$) (mg/L as N).
t = Time (d).
Q = Flow rate (m^3/d).
C_0 = Influent concentration of ($NH_4^+ + NH_3$) (mg/L as N).
C_e = Effluent concentration of ($NH_4^+ + NH_3$) (mg/L as N).
k = Mass-transfer coefficient (m/d).
A = Surface area of the pond (m^3).

The equilibrium equation for ammonia dissociation may be expressed as:

$$K_b = \frac{[NH_4^+][OH^-]}{[NH_3]} \tag{4.26}$$

where K_b is an ammonia dissociation constant.

By modifying Equation 4.26, gaseous ammonia concentration may be expressed as a function of the pH value and total ammonia concentration (NH_4^+ + NH_3) as follows:

$$\left[H^+\right] = \frac{K_W}{\left[OH^-\right]} \tag{4.27}$$

$$C = NH_4^+ + NH_3 \tag{4.28}$$

$$NH_3 = \frac{C}{1 + 10^{pK_W - pK_b - pH}} \tag{4.29}$$

where $pK_W = -\log K_W$, and $pK_b = -\log K_b$.

Assuming steady-state conditions and a completely mixed lagoon where $C_e = C$, Equation 4.28 and Equation 4.29 will yield the following relationship:

$$\frac{C_e}{C_o} = \frac{1}{1 + \dfrac{A}{Q}k\left[\dfrac{1}{1 + 10^{pK_W - pK_b - pH}}\right]} \tag{4.30}$$

This relationship emphasizes the effect of pH, temperature (pK_W and pK_b are functions of temperature), and hydraulic loading rate on ammonia nitrogen removal.

Experiments on ammonia stripping conducted by Stratton (1968, 1969) showed that the ammonia loss-rate constant was dependent on the pH value and temperature (°C) as shown in the following relationships:

$$\text{Ammonia loss rate constant} \times e^{1.57(pH-8.5)} \tag{4.31}$$

$$\text{Ammonia loss rate constant} \times e^{0.13(T-20)} \tag{4.32}$$

King (1978) reported that only 4% nitrogen removal was achieved by harvesting floating *Cladophora fracta* from the first lagoon in a series of four receiving secondary effluents. The major nitrogen removal in the lagoons was attributable to ammonia gas stripping. The removal of total nitrogen was described by first-order kinetics using a plug-flow model: $N_t = N_0 e^{-0.03t}$, where N_t is the total nitrogen concentration (mg/L), N_0 is the initial total nitrogen concentration (mg/L), and t is time (d).

It is well understood that large-scale facultative wastewater stabilization lagoon systems only approach steady-state conditions, and only during windy seasons will well-designed lagoons completely achieve mixed conditions. Moreover, when ammonia removal through biological activity becomes significant or ammonia is released into the contents of the lagoon from anaerobic activity at the bottom of the lagoon, the expressions for ammonia removal in the system must include these factors along with the theoretical consideration of ammonia stripping as shown in Equation 4.30.

In the following text, mathematical relationships for total nitrogen removal based on the performance of three full-scale facultative wastewater stabilization lagoons are developed taking into consideration the theoretical approach and incorporating temperature, pH value, and hydraulic loading rate as variables. Therefore, rather than using the theoretical expression for ammonia nitrogen stripping (Equation 4.30), the following equation is considered for TKN removal in facultative lagoons:

$$\frac{C_e}{C_0} = \frac{1}{1 + \frac{A}{Q} K \cdot f(\text{pH})} \tag{4.33}$$

where K is a removal rate coefficient (L/t), and $f(\text{pH})$ is a function of pH.

The K values are considered to be a function of temperature and mixing conditions. For a similar lagoon configuration and climatic region, the K values may be expressed as a function of temperature only. The function of pH, which is considered to be dependent on temperature, affects the pK and pK_b values, as well as the biological activity in the lagoon. When the effect of the pH function on ammonia nitrogen stripping was incorporated (Equation 4.33), the pH function was found to be an exponential relationship; the selection of an exponential function to describe the pH function was based on statistical analyses indicating that an exponential relationship best described the data. Also, most reaction rate and temperature relationships are described by exponential functions such as the Van't Hoff–Arrhenius equation; therefore, it is logical to assume that such a relationship would apply in the application of the theoretical equation to a practical problem.

4.11.2.2 Design Models

Data were collected on a frequent schedule from every cell at all of the lagoon systems shown in Table 4.17 for at least a full annual cycle. This large body of data allowed quantitative analysis that included all major variables, and several design models were independently developed. The two models discussed here have been shown to be the most accurate in predicting nitrogen removal in facultative lagoon systems. These have been validated using data from sources not used in model development. The two models are summarized in Table 4.18 and Table 4.19, and details on the theoretical development of the models were presented above. Further validation of the two models can be found in Reed et al. (1995), Reed (1984, 1985), and USEPA (1983). Both are first-order models and both depend on pH, temperature, and detention time in the system. Although they both predict the removal of total nitrogen, it is implied in the development of each that volatilization of ammonia is the major pathway for nitrogen removal from wastewater stabilization lagoons. The application of the two models is shown in Figure 4.29, and the predicted total nitrogen in the effluent is compared to the actual monthly average values measured at Peterborough, New Hampshire. Both

TABLE 4.18
Model 1. Nitrogen Removal in Facultative Lagoons
(Plug-Flow Model)

$$N_e = N_0 e^{-K_T[t+60.6\,(pH-6.6)]}$$

where:

N_e = effluent total nitrogen (mg/L)

N_0 = influent total nitrogen (mg/L)

K_T = temperature dependent rate constant = $K_{20}(\theta)^{(T-20)}$ = rate constant at 20°C = 0.0064, where θ = 1.039

t = detention time in system (d)

pH = pH of near-surface bulk liquid

Note: See USEPA (1983) or Reed (1984) for typical pH values or estimate using pH = $7.3e^{0.0005ALK}$, where ALK = expected influent alkalinity (mg/L) (EPA, 1983; Reed, 1984).

Use the Mancini and Barnhart (1976) equation to determine lagoon water temperature:

$$T = \frac{0.5AT_a + QT_i}{0.5A + Q}$$

where:

A = surface area of pond (m³)
T_a = ambient air temperature (°C)
T_i = influent temperature (°C)
Q = influent flow rate (m³/d)

Source: Reed, S.C., *J. WPCF*, 57(1), 39–45, 1985. With permission.

of these models are written in terms of total nitrogen, and they should not be confused with the still-valid equations reported by Pano and Middlebrooks (1982) that are limited to the ammonia fraction. Calculations and predictions based on total nitrogen should be even more conservative.

A high rate of ammonia removal by air stripping in advanced wastewater treatment depends on a high (>10) chemically adjusted pH. Algae–carbonate interactions in wastewater lagoons can elevate the pH to similar levels for brief periods. At other times, at moderate pH levels, the rate of nitrogen removal may be low, but the long detention time in the lagoon compensates.

Figure 4.30 illustrates the validation of both models using data from lagoon systems not used previously. The diagonal line on the figure represents a perfect fit of predicted vs. actual values. The close fit and consistent trends verify that either model can be used to estimate nitrogen removal. In addition, the models have been used in the design of several lagoons systems and have been found to work well.

TABLE 4.19
Model 2. Nitrogen Removal in Facultative Lagoons
(Complete-Mix Model)

$$N_e = \frac{N_0}{1 + t(0.000576T - 0.00028)e^{(1.080 - 0.042T)(pH - 6.6)}}$$

where:

N_e = effluent total nitrogen (mg/L)
N_0 = influent total nitrogen (mg/L)
t = detention time (d)
T = temperature of pond water (°C)
pH = pH of near-surface bulk liquid

Use the Mancini and Barnhart (1976) equation to determine lagoon water temperature:

$$T = \frac{0.5AT_a + QT_i}{0.5A + Q}$$

where:

A = surface area of pond (m³)
T_a = ambient air temperature (°C)
T_i = influent temperature (°C)
Q = influent flow rate (m³/d)

Source: Middlebrooks, E.J., Nitrogen removal model developed for U.S. Environmental Protection Agency, Washington, D.C., 1985.

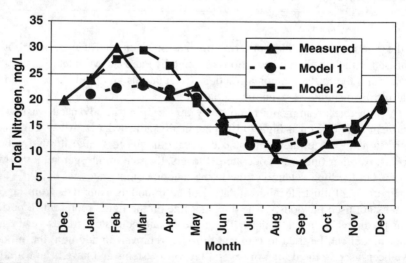

FIGURE 4.29 Predicted vs. actual effluent nitrogen for Peterborough, New Hampshire.

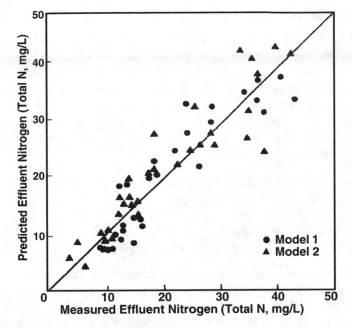

FIGURE 4.30 Verification of design models.

4.11.2.3 Applications

These models should be useful for new or existing wastewater lagoons when nitrogen removal or ammonia conversion is required. The design of new systems would typically base detention time on the BOD removal requirements. The nitrogen removal that will occur during that time can then be calculated with either model. It is prudent to assume that the remaining nitrogen in the effluent will be ammonia and then design any further removal or conversion for that amount. If additional land is available, a final step can be used to compare the provision of additional detention time in the lagoon for nitrogen removal with the costs for other removal alternatives. Use of these models is particularly important when lagoons are used as a component in land treatment systems because nitrogen is often the controlling design parameter. A reduction in lagoon effluent nitrogen will often permit a very significant reduction in the land area needed and, therefore, the costs for land treatment.

4.11.2.4 Summary

Nitrogen removal occurs in facultative wastewater stabilization lagoons, and it can be reliably predicted for design purposes with either of the two models presented above. Nitrogen removal in lagoons may be more cost effective than other alternatives for removal or ammonia conversion. Nitrogen removal in lagoons used as a component in land treatment systems can influence the cost effectiveness of the project.

4.11.3 AERATED LAGOONS

At a pH value of 8.0, approximately 95% of the ammonia nitrogen is in the form of ammonium ion; therefore, in biological systems, such as aerated lagoons where the pH values are usually less than 8.0, the majority of the ammonia nitrogen is in the form of ammonium ion. Total Kjeldahl nitrogen (TKN) is composed of the ammonia nitrogen and organic nitrogen. Organic nitrogen is a potential source of ammonia nitrogen because of the deamination reactions during the metabolism of organic matter in wastewater. Ammonia and TKN reduction in aerated lagoons can occur through several processes:

- Gaseous ammonia stripping to the atmosphere
- Ammonia assimilation in biomass
- Biological nitrification
- Biological denitrification
- Sedimentation of insoluble organic nitrogen

The rate of gaseous ammonia losses to the atmosphere depends primarily on the pH value, temperature, hydraulic loading rate, and mixing conditions in the lagoon. The equilibrium equation $NH_3 + H_2O \leftrightarrow NH_4^+ + OH^-$ is shifted toward gaseous ammonia by an alkaline pH value, while the mixing conditions affect the magnitude of the mass-transfer coefficient. Temperature affects both the equilibrium constant and mass-transfer coefficient.

Ammonia nitrogen assimilation into the biomass depends on the biological activity in the system and is affected by several factors such as temperature, organic load, detention time, and wastewater characteristics. Biological nitrification depends on adequate environmental conditions for nitrifiers to grow and is affected by several factors such as temperature, dissolved oxygen concentration, pH value, detention time, and wastewater characteristics.

Denitrification can take place in bottom sediments under anoxic conditions, and temperature, redox potential, and sediment characteristics affect the rate of denitrification. In well-designed aerated lagoons with good mixing conditions and distribution of dissolved oxygen, denitrification will be negligible.

The U.S. Environmental Protection Agency sponsored comprehensive studies of aerated wastewater lagoon systems between 1978 and 1980 that provided information about nitrogen removal in aerated lagoon systems (Earnest et al., 1978; Englande, 1980; Gurnham et al., 1979; Polkowski, 1979; Reid and Streebin, 1979; Russel et al., 1980). Table 4.20 and Table 4.21 summarize the key findings from those studies. These results verify the consensus of previous investigators that nitrogen removal was in some way related to pH, detention time, and temperature in the lagoon system.

4.11.3.1 Comparison of Equations

Table 4.22 contains a summary of selected equations developed to predict ammonia nitrogen and TKN removal in diffused-air aerated lagoons (Middlebrooks and

TABLE 4.20
Wastewater Characteristics and Operating Conditions for the Five Aerated Lagoons

	System				
Parameter	Pawnee	Bixby	Koshkonong	Windber	North Gulfport
BOD (mg/L)	473	368	85	173	178
COD (mg/L)	1026	635	196	424	338
TKN (mg/L)	51.41	45.04	15.3	24.33	26.5
NH_3–N (mg/L)	26.32	29.58	10.04	22.85	15.7
Alkalinity (mg/L)	242	154	397	67	144
pH	6.8–7.4	6.1–7.1	7.2–7.4	5.6–6.9	6.7–7.5
Hydraulic loading rate (m/d)	0.0213	0.0285	0.0423	0.0663	0.0873
Organic loading rate (kg BOD_5 ha^{-1} d^{-1})	151	161	87	285	486
Detention time (d)	143	107	72	46	22

Source: Data from Earnest et al. (1978), Englande (1980), Gurnham et al. (1979), Polkowski (1979), Reid and Streebin (1979).

Pano, 1983). All of the equations have a common database; however, the data were used differently to develop several of the equations. The "System" column in Table 4.22 describes the lagoons or series of lagoons that were used to develop the equation. An explanation of the system combinations was presented above. These combinations of data were analyzed statistically, and the equations presented in Table 4.22 were selected based on the best statistical fit of the data for the various combinations that were tried. The combinations of data are not directly comparable, but the presentation in Table 4.22 takes into account the best statistical fit of the data.

A comparison of the hydraulic detention times calculated using the various formulas for TKN removal show that the maximum deviation between the maximum and minimum detention times calculated from the equation is 14%. In view of the wide variation in methods used to develop the various relationships, this is a very small deviation. All of the relationships are statistically significant at levels higher than 1%. The small difference in detention times calculated using all of the expressions establishes a good basis to apply any of the relationships in the design of lagoons to estimate TKN removal. Because of the simplicity of the plug-flow model and the fraction-removed model, it is recommended that these two be employed, with the others used as a check to ensure adequate removal in the event that unusual loading rates or BOD_5 loading rates are encountered.

TABLE 4.21
Nitrogen Removal in Aerated Lagoons

Parameter (mg/L)	Pawnee		Bixby		Koshkonong	
	Influent	Effluent	Influent	Effluent	Influent	Effluent
TKN	51.41	5.04	45.04	8.44	15.3	7.6
Range	24.93–80.20	2.21–12.74	36.33–64.80	3.04–22.20	6.37–21.34	3.38–13.83
Ammonia–N	26.32	1.27	29.59	3.46	10.04	5.26
Range	12.00–37.00	0.19–5.47	23.71–40.35	0.11–14.76	4.40–16.12	0.66–12.51
Nitrate–N	—	0.81	—	—	1.66	4.35
Range	—	0.15–1.54	—	—	0.18–5.78	1.14–9.13
Nitrite–N	—	0.13	—	—	0.08	—
Range	—	0.02–0.55	—	—	0.02–0.17	0.03–1.05
Alkalinity	242	161	154	70	397	382
pH	6.8–7.4	7.8–9.3	6.1–7.1	6.7–9.2	7.2–7.4–7.9	—
Temperature (°C)	—	11.3	—	16.3	—	11.6
Range	—	3–22	—	5–29–	1–25	—
Dissolved oxygen	—	1.9–16.0	—	3.9–13.5	—	7.6–15.3
Operating conditions						
Hydraulic loading rate (m/d)	0.02	—	0.03	—	0.04	—
Organic loading rate (kg BOD ha^{-1} d^{-1})	151	—	161	—	87	—
Hydraulic detention time (d)	143	—	107	—	72	—
Power level (CFM/MG)	—	—	29.8, 17.0	—	68, 28, 16	—
TKN	24.33	23.57	26.5	10.8	15.7	11.1

Parameter (mg/L)	Windber		North Gulfport		Mt. Shasta	
	Influent	Effluent	Influent	Effluent	Influent	Effluent
Range	13.21–46.00	14.43–34.11	20.6–30.9	7.2–13.3	10.1–20.9	6.8–14.2
Ammonia–N	22.85	22.92	15.73	5.10	10.30	5.40
Range	12.32–37.24	12.04–32.75	11.6–20.0	0.9–9.7	4.5–17.5	0.5–12.0
Nitrate–N	—	0.72	—	2.36	0.30	0.73
Range	—	0.11–2.63	—	0.12–6.46	0.01–0.86	0.04–2.32
Nitrite–N	—	0.24	—	0.64	0.15	0.49
Range	—	0.10–0.66	—	0.04–1.76	0.01–0.95	0.01–2.06
Alkalinity	67	82	144	102	93	74
pH	5.6–6.9	6.8–8.5	6.7–7.5	6.8–7.5	6.5–7.6	7.4–9.7
Temperature (°C)	—	13.9	—	21.5	—	13.7
Range	—	2–24	—	11–29	—	2–27
Dissolved oxygen	—	5.7–15.0	—	0.8–9.3	—	10.9–14.0
Operating conditions						
Hydraulic loading rate (m/d)	0.07	—	0.09	—	0.08	—
Organic loading rate (kg BOD ha^{-1} d^{-1})	285	—	486	—	202	—
Hydraulic detention time (d)	46	—	22	—	21 + 10 Fac.	—
Power level (CFM/MG)	34, 14, 6	—	7.7–8.5 hp/MG	—	—	—

Source: USEPA, *Technology Transfer Process Design Manual for Municipal Wastewater Stabilization Ponds,* EPA 625/1-83-015, Center for Environmental Research Information, U.S. Environmental Protection Agency, Cincinnati, OH, 1983.

TABLE 4.22
Comparisons of Various Equations Developed To Predict Ammonia–Nitrogen and Total Kjeldahl Nitrogen (TKN) Removal in Diffused-Air Aerated Lagoons

Equation Used To Estimate Correlation Coefficient	Correlation Coefficient	Hydraulic Detention Time (d)	Comparison with Maximum Detention Time (% Difference)	System
TKN Removal				
$\ln C_e/C_0 = -0.0129$ (detention time)	0.911	125	5.3	Ponds 1, 2, and 3 (mean monthly data)
TKN removal rate = 0.809 (TKN loading rate)	0.983	132	0.0	Total system (mean monthly data)
TKN removal rate = 0.0946 (BOD_5 loading rate)	0.967	113	14.4	Total system (mean monthly data)
TKN fraction removed = 0.0062 (detention time)	0.959	129	2.3	Ponds 1, 2, and 3 (mean monthly data)
Ammonia Nitrogen Removal				
$\ln C_e/C_0 = -0.0205t$	0.798	79	40.2	All data (mean monthly data)
NH_3–N removal rate = 0.869 (NH_3–N loading rate)	0.968	92	30.3	Total system (mean monthly data)
NH_3–N removal rate = 0.0606 (BOD_5 loading rate)	0.932	132	0.0	Total system (mean monthly data)
NH_3–N fraction removed = 0.0066 (detention time)	0.936	121	8.3	Ponds 1, 2, and 3

Source: Middlebrooks, E.J. and Pano, A., *Water Res.*, 17(10), 1369–1378, 1983. With permission.

Using any of these expressions will result in a good estimate of the TKN removal that is likely to occur in diffused-air aerated lagoons. Unfortunately, data are not available to develop relationships for surface aerated lagoons. The relationships developed to predict ammonia nitrogen removal yielded highly significant (1% level) relationships for all of the equations presented in Table 4.22; however, the agreement between the calculated detention times for ammonia nitrogen removal differed significantly from that observed for the TKN data. This variation is not surprising in view of the many mechanisms involved in ammonia nitrogen production and removal in wastewater lagoons, but this variation in results does complicate the use of the equations to estimate ammonia nitrogen removal in aerated lagoons.

Statistically, a justification exists to use either of the expressions in Table 4.22 to calculate the detention time required to achieve a given percentage reduction in ammonia nitrogen. Perhaps the best equation to use during design to predict ammonia nitrogen removal is the relationship between the fraction removed and the detention time. The correlation coefficient for this relationship is higher than the correlation coefficient for the plug-flow model, and both equations are equally simple.

Rich (1996, 1999) has proposed continuous-feed, intermittent-discharge (CFID) basins for use in aerated lagoon systems for nitrification and denitrification. The systems are designed to use in-basin sedimentation to uncouple the solids retention time from the hydraulic retention time. Unlike sequencing batch reactor (SBR) systems, the influent flow is continuous. A single basin with a dividing baffle to prevent short-circuiting is frequently used. Some CFID systems have experienced major operational problems with short-circuiting and sludge bulking; however, by minimizing these problems with design changes the systems can be made to function properly. CFID design modifications can be made to overcome most difficulties, and details are presented by Rich (1999). The basic CFID system consists of a single reactor basin divided into two cells with a floating baffle. The two cells are referred to as the influent (cell 1) and effluent cell (cell 2). Mixed liquor is recycled from cell 2 to the headworks to provide a high ratio of soluble biodegradable organics to organisms, and the oxygen source is primarily nitrates. This approach is used to control bulking. Although some nitrification will occur in the influent cell, the system is designed for nitrification to occur in the effluent cell. To learn more about the operation of the CFID systems, consult Rich (1999).

4.11.3.2 Summary

Rich's (1999) method provides a way to design for nitrification in an aerated lagoon. The equations in Table 4.22 are empirical and may or may not apply to a general design; however, these equations will serve as an estimate of what might be expected in terms of nitrogen removal. Designing a lagoon system to nitrify a wastewater is not difficult if the water temperature and detention time are adequate to support nitrifiers and adequate dissolved oxygen is supplied.

Obviously, providing recycle of the mixed liquor is a significant benefit. As with all treatment methods, an economic analysis should be performed to determine the choice of a system.

4.11.4 PUMP SYSTEMS, INC., BATCH STUDY

In 1998 a solar-powered circulator (equivalent to the SolarBee® Model SB2500) was installed in a 29-acre pond with a depth of 15 feet at Dickinson, North Dakota, with no incoming wastewater. The circulator flow rate was 2500 gpm. The ammonia nitrogen concentration at the beginning of the experiment was approximately 20 mg/L. Dissolved oxygen, pH, BOD, TSS, ammonia nitrogen, water temperature, and various other parameters were measured over a 90-day period at various locations and depths. Over 1500 samples were collected over the 90-day testing period. Average data for the various locations and depths are shown in Table 4.23. The average water temperature during the 90 days of testing was 20.5°C. Dissolved oxygen was present throughout the pond at all depths but on occasion dropped to 0.4 mg/L at the bottom. These occasional low DO concentrations may have had an adverse effect on the results presented below, but the results do provide some guidance as to how to estimate the expected conversion of ammonia nitrogen in a partial-mix aerated lagoon system.

A plot of the data for complete-mix and plug-flow models was prepared, but little difference in the fit of the data was observed. The plug-flow plot is shown in Figure 4.31. The plug-flow model and the model for a batch test are the same and should fit the data best. The reduction in NH_4–N with time was directly related to the variation in pH value (Figure 4.31). When the pH exceeded 8.0, the reduction in NH_4–N increased, resulting in a greater loss of the ammonia gas to the atmosphere.

The results of this study are useful for revealing the very low reaction rate for nitrification that occurs in partial mix aerated lagoons. The reaction rate of 0.0107 d^{-1} obtained at an average temperature of 20.5°C in the Dickinson experiments agrees with results obtained with data collected in an aerated lagoon located in Wisconsin (Middlebrooks, 1982). At 1°C, the ammonia nitrogen conversion reaction rate for the Wisconsin partial-mix lagoon ranged between 0.0035 and 0.0070 d^{-1}. Using an average value of 0.005 d^{-1} at 1°C and the value of 0.0107 d^{-1} obtained at Dickinson at 20.51°C, an approximate value of 1.04 results for θ in the classical temperature correction equation; thus, $k_T = k_{20}(1.04)^{(T-20)}$. Example 4.8 illustrates the effects of reaction rates and temperature on the performance of partial-mix lagoon systems.

Example 4.8

Estimate the expected NH_4–N conversion in a partial mix aerated lagoon receiving adequate DO and alkalinity to nitrify an ammonia-N concentration of 20 mg/L at a water temperature of 10°C. Determine the effluent concentration at a detention time of 30 days and at a desired effluent concentration of 10 mg/L.

Input data:
Temperature = 1°C.
Influent $NH_4–N = C_0 = 20$ mg/L.
Temperature correction factor = $\theta = 1.04$.
Reaction rate = $k_{20} = 0.0107$ d^{-1}.
Known detention time = 30 d.
Desired effluent $NH_4–N = 10$ mg/L.

Solution

1. Correct reaction rate for temperature:
   $k_T = k_{20}(\theta)^{(T-20)} = 0.005079$ d^{-1}
2. Determine the effluent concentration with the detention time known:
   $C_e/C_0 = e^{-kt}$
   $C_e = 17.17$ mg/L
3. Determine the detention time required to achieve the desired effluent concentration:
   $t = (\ln(C_e/C_0))/–k = 136.48$ d

4.11.5 COMMERCIAL PRODUCTS

Numerous products and processes are available to improve lagoon performance and remove nitrogen. Several of these options are presented below; information about them was extracted from Burnett et al. (2004).

4.11.5.1 Add Solids Recycle

The addition of solids recycle can be a reliable method of producing an effluent meeting stringent ammonia limits. With the addition of recycle, a lagoon is converted to a low mixed liquor suspended solids (MLSS) activated sludge system. This can be accomplished using an external clarifier and adding a pump to return solids to the headworks. The BIOLAC® process uses an internal clarifier. Effluent from the clarifier is discharged to disinfection or routed through subsequent cells of the lagoon system.

Successful operation of a low-MLSS activated sludge system requires that the recycled solids be kept in suspension. The aerated lagoon must be kept completely mixed. In most cases, a portion of the existing lagoon is partitioned into a complete mix cell because the power required to mix the cell is far greater than that required to reduce the BOD or nitrify the ammonia. The remaining portion of the system is used for polishing the effluent or to store the water before discharge.

Because the recycle system is an activated sludge variation, it can be designed and operated with traditional activated sludge design methods. Floating baffle curtains with exit ports are frequently used for cell partitioning. Excess sludge wasting can be accomplished in a separate holding pond, or downstream cells of the existing lagoon can be used to store and treat sludge for disposal.

TABLE 4.23
Average Values for Batch Test in Pond 4 at Dickinson, North Dakota

Sampling Date	Days	Temperature	DO	pH	NH_4-N	C_e/C_o	ln C_e/C_o	k_{CM}	k_{PM}	$C_o/C_e - 1$
6/3/98 averages	0.00	15.92	0.83	7.74	19.52	1	0	—	—	0
6/4/98 averages	1.00	16.12	1.00	7.72	22.74	1.1652	0.1529	-0.1418	0.152877	-0.1418
6/10/98 averages	7.00	15.26	1.49	7.74	24.02	1.2309	0.2077	-0.0268	0.029673	-0.1876
6/15/98 averages	12.00	17.80	7.05	8.02	22.99	1.1778	0.1636	-0.0126	0.013634	-0.1509
6/16/98 averages	13.00	17.38	7.58	8.06	21.73	1.1134	0.1074	-0.0078	0.008264	-0.1019
6/22/98 averages	19.00	17.30	12.10	8.41	18.43	0.9440	-0.0576	0.0031	-0.00303	0.0593
6/29/98 averages	26.00	18.07	14.85	8.81	13.59	0.6965	-0.3618	0.0168	-0.01391	0.4358
7/2/98 averages	29.00	20.24	15.70	8.73	13.62	0.6979	-0.3597	0.0149	-0.0124	0.4329
7/9/98 averages	36.00	23.32	14.32	8.78	10.71	0.5489	-0.5999	0.0228	-0.01666	0.8219
7/15/98 averages	42.00	23.95	3.88	8.59	9.95	0.5096	-0.6740	0.0229	-0.01605	0.9621
7/21/98 averages	48.00	24.77	7.58	8.50	8.88	0.4550	-0.7874	0.0249	-0.0164	1.1976
7/23/98 averages	50.00	24.23	5.72	8.47	8.62	0.4414	-0.8177	0.0253	-0.01635	1.2653
7/30/98 averages	57.00	22.87	0.85	7.72	9.10	0.4663	-0.7629	0.0201	-0.01338	1.1444
8/4/98 averages	62.00	21.91	1.30	7.95	9.20	0.4716	-0.7517	0.0181	-0.01212	1.1206

8/12/98 averages	70.00	24.36	8.36	8.14	12.07	0.6184	-0.4807	0.0088	-0.00687	0.6172
8/18/98 averages	76.00	22.54	2.82	7.93	10.46	0.5360	-0.6237	0.0114	-0.00821	0.8657
8/26/98 averages	84.00	21.89	4.86	7.79	9.35	0.4793	-0.7355	0.0129	-0.00876	1.0864
9/1/98 averages	90.00	22.37	9.12	8.14	7.88	0.4035	-0.9076	0.0164	-0.01008	1.4783
Average	—	20.57	6.63	8.17	—	—	—	0.0017	0.0030	—
Minimum	—	15.26	0.83	7.72	—	—	—	-0.1418	-0.0167	—
Maximum	—	24.77	15.70	8.81	—	—	—	0.0253	0.1529	—

Note: Area = 29 acres; no inflow.

Source: Courtesy of Pump Systems, Inc., Dickinson, ND.

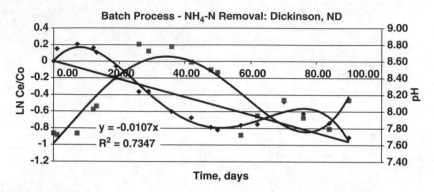

FIGURE 4.31 Plot of NH_4–N data using plug-flow model.

A complete-mix section can be located anywhere in the flow train of an aerated lagoon system. Locating the complete-mix cell near the end of the flow train has the advantage of nitrifying after carbonaceous BOD has been removed in the lagoon system. When the complete-mix zone is first in the process, sludge can easily be returned to a manhole or other suitable location upstream of the plant influent. By recycling sludge to the headworks, anoxic conditions and a high food to microorganism ratio will help control sludge bulking, provide some denitrification, and recover alkalinity.

4.11.5.2   Convert to Sequencing Batch Reactor Operation

Converting an aerated lagoon to an activated sludge system can be accomplished by operating the aerated lagoon as a sequencing batch reactor. A portion of the aerated lagoon is partitioned into two or more complete-mix SBR zones. SBRs operate in a sequence of fill, react, settle, and decant. In a single-train lagoon SBR, flow into the basin will continue through all four cycles. Where parallel systems exist, the SBR can be operated as a typical SBR system; however, the construction costs will be greater. As noted previously, Rich (1999) refers to this operation as a continuous-feed, intermittent discharge process, but it is the same as the commercial SBR system marketed by Austgen Biojet. In SBR mode, intermittent aeration is used, and a decanting process is used to transport the settled wastewater to downstream facultative cells or disinfection before discharge. Decant is accomplished with pumps, surface weirs, or floating decant devices. A portion of the MLSS must be wasted during the react (mixing and aeration) phase to keep the process in balance. Rich (1999) suggested adding a recycle pump station and returning the mixed liquor to the influent sewer to provide an anoxic environment for control of sludge bulking.

4.11.5.3   Install Biomass Carrier Elements

The addition of baffles and suspended fabrics for attached growth to accumulate and reduce pollutants has been suggested for many years (Polprasert and Agarwalla,

1995; Reynolds et al., 1975). Commercial fabrics are relatively new for the removal of ammonia. The carriers are plastic ribbons or wheels that are installed in the aerated zone to provide surface area for the growth of microorganisms. Given adequate surface area, nitrifying microorganisms can grow and multiply on the plastic surfaces and achieve ammonia removal. The aerated cell does not have to be completely mixed, which is required in the recycle and SBR approaches. The increase in oxygen demand exerted by the attached growth microorganisms must be provided. Solids that drop from the biomass carriers settle or pass to following lagoon cells. Sludge buildup will increase, but the transfer of solids to subsequent facultative lagoons will anaerobicly reduce the volume, and the need for sludge removal will be affected very little.

4.11.5.4 Commercial Lagoon Nitrification Systems

Lagoon nitrification systems offered commercially include:

- ATLAS-IS™ — Internal clarifier system by Environmental Dynamics, Inc.
- CLEAR™ process — SBR variant by Environmental Dynamics, Inc.
- Ashbrook SBR — SBR system by Ashbrook Corporation
- AquaMat® process — Plastic biomass carrier ribbons by Nelson Environmental, Inc.
- MBBR™ process — Plastic biomass carrier elements by Kaldnes North America, Inc.
- Zenon membrane process

4.11.5.4.1 ATLAS-IS™

Environmental Dynamics, Inc. (EDI; Columbia, MO) offers their Advanced Technology Lagoon Aeration System with Internal Separator (ATLAS-IS™), which is designed to provide a high level of treatment with minimal operation and maintenance requirements. The process consists of a fine-bubble, floating, lateral aeration system that contains a series of internal clarifiers or settlers. The settlers are constructed of a plastic material and may contain lamella baffles. The units are installed within a complete-mix zone of the aerated pond system. Mixed liquor enters the settling chamber through the bottom. A slight concentration of the MLSS takes place in the settler as the mixed liquor rises and spills over a weir into an effluent pipe. No return-activated sludge (RAS) or waste-activated sludge (WAS) is required. Over time the MLSS will build up to a level adequate to grow nitrifying microorganisms. Some solids are carried downstream so no separate sludge wasting is necessary. The ATLAS-IS™ system has been tested at Ashland, Missouri, and has been successful in building up MLSS and in achieving nitrification. A layout of the system is shown in Figure 4.32.

4.11.5.4.2 CLEAR™ Process

Environmental Dynamics, Inc., also offers an SBR variant known as the Cyclical Lagoon Extended Aeration Reactor (CLEAR™). A completely mixed aerated

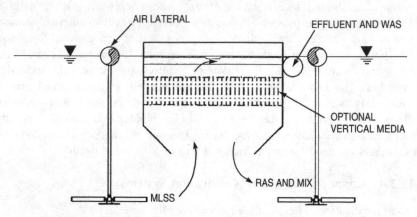

FIGURE 4.32 EDI's ATLAS-IS™ internal lagoon settler. (From Burnett, C.H. et al., Ammonia Removal in Large Aerated Lagoons, paper presented at WEFTEC Annual Meeting, Water Environment Federation, New Orleans, LA, October, 2004.)

lagoon cell is partitioned into three zones using floating baffle curtains. Influent is fed to each of the three zones sequentially. Aeration is applied to the zone receiving influent wastewater and, for part of this cycle, one of the other two zones. While the inflowing zone is aerated, the other two zones cycle between settling and decanting. WAS is removed using airlift pumps, either to downstream facultative ponds for storage or to further processing and disposal. A control system is provided to operate the motorized wastewater influent valves and decanters. Currently, no full-scale installations of the CLEAR™ process are in operation, but it appears the process should function as claimed. A depiction of the process is shown in Figure 4.33.

4.11.5.4.3 Ashbrook SBR

The Ashbrook Corporation (Houston, TX) SBR system consists of decanters, motorized valves, and a control system. A facility has been installed in a lagoon system in Quincy, Washington, where the aerated lagoon has been portioned into sections and air is provided for complete mixing in two or more SBR cells. Operation is similar to a conventional SBR process, and the system in Quincy has been working well. The facility is shown in Figure 4.34, and effluent data are provided in Table 4.24.

4.11.5.4.4 AquaMat® Process

The biomass carrier system AquaMat® is marketed by Nelson Environmental, Inc. (Winnipeg, Manitoba). Plastic ribbons slightly more dense than water are connected to a plastic float; these ribbons extend into the waste stream 3 feet or more and provide additional surface area for bacteria to grow. When used with lagoon systems, the application is referred to as the Advanced Microbial Treatment System (AMTS). Year-round nitrification has been achieved in an aerated

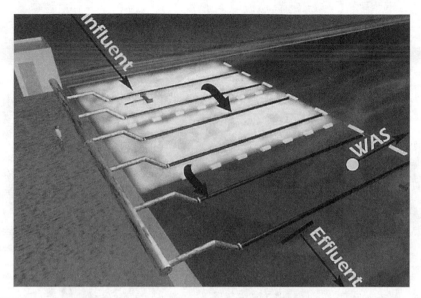

FIGURE 4.33 CLEAR™ process. (From Burnett, C.H. et al., Ammonia Removal in Large Aerated Lagoons, paper presented at WEFTEC Annual Meeting, Water Environment Federation, New Orleans, LA, October, 2004.)

FIGURE 4.34 Ashbrook Lagoon sequencing batch reactor (SBR), Quincy, Washington. (From Burnett, C.H. et al., Ammonia Removal in Large Aerated Lagoons, paper presented at WEFTEC Annual Meeting, Water Environment Federation, New Orleans, LA, October, 2004.)

TABLE 4.24
Ashbrook Lagoon SBR
(Quincy, Washington)

	2000/2003 Average
Flow (mgd)	0.78
Influent:	
BOD (mg/L)	145
TSS (mg/L)	159
NH_3 (mg/L)	19
Effluent:	
BOD (mg/L)	14
TSS (mg/L)	6
NH_3 (mg/L)	1.7

Source: Burnett, C.H. et al., Ammonia Removal in Large Aerated Lagoons, paper presented at WEFTEC Annual Meeting, Water Environment Federation, New Orleans, LA, October, 2004.

lagoon in Laurelville, Ohio, and other locations in Canada. Performance data are shown in Table 4.25, and an example of the AquaMat® process is shown in Figure 4.35.

4.11.5.4.5 MBBR™ Process

The Moving Bed™ Biofilm Reactor (MBBR™) is marketed by Kaldnes North America, Inc. (Providence, RI). The process is similar to the AquaMat® process except that thousands of small polyethylene wheels, as shown in Figure 4.36, are suspended in the lagoon. A sufficient number of these wheels provides adequate surface area for the growth of nitrifiers. An aerated lagoon in Johnstown, Colorado, has been successfully upgraded using the MBBR™ process.

4.11.5.5 Other Process Notes

Partial denitrification has been achieved by most of the systems described above, although the nitrogen removal pathways are not well understood. Several other commercial SBR systems and biomass carrier systems are available; however, experience with these in lagoons appears to be limited. The principle is the same and it appears reasonable to expect these proprietary systems would work. The manufacturers of the products mentioned here have unique experience working

TABLE 4.25
Nelson AquaMat Biomass
Carrier (Larchmont, Georgia)

Parameter	Reported Average Effluent Quality (mg/L)
BOD	6
TSS	10
NH_3	0.1

Source: Burnett, C.H. et al., Ammonia Removal in Large Aerated Lagoons, paper presented at WEFTEC Annual Meeting, Water Environment Federation, New Orleans, LA, October, 2004.

FIGURE 4.35 AquaMat® process, Nelson Environmental, Inc., Winnipeg, Manitoba. (From Burnett, C.H. et al., Ammonia Removal in Large Aerated Lagoons, paper presented at WEFTEC Annual Meeting, Water Environment Federation, New Orleans, LA, October, 2004.)

FIGURE 4.36 MBBR™ process (Moving Bed™ Biofilm Reactor; Kaldnes North America, Inc.). (From Burnett, C.H. et al., Ammonia Removal in Large Aerated Lagoons, paper presented at WEFTEC Annual Meeting, Water Environment Federation, New Orleans, LA, October, 2004.)

with lagoon systems. In addition to the techniques discussed here, the companies have experience with floating baffle curtains for partitioning, installation of equipment without removing existing lagoons from service, cost-effective and efficient aeration systems for large surface area installations, and optimizing complete-mix and partial-mix aeration regimes.

4.11.5.6 Ultrafiltration Membrane Filtration

As noted earlier, John Thompson Engineering of New South Wales, Australia, designed and installed a 0.264-mgd (1000 m^3/d) zenon membrane facility to polish an effluent from a lagoon facility in 2001. Operating data are not available, but all indications are that the facility is functioning well. Very little operating experience is available with membranes in lagoons, but it is an option that should be evaluated because of the potential for production of an excellent effluent in a confined space.

4.11.5.7 BIOLAC® Process (Parkson Corporation)

The BIOLAC® process is an activated sludge system contained within an earthen dike. The process is described in detail in Section 4.10.2 of this chapter.

4.12 MODIFIED HIGH-PERFORMANCE AERATED POND SYSTEMS FOR NITRIFICATION AND DENITRIFICATION

Continuous-feed, intermittent-discharge (CFID) basins have been in use for many years in Australia and the United States (Rich, 1999). In-basin sedimentation is used rather than a secondary clarifier, and sludge recycle is used. Discharge is intermittent, but flow into the basin is continuous which differs from sequencing batch reactors. These systems generally have been successful but have had problems with short-circuiting and sludge bulking; however, modifications to the system can be made to overcome these difficulties (Rich, 1999). Rich (1996, 1999) has proposed a modified CFID basin for use in aerated lagoon systems for nitrification and denitrification. The systems are designed to use in-basin sedimentation to uncouple the solids retention time from the hydraulic retention time, and the influent flow is continuous. He also has proposed a nitrification system consisting of a combination of a modified CFID with a sludge basin. The modified CFID basin serves as the reactor basin, and the sludge basin stabilizes and stores the sludge. This arrangement provides flexibility in operation. To learn more about the operation of CFID systems, consult Rich (1999).

4.13 NITROGEN REMOVAL IN PONDS COUPLED WITH WETLANDS AND GRAVEL BED NITRIFICATION FILTERS

See Section 7.9 of this book for design detail. This system was developed as a retrofit for free water surface (FWS) and subsurface flow (SSF) existing wetland systems having trouble meeting ammonia effluent standards. NFB units can be located at the front or near the end of the wetland where wetland effluent is pumped to the top of the NFB and distributed evenly over the surface. Introducing the wetland effluent to the NFB at the head of the system has the advantage of mixing the influent wastewater with the highly nitrified NFB effluent which results in denitrification and removal of nitrogen from the system. In addition, the BOD will be reduced, and some of the alkalinity lost during nitrification will be recovered. By locating the NFB at the end of the wetland, nitrification will occur but denitrification will be limited and the nitrates will pass out of the system. Locating the NFB at the end of the wetland requires less pumping capacity, but the advantages of denitrification obtained at the front end could easily offset the pumping advantage. When retrofitting an existing lagoon-wetland for nitrification and nitrogen removal, the NFB appears to have economic advantages and simplicity of construction and operation. It also is likely that the NFB would be a viable alternative for nitrogen removal in the initial design of a lagoon–wetland system.

4.14 CONTROL OF ALGAE AND
DESIGN OF SETTLING BASINS

Control of algae in wastewater stabilization pond effluents has been a major concern throughout the history of the use of the systems. The use of maturation ponds and polishing ponds following all types of treatment processes has resulted in a need to control algae in the effluent. State design standards have contributed to the problem by requiring long detention times in the final cell in a lagoon system. It has been established that few, if any, of the solids in lagoon effluents are fecal matter or material entering the lagoon system. This fact led to much discussion about the necessity to remove algae from lagoon effluents; however, it was pointed out that the algae die, settle out, and decay, thus inducing an oxygen demand on the receiving stream. This concern about decay and oxygen consumption resulted in investigations into the most effective methods to remove algae and ways to design systems to minimize the growth of algae in settling basins. Methods for removing algae are discussed in Chapter 5, and the design of settling basins is discussed in the following text.

For 18 months, Toms et al. (1975) studied algae growth rates in full-scale polishing lagoons receiving activated sludge effluents. It was concluded that growth rates for the dominant species always were less than 0.48 d^{-1}, and, if the hydraulic retention time (HRT) was less than 2 d, algae growth would not become a problem. At a HRT of less than 2.5 d, the effluent TSS decreased, and beyond this HRT the TSS increased. Uhlmann (1971) reported no algae growth in hyper-fertilized ponds when the detention times were less than 2.5 d. Toms et al. (1975) evaluated one-cell and four-cell polishing lagoons and found that, for HRTs beyond 2.5 d, the TSS increased in both lagoons, but significant growth did not occur until after 4 to 5 d in the four-cell lagoon.

Light penetration is reduced as the depth of a lagoon is increased; however, because of the trapezoidal shape of most lagoon cells, little advantage is achieved by increasing the depth beyond 3 or 4 m. Thermal stratification occurs in lagoons without mechanical mixing and provides an excellent environment for algae growth. Disturbing stratification will reduce algae growth. Rich (1999) recommends some degree of aeration for lagoon cells to control algae. The intensity of aeration also has an influence on algae growth by suspending more and more solids as the intensity increases. This results in a reduction in light transmission and consequently fewer algae.

4.15 HYDRAULIC CONTROL OF PONDS

In the past, the majority of ponds were designed to receive influent wastewater through a single pipe, usually located toward the center of the first cell in the system. Hydraulic and performance studies have shown that the center discharge point is not the most efficient method of introducing wastewater to a pond (Finney and Middlebrooks, 1980; Mangelson an dWatters, 1972). Multiple inlet arrangements are preferred even in small ponds (<0.5 ha [<1.2 ac]). The inlet points should be as far apart as possible, and the water should preferably be introduced by means of a long diffuser. The inlets and outlets should be placed so flow through the pond is uniform between successive inlets and outlets.

Single inlets can be used successfully if the inlet is located at the greatest distance possible from the outlet structure and is baffled, or the flow is otherwise directed to avoid currents and short-circuiting. Outlet structures should be designed for multiple-depth withdrawal, and all withdrawals should be a minimum of 0.3 m (1 ft) below the water surface to reduce the potential impact of algae and other surface detritus on effluent quality.

Analysis of the performance data from selected aerated and facultative ponds indicates that four cells in series are desirable to give the best BOD and fecal coliform removals for ponds designed as plug flow systems. Good performance can also be obtained with a smaller number of cells if baffles or dikes are used to optimize the hydraulic characteristics of the system.

Better treatment is obtained when the flow is guided more carefully through the pond. In addition to treatment efficiency, economics and esthetics play an important role in deciding whether or not baffling is desirable. In general, the more baffling is used, the better are the flow control and treatment efficiency. The lateral spacing and length of the baffle should be specified so the cross-sectional area of flow is as close to a constant as possible.

Wind generates a circulatory flow in bodies of water. To minimize short-circuiting due to wind, the pond inlet–outlet axis should be aligned perpendicular to the prevailing wind direction if possible. If this is not possible, baffling can be used to control wind-induced circulation to some extent. In a constant-depth pond, the surface current will be in the direction of the wind, and the return flow will be in the upwind direction along the bottom.

Ponds that are stratified because of temperature differences between the inflow and the pond contents tend to behave differently in winter and summer. In summer the inflow is generally colder than the pond, so it sinks to the pond bottom and flows toward the outlet. In the winter, the reverse is generally true, and the inflow rises to the surface and flows toward the outlet. A likely consequence is that the effective treatment volume of the pond is reduced to that of the stratified inflow layer (density current). The result can be a drastic decrease in detention time and an unacceptable level of treatment.

4.16 REMOVAL OF PHOSPHORUS

In general, removal of phosphorus is not often required for wastewaters that receive lagoon treatment, although a number of exceptions can be found for systems in the northcentral United States and Canada. If such a requirement is imposed, the experiences described in the following text will provide some guidance.

4.16.1 Batch Chemical Treatment

In order to meet a phosphorus requirement of 1 mg/L for discharge to the Great Lakes, an approach using in-pond chemical treatment in controlled-discharge ponds was developed in Canada. Alum, ferric chloride, and lime were all tested by using a motorboat for distribution and mixing of the chemical. A typical alum dosage might be 150 mg/L, which should produce an effluent from the controlled-discharge pond that contains less than 1 mg/L of phosphorus and less than 20 mg/L BOD and SS.

The sludge buildup from the additional chemicals is insignificant and would allow years of operation before requiring cleaning. The costs for this method were very reasonable and much less than those for conventional phosphorus removal methods. This method has been applied successfully in several midwestern states.

4.16.2 Continuous-Overflow Chemical Treatment

Studies of in-pond precipitation of phosphorus, BOD, and SS were conducted over a 2-year period in Ontario, Canada. The primary objective of the chemical dosing process was to test removal of phosphorus with ferric chloride, alum, and lime. Ferric chloride doses of 20 mg/L and alum doses of 225 mg/L, when added continuously to the pond influent, effectively maintained pond effluent phosphorus levels below 1 mg/L over a 2-year period. Hydrated lime at dosages up to 400 mg/L was not effective in consistently reducing phosphorus below 1 mg/L (1 to 3 mg/L was achieved) and produced no BOD reduction while slightly increasing the SS concentration. Ferric chloride reduced effluent BOD from 17 to 11 mg/L and SS from 28 to 21 mg/L; alum produced no BOD reduction and a slight SS reduction (from 43 to 28–34 mg/L). Consequently, direct chemical addition appears to be effective only for phosphorus removal. A six-cell pond system located in Waldorf, Maryland, was modified to operate as two three-cell units in parallel. One system was used as a control, and alum was added to the other for phosphorus removal. Each system contained an aerated first cell. Alum addition to the third cell of the system proved to be more efficient in removing total phosphorus, BOD, and SS than alum addition to the first cell. Total phosphorus reduction averaged 81% when alum was added to the inlet to the third cell and 60% when alum was added to the inlet of the first cell. Total phosphorus removal in the control ponds averaged 37%. When alum was added to the third cell, the effluent total phosphorus concentration averaged 2.5 mg/L, with the control units averaging 8.3 mg/L. Improvements in BOD and SS removal by alum addition were more difficult to detect, and at times increases in effluent concentrations were observed.

REFERENCES

Agunwamba, J.C., Egbuniwe, N., and Ademiluyi, J.O. (1992). Prediction of the dispersion number in waste stabilization, *Water Res.*, 26, 85.

Benefield, L.D. and Randall C.W. (1980). *Biological Process Design for Wastewater Treatment*, Prentice-Hall, Englewood Cliffs, NJ.

Bhagat, S.K. and Proctor, D.E. (1969). Treatment of dairy manure by lagooning, *J. Water Pollut. Control Fed.*, 41, 5.

Boulier, G.A. and Atchinson, T.J. (1975). *Practical Design and Application of the Aerated-Facultative Lagoon Process*, Hinde Engineering Company, Highland Park, IL.

Bowen, S.P. (1977). *Performance Evaluation of Existing Lagoons, Peterborough, NH*, EPA-600/2-77-085, Municipal Environmental Laboratory, U.S. Environmental Protection Agency, Cincinnati, OH.

Bowman, R.H. (pers. comm., 2000). West Slope Unit Leader, Water Quality Division, Colorado Department of Public Health and Environment.

Buddhavarapu, L.R. and Hancock, S.J. (1989). Advanced Treatment for Lagoons Using Lemna Technology, paper presented at the 62nd Annual Water Pollution Control Federation Conference, San Francisco, CA, October.

Burnett, C.H., Featherston, B., and Middlebrooks, E.J. (2004). Ammonia Removal in Large Aerated Lagoons, paper presented at WEFTEC Annual Meeting, Water Environment Federation, New Orleans, LA, October, 2004.

Canter, L.W. and Englande, A.J. (1970). States' design criteria for waste stabilization ponds, *J. Water Pollut. Control Fed.*, 42(10), 1840–1847.

Chan, D.B. and Pearson, E.A. (1970). *Comprehensive Studies of Solid Waste Management: Hydrolysis Rate of Cellulose in Anaerobic Digesters*, SERL Report No. 70-3, University of California, Berkeley.

City of Hotchkiss (2000). *Wastewater Treatment Plant*, Hotchkiss, CO.

Cooper, R.C. (1968). Industrial waste oxidation ponds, *Southwest Water Works J.*, 5, 21.

Culley, D.D. and Epps, E.A. (1973). Use of duckweed for waste treatment and animal feed, *J. Water Pollut. Control Fed.*, 45, 337.

Downing, J.B., Bracco, E., Green, F.B., Ku, A.Y., Lundquist, T.J., Zubieta, I.X., and Oswald, W.J. (2002). Low cost reclamation using the Advanced Integrated Wastewater Pond Systems® technology and reverse osmosis, *Water Sci. Technol.*, 45(1), 117–125.

Earnest, C.M., Vizzini, E.A., Brown D.L., and Harris J.L. (1978). *Performance Evaluation of the Aerated Lagoon System at Windber, Pennsylvania*, EPA-600/2-78-023, Municipal Environmental Research Laboratory, U.S. Environmental Protection Agency, Cincinnati, OH.

Eckenfelder, W.W. (1961). *Biological Waste Treatment*, Pergamon Press, London.

Englande A.J., Jr. (1980). *Performance Evaluation of the Aerated Lagoon System at North Gulfport, Mississippi*, EPA-600/2-80-006, Municipal Environmental Research Laboratory, U.S. Environmental Protection Agency, Cincinnati, OH.

Everall, N.C. and Lees, D.R. (1997). The identification and significance of chemicals released from decomposing barley straw during reservoir algal control, *Water Res.*, 31, 614-620.

Fagan, J. (pers. comm., 2000). Consolidated Consulting Services, Delta, CO.

Finney, B.A. and Middlebrooks, E.J. (1980). Facultative waste stabilization pond design, *J. Water Pollut. Control Fed.*, 52(1), 134–147.

Fritz, J.J., Middleton, A.C., and Meredith, D.D. (1979). Dynamic process modeling of wastewater stabilization ponds, *J. Water Pollut. Control Fed.*, 51(11), 2724–2743.

Ghosh, S. and Klass, D.L. (1974). Conversion of urban refuse to substitute natural gas by the BIOGAS® process, in *Proceedings of the 4th Mineral Waste Utilization Symposium*, Chicago, IL, May 7-8.

Ghosh, S., Conrad, J.R., and Klass, D.L. (1974). Development of an Anaerobic Digestion-Based Refuse Disposal Reclamation System, paper presented at the 47th Annual Conference of the Water Pollution Control Federation, Denver, CO, October 6–11.

Gloyna, E.F. (1971). *Waste Stabilization Ponds*, Monograph Series No. 60, World Health Organization, Geneva, Switzerland.

Gloyna, E.F. (1976). Facultative waste stabilization pond design, in *Ponds as a Waste Treatment Alternative*, Gloyna, E.F., Malina, J.F., Jr., and Davis, E.M., Eds., Water Resources Symposium No. 9, University of Texas Press, Austin, TX.

Great Lakes–Upper Mississippi River Board of State Sanitary Engineers. (1990). *Recommended Standards for Sewage Works (Ten-States Standards)*, Health Education Services, Inc., Albany, NY.

Green, F.B., Lundquist, T.J., and Oswald, W.J. (1995a). Energetics of advanced integrated wastewater pond systems, *Water Sci. Technol.*, 31(12) 9–20.

Green, F.B., Bernstone, L.S., Lundquist, T.J., Muir, J., Tresan, R.B., and Oswald, W.J. (1995b). Methane fermentation, submerged gas collection, and the fate of carbon in advanced integrated wastewater pond systems, *Water Sci. Technol.*, 31(12) 55–65.

Green, F.B., Bernstone, L.S., Lundquist, T.J., and Oswald, W.J. (1996). Advanced integrated wastewater pond systems for nitrogen removal, *Water Sci. Technol.*, 33(7) 207–217.

Green, F.B., Liundquist, T.J., Quinn, N.W.T., Zarate, M.A., Zubieta, I.X., and Oswald, W.J. (2003). Selenium and nitrate removal from agricultural drainage using the AIWPS® technology, *Water Sci. Technol.*, 48(2), 299–305.

Gurnham C.F., Rose, B.A., and Fetherston, W.T. (1979). *Performance Evaluation of the Existing Three-Lagoon Wastewater Treatment Plant at Pawnee, Illinois*, EPA-600/2-79-043, Municipal Environmental Research Laboratory, U.S. Environmental Protection Agency, Cincinnati, OH.

Hill, D.O. and Shindala, A. (1977). *Performance Evaluation of Kilmichael Lagoon*, EPA-600/2-77-109, Municipal Environmental Research Laboratory, U.S. Environmental Protection Agency, Cincinnati, OH.

Hobson, P.N., Bousfield, S., and Summers, R. (1974). Anaerobic digestion of organic matter, *CRC Crit. Rev. Environ. Control*, 4, 131–191.

IACR–Centre for Aquatic Plant Management (1999). *Information Sheet 3: Control of Algae Using Straw*, Centre for Aquatic Plant Management, Reading, Berkshire, U.K.

King, D.L. (1978). The role of ponds in land treatment of wastewater, in *Proceedings of the International Symposium on Land Treatment of Wastewater*, Hanover, NH, 191.

Kotze, J.P., Thiel, P.G., Toerien, D.F., Attingh, W.H.J., and Siebert, M.L. (1968). A biological and chemical study of several anaerobic digesters, *Water Res.*, 2(3), 195–213.

Larson, T.B. (1974). A Dimensionless Design Equation for Sewage Lagoons, Ph.D. dissertation, University of New Mexico, Albuquerque.

Lemna Technologies, Inc. (1999a). News release, 2445 Park Avenue, Minneapolis, MN.

Lemna Technologies, Inc. (1999b). Brochures, 2445 Park Avenue, Minneapolis, MN.

Malina, J.F., Jr., and Rios, R.A. (1976). Anaerobic ponds, in *Ponds as a Wastewater Alternative*, Gloyna, E.F., Malina, J.F., Jr., and Davis, E.M., Eds., Water Resources Symposium No. 9, University of Texas Press, Austin, TX.

Mancini, J.L., and Barnhart, E.L. (1976). Industrial waste treatment in aerated lagoon, in *Ponds as a Wastewater Alternative*, Gloyna, E.F., Malina, J.F., Jr., and Davis, E.M., Eds., Water Resources Symposium No. 9, University of Texas Press, Austin, TX.

Mangelson, K.A. and Watters, G.Z. (1972). Treatment efficiency of waste stabilization ponds, *J. Sanit. Eng. Div. ASCE*, 98(SA2), 407–425.

Mara, D.D. (1975). Discussion, *Water Res.*, 9, 595.

Mara, D.D. (1976). *Sewage Treatment in Hot Climates*, John Wiley & Sons, New York.

Marais, G.V.R. (1970). Dynamic behavior of oxidation ponds, in *Proceedings of Second International Symposium for Waste Treatment Lagoons*, Kansas City, MO, June 23–25.

Marais, G.V.R. and Shaw, V.A. (1961). A rational theory for the design of sewage stabilization ponds in Central and South Africa, *Trans. South African Inst. Civil Eng.*, 3, 205.

McAnaney, D.W. and Poole, W.D. (1997). *Cold Weather Nitrification in Pond-Based Systems*, Lemna USA, Inc., 2445 Park Avenue, Minneapolis, MN.

McGarry, M.C. and Pescod, M.B. (1970). Stabilization pond design criteria for tropical Asia, in *Proceedings of Second International Symposium for Waste Treatment Lagoons*, Kansas City, MO, June 23–25.

McKinney, R.E. (1977). *Performance Evaluation of an Existing Lagoon System at Eudora, Kansas*, EPA-600/2-77-167, Municipal Environmental Research Laboratory, U.S. Environmental Protection Agency, Cincinnati, OH.

Meron, A. (1970). Stabilization Pond Systems for Water Quality Control, Ph.D. dissertation, University of California, Berkeley.

Metcalf & Eddy (1991). *Wastewater Engineering Treatment Disposal Reuse*, 3rd ed., McGraw-Hill, New York.

Metcalf & Eddy (2003). *Wastewater Engineering Treatment Disposal Reuse*, 4th ed., McGraw-Hill, New York.

Middlebrooks, E.J. (1985). Nitrogen removal model developed for U.S. Environmental Protection Agency, Washington, D.C.

Middlebrooks, E.J. (1987). Design equations for BOD removal in facultative ponds, *Water Sci. Technol.*, 19, 12.

Middlebrooks, E.J. (2000). Joemiddle@aol.com; (fax) 303-664-5651.

Middlebrooks, E.J. and Pano, A. (1983). Nitrogen removal in aerated lagoons, *Water Res.*, 17(10), 1369–1378.

Middlebrooks, E.J. and Procella, D.B. (1971). Rational multivariate algal growth kinetics, *J. Sanit. Eng. Div. Am. Soc. Civ. Eng.*, SA1, 135–140.

Middlebrooks, E.J., Middlebrooks, C.H., Reynolds, J.H., Watters, G.Z., Reed, S.C., and George, D.B. (1982). *Wastewater Stabilization Lagoon Design, Performance, and Upgrading*, Macmillan, New York.

Monod J. (1950). La technique de culture continue: theorie et application, *Ann. Inst. Pasteur*, 79, 390.

Neel, J.K., McDermott, J.H., and Monday, C.A. (1961). Experimental lagooning of raw sewage, *J. Water Pollut. Control Fed.*, 33(6), 603–641.

Nurdogan, Y. and Oswald, W.J. (1995). Enhanced nutrient removal in high-rate ponds, *Water Sci. Technol.*, 31(12), 33–43.

Oleszkiewicz, J.A. (1986). Nitrogen transformations in an aerated lagoon treating piggery wastes, *Agric. Wastes AGWADL*, 16(3), 171–181.

Oswald, W.J. (1968). Advances in anaerobic pond systems design, in *Advances in Water Quality Improvement*, Gloyna, E.F. and Eckenfelder, W.W., Jr., Eds., University of Texas Press, Austin, TX.

Oswald, W.J. (1990a). Advanced integrated wastewater pond systems: supplying water and saving the environment for six billion people, in *Proceedings of the ASCE Convention, Environmental Engineering Division*, San Francisco, CA, November 5–8.

Oswald, W.J. (1990b). Sistemas Avanzados De Lagunas Integradas Para Tratamiento De Aguas Servidas (SALI), in *Proceedings of the ASCE Convention, Environmental Engineering Division*, San Francisco, CA, November 5–8.

Oswald, W. J. (1995). Ponds in the twenty-first century, *Water Sci. Technol.*, 31(12), 1–8.

Oswald, W.J. (1996). *A Syllabus on Advanced Integrated Pond Systems®*, University of California, Berkeley.

Oswald, W. J. (2003). My sixty years in applied algology, *J. Appl. Phycol.*, 15, 99–106.

Oswald, W.J., Golueke, C.G., and Tyler, R.W. (1967). Integrated pond systems for subdivisions, *J. Water Pollut. Control Fed.*, 39(8), 1289.

Oswald, W.J., Meron, A., and Zabat, M.D. (1970). Designing waste ponds to meet water quality criteria, in *Proceedings of Second International Symposium for Waste Treatment Lagoons*, Kansas City, MO, June 23–25.

Oswald, W.J., Green, F.B., and Lundquist, T.J. (1994). Performance of methane fermentation pits in advanced integrated wastewater pond systems, *Water Sci. Technol.*, 30(12), 287–295.

Pano, A. and Middlebrooks, E.J. (1982). Ammonia nitrogen removal in facultative wastewater stabilization ponds, *J. WPCF*, 54(4), 2148.

Parker, C.D. (1970). Experiences with anaerobic lagoons in Australia, in *Proceedings of the Second International Symposium for Waste Treatment Lagoons*, Kansas City, MO, June 23–25.

Parker, C.D., Jones, H.L., and Greene, N.C.(1959). Performance of large sewage lagoons at Melbourne, Australia, *Sewage Indust. Wastes*, 31(2), 133.

Parkson Corp. (pers. comm., 2004). Personal communication with Chuck Morgan and case studies from website (www.parkson.com), Ft. Lauderdale, FL.

Pearson, H.W. and Green, F.B., Eds. (1995). Waste stabilisation ponds and the reuse of pond effluents, *Water Sci. Technol.*, 31, 12.

Polkowski, L.B. (1979). *Performance Evaluation of Existing Aerated Lagoon System at Consolidated Koshkonong Sanitary District, Edgerton, Wisconsin*, EPA-600/2-79-182, Municipal Environmental Research Laboratory, U.S. Environmental Protection Agency, Cincinnati, OH.

Polprasert, C. and Agarwalla, B.K. (1995). Significance of biofilm activity in facultative pond design and performance, *Water Sci. Technol.*, 31(12), 119–128.

Polprasert, C. and Bhattarai, K.K. (1985). Dispersion model for waste stabilization ponds, *J. Environ. Eng. Div. ASCE*, 111(EE1), 45–59.

Pump Systems, Inc. (2004). Dickinson, ND.

Racault, Y., Boutin, C., and Seguin, A. (1995). Waste stabilization ponds in France: a report on fifteen years experience, *Water Sci. Technol.*, 31(12), 91–101.

Ramani, R. (1976). Design criteria for polishing ponds, in *Advances in Water Quality Improvement*, Gloyna, E.F. and Eckenfelder, W.W., Jr., Eds., University of Texas Press, Austin, TX.

Reed, S.C. (1984). *Nitrogen Removal in Wastewater Ponds*, CRREL Report 84-13, Cold Regions Engineering and Research Laboratory (CRREL), Hanover, NH.

Reed, S.C. (1985). Nitrogen removal in wastewater stabilization ponds, *J. WPCF*, 57(1), 39–45.

Reed, S.C., Crites, R.W., and Middlebrooks, E.J. (1995). *Natural Systems for Waste Management and Treatment*, 2nd ed., McGraw-Hill, New York.

Reid G.W. and Streebin, L. (1979). *Performance Evaluation of Existing Aerated Lagoon System at Bixby, Oklahoma*, EPA-600/2-79-014, Municipal Environmental Research Laboratory, U.S. Environmental Protection Agency, Cincinnati, OH.

Reid, L.D., Jr. (1970). *Design and Operation for Aerated Lagoons in the Arctic and Subarctic*, Report 120, U.S. Public Health Service, Arctic Health Research Center, Fairbanks, AK.

Reynolds, J.H. and Middlebrooks, E.J. (1990). Aerated lagoon design equation and performance evaluation (poster presentation), *Water Pollut. Res. Control*, 23, 10–12.

Reynolds, T.D. and Richards, P.A. (1996). *Unit Operations and Processes in Environmental Engineering*, 2nd ed., PWS Publishing, New York.

Reynolds, J.H., Nielson, S.B., and Middlebrooks, E.J. (1975). Biomass distribution and kinetics of baffled lagoons, *J. Environ. Eng. Div. ASCE*, 101(EE6), 1005–1024.

Reynolds, J.H., Swiss, R.E., Macko, C.A., and Middlebrooks, E.J. (1977). *Performance Evaluation of an Existing Seven Cell Lagoon System*, EPA-600/2-77-086, Municipal Environmental Research Laboratory, U.S. Environmental Protection Agency, Cincinnati, OH.

Rich, L.G. (1996). Nitrification systems for small and intermediate size communities, *South Carolina Water Pollut. Control J.*, 26(3), 14–15.

Rich, L.G. (1999). *High-Performance Aerated Lagoon Systems*, American Academy of Environmental Engineers, Annapolis, MD.

Rich, L.G. (pers. comm., 2000). Department of Environmental Engineering and Science, Clemson University, Clemson, SC (lrich@clemson.edu).

Russel, J.S., Reynolds, J.H., and Middlebrooks, E.J. (1980). *Wastewater Stabilization Lagoon–Intermittent Sand Filter Systems*, EPA 600/2-80-032, Municipal Environmental Research Laboratory, U.S. Environmental Protection Agency, Cincinnati, OH.

Shephard-Wesnitzer, Inc. (1998). Preliminary Engineering Report for Dove Creek (Colorado) Constructed Wetlands, SWI, Sedona, AZ.

Shilton, A. (1995). Ammonia volatilization from a piggery pond, in *Proceedings of Symposium on Waste Stabilization Ponds: Technology and Applications*, Joao Pessoa, Paraiba, Brazil.

Soares, J., Silva, S.A., De-Oliveira, R., Araujo, A.L.C., Mara, D.D., and Pearson, H.W. (1995). Ammonia Removal in a Pilot-Scale WSP Complex in Northeast Brazil, in *Proceedings of Symposium on Waste Stabilization Ponds: Technology and Applications*, Joao Pessoa, Paraiba, Brazil.Stratton, F.E. (1968). Ammonia nitrogen losses from streams, *J. Sanit. Eng. Div. ASCE*, 94, 1085–1092.

Stratton, F.E. (1969). Nitrogen losses from alkaline water impoundments, *J. Sanit. Eng. Div. ASCE*, 95, 223–231.

Thirumurthi, D. (1974). Design criteria for waste stabilization ponds, *J. Water Pollut. Control Fed.*, 46, 2094–2106.

Toms, I.P., Owens, M., Hall, J.A., and Mindenhall, M.J. (1975). Observations on the performance of polishing ponds at a large regional works, *Water Pollut. Control*, 74, 383–401.

Uhlmann, D. (1971). Influence of dilution, sinking, and grazing rates on phytoplankton populations of hyper-fertilized ponds and ecosystems, *Mitt. Internat. Verein Limnol.*, 19, 100–124.

USEPA. (1975). *Technology Transfer Process Design Manual for Nitrogen Control*, Center for Environmental Research Information, U.S. Environmental Protection Agency, Cincinnati, OH.

USEPA. (1983). *Design Manual: Municipal Wastewater Stabilization Ponds*, EPA 625/1-83-015, Center for Environmental Research Information, U.S. Environmental Protection Agency, Cincinnati, OH.

USEPA. (1985). *Wastewater Stabilization Ponds: Nitrogen Removal*, U.S. Environmental Protection Agency, Washington, D.C.

USEPA. (1990). *Assessment of the BIOLAC Technology*, EPA 430/09-90-013, Office of Water (WH-595), U.S. Environmental Protection Agency, Washington, D.C.

Walter, C.M. and Bugbee, S.L. (1974) Progress Report, Blue Springs Lagoon Study, Blue Springs, Missouri, in *Upgrading Wastewater Stabilization Ponds To Meet New Discharge Standards*, Middlebrooks, E.J., Ed., Utah State University Press, Logan.

WEF/ASCE. (1991). *Design of Municipal Wastewater Treatment Plants*, Vols. 1 and 2, Water Environment Federation and American Society of Civil Engineers Washington, D.C.

Wehner, J.F. and Wilhelm, R.H. (1956). Boundary conditions of flow reactor, *Chem. Eng. Sci.*, 6, 89–93.

WHO. (1987). *Wastewater Stabilization Ponds: Principles of Planning and Practice*, WHO Tech. Publ. 10, Regional Office for the Eastern Mediterranean, World Health Organization, Alexandria.

Wolverton, B.C., and McDonald, R.C. (1979). Upgrading facultative wastewater lagoons with vascular aquatic plants, *J. Water Pollut. Control Fed.*, 51(2), 305.

Wrigley, J.J. and Toerien, D.F. (1990). Limnological aspects of small sewage ponds, *Water Res.*, 24(1), 83–90.

Zirschky, J. and Reed, S.C. (1988). The use of duckweed for wastewater treatment, *J. Water Pollut. Control Fed.*, 60, 1253.

5 Pond Modifications for Polishing Effluents

The two principal ways to upgrade lagoon effluents are solids removal methods and modifications to the lagoon process. The selection of the best method to achieve a desired effluent quality depends on the design conditions and effluent limits imposed on the facility. The advantages and limitations of the various methods are discussed in Section 5.1 and Section 5.2.

5.1 SOLIDS REMOVAL METHODS

5.1.1 INTRODUCTION

The occasional high concentration of total suspended solids (TSS), which can exceed 100 mg/L, in the effluent is the major disadvantage of pond systems. The solids are primarily composed of algae and other pond detritus, not wastewater solids. These high concentrations are usually limited to 2 to 4 months during the year. Solids removal methods that are discussed in this chapter include intermittent sand filters, recirculating sand filters, rock filters, coagulation–flocculation, and dissolved-air flotation. The rock filter is not a true filter but is included because of its association with filters when discussing solids removal from lagoon effluents. Further details for all methods can be found in the references at the end of the chapter. Although slightly dated, an excellent introduction to the design and performance of intermittent sand filters and rock filters is presented in the U.S. Environmental Protection Agency's *Design Manual: Municipal Wastewater Stabilization Ponds* (USEPA, 1983). Information on both processes can also be found in a document published by the Water Environment Federation (2001). A literature search on the application of recirculating sand filters to the removal of TSS from lagoon effluents was unsuccessful, but several references to their application in treating septic tank effluents were found. This lack of information may be attributable to concern about the accumulation of algae in the media. Nolte & Associates (1992) conducted a review of the literature covering recirculating sand filters and intermittent sand filters.

5.1.2 INTERMITTENT SAND FILTRATION

Intermittent sand filters have a long and successful history of treating wastewaters (Furman et al., 1955; Grantham et al., 1949; Massachusetts Board of Health, 1912). Table 5.1 presents a summary of the design characteristics and performance of several systems employed in Massachusetts around 1900. These systems

TABLE 5.1
Design and Performance of Early Massachusetts Intermittent Sand Filters

Location	Year Started	Loading Rate (gal/d/ac)	Filter Depth (in.)	Sand Size (mm)	Ammonia Removal		BOD$_5$ Removal	
					Influent (mg/L)	Effluent (mg/L)	Influent (mg/L)	Effluent (mg/L)
Andover	1902	35,000	48–60	0.15–0.2	–	–	–	–
Brockton	–	–	–	–	40.7	1.5	314	6.2
Concord	1899	83,000	–	–	–	2.7	–	–
Farmington	–	–	70	0.06–0.12	27.3	2.7	–	–
Gardner	1891	122,000	60	0.12–0.18	21.2	7.5	139	9.5
Leicester	–	–	–	–	–	–	321	13.1
Natick	–	–	–	–	12.4	2.3	–	–
Spencer	1897	61,000	48	0.18–0.34	16	2.1	116	6.9

Source: Data from the Massachusetts Board of Health (1912) and Mancl and Peeples (1991).

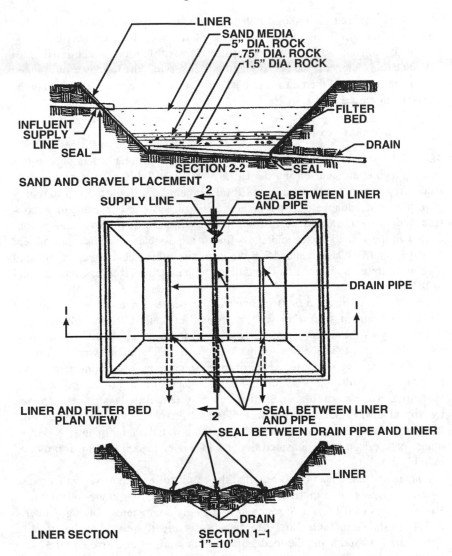

FIGURE 5.1 Cross-sectional and plan views of typical intermittent sand filter. (From Middlebrooks, E.J. et al., *Wastewater Stabilization Lagoon Design, Performance, and Upgrading*, Macmillan, New York, 1982. With permission.)

treated raw or primary effluent wastewater and produced an excellent effluent. A typical intermittent sand filter is shown in Figure 5.1. Intermittent sand filtration is capable of polishing pond effluents at relatively low cost and is similar to the practice of slow sand filtration in potable water treatment. Intermittent sand filtration of pond effluents is the application of pond effluent on a periodic or intermittent basis to a sand filter bed. As the wastewater passes through the bed, suspended solids and other organic matter are removed through a combination of physical straining and biological degradation processes. The particulate matter collects in the top 2 to 3 in. (5 to 8 cm) of the filter bed; this accumulation eventually clogs the surface and prevents effective infiltration of additional effluent. When this happens, the bed is taken out of service, the top layer of clogged sand is removed, and the unit is put back into service. The removed sand can be washed and reused or discarded.

5.1.2.1 Summary of Performance

Table 5.2 and Table 5.3 summarize the performance of intermittent sand filters treating lagoon effluents during the 1970s and 1980s. Table 5.2 is a summary of studies reported in the literature and EPA documents, and Table 5.3 is a summary of results from field investigations at three full-scale systems consisting of lagoons followed by intermittent sand filters. These are the most extensive studies conducted in the United States and show that it is possible to produce an effluent with TSS and BOD_5 less than 15 mg/L from anaerobic, facultative, and aerated lagoons followed by intermittent sand filters with effective sizes less than or equal to 0.3 mm.

Rich and Wahlberg (1990) evaluated the performance of five facultative lagoon–intermittent sand filter systems located in South Carolina and Georgia. A summary of the design characteristics and performance of these systems is shown in Table 5.4. The systems provided superior performance when compared with ten aerated lagoon systems. Six of the systems consisted of one aerated cell followed by a polishing pond; three were designed as dual-power, multicellular systems, and one was a single-cell, dual-power system. Based on the data reported by Niku et al. (1981), the performance of the facultative lagoon–intermittent sand filter systems compared most favorably with activated sludge plants. A performance comparison of the activated sludge and aerated lagoon systems is provided in Table 5.5.

Truax and Shindala (1994) reported the results of an extensive evaluation of facultative lagoon–intermittent sand filter systems using four grades of sand with effective sizes of 0.18 to 0.70 mm and uniformity coefficients ranging from 1.4 to 7.0. As shown in Table 5.6, performance was directly related to the effective size of the sand and hydraulic loading rate. With sands of effective size 0.37 mm or less and hydraulic loading rates of 4.9 gal/ft²·d (0.2 m³/m²·d), effluents with biological oxygen demand (BOD_5) and total suspended solids (TSS) of less than 15 mg/L were obtained. Total Kjeldahl nitrogen (TKN) concentrations were reduced from 11.6 to 4.3 mg/L at the 4.9-gal/ft²·d (0.2-m³/m²·d) loading rate. The experiments were conducted in a mild climate, and it is unlikely that similar nitrogen removal will occur during cold months of more severe climates.

Melcer et al. (1995) reported the performance of a full-scale aerated lagoon–intermittent system located in New Hamburg, Ontario, Canada, that had been in operation since 1980. Results for 1990 and from January to August of 1991 are presented in Table 5.7. Surface loading rates for both periods were 79.6 gal/ft²·d (3.24 m³/m²·d) with influent BOD_5, TSS, and TKN concentrations of 12,

16, and 19 mg/L, respectively. Filter effluent quality was exceptional, with BOD_5, TSS, and TKN concentrations less than 2 mg/L.

5.1.2.2 Operating Periods

The length of filter run is a function of the effective size of the sand and the quantity of solids deposited on the surface of the filter. The EPA's *Design Manual: Municipal Wastewater Stabilization Ponds* (USEPA, 1983) and several other publications (Bishop et al., 1977; Harris et al., 1978; Hill et al., 1977; Marshall and Middlebrooks, 1974; Messinger, 1976; Russell et al., 1983; Tupyi et al., 1979) contain extensive information on the relationship between solids deposited on the surface of a filter and the length of run time. Truax and Shindala (1994) also reported run times very similar to those in the above studies, and their results are presented in Table 5.8.

5.1.2.3 Maintenance Requirements

Maintenance is directly related to the quantity of solids applied to the surface of the filter which is related to the concentration of solids in the influent to the filter and the hydraulic loading rate. Filters with low hydraulic loading rates tend to operate for extended periods as shown in the above references and Table 5.8. With such extended operating periods, maintenance consists of routine inspection of the filter, removing weeds, and an occasional cleaning by removing the top 5 to 8 cm of sand after allowing the filter to dry out. A summary of reported annual maintenance for three field-scale, lagoon–intermittent sand filter facilities is shown in Table 5.9.

5.1.2.4 Hydraulic Loading Rates

Typical hydraulic loading rates on a single-stage filter range from 0.4 to 0.6 MG/ac·d (0.37 to 0.56 m^3/m^2·d). If the SS in the influent to the filter will routinely exceed 50 mg/L, the hydraulic loading rate should be reduced to 0.2 to 0.4 MG/ac·d (0.19 to 0.37 m^3/m^2·d) to increase the filter run. In cold weather locations, the lower end of the range is recommended during winter operations to avoid the possible need for bed cleaning during the winter months.

5.1.3.5 Design of Intermittent Sand Filters

Algae removal from lagoon effluent is almost totally a function of the sand gradation used. When BOD_5 and SS below 30 mg/L will satisfy requirements, a single-stage filter with medium sand (effective size of 0.3 mm) will produce a reasonable filter run. If better effluent quality is necessary, finer sand (effective size of 0.15 to 0.2 mm) is necessary or the use of a two-stage filtration system with the finer sand in the second stage.

The total filter area required for a single-stage operation is obtained by dividing the anticipated influent flow rate by the hydraulic loading rate selected

TABLE 5.2
Intermittent Sand Filter Performance Treating Lagoon Effluents[a]

Pond Type	u[b]	Loading Rate (mgd/ac)	TSS Influent (mg/L)	TSS Effluent (mg/L)	TSS Removal (%)	VSS Influent (mg/L)	VSS Effluent (mg/L)	VSS Removal (%)	BOD Influent (mg/L)	BOD Effluent (mg/L)	BOD Removal (%)	Ref.
Facultative	5.8	0.1	13.7	4.0	71	9.2	2.0	78	6.3	1.2	82	Marshall and Middlebrooks (1974)
		0.2	13.7	4.8	65	9.2	2.1	77	6.3	1.3	80	
		0.3	13.7	6.0	56	9.2	2.3	75	6.3	2.0	69	
Facultative	9.74	0.2	30.3	3.5	88	23.0	1.3	94	19.5	1.9	90	Harris et al. (1978)
		0.4	30.1	2.9	90	22.5	3.4	85	20.6	2.5	88	
		0.6	34.0	5.9	83	25.9	3.1	88	25.6	4.2	84	
		0.8	23.9	4.7	80	15.2	1.2	92	2.8	1.8	36	
		1.0	28.5	5.1	82	21.5	2.5	88	13.5	2.6	81	
		1.0	24.3	3.7	85	18.6	1.6	91	6.1	2.2	64	
Facultative	6.2	0.5	32.4	8.6	74	21.9	3.3	85	10.7	1.8	83	Hill et al. (1977)
		1.0	32.4	7.8	76	21.9	3.2	85	10.7	2.0	82	
		1.5	32.4	6.4	80	21.9	3.3	85	10.7	2.3	79	
Facultative	9.73	0.25	70.7	10.1	86	38.8	6.5	83	20.2	6.6	67	Bishop et al. (1977)
		0.5	197	15.6	92	155	11.9	92	71.4	9.4	87	
		1.0	108	11.8	89	83.0	8.8	89	34.0	13.0	62	

Aerated[c]	9.73	0.5	158	52.5	67	71.1	13.2	81	34.4	5.1	85	Bishop et al. (1977)
		1.0	68.7	32.9	52	36.6	11.3	69	19.6	11.7	40	
Anaerobic	NA	0.1	353	45.5	87	264	28.1	84	123	19.5	84	Messinger (1976)
		0.35	208	46.5	78	162	35.3	78	108	43.7	60	
		0.5	194	45.1	77	175	35.7	80	107	67.6	37	
Facultative	9.7	0.2	23.0	2.7	88	17.8	1.0	95	10.9	1.1	90	Tupyi et al. (1979)
		0.4	20.8	3.5	83	18.5	2.3	88	11.5	2.6	77	

[a] Results for best overall performing 0.17-mm e.s. filters.
[b] Uniformity coefficient.
[c] Dairy waste.

TABLE 5.3
Mean Performance Values for Three Full-Scale Lagoon–Intermittent Sand Filter Systems

Parameter	Mt. Shasta Facility				Moriarty Facility				Ailey Facility			
	Facility Influent	Lagoon Effluent	Filter Effluent	Facility Effluent	Facility Influent	Lagoon Effluent	Filter Effluent	Facility Effluent	Facility Influent	Lagoon Effluent	Filter Effluent	Facility Effluent
BOD (mg/L)	114	22	11	8	148	30	17	17	67	22	8	6
Soluble BOD (mg/L)	41	7	4	5	74	17	16	16	17	10	6	5
TSS (mg/L)	83	49	18	16	143	81	13	13	109	43	15	13
VSS (mg/L)	70	34	13	10	118	64	9	9	87	32	8	6
FC (number/100 mL)	1.16×10^6	292	30	<2	4.24×10^6	290	18	34	2.17×10^6	55	8	<1
pH	6.9	8.7	6.8	6.6	8.0	8.9	8.0	8.0	7.3	8.9	7.1	6.8
DO (mg/L)	4.8	12.4	5.5	5.3	1.8	10.9	8.3	8.3	6.7	10.2	7.4	7.9
COD (mg/L)	244	100	87	68	305	84	43	43	160	57	32	25
Soluble COD (mg/L)	159	71	64	50	197	67	34	34	82	41	23	16
Alk (mg/L as $CaCO_3$)	95	75	51	42	436	293	260	260	93	84	76	69
TP (mg-P/L)	4.68	3.88	3.09	2.72	10.3	4.02	2.8	2.8	4.96	3.10	2.67	2.45
TKN (mg-N/L)	15.5	11.1	7.5	5.2	60	22	12.1	12.1	14.2	7.3	4.1	2.2
NH_3 (mg-N/L)	10.8	5.56	1.83	1.76	38	16	9.16	9.16	5.5	0.658	0.402	0.31

Organic nitrogen (mg-N/L)	4.8	5.6	5.7	3.4	22	5.7	3.3	3.3	8.7	6.7	3.8	1.9
NO_2 (mg-N/L)	0.16	0.56	77	0.020	0.05	159	1.66	1.66	0.479	0.028	73	0.010
NO_3 (mg-N/L)	0.28	0.78	4.3	4.5	0.05	0.09	4.09	4.09	1.6	0.15	2.36	2.14
Total algal count (cells/mL)	NA	398,022	144,189	141,305	NA	756,681	32,417	32,417	NA	349,175	21,583	29,360
Flow (mgd)	0.637	NA.	NA.	0.488	0.096	NA	0.046	NA	NA	NA	NA	0.070

Note: NA = not available.

Source: Data from Russell et al. (1980, 1983).

TABLE 5.4
Design Characteristics and Performance of Facultative Lagoon–Intermittent Sand Filter Systems

Design Flow (m³/L)	Present Flow (% Design)	Hydraulic Retention Time[a]	Filter Dosing (m³/m²·d)[a]	BOD$_5$ (g/m³)		TSS (g/m³)		NH$_3$–N (g/m³)	
				50%	95%	50%	95%	50%	95%
303	56	93	0.03	9	28	12	41	0.9	4
303	79	70	0.37	6	22	7	29	0.4	1.2
568	48	59	0.47	7	17	11	30	–	–
378	66	52	0.37	9	21	11	25	0.9	2.4
568	37	55	0.31	6	17	6	16	1.3	5.4

[a] Based on design flow rate.

Source: Rich, L.G. and Wahlberg, E.J., *J. Water Pollut. Control Fed.*, 62, 697–699, 1990. With permission.

TABLE 5.5
**Performance Comparison of Lagoon–Intermittent Sand Filters
with Aerated Lagoons and Activated Sludge Plants**

Process	Number of Systems	BOD$_5$ (g/m^3)		TSS (g/m^3)	
		Mean	SD	Mean	SD
Conventional activated sludge	18	12.8	6.85	14.92	10.53
Step-feed activated sludge	13	10.84	7.68	16.23	16.65
Aerated lagoons[a]	6	28	71	50	129
DPMC upgrades[b]	3	18	39	13	36
DPMC new	1	14	37	11	31
Lagoon–intermittent sand filter	9	8.35	3.07	9.88	3.84

[a] One aerated cell followed by a polishing pond.
[b] Facultative lagoons upgraded to dual-power, multicellular aerated lagoons.

Source: Rich, L.G. and Wahlberg, E.J., *J. Water Pollut. Control Fed.*, 62, 697–699, 1990. With permission.

for the system. One spare filter unit should be included to permit continuous operation as the cleaning operation may require several days. An alternative approach is to provide temporary storage in the pond units. Three filter beds are the preferred arrangement to permit maximum flexibility. In small systems that depend on manual cleaning, the individual bed should not be bigger than about 1000 ft^2 (90 m^2). Larger systems with mechanical cleaning equipment might have individual filter beds up to 55,000 ft^2 (5000 m^2) in area.

Selected sand is usually used as the filter media. These are generally described by their effective size (e.s.) and uniformity coefficient (U.C.). The e.s. is the 10 percentile size; that is, only 10% of the filter sand, by weight, is smaller than that size. The uniformity coefficient is the ratio of the 60-percentile size to the 10-percentile size. The sand for single-stage filters should have an e.s. ranging from 0.20 to 0.30 mm and a U.C. of less than 7.0, with less than 1% of the sand smaller than 0.1 mm. The U.C. value has little effect on performance, and values ranging from 1.5 to 7.0 are acceptable. Generally, clean, pit-run concrete sand is suitable for use in intermittent sand filters if the e.s., U.C., and minimum sand size are suitable.

The design depth of sand in the bed should be at least 45 cm (18 in.) plus a sufficient depth for at least 1 year of cleaning cycles. A single cleaning operation may remove 1 to 2 in. (2.5 to 5 cm) of sand. A 30-d filter run would then require an additional 12 in. (30 cm) of sand. In the typical case, an initial bed depth of about 36 in. (90 cm) of sand is usually provided. A graded gravel layer, 12 to 18 in. (30 to 45 cm), separates the sand layer from the underdrains. The bottom layer is graded so its e.s. is four times as great as the openings in the underdrain piping. The successive layers of gravel are progressively finer to prevent intrusion of

TABLE 5.6
Lagoon–Intermittent Sand Filtration Performance Data

Parameter	Filter Influent (mg/L)	Sand 1 (e.s. = 0.70 mm; U.C. = 21) Hydraulic Loading Rate (m³/m²/d)		Sand 2 (e.s. = 0.35 mm; U.C. = 14) Hydraulic Loading Rate (m³/m²/d)		Sand 3 (e.s. = 0.37 mm; U.C. = 7.0) Hydraulic Loading Rate (m³/m²/d)		Sand 4 (e.s. = 0.18 mm; U.C. = 2.7) Hydraulic Loading Rate (m³/m²/d)	
		0.2	1.2	0.2	1.2	0.2	1.2	0.2	1.2
TSS (mg/L)									
Mean	100	18	32	13	39	13	41	13	25
SD	38	9	17	5	12	6	15	4	8
VSS (mg/L)									
Mean	86	13	22	9	26	10	27	7	17
SD	32	5	12	4	5	4	9	2	4
COD (mg/L)									
Mean	156	79	109	73	105	59	91	48	75
SD	38	21	35	21	31	19	32	12	32
BOD$_5$ (mg/L)									
Mean	31	16	20	13	20	11	16	6	14
SD	7	6	7	5	8	4	5	4	7
TKN (mg/L)									
Mean	11.6	5.6	7.9	4.6	7.3	6.0	7.0	4.3	7.8
SD	4.2	1.7	3.2	2.1	2.2	2.2	2.8	2.2	2.3

	1	2	3	4	5	6	7	8	9
NH_3–N (mg/L)									
Mean	3.1	0.4	1.6	0.5	1.2	0.7	1.7	0.5	1.2
SD	1.5	0.6	0.9	0.7	0.7	0.5	1.1	0.8	0.9
NO_3–N (mg/L)									
Mean	0.3	0.6	1.1	2.9	2.6	2.8	1.1	3.1	1.5
SD	0.2	1.9	1.0	2.0	2.4	2.4	1.8	2.6	1.7
PO_4–P (mg/L)									
Mean	5.7	3.5	4.9	5.5	4.9	4.0	4.9	4.0	3.9
SD	2.6	2.6	3	2.3	1.4	3.1	3.3	2.8	3.1
TC (number/100 mL)									
Mean	1.65×10^5	0.92×10^5	0.59×10^5	0.71×10^5	0.83×10^5	0.47×10^5	0.48×10^5	0.38×10^5	0.28×10^5
SD	3.21×10^5	2.57×10^5	1.68×10^5	1.49×10^5	2.74×10^5	1.82×10^5	1.17×10^5	2.15×10^5	0.41×10^5
pH									
Mean	7.8	7.2	7.2	7.2	7.0	6.9	7.0	7.0	7.0
SD	0.5	0.3	0.4	0.3	0.4	0.4	0.4	0.4	0.4

Source: Truax, D.D. and Shindala, A., Water Environ. Res., 66(7), 894–898, 1994. With permission.

TABLE 5.7
Performance of Aerated Lagoon–Intermittent Sand Filter, Hamburg Plant

Parameter	1990	1991 (January to August)
Average flow rate (m³/d)	1676	1673
Maximum flow rate (m³/d)	4530	3990
Raw Sewage		
BOD (mg/L)	186	120
TSS (mg/L)	314	171
TKN (mg/L)	45	44
TP (mg/L)	9.3	9.5
Aerated Cell		
HRT (d)	7	7
BOD loading (kg/m³·d)	0.03	0.02
Aerated Cell Effluent		
BOD (mg/L)	34	36
TSS (mg/L)	44	44
TP (mg/L)	6	5
Facultative Lagoon		
HRT (d)	165	165
Average BOD loading (kg/1000 m²·d)	0.51	0.55
Cell 2 Effluent		
BOD (mg/L)	12	11
TSS (mg/L)	16	18
TKN (mg/L)	19	18
NH₃–N (mg/L)	15	14
NO(T)–N (mg/L)	1.1	0.8
TP (mg/L)	1.2	0.7
Filter		
Annual surface loading (m³/m²)	195	153
Surface loading (L/m²·d)	3240	3240
Filter Effluent	**March to December**	**March to August**
BOD (mg/L)	2	2
TSS (mg/L)	1.7	1.1
TKN (mg/L)	2	1.1
NH₃–N (mg/L)	1.2	0.6
NO(T)–N (mg/L)	7	9
TP (mg/L)	0.5	0.4

Source: Melcer, H. et al., *Water Sci. Technol.*, 31(12), 379–387, 1995. With permission.

TABLE 5.8
Lagoon–Intermittent Sand Filtration Run Lengths

Sand Characteristics	Hydraulic Loading Rate (m³/m²/d)	Days of Filter Operation Before Initial Clogging
Sand 1 (e.s. = 0.70 mm; U.C. = 2.1)	0.2	469
	0.9	335
	1.1	106
Sand 2 (e.s. = 0.35 mm; U.C. = 1.4)	0.2	468
	0.7	259
	0.9	16
Sand 3 (e.s. = 0.37 mm; U.C. = 7.0)	0.2	130
	0.4	305
	0.6	159
	0.7	27
	0.9	9
Sand 4 (e.s. = 0.18 mm; U.C. = 2.7)	0.2	131
	0.4	130
	0.7	35
	0.9	5

Source: Truax, D.D. and Shindala, A., *Water Environ. Res.*, 66(7), 894–898, 1994. With permission.

sand. An alternative is to use gravel around the underdrain piping and then a permeable geotextile membrane to separate the sand from the gravel. Further details on design and performance are presented in USEPA (1983). A design example for an intermittent sand filter treating a lagoon effluent is presented in Example 5.1.

Example 5.1. Typical Design of Intermittent
Sand Filter Treating Lagoon Effluent

Design data and assumptions:
1. Design flow = Q = 379 m³/d (0.100 MG/d).
2. Hydraulic loading rate = HLR = 0.29 m³/m²·d (0.310 MG/ac·d).
3. Minimum number of filters = 2.
4. Design to minimize operation and maintenance.
5. Gravity flow is possible.
6. Topography and location satisfactory.
7. Adequate land is available at reasonable cost.
8. Filter sand is locally available.
9. Filters are considered plugged when, at the time of dosing, the water from the previous dose has not dropped below the filter service.

TABLE 5.9
Summary of Reported Annual Maintenance For Field-Scale Facilities

Job Description	Mt. Shasta Facility	Moriarty Facility	Ailey Facility
Daily operation and maintenance (daily monitoring)	(1.0 hr) × 7 d × 52 wk = 364	(1.0 hr) × 7 d × 52 wk = 364	(0.5 hr) × 5 d × 52 wk = 130
Filter cleaning	54[a]	28[a]	None
Filter raking	12 raking; 16 mixing	13	22
Filter weed control	NA	None	26
Miscellaneous maintenance	NA	11	None
Grounds maintenance	42	8	28
Total reported man-hr/yr	488+	424	206
Computed manpower requirements	2.4 man-yr[b]	1 man-yr[b]	1 man-yr[b]
Actual reported manpower input	2.0 man-yr[c]	0.28 man-yr[b]	0.14 man-yr[b]

[a] Man-hours with mechanical assistance.
[b] Assuming 1500 man-hr per 1 man-yr.
[c] Considering extra assistance for filter cleaning and weekend monitoring.

Source: Data from Russell et al. (1980, 1983).

Design

Determine dimensions of filters:

> Area of each filter = Q/HLR.
> Area = 1307 m^2 (0.323 ac).
> Length-to-width radio = 2:1.
> W = 25.56 m (83.87 ft).
> L = 51.13 m (167.7 ft).

A minimum of two filters is required.

Influent distribution system

Design assumptions:

1. Dosing syphon will be used to gravity feed filters. Electric activated valves also may be used.
2. Loading sequence will be designed to deliver one half the daily flow rate to one filter unit per day in two equal doses. More frequent dosing is acceptable.
3. Pipe sizes are selected to avoid clogging and to make cleaning convenient. Hydraulics do not control.

Dosing basin sizing:
  Number of dosings per day = 2.
  Q = (design Q)/(number of dosings) = 189.5 m^3/d.
  Volume = 189.5 m^3.

Install overflow pipe to filters. Distribution manifold from dosing siphon should be designed to minimize the velocity of water entering the filter. Use 10-in.-diameter pipe in this design. Each of the outlets from the manifold will be spaced 10 ft from each end and 21 ft on-centers on the long side of the filter. Manifold outlets will discharge onto 3-ft × 3-ft splash pads constructed of gravel 1.5 to 3 in. in diameter.

Filter containment and underdrain system

The filter may be contained in a reinforced concrete structure or a synthetic liner to prevent groundwater contamination. The slope of the filter bottom is dependent on the slope of the drain pipe configuration. Use a slope of 0.025% with lateral collection lines 15 ft on-center. Six-inch lateral collecting pipe and 8-in. collection manifolds will provide adequate hydraulic capacity and ease of maintenance.

  Minimum freeboard required for filters: Must be adequate to receive one dosing × safety factor.
  Safety factor = 3.
  Depth = (SF × Q_{dosing})/(L × W) = (3 × 189.5)/(25.56 × 51.13).
  Water depth assuming no passage through filter = 0.435 m (1.47 ft).

5.1.3 ROCK FILTERS

In a rock filter, pond effluent travels through a submerged porous rock bed, causing algae to settle out on the rock surfaces as the liquid flows through the void spaces. The accumulated algae are then biologically degraded. Algae removal with rock filters has been studied extensively at Eudora, Kansas; California; Missouri; and Veneta, Oregon (USEPA, 1983). Rock filters have been installed throughout the United States and the world, and performance has varied (Middlebrooks, 1988; Saidam et al., 1995; USEPA, 1983). A diagram of the Veneta rock filter can be seen in Figure 5.2. The West Monroe, Louisiana, rock filters were essentially the same as the one in Veneta, but the filters received higher loading rates than those employed at the Veneta system. Several rock filters of various designs have been constructed in Illinois with varied success. Many of the Illinois filters produced an excellent effluent, but the designs varied widely (Adam, 1986; Menninga, 1986); see Figure 5.3 for diagrams of the various types of rock filters applied in Illinois. Snider (1998) designed a rock filter for Prineville, Oregon, and knew of one built at Harrisburg, Oregon. Performance and design details are not available; however, Snider indicated that the systems were designed using information from the Veneta system. The principal advantages of the rock filter are its relatively low construction costs and simple operation. Odor problems can occur, and the design lives of the filters and the cleaning procedures have not yet been firmly established; however, several units have operated successfully for over 20 years.

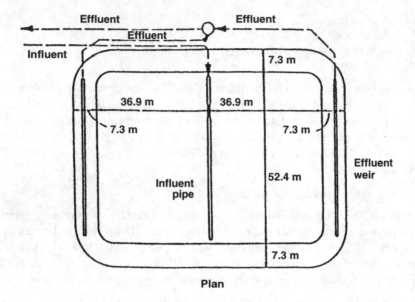

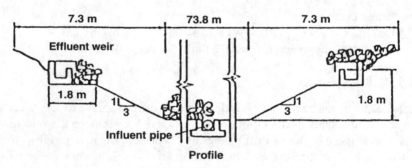

FIGURE 5.2 Rock filter at Veneta, Oregon. (From Swanson, G.R. and Williamson, K.J., *J. Environ. Eng. Div. ASCE*, 106(EE6), 1111–1119, 1980. With permission.)

5.1.3.1 Performance of Rock Filters

Mixed results have been obtained with rock filters. The most successful ones have been located in Veneta, Oregon, and West Monroe, Louisiana. Performance data from a study by Swanson and Williamson (1980) for the Veneta system are shown in Figure 5.4. Performance data for 1994 are shown in Table 5.10. After approximately 20 years of operation, the system was producing an effluent meeting secondary standards with regard to BOD_5, TSS, and fecal coliforms (FC). During the winter months, high ammonia nitrogen concentrations were observed in the effluent. Stamberg et al. (1984) presented performance results for the two rock filters operating in West Monroe (Figure 5.5). The systems were loaded at higher hydraulic loading rates than that used at the Veneta facility (<0.3 m³ of

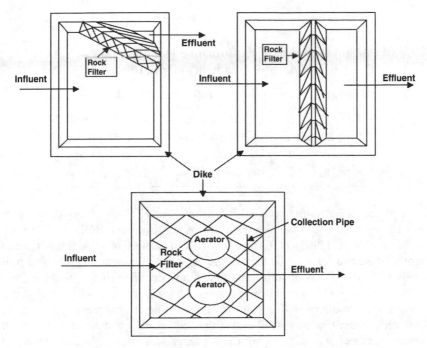

FIGURE 5.3 State of Illinois rock filter configurations.

wastewater per d per m³ of rock), and the TSS removals were less than those reported for the Veneta system. In general, the West Monroe systems produced effluent BOD₅ and TSS concentrations less than 30 mg/L, but these concentrations

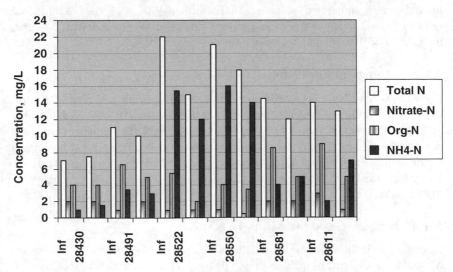

FIGURE 5.4 Performance of Veneta, Oregon, rock filter. (From Swanson, G.R. and Williamson, K.J., *J. Environ. Eng. Div. ASCE*, 106(EE6), 1111–1119, 1980. With permission.)

TABLE 5.10
Mean and Range of Performance Data For Veneta,
Oregon, Wastewater Treatment Plant (1994)

Constituent	Influent	Effluent
BOD5 (mg/L)	138 (50–238)	17 (5–30)
TSS (mg/L)	124 (50–202)	9 (2–27)
FC (number/100 mL)	—	<10 (<10–20)
Flow (mgd)	0.251 (0.159–0.452)	0.309 (0.079–0.526)

deviated occasionally, up to a BOD_5 of 40 mg/L and TSS of 50 mg/L; however, only 12 out of over 100 samples exceeded 30 mg/L in both BOD_5 and TSS.

Saidam et al. (1995) performed a series of studies of rock filters treating lagoon effluent in Jordan. The filters were arranged in three trains. The first train consisted of two filters in series; the first filter contained rock with an average diameter of 7 in. (18 cm) and was followed by a filter containing local gravel with an average diameter of 4.6 in. (11.6 cm). The second train contained the same rock as used in the first filter, but the second filter contained rock with an average diameter of 1 in. (2.4 cm). Local gravel with an average diameter of 4.6 in. (11.6 cm) was used in the first filter of the third train, and the second filter contained an aggregate with an average diameter of 0.5 in. (1.27 cm). The filters in the three trains were operated in series. The characteristics of the wastewater, hydraulic loading rates, and effluents from the various filters are shown in Table 5.11. The removal efficiencies obtained in the first run for the various filters and the trains are summarized in Table 5.12. Even though the rock sizes and loading rates of several of the filters were similar to those at Veneta and West Monroe, the quality of the effluents was much lower.

5.1.3.2 Design of Rock Filters

Rock filters have been designed using a number of varying parameters. A summary of the design parameters used for several locations is provided in Table 5.13. The parameters shown for Illinois are the current standards and were not necessarily used to design the systems diagrammed in Figure 5.3. The critical factor in the design of rock filters appears to be the hydraulic loading rate. Rates less than 0.3 $m^3/m^3{\cdot}d$ appear to give the best results with rocks in the range of 3 to 8 in. (8 to 20 cm), a depth of 6.6 ft (2 m), and water applied in an upflow pattern.

5.1.4 NORMAL GRANULAR MEDIA FILTRATION

Granular media filtration (rapid sand filters) has proven very successful as a means of liquid–solids separation. The simple design and operation processes make it applicable to wastewater streams containing up to 200 mg/L suspended solids. Automation based on easily measured parameters results in minimum operation

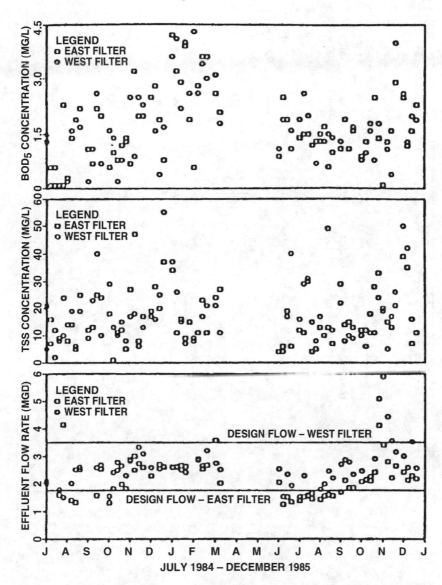

FIGURE 5.5 Rock filter performance, West Monroe, Louisiana. (From Stamberg, J.B. et al., Simple Rock Filter Upgrades Lagoon Effluent to AWT Quality in West Monroe, LA, paper presented at 57th Conf. Water Pollution Control Federation, New Orleans, LA, 1984.)

and maintenance costs; however, when regular granular media filtration has been applied to the removal of algae from wastewater stabilization effluents, very poor results have been obtained. Good efficiencies can be obtained when chemicals are added prior to filtration or when the wastewater is treated by coagulation and flocculation prior to filtration. Table 5.14 contains a summary of the results with

TABLE 5.11
Performance of Rock Filters

Unit	Hydraulic Loading Rate (m³/m³·d)	Run	T(°C)	pH	DO (mg/L)	H_2S (mg/L)	NH_4–N (mg/L)	TSS (mg/L)	TVSS (mg/L)	TP (mg/L)	BOD_5 (mg/L)	COD (mg/L)	TFCC (MPN/ 100 mL)
Influent													
	—	1	25.7	7.7–8.3	3.2	0.02	85	201	161	23	95	334	1.10E+04
	—	2	21.0	7.4–8.1	4.8	0.02	93	234	182	21	105	341	6.3E+04
	—	3	14.0	7.9–8.0	4.0	0.30	97	213	179	19	122	398	9.6E+05
	—	4	15.0	7.3–7.9	3.5	0.02	71	101	88	14	108	323	1.6E+04
First Train													
Rock filter 1	0.498	1	25.1	7.3–8.3	1.2	0.18	89	131	113	21	61	272	2.2E+04
Average diameter = 18 cm	0.634	2	20.0	7.0–7.9	1.5	0.02	96	200	146	20	81	300	5.7E+04
Voids = 49%	0.5–0.58	3	13.4	7.8–8.0	1.0	0.30	96	156	133	18	100	317	8.1E+05
Surface area = 17 m²/m³	0.5–0.58	4	13.0	7.0–7.9	2.1	0.02	72	76	68	13	77	281	1.4E+04
Wadi gravel filter 1	0.386	1	25.2	7.5–8.5	1.9	0.63	91	78	80	17	36	223	1.00E+03
Average diameter = 11.6 cm	0.634	2	19.9	7.1–7.8	1.4	0.02	97	161	118	18	66	263	4.2E+04

Voids = 41%	0.5–0.58	3	13.4	7.8–8.0	1.0	0.40	97	129	114	16	77	306	4.7E+05
Surface area = 25 m²/m³	0.5–0.58	4	13.0	7.4–8.6	1.9	0.02	71	66	60	13	74	263	1.10E+04
Second Train													
Rock filter 2	0.311	1	25.3	7.7–8.2	1.1	0.43	89	130	110	21	53	263	1.9E+03
Average diameter = 18 cm	0.634	2	19.7	7.2–7.9	1.4	0.02	98	203	144	20	79	301	5.00E+04
Voids = 49%	0.5–0.58	3	13.3	7.7–8.1	1.0	0.40	98	164	137	17	87	336	8.6E+05
Surface area = 17 m²/m³	0.5–0.58	4	13.7	7.6–8.0	1.9	0.02	71	88	79	13	92	293	1.00E+04
Coarse aggregate filter 2	0.333	1	25.6	7.5–8.5	1.7	0.47	89	102	92	15	51	223	1.5E+03
Average diameter = 2.4 cm	0.634	2	19.9	7.1–7.9	1.4	1.5	98	154	117	16	65	257	3.2E+04
Voids = 40%	0.5–0.58	3	13.4	7.7–8.1	1.0	0.30	96	134	115	16	73	285	5.4E+05
Surface area = 150 m²/m³	0.5–0.58	4	13.6	7.3–8.0	2.0	0.02	71	60	52	11	87	246	6.5E+03

TABLE 5.11 (cont.)
Performance of Rock Filters

Unit	Hydraulic Loading Rate (m³/m³·d)	Run	T(°C)	pH	DO (mg/L)	H₂S (mg/L)	NH₄–N (mg/L)	TSS (mg/L)	TVSS (mg/L)	TP (mg/L)	BOD₅ (mg/L)	COD (mg/L)	TFCC (MPN/100 mL)
							Parameter						
Third Train													
Wadi gravel filter 3	0.274	1	25.7	7.1–8.5	1.6	0.37	91	109	101	18	48	255	1.6E+03
Average diameter = 11.6 cm	0.634	2	20.2	7.2–7.9	1.4	0.77	96	206	151	19	76	304	6.8E+04
Voids = 41%	0.5–0.58	3	13.3	7.9–8.0	1.0	0.3	97	150	131	18	86	312	3.2E+05
Surface area = 25 m²/m³	0.5–0.58	4	15.0	7.6–8.0	1.9	0.02	71	81	71	13	76	272	6.3E+03
Medium aggregate filter 3	0.442	1	25.9	7.3–8.6	2.0	0.35	92	82	79	12	42	188	6.4E+02
Average diameter = 1.27 cm	0.634	2	19.7	7.3–8.2	1.5	0.88	96	162	121	16	72	249	3.3E+04
Voids = 28%	0.5–0.58	3	13.4	7.7–8.1	1.0	0.02	100	132	108	15	66	257	4.4E+05
Surface area = 327 m²/m³	0.5–0.58	4	13.0	7.3–7.8	1.9	0.02	71	52	45	10	59	206	3.3E+03

Source: Saidam, M.Y. et al., Water Sci. Technol., 31(12), 369–378, 1995. With permission.

TABLE 5.12
Summary of the Removal Efficiency in the First Run

Parameter	Percent Removal of Individual Filters						Percent Removal Per Train		
	Rock Filter 1	Wadi Gravel Filter 1	Rock Filter 2	Coarse Aggregate Filter 2	Wadi Gravel Filter 3	Medium Aggregate Filter 3	First Train	Second Train	Third Train
TSS	34	41	35	22	46	25	61	49	59
BOD$_5$	36	41	44	4	49	13	62	46	56
COD	19	18	21	15	24	25	33	33	44
Total phosphorus	9	15	9	30	18	33	24	35	46
Total fecal coliform	80	55	83	21	85	60	90	86	94
Color	25	34	28	20	30	36	51	42	55
Hydraulic loading rate (m^3/m^3·d)	0.498	0.386	0.311	0.333	0.274	0.442	—	—	—

Source: Saidam, M.Y. et al., *Water Sci. Technol.,* 31(12), 369–378, 1995. With permission.

TABLE 5.13
Rock Filter Design Parameters

Parameter	Veneta, Oregon	W. Monroe, Louisiana	State of Illinois	Eudora, Kansas	California, Missouri
Hydraulic loading rate ($m^3/m^3 \cdot d$)	0.3	0.36	0.8	Up to 1.2 in the summer; 0.4 in winter and spring	0.4
Rock (cm)	7.5–20	5–13	8–15, free of fines, soft weathering stone, and no flat rock.	2.5	6–13
Aeration	None	None	Post-aeration ability is necessary.	None	None
Depth (m)	2	1.8	Rock media must extend 0.3 m above water surface.	1.5	1.68
Disinfection	Yes	Yes	Chlorination of post-aeration cell is encouraged.	Not applicable	Yes

TABLE 5.14
Summary of Direct Filtration With Rapid Sand Filters

Investigator	Coagulant Aid and Dose (mg/L)	Filter Loading (gpm/ft²)	Filter Depth (ft)	Sand Size (mm)	Findings
Borchardt and O'Melia (1961)[a]	None	0.2–2	2	$d_{50} = 0.32$	Removal declines to 21–45% after 15 hr
Davis and Borchardt (1966)	Fe (7 mg/L)	2.1	2	$d_{50} = 0.40$	50% algae removal
	None	0.49	NA	$d_{50} = 0.75$	22% algae removal
	None	0.49		$d_{50} = 0.29$	34% algae removal
	None	1.9		$d_{50} = 0.75$	10% algae removal
	None/Fe	1.9		$d_{50} = 0.29$	2% algae removal
		NA	NA	$d_{50} = 0.75$	45% algae removal
Foess and Borchardt (1969)	None	2	2	$d_{50} = 0.71$	pH 2.5; 90% algae removal pH 8.9; 14% removal
Lynam et al. (1969)	None	1.1	0.92	$d_{50} = 0.55$	62% SS removal
Kormanik and Cravens (1978)	None	—	—	—	11–45% SS removal

a Lab culture of algae.

direct granular media filtration. Diatomaceous earth filtration is capable of producing a high-quality effluent when treating wastewater stabilization pond water, but the filter cycles are generally less than 3 hr. This results in excessive usage of backwash water and diatomaceous earth, which leads to very high costs and eliminates this method of filtration as an alternative for polishing wastewater stabilization pond effluents.

5.1.5 COAGULATION–FLOCCULATION

A process of coagulation followed by sedimentation has been applied extensively for the removal of suspended and colloidal materials from water. Lime, alum, and ferric salts are the most commonly used coagulating agents. Floc formation is sensitive to parameters such as pH, alkalinity, turbidity, and temperature. Most of these variables have been studied, and their effects on the removal of turbidity of water supplies have been evaluated. In the case of the chemical treatment of wastewater stabilization pond effluents, however, the data are not comprehensive.

Shindala and Stewart (1971) investigated the chemical treatment of stabilization pond effluents as a post-treatment process to remove the algae and improve the quality of the effluent. They found that the optimum dosage for best removal o£ the parameters studied was 75 to 100 mg/L of alum. When this dosage was used, the removal of phosphate was 90% and the chemical oxygen demand (COD) was 70%.

Tenney (1968) has shown that, at a pH range of 2 to 4, algal flocculation was effective when a constant concentration of a cationic polyelectrolyte (10 mg/L of C-31) was used. Golueke and Oswald (1965) conducted a series of experiments to investigate the relation of hydrogen ion concentrations to algal flocculation. In this study, only H_2SO_4 was used, and only to lower the pH. Golueke and Oswald (1965) found that flocculation was most extensive at a pH value of 3, with which Tenney's results agree. They obtained algal removals of about 80 to 90%. Algal removal efficiencies were not affected in the pH range of 6 to 10 by cationic polyelectrolytes.

The California Department of Water Resources (1971) reported that of 60 polyelectrolytes tested, 17 compounds were effective in the coagulation of algae and were economically competitive when they were compared with mineral coagulation used alone. Generally, less than 10 mg/L of the polyelectrolytes was required for effective coagulation. A daily addition of 1 mg/L of ferric chloride to the algal growth pond resulted in significant reductions in the required dosage of both organic and inorganic coagulants.

McGarry (1970) studied the coagulation of algae in stabilization pond effluents and reported the results of a complete factorial-designed experiment using the common jar test. Tests were performed to determine the economic feasibility of using polyelectrolytes as primary coagulants alone or in combination with alum. He also investigated some of the independent variables that affected the flocculation process, such as concentration of alum, flocculation turbulence, concentration of polyelectrolyte, pH after the addition of coagulants, chemical dispersal conditions, and high-rate oxidation pond suspension characteristics. Alum was found to be effective for coagulation of algae from high-rate oxidation pond effluents, and the polyelectrolytes used did not reduce the overall costs of algal removal. The minimum cost per unit algal removal was obtained with alum alone (75 to 100 mg/L). The most significant effects occurred with alum and polyelectrolyte concentrations. The time of polyelectrolyte addition alone had no significant effect. The more important interactions occurred between alum and polyelectrolyte, alum and polyelectrolyte concentrations, time of polyelectrolyte addition and alum concentration, and time of polyelectrolyte addition and polyelectrolyte concentration.

Al-Layla and Middlebrooks (1975) evaluated the effects of temperature on algae removal using coagulation–flocculation–sedimentation. Algae removal at a given alum dosage decreased as the temperature increased. Maximum algae removal generally occurred at an alum dosage of approximately 300 mg/L at 10°C. At higher temperatures, alum dosages as high as 600 mg/L did not produce removals equivalent to the results obtained at 10°C with 300 mg/L of alum. The settling time required to achieve good removals, flocculation time, organic carbon removal, total phosphorus removal, and turbidity removal were found to vary adversely as the temperature of the wastewater increased.

Dryden and Stern (1968) and Parker (1976) reported on the performance and operating costs of a coagulation–flocculation system followed by sedimentation, filtration, and chlorination with discharge to recreational lakes. This system probably has the longest operating record of any coagulation–flocculation system treating wastewater stabilization pond effluent. The TSS concentrations applied to the plant have ranged from about 120 to 175 mg/L, and the plant has produced an effluent with a turbidity of less than 1 Jackson turbidity unit (JTU) most of the time. Aluminum sulfate dosages have ranged from 200 to 360 mg/L. The design capacity is 0.5 mgd (1893 m³/d), and the plant was constructed in 1970 at a cost of $243,000. Operating and maintenance costs for 1973–1974 were $304/mg ($0.08/m³). Because of seasonal flow variations, operations and maintenance costs ranged from $200 to $800 per mg ($0.053 to $0.21 per m³).

Coagulation–flocculation is not easily controlled and requires expert operating personnel at all times. A large volume of sludge is produced, which introduces an additional operating problem that would very likely be ignored in a small community that is accustomed to a minimum of operation and maintenance of a wastewater lagoon. Therefore, coagulation–flocculation does not seem feasible for application in small communities.

5.1.6 DISSOLVED-AIR FLOTATION

Several studies have shown the dissolved-air flotation process to be an efficient and a cost-effective means of algae removal from wastewater stabilization lagoon effluents. The performance data for several of these studies are summarized in Table 5.15. Three basic types of dissolved-air flotation are employed to treat wastewaters: total pressurization, partial pressurization, and recycle pressurization.

TABLE 5.15
Summary of Typical Dissolved Air Flotation Performance (University of Texas at Austin, 1976)

Investigator/Location	Coagulant and Dose (mg/L)	Rate (gpm/f²)	Detention Time (min)	BOD$_5$			Suspended Solids		
				Influent (mg/L)	Effluent (mg/L)	% Removed	Influent (mg/L)	Effluent (mg/L)	% Removed
Parker (1976) Stockton, California	Alum (225 mg/L); acid added to pH 6.4	2.7[a]	17[a]	46	5	89	104	20	81
Ort (1972) Lubbock, Texas	Lime (150 mg/L)[c]	NA	12[b]	280–450	0–3	>99	240–360	0–50	>79
Komline-Sanderson (1972) El Dorado, Arkansas	Alum (200 mg/L)	4.0[c]	8[c]	93	<3	>97	450	36	92
Bore et al. (1975) Logan, UT	Alum (300 mg/L)	1.3–2.4[d]	NA	NA	NA	NA	125	4	96
Stone et al. (1975) Sunnyvale, California	Alum (175 mg/L); acid added to pH 6.0 to 6.3	2.0[e]	11[e]	NA	NA	NA	150	30	80

[a] Including 33% pressurized (35–60 psig) recycle.
[b] Including 30% pressurized (50 psig) recycle.
[c] Including 100% pressurized recycle.
[d] Including 25% pressurized (45 psig) recycle.
[e] Including 27% pressurized (55–70 psig) influent.

FULL FLOW PRESSURIZATION

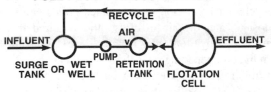

PARTIAL PRESSURIZATION

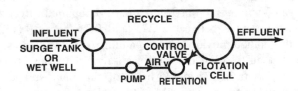

RECYCLE PRESSURIZATION

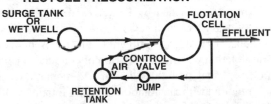

FIGURE 5.6 Types of dissolved-air flotation systems. (From Snider, Jr., E.F., in *Ponds as a Wastewater Treatment Alternative*, Gloyna, E.F. et al., Eds., Center for Research in Water Resources, College of Engineering, University of Texas, Austin, 1976.)

These three types of dissolved-air flotation are illustrated by flow diagrams in Figure 5.6. In the total pressurization system, the entire wastewater stream is injected with air and pressurized and held in a retention tank before entering the flotation cell. The flow is direct, and all recycled effluent is repressurized. In partial pressurization, only part of the wastewater stream is pressurized, and the remainder of the flow bypasses the air dissolution system and enters the separator directly. Recycling serves to protect the pump during periods of low flow, but it does hydraulically load the separator. Partial pressurization requires a smaller pump and a smaller pressurization system. In recycle pressurization, clarified effluent is recycled for the purpose of adding air and then is injected into the raw wastewater. Approximately 20 to 50% of the effluent is pressurized in this system. The recycle flow is blended with the raw water flow in the flotation cell or in an inlet manifold.

Important parameters in the design of a flotation system are hydraulic loading rate, including recycle, concentration of suspended solids contained within the flow, coagulant dosage, and the air-to-solids ratio required to effect efficient removal. Pilot-plant studies by Stone et al. (1975), Bare (1971), and Snider (1976) reported maximum hydraulic loading rates that ranged between 2 and 2.5 gpm/ft² (81.5 and 101.8 L/min·m²). A most efficient air-to-solids ratio was found to be 0.019 to 1.0 by Bare (1971). Solids concentrations during Bare's studies were 125 mg/L. Experimental results with the removal of algae indicate that lower hydraulic rates and air-to-solids ratios than those recommended by the manufacturers of industrial equipment should be utilized when attempting to remove algae.

In combined sedimentation flotation pilot-plant studies at Windhoek, Southwest Africa, van Vuuren and van Duuren (1965) reported effective hydraulic loading rates that ranged between 0.275 and 0.75 gpm/ft² (11.2 and 30.5 L/min·m²), with flotation provided by the naturally dissolved gases. Because air was not added, the air-to-solids ratios were not reported. They also noted that it was necessary to use from 125 to 175 mg/L of aluminum sulfate to flocculate the effluent containing from 25 to 40 mg/L of algae. Subsequent reports on a total flotation system by van Vuuren et al. (1965) stated that a dose of 400 mg/L of aluminum sulfate was required to flocculate a 110-mg/L algal suspension sufficiently to obtain a removal that was satisfactory for consumptive reuse of the water. Based on data provided by Parker et al. (1973), Stone et al. (1975), Bare (1971), and Snider (1976), it appears that a much lower dose of alum would be required to produce a satisfactory effluent to meet present discharge standards.

Dissolved-air flotation with the application of coagulants performs essentially the same function as coagulation–flocculation–sedimentation, except that a much smaller system is required with the flotation device. Flotation will occur in shallow tanks with hydraulic residence times of 7 to 20 min, compared with hours in deep sedimentation tanks. Overflow rates of 2 to 2.5 gpm/ft² (81.5 to 101.8 L/min·m²) can be employed with flotation, whereas a value of less than 1 gpm/ft² (40.7 L/min·m²) is recommended with sedimentation. However, it must be pointed out that the sedimentation process is much simpler than the flotation process, and, when applied to small systems, consideration must be given to this factor.

The flotation process does not require a separate flocculation unit, and this has definite advantages. It has been shown that the introduction of a flocculation step after chemical addition in the flotation system is detrimental. It is best to add alum at the point of pressure release where mixing occurs and a good dispersion of the chemicals occurs. Brown and Caldwell (1976) have designed two tertiary treatment plants that employ flotation, and they have developed design considerations that should be applied when employing flotation. These features are not included in standard flotation units and should be incorporated to ensure good algae removal (Parker, 1976). In addition to incorporating various mechanical improvements, the Brown and Caldwell study recommended that the tank surface be protected from excessive wind currents to prevent float movement to one side of the tank. They found that the relatively light float is easily moved across the water surface by wind action. It was also recommended that the flotation tank be covered in rainy climates to prevent breakdown of the floc by rain. Another alternative proposed has been to store the wastewater in stabilization ponds during the rainy season and then operate the flotation process at a higher rate during dry weather.

Alum–algae sludge was returned to the wastewater stabilization ponds for over 3 years at Sunnyvale, California, with no apparent detrimental effect (Farnham, 1981). Sludge banks, floating mats of material, and increased TSS concentrations in the pond effluent were not observed. Return of the float to the pond system is an alternative at least for a few years. Most estimates of a period of time that sludge can be returned range from 10 to 20 years.

Sludge disposal from a dissolved-air flotation system can impose considerable difficulties. Alum–algae sludge is very difficult to dewater and discard. Centrifugation and vacuum filtration of unconditioned algae–alum sludge have produced marginal results. Indications are that lime coagulation may prove to be as effective as alum and produce sludge more easily dewatered.

Brown and Caldwell (1976) evaluated heat treatment of alum–algae sludges using the Porteous, Zimpro® low-oxidation, and Zimpro® high-oxidation processes and found relatively inefficient results. The Purifa process, using chlorine to stabilize the sludge, produced a sludge dewaterable on sand beds or in a lagoon; however, the high cost of chlorine eliminates this alternative. If algae are killed before entering an anaerobic digester, volatile matter destruction and dewatering results are reasonable. But, as with the other sludge treatment and disposal processes, additional operations and costs are incurred, and the option of dissolved-air flotation loses its competitive position.

5.2 MODIFICATIONS AND ADDITIONS TO TYPICAL DESIGNS

5.2.1 CONTROLLED DISCHARGE

Controlled discharge is defined as limiting the discharge from a lagoon system to those periods when the effluent quality will satisfy existing discharge requirements. The usual practice is to prevent discharge from the lagoon during the winter period and during the spring and fall overturn periods and algal bloom periods. Many countries currently do not permit lagoon discharges during winter months.

Pierce (1974) reported on the quality of lagoon effluent obtained from 49 lagoon installations in Michigan that practice controlled discharge. Of these 49 lagoon systems, 27 have two cells, 19 have three cells, 2 have four cells, and 1 has five cells. Discharge from these systems is generally limited to late spring and early fall; however, several of the systems discharged at various times throughout the year. The period of discharge varied from fewer than 5 d to more than 31 d. The lagoons were emptied to a minimum depth of approximately 0.46 m (18 in.) during each controlled discharge to provide storage capacity for the non-discharge periods.

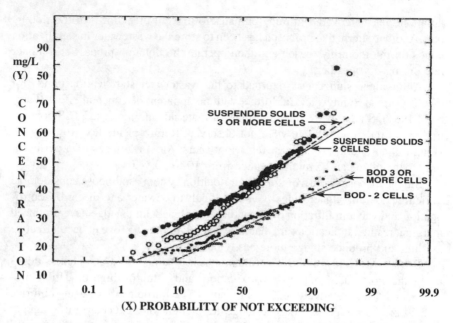

FIGURE 5.7 Comparison of the effluent quality of two-cell systems vs. lagoon systems with three or more cells with long storage periods before discharge (Michigan). (Pierce, D.M., in *Upgrading Wastewater Stabilization Ponds To Meet New Discharge Standards*, PRWG151, Utah Water Research Laboratory, Utah State University, Logan, 1974.)

During the discharge period, the lagoon effluent was monitored for BOD_5, TSS, and FC. The effluent BOD_5 and TSS concentrations measured during the study are illustrated in terms of probability of occurrence in Figure 5.7. All values are arranged in order of magnitude and plotted on normal probability paper with concentration (mg/L) plotted against the probability that the value would not be exceeded under similar conditions. The plot compares the performance of two-cell lagoon systems vs. lagoon systems with three or more cells. The results shown in Figure 5.7 are summarized in Table 5.16.

The results of the study indicated that the most probable effluent BOD_5 concentration for controlled discharge systems was 17 mg/L for the two-cell lagoon systems and 14 mg/L for lagoon systems with three or more cells. There was a 90% probability that the effluent BOD_5 concentration from two-cell systems and lagoon systems with three or more cells would not exceed 27 mg/L. This value was slightly less than the 30-mg/L BOD_5 U.S. Federal Secondary Treatment Standard. The most probable effluent TSS concentration was found to be 30 mg/L for two-cell lagoon systems and 27 mg/L for lagoon systems with three or more cells. The 90% probability levels for effluent TSS concentrations were 46 mg/L for two-cell lagoon systems and 47 mg/L for lagoon systems with three or more cells. The results of the study also indicate that the FC levels were generally less than 200/100 mL, although this standard was exceeded on several occasions when chlorination was not employed.

TABLE 5.16
Effluent Quality Resulting from Controlled Discharge
Operations of 49 Michigan Lagoon Installations

Percent Probability of Occurrence	Effluent BOD$_5$ Concentration (mg/L)		Effluent SS Concentration (mg/L)	
	Two Cells	Three or More Cells	Two Cells	Three or More Cells
50% probability (most probable)	17	14	30	27
90% probability (will not be exceeded 9 out of 10 samples)	27	27	27	47

Source: Pierce, D.M., in *Upgrading Wastewater Stabilization Ponds To Meet New Discharge Standards*, PRWG151, Utah Water Research Laboratory, Utah State University, Logan, 1974.

A similar study of controlled discharge lagoon systems was conducted in Minnesota (Pierce, 1974). The discharge practices of the 39 installations studied were similar to those employed in Michigan. The results of that study from the fall discharge period indicated that the effluent BOD$_5$ concentrations for 36 of 39 installations sampled were less than 25 mg/L, and the effluent TSS concentrations were less than 30 mg/L. In addition, effluent FC concentrations were measured at 17 of the lagoon installations studied. All the installations reported effluent FC concentrations of less than 200/100 mL.

During the spring discharge period, 49 municipal lagoon installations were monitored. Effluent BOD$_5$ concentrations exceeded 30 mg/L at only three installations, while the maximum effluent BOD$_5$ concentration reported was only 39 mg/L. Effluent TSS concentrations ranged from 7 to 128 mg/L, with 16 of the 49 installations reporting effluent TSS concentrations greater than 30 mg/L. Only 3 of the 45 installations monitored for effluent FC concentrations exceeded 200/100 mL.

The controlled discharge of lagoon effluent is a simple, economical, and practical method of achieving a high degree of treatment. Experience indicates that routine monitoring of the lagoon effluent is necessary to determine the proper discharge period; however, these discharge periods may extend throughout the major portion of the year. It will be necessary to increase the storage capacity of certain lagoon systems that employ controlled discharge. Many lagoon systems already have additional freeboard and storage capacity that could be utilized without significant modification.

5.2.2 HYDROGRAPH CONTROLLED RELEASE

The hydrograph controlled release (HCR) pond is a variation of the controlled discharge pond. This concept was developed in the southern United States but can be used effectively in most areas of the world. In this case, the discharge

TABLE 5.17
Hydrograph-Controlled Release Pond
Design Basics Used in United States

a.   Basic principle: At critically low river flow, BOD and SS loadings
      are reduced by restricting effluent discharge rates rather than
      decreasing concentration of pollutants.

b.   Must be equipped to retain wastewater during low flow ($Q_{10/7}$). Use
      existing ponds or build storage ponds. ($Q_{10/7}$ = once-in-10-year low
      flow rate for 7-day period.)

c.   Assimilative capacity of receiving stream must be established by
      studying historical data or estimated using techniques available in
      the literature (Zirschsky and Thomas, 1987).

periods are controlled by a gauging station in the receiving stream and are allowed
to occur during high-flow periods. During low-flow periods, the effluent is stored
in the HCR pond. The process design uses conventional facultative or aerated
ponds for the basic treatment, followed by the HCR cell for storage and discharge.
No treatment allowances are made during design for the residence time in the
HCR cell; its sole function is storage. Depending on stream flow conditions,
storage needs may range from 30 to 120 d. The design maximum water level in
the HCR cell is typically about 8 ft (2.4 m), with a minimum water level of 2 ft
(0.6 m). Other physical elements are similar to conventional pond systems. The
major advantage of HCR systems is the possibility of utilizing lower discharge
standards during high-flow conditions as compared to a system designed for very
stringent low-flow requirements and then operated in that mode on a continuous
basis. A summary of the design approach used in the United States is shown in
Table 5.17. Zirschsky and Thomas (1987) performed an assessment of HCR
systems in the United States that demonstrated that the HCR system is an effective, economical, and easily operated system. It was also found to be an effective
means of upgrading a lagoon system. Several simple effluent release structures
are illustrated in the article.

5.2.3 COMPLETE RETENTION PONDS

In areas of the world where the moisture deficit (evaporation minus rainfall)
exceeds 30 in. (75 cm) annually, a complete retention wastewater pond may prove
to be the most economical method of disposal if low-cost land is available. The
pond must be sized to provide the necessary surface area to evaporate the total
annual wastewater volume plus the precipitation that would fall on the pond. The
system should be designed for the maximum wet year and minimum evaporation
year of record if overflow is not permissible under any circumstance. Less stringent design standards may be appropriate in situations where occasional overflow

is acceptable or an alternative disposal area is available under emergency conditions. Monthly evaporation and precipitation rates must be known in order to size the system properly. Complete retention ponds usually require large land areas, and these areas are not productive once they have been committed to this type of system. Land for this system must be naturally flat or must be shaped to provide ponds that are uniform in depth and have large surface areas. The design procedure for a complete retention wastewater pond system is relatively detailed; the procedure can be obtained from the National Technical Information Service, U.S. Department of Commerce, Springfield, VA 22161 (ask for the EPA-625/1-83-015 document).

5.2.4 AUTOFLOCCULATION AND PHASE ISOLATION

Autoflocculation of algae has been observed during some studies (Golueke and Oswald, 1965; Hill et al., 1977; McGriff and McKinney, 1971; McKinney, 1971). *Chlorella* was the predominant alga occurring in most of the cultures. Laboratory-scale continuous experiments with mixtures of activated sludge and algae have produced large bacteria–algae flocs with good settling characteristics (Hill et al., 1977; Hill and Shindala, 1977). Floating algae blankets have been reported in some cases in the presence of chemical coagulants (Shindala and Stewart, 1971; van Vuuren and van Duuren, 1965). The phenomenon may be caused by the entrapment of gas bubbles produced during metabolism or by the fact that in a particular physiological state the algae have a neutral buoyancy. In a 3000-gph (11,355 L/hr) pilot plant (combined flocculation and sedimentation), a floating algal blanket occurred with alum doses of 125 to 170 mg/L. About 50% of the algae removed were skimmed from the surface (van Vuuren and van Duuren, 1965). Because of the infrequent occurrence of conditions necessary for autoflocculation, it is not a viable alternative for removing algae from wastewater stabilization ponds. Phase isolation experiments to remove algae from lagoon effluents are based on this concept, and some success has been reported; however, full-scale operation of a phase isolation system did not produce consistent results (McGriff, 1981).

5.2.5 BAFFLES AND ATTACHED GROWTH

The encouragement of attached microbial growth in oxidation ponds is an apparent practical solution for maintaining biological populations while still obtaining the treatment desired. Although baffles are considered useful primarily to ensure good mixing and to eliminate the problem of short-circuiting, they behave similarly to the biological disks in that they provide a substrate for bacteria, algae, and other microorganisms to grow (Polprasert and Agarwalla, 1995; Reynolds et al., 1975). In general, attached growth surpasses suspended growth if sufficient surface area is available. In anaerobic or facultative ponds with baffling or biological disks, the microbiological community consists of a gradient of algae to photosynthetic, chromogenic bacteria and, finally, to nonphotosynthetic, nonchromogenic bacteria

(Reynolds et al., 1975). In these baffle experiments, the presence of attached growth on the baffles has been the reason for the higher efficiency of treatment than that in the nonbaffled systems. Polprasert and Agarwalla (1995) demonstrated the significance of biofilm biomass growing on the sidewalls and bottoms of ponds. A model for substrate utilization in facultative ponds was presented using first-order reactions for both suspended and biofilm biomass.

5.2.6 LAND APPLICATION

The design and operation of land treatment systems is described in detail in Reed et al. (1995), Crites et al. (2000), and Chapter 8. Three types of land application — slow rate, soil aquifer treatment, and overland flow — are discussed in Chapter 8.

5.2.7 MACROPHYTE AND ANIMAL SYSTEMS

It is recommended that the references in this chapter be consulted before attempting to design a macrophyte system. Detailed design information can be obtained in Mara et al. (1996, 2000), Pearson and Green (1995), Reed et al. (1995), and others in various volumes of the *Water Science & Technology* journal.

5.2.7.1 Floating Plants

Water hyacinths, duckweeds, pennywort, and water ferns appear to offer the greatest potential for wastewater treatment, and each has its own environmental requirements. Hyacinths, pennywort, and duckweeds are the only floating plants that have been evaluated in pilot- or full-scale systems. Detailed design considerations are presented in Reed et al. (1995).

5.2.7.2 Submerged Plants

Submerged aquatic macrophytes for treatment of wastewaters have been studied extensively in the laboratory, in greenhouses, in a pilot study by McNabb (1976), and in large-scale wetland stormwater treatment systems designed to remove phosphorus to levels less than 20 µm/L (SFWMD, 2003).

5.2.7.3 *Daphnia* and Brine Shrimp

Daphnia are filter feeders, and their main contribution to wastewater treatment is the removal of suspended solids. *Daphnia* cultured in wastewater is very sensitive to pH because high pH values result in the production of un-ionized ammonia, which is toxic to the *Daphnia*. To be effective, shading is required to prevent the growth of algae that will result in high pH values during the daytime. The addition of acid and gentle aeration may be necessary. Brine shrimp survive only in saline waters, which limits their application in wastewater treatment. Laboratory- and pilot-scale experiments have been conducted, but the environmental and management requirements make the process uneconomical.

5.2.7.4 Fish

Fish have been grown in treated wastewaters for centuries, and where toxics are not encountered the process has been successful. Many species of fish have been used in wastewater treatment, but fish activity is temperature dependent. Most grow successfully in warm water, but catfish and minnows are exceptions. Dissolved oxygen concentrations are critical and the presence of un-ionized ammonia is toxic to the young of larger species. Detailed studies of fish in wastewater stabilization ponds have been conducted by Coleman (1974) and Henderson (1979). Numerous studies of fish culture have been conducted around the world (Reed et al., 1995).

5.2.8 CONTROL OF ALGAE GROWTH BY SHADING AND BARLEY STRAW

5.2.8.1 Dyes

Dyes have been applied to small ponds to control algae growth; however, the USEPA has not approved dyes for use in municipal or industrial wastewater lagoons. Aquashade®, a mixture of blue and yellow dyes, is marketed as a means of controlling algae in backyard garden pools and large business park and residential development ponds. The product is registered with the USEPA for these uses. The future approval of the use of dyes in wastewater lagoons is unknown.

5.2.8.2 Fabric Structures

Lagoons in Colorado and other locations have constructed structures suspending greenhouse fabrics of various light transmittance and opaque materials to reduce or eliminate light transmittance in small wastewater ponds. Figure 5.8 shows a partially covered lagoon located in Naturita, Colorado, that uses fabrics. The screening has been successful, but in some cases the fabrics were not fastened adequately to prevent wind damage. A cover adequately protected from the wind should be successful in reducing or eliminating algae growth. With full coverage of the surface, anaerobic conditions are possible, and aeration of the effluent may be necessary to meet discharge standards. Partial shading in correct proportions should avoid anaerobic conditions.

5.2.8.3 Barley Straw

In 1980, a farmer observed that the accidental addition of barley straw to a lake reduced the algae concentration. Allowing barley straw to decompose in ponds has been proposed as a means of controlling algae growth in ponds. Details regarding the application of barley straw are provided in IACR-Centre for Aquatic Plant Management (1999). During decomposition, the chemicals listed in Table 5.18 are released to the water and inhibit the growth of algae (Everall and Lees, 1997). The acceptability of this method of algae control by regulatory agencies has not been resolved.

FIGURE 5.8 Photograph of shading for algae control at Naturita, Colorado. (R.H. Bowman, West Slope Unit Leader, Water Quality Division, Colorado Department of Public Health and Environment, personal communication, 2000.)

5.2.8.4 Lemna Systems

The various uses of Lemna processes are described in Chapter 4.

5.3 PERFORMANCE COMPARISONS WITH OTHER REMOVAL METHODS

Designers and owners of small systems are strongly encouraged to use technology that is as simple as feasible. Experience has shown that small communities, or large ones without properly trained operating personnel and access to spare parts, that use sophisticated technology inevitably encounter serious maintenance problems and frequently fail to meet effluent standards. Methods discussed in this chapter that require good maintenance and operator skills are dissolved-air flotation, centrifugation, coagulation–flocculation, and granular media filtration (rapid sand or mixed media filters with chemical addition). At locations where operation and maintenance are available, these processes can be made to work well.

From the preceding sections, it is obvious that many methods of removing or controlling algae concentrations in lagoon effluents are available; however, selection of the proper method for a particular site is dependent on many variables. Small communities with limited resources and untrained operating personnel should select as simple a system as applicable to their site situation.

In rural areas with adequate land, lagoons such as controlled discharge lagoons or hydrograph controlled release lagoons are an excellent choice. In arid

TABLE 5.18
List of Chemicals Produced by Decomposing Straw

Acetic acid

3-Methylbutanoic acid

2-Methylbutanonic acid

Hexanoic acid

Octanoic acid

Nonanoic acid

Decanoic acid

Dodecanoic acid

Tetradecanoic acid

Hexadecanoic acid

1-Methylnaphthalene

2-(1,1-Dimethlyethyl phenol)

2,6-Dimethoxy-4-(2-propenyl) phenol

2,3-Dihydrobenzofuron

5,6,7,7A-Tetrahydro-4,4,7A-trimethyl-2(4H)benzofuranone

1,1,4,4-Tetramethyl-2,6-*bis*(methylene) cyclohexone

1-Hexacosene

11 Unidentified

Source: Everall, N.C. and Lees, D.R., *Water Res.*, 31, 614–620, 1997.
With permission

areas, the total containment lagoon should be considered. Performance by these types of treatment is controlled by selecting the time of discharge and can be controlled to produce an excellent effluent (BOD_5 and TSS < 30 mg/L).

Where land is limited but resources and personnel are unavailable, it is again best to utilize relatively simple methods to control algae in effluents. Intermittent sand filters, application of effluent to farm lands, overland flow, rapid infiltration, constructed wetlands, and rock filters may serve well. Intermittent sand filters with low application rates and a warm climate will provide nitrification. Land application to farm land will reduce both nitrogen and phosphorus while producing an excellent effluent.

Energy savings with these type processes are substantial, and an excellent effluent can be produced (Middlebrooks et al., 1981). Table 5.19 provides a comparison of expected effluent quality and energy consumption in relatively simple processes up to the most sophisticated used in wastewater treatment. Obviously, many variables must be considered in design, but energy consumption and sophistication must receive due consideration.

TABLE 5.19
Total Annual Energy for Typical 1-mgd System
Including Electrical and Fuel

Treatment System	Effluent Quality				Energy (1000 kWh/yr)
	BOD	SS	P	N	
Rapid infiltration (facultative lagoon)	5	1	2	10	150
Slow rate, ridge + furrow (facultative lagoon)	1	1	0.1	3	181
Overland flow (facultative lagoon)	5	5	5	3	226
Facultative lagoon + intermittent sand filter	15	15	—	10	241
Facultative lagoon + microscreens	30	30	—	15	281
Aerated lagoon + intermittent sand filter	15	15	—	20	506
Extended aeration + sludge drying	20	20	—	—	683
Extended aeration + intermittent sand filter	15	15	—	—	708
Trickling filter + anaerobic digestion	30	30	—	—	783
RBC + anaerobic digestion	30	30	—	—	794
Trickling filter + gravity filtration	20	10	—	—	805
Trickling filter + N removal + filter	20	10	—	5	838
Activated sludge + anaerobic digestion	20	20	—	—	889
Activated sludge + anaerobic digestion + filter	15	10	—	—	911
Activated sludge + nitrification + filter	15	10	—	—	1051
Activated sludge + sludge incineration	20	20	—	—	1440
Activated sludge + AWT	<10	5	<1	<1	3809
Physical chemical advanced secondary	30	10	1	—	4464

Source: Middlebrooks, E.J. et al., *J. Water Pollut. Control Fed.*, 53(7), 1172–1198, 1981. With permission.

REFERENCES

Adam, J.M. (1986). Investigation of Rock Filters in Northwestern Illinois, paper presented at the 7th Annual Conference, Illinois Water Pollution Control Association, June 24–26, 1986.

Al-Layla, M.A. and Middlebrooks, E.J. (1975). Effect of temperature on algal removal from wastewater stabilization ponds by alum coagulation, *Water Res.*, 9, 873–879.

Al-Layla, M.A., Ahmad, S., and Middlebrooks, E.J. (1980). *Handbook of Wastewater Collection and Treatment: Principles and Practices*, Garland STPM Press, New York.

Bare, W.F.R., Jones, N.B., and Middlebrooks, E.J. (1975) Algae removal using dissolved-air flotation, *J. Water Pollut. Control Fed.*, 47(1), 153–169.

Benefield, L.D., Randall, C.W. (1980). *Biological Process Design for Wastewater Treatment*, Prentice-Hall, Englewood Cliffs, NJ.

Bishop, R.P., Reynolds, J.H., Filip, D.S., and Middlebrooks, E.J. (1977). *Upgrading Aerated Lagoon Effluent with Intermittent Sand Filtration*, PRWR&T 167-1, Utah Water Research Laboratory, Utah State University, Logan.

Borchardt, J.A. and O'Melia, C.R. (1961). Sand filtration of algae suspensions, *J. Am. Water Works Assoc.*, 53(12), 1493–1502.

Boulier, G.A. and Atchinson, T.J. (1975). *Practical Design and Application of the Aerated–Facultative Lagoon Process*, Hinde Engineering Co., Highland Park, IL.

Brown and Caldwell. (1976). *Draft Project Report*, City of Davis–Algae Removal Facilities, Walnut Creek, CA.

California Department of Water Resources. (1971). *Removal of Nitrate by an Algal System: Bioengineering Aspects of Agricultural Drainage*, San Joaquin Valley, CA.

Chan, D.B. and Pearson, E.A. (1970). *Comprehensive Studies of Solid Waste Management: Hydrolysis Rate of Cellulose in Anaerobic Digesters*, SERL Report No. 70-3, University of California, Berkeley.

Coleman, M.S. (1974). Aquaculture as a means to achieve effluent standards, in *Wastewater Use in the Production of Food and Fiber*, EPA 660/2-74-041, U.S. Environmental Protection Agency, Washington, D.C., 199–214.

Crites, R.W. and Tchobanoglous, G. (1998). *Small and Decentralized Wastewater Management Systems*, McGraw-Hill, New York.

Crites, R.W., Reed, S.C., and Bastian, R.K. (2000). *Land Treatment Systems for Municipal and Industrial Wastes*, McGraw Hill, New York.

Davis, E. and Borchardt, J.A. (1966). Sand filtration of particulate matter, *Proc. Am. Soc. Civil Eng. J. Sanit. Eng. Div.*, 92(SA5), 47–60.

Dryden, F.D. and Stern, G. (1968). Renovated wastewater creates recreational lake, *Environ. Sci. Technol.*, 2(4), 266–278.

Everall, N.C. and Lees, D.R. (1997). The identification and significance of chemicals released from decomposing barley straw during reservoir algal control, *Water Res.*, 31, 614-620.

Farnham, H. (pers. comm., 1981). Sunnyvale, CA, Wastewater Treatment Plant, Sunnyvale, CA.

Foess, G.W. and Borchardt, J.A. (1969). Electrokinetic phenomenon in the filtration of algae suspensions, *J. Am. Water Works Assoc.*, 61(7), 333–337.

Furman, T. deS., Calaway, W. T., and Grantham, G.R. (1955). Intermittent sand filter multiple loadings, *Sewage Indust. Wastes*, 27(3), 261–276.

Ghosh, S. and Klass, D.L. (1974). Conversion of urban refuse to substitute natural gas by the BIOGAS® process, in *Proceedings of the Fourth Mineral Waste Utilization Symposium*, Chicago, IL, May 7–8.

Ghosh, S., Conrad, J.R., and Klass, D.L. (1974). Development of an Anaerobic Digestion-Based Refuse Disposal Reclamation System, paper presented at the 47th Annual Conference of the Water Pollution Control Federation, Denver, CO, October 6–11.

Gloyna, E.F. (1971). *Waste Stabilization Ponds*, Monograph Series No. 60, World Health Organization, Geneva, Switzerland.

Golueke, C. and Oswald, W.J. (1965). Harvesting and processing sewage-grown planktonic algae, *J. Water Pollut. Control Fed.*, 37(4), 471–498.

Grantham, G.R., Emerson, D.L., and Henry, A.K. (1949). Intermittent sand filter studies, *Sewage Works J.*, 21(6), 1002–1015.

Great Lakes–Upper Mississippi River Board of State Sanitary Engineers. (1973). *Recommended Standards for Sewage Works (Ten-States Standards)*, Health Education Services, Inc., Albany, NY.

Harris, S.E., Filip, D.S., Reynolds, J.H., and Middlebrooks E.J. (1978). *Separation of Algal Cells from Wastewater Lagoon Effluents*. Vol. I. *Intermittent Sand Filtration To Upgrade Waste Stabilization Lagoon Effluent*, EPA-600/2-78-033, NTIS No. PB 284925, Municipal Environmental Research Laboratory, U.S. Environmental Protection Agency, Cincinnati, OH.

Henderson, S. (1979). Utilization of silver and bighead carp for water quality improvement, in *Aquaculture Systems for Wastewater Treatment Seminar Proceedings and Engineering Assessment*, EPA 430/9-80-006, U.S. Environmental Protection Agency, Washington, D.C., 309–350.

Hill, D.O. and Shindala, A. (1977). *Performance Evaluation of Kilmichael Lagoon*, EPA-600/2-77-109, Municipal Environmental Research Laboratory, U.S. Environmental Protection Agency, Cincinnati, OH.

Hill, F.E., Reynolds, J.H., Filip, D.S., and Middlebrooks, E.J. (1977). *Series Intermittent Sand Filtration to Upgrade Wastewater Lagoon Effluents*, PRWR 153-1, Utah Water Research Laboratory, Utah State University, Logan.

Hobson, P., Bousfield, N.S., and Summers, R. (1974). Anaerobic digestion of organic matter, *CRC Crit. Rev. Environ. Control*, 4, 131–191.

IACR–Centre for Aquatic Plant Management (1999). *Information Sheet 3: Control of Algae Using Straw*, Centre for Aquatic Plant Management, Reading, Berkshire, U.K.

Kormanik, R.A. and Cravens, J.B. (1978). Microscreening and other physical–chemical techniques for algae removal, in *Proceedings of Performance and Upgrading of Wastewater Stabilization Ponds Conference*, Utah State University, Logan, August 23–25, 1978.

Kotze, J.P., Thiel, P.G., Toerien, D.F., Attingh, W.H.J., and Siebert, M.L. (1968). A biological and chemical study of several anaerobic digesters, *Water Res.*, 2(3), 195–213.

Lynam, G., Ettelt, G., and McAllon, T. (1969). Tertiary treatment of metro Chicago by means of rapid sand filtration and microstrainers, *J. Water Pollut. Control Fed.*, 41(2), 247–279.

Malina, Jr., J.F. and Rios, R.A. (1976). Anaerobic ponds, in *Ponds as a Wastewater Treatment Alternative*, Gloyna, E.F., Malina, Jr., J.F., and Davis, E.M., Eds., Water Resources Symposium No. 9, University of Texas Press, Austin, TX.

Mancini, J.L. and Barnhart, E.L. (1976). Industrial waste treatment in aerated lagoons, in *Ponds as a Wastewater Treatment Alternative*, Gloyna, E.F., Malina, Jr., J.F., and Davis, E.M., Eds., Water Resources Symposium No. 9, University of Texas Press, Austin, TX.

Mancl, K.M. and Peeples, J.A. (1991). One hundred years later: reviewing the work of the Massachusetts State Board of Health on the intermittent sand filtration of wastewater from small communities, in *Proceedings of the 6th National Symposium on Individual and Small Community Sewage Systems*, Chicago, American Society of Agricultural Engineers (ASAE), December 16–17, 155.

Mara, D.D. (1996). *Sewage Treatment in Hot Climates*, John Wiley & Sons, New York, 1976.

Mara, D.D., Pearson, H.W., and Silva, S.A., Eds. (1996). Waste stabilization ponds: technology and applications, *Water Sci. Technol.*, 33, 7.

Mara, D.D., Azov, Y., and Pearson, H.W., Eds. (2000). Waste stabilization ponds: technology and the environment, *Water Sci. Technol.*, 42,10-11.

Marshall, G.R. and Middlebrooks, E.J. (1974). *Intermittent Sand Filtration To Upgrade Existing Wastewater Treatment Facilities*, PRJEW 115-2, Utah Water Research Laboratory, Utah State University, Logan.

Massachusetts Board of Health. (1912). The condition of an intermittent sand filter for sewage after twenty-three years of operation, *Eng. Contracting*, 37, 271.

McGarry, M.G. (1970). Algal flocculation with aluminum sulfate and polyelectrolytes, *J. Water Pollut. Control Fed.*, 42, 5, R191.

McGriff, E.C. (1981). *Facultative Lagoon Effluent Polishing Using Phase Isolation Ponds*, EPA-600/2-81-084, NTIS No. PB 81-205965, Municipal Environmental Research Laboratory, U.S. Environmental Protection Agency, Cincinnati, OH.

McGriff, E.C. and McKinney, R.E. (1971). Activated algae? A nutrient removal process, *Water Sewage Works*, 118, 337.

McKinney, R.E. (1971). Ahead: activated algae?, *Water Wastes Eng.*, 8, 51.

McNabb, C.D. (1976). The potential of submerged vascular plants for reclamation of wastewater in temperate zone ponds, in *Biological Control of Water Pollution*, University of Pennsylvania Press, Philadelphia, PA, 123–132.

Melcer, H., Evans, B., Nutt, S.G., and Ho, A. (1995). Upgrading effluent quality for lagoon-based systems, *Water Sci. Technol.*, 31(12), 379–387.

Menninga, N. (pers. comm., 1986). Illinois Environmental Protection Agency, Springfield.

Messinger, S.S. (1976). Anaerobic Lagoon–Intermittent Sand Filter System for Treatment of Dairy Parlor Wastes, M.S. thesis, Utah State University, Logan.

Metcalf & Eddy (2003). *Wastewater Engineering Treatment Disposal Reuse*, 4th. ed., McGraw-Hill, New York.

Middlebrooks, E.J. (1988). Review of rock filters for the upgrade of lagoon effluents, *J. Water Pollut. Control Fed.*, 60, 1657–1662.

Middlebrooks, E.J., Middlebrooks, C.H., and Reed, S.C. (1981). Energy requirements for small wastewater treatment systems, *J. Water Pollut. Control Fed.*, 53(7), 1172–1198.

Middlebrooks, E.J., Middlebrooks, C.H., Reynolds, J.H., Watters, G.Z., Reed, S.C., and George, D.B. (1982). *Wastewater Stabilization Lagoon Design, Performance, and Upgrading*, Macmillan, New York.

Niku, S., Schroeder, E.D., Tchobanoglous, G., and Samaniego, F.J. (1981). *Performance of Activated Sludge Processes: Reliability, Stability, and Variability*, EPA-600/52-81-227, U.S. Environmental Protection Agency, Cincinnati, OH.

Nolte & Associates (1992). *Literature Review of Recirculating and Intermittent Sand Filters: Operation and Performance, Town of Paradise*, prepared for the California Regional Water Quality Control Board, Sacramento.

Oswald, W.J. (1968). Advances in anaerobic pond systems design, in *Advances in Water Quality Improvement*, Gloyna, E.F. and Eckenfelder, Jr., W.W., Eds., University of Texas Press, Austin, TX, 409.

Oswald, W.J. (1996). *A Syllabus on Advanced Integrated Pond Systems*, University of California, Berkeley.

Oswald, W.J., Golueke, C.G., and Tyler, R.W. (1967). Integrated pond systems for subdivisions, *J. Water Pollut. Control Fed.*, 39(8), 1289.

Parker, C.D. (1970). Experiences with anaerobic lagoons in Australia, in *Proceedings of the Second International Symposium for Waste Treatment Lagoons*, Kansas City, MO, June 23–25, 334.

Parker, C.D., Jones, H.L., and Greene, N.C. (1959). Performance of Large Sewage Lagoons at Melbourne, Australia, *Sewage Indust. Wastes*, 31(2), 133.

Parker, D.S. (1976). Performance of alternative algae removal systems, in *Ponds as a Wastewater Treatment Alternative*, Gloyna, E.F., Malina, Jr., J.F., and Davis, E.M., Eds., Water Resources Symposium No. 9, University of Texas Press, Austin, TX.

Parker, D.S., Tyler, J.B., and Dosh, T.J. (1973). Algae removal improves pond effluent, *Water Wastes Eng.*, 10, 1.

Pearson, H.W. and Green, F.B., Eds. (1995). Waste stabilization ponds and the reuse of pond effluents, *Water Sci. Technol.*, 31, 12.

Pierce, D.M. (1974). Performance of raw waste stabilization lagoons in Michigan with long period storage before discharge, in *Upgrading Wastewater Stabilization Ponds To Meet New Discharge Standards*, PRWG151, Utah Water Research Laboratory, Utah State University, Logan.

Polprasert, C. and Agarwalla, B.K. (1995). Significance of biofilm activity in facultative pond design and performance, *Water Sci. Technol.*, 31(12), 119–128.

Reed, S.C., Crites, R.W., and Middlebrooks, E.J. (1995). *Natural Systems for Waste Management and Treatment*, 2nd ed., McGraw-Hill, New York.

Reid, Jr., L.D. (1970). *Design and Operation for Aerated Lagoons in the Arctic and Subarctic*, Report 120, U.S. Public Health Service, Arctic Health Research Center, Fairbanks, AK.

Reynolds, J.H., Nielson, S.B., and Middlebrooks, E.J. (1975). Biomass distribution and kinetics of baffled lagoons, *J. Environ. Eng. Div. ASCE*, 101(EE6), 1005–1024.

Rich, L.G. and Wahlberg, E.J. (1990). Performance of lagoon-intermittent sand filter systems, *J. Water Pollut. Control Fed.*, 62, 697–699.

Russell, J.S., Middlebrooks, E.J., and Reynolds, J.H. (1980). *Wastewater Stabilization Lagoon–Intermittent Sand Filter Systems*, EPA 600/2-80-032, Municipal Engineering Research Laboratory, U.S. Environmental Protection Agency, Cincinnati, OH.

Russell, J.S., Middlebrooks, E.J. Lewis, R.F., and Barth, E.F. (1983). Lagoon effluent polishing with intermittent sand filters, *J. Environ. Eng. Div. ASCE*, 109(6), 1333–1353.

Saidam, M.Y., Ramadan, S.A., and Butler, D. (1995). Upgrading waste stabilization pond effluent by rock filters, *Water Sci. Technol.*, 31(12), 369–378.

SFWMD. (2003). *2003 Everglades Consolidated Report*, South Florida Water Management District, West Palm Beach, FL.

Shindala, A. and Stewart, J.W. (1971). Chemical coagulation of effluents from municipal waste stabilization ponds, *Water Sewage Works*, 118(4), 100–103.

Snider, Jr., E.F. (1976). Algae removal by air flotation, in *Ponds as a Wastewater Treatment Alternative*, Gloynd, E.F., Malina, Jr., J.F., and Davis, E.M., Eds., Water Resources Symposium No. 9, University of Texas Press, Austin, TX.

Snider, K.E. (pers. comm., 1998). The Engineering Co., Fort Collins, CO.

Stamberg, J.B. et al. (1984). Simple Rock Filter Upgrades Lagoon Effluent to AWT Quality in West Monroe, Louisiana, paper presented at 57th Conf. Water Pollution Control Federation, New Orleans, LA.

Stone, R.W., Parker, D.S., and Cotteral, J.A. (1975). Upgrading lagoon effluent to meet best practicable treatment, *J. Water Pollut. Control Fed.*, 47(8), 2019–2042.

Swanson, G.R. and Williamson, K.J. (1980). Upgrading lagoon effluents with rock filters, *J. Environ. Eng. Div. ASCE*, 106(EE6), 1111–1119.

Tenney, M.W. (1968). Algal flocculation with aluminum sulfate and polyelectrolytes, *Appl. Microbiol.*, 18(6), 965.

Truax, D.D. and Shindala, A. (1994). A filtration technique for algal removal from lagoon effluents, *Water Environ. Res.*, 66(7), 894–898.

Tupyi, B., Reynolds, J.H., Filip, D.S., and Middlebrooks, E.J. (1979). *Separation of Algal Cells from Wastewater Lagoon Effluents*. Vol. II. *Effect of Sand Size on the Performance of Intermittent Sand Filters*, EPA-600/2-79-152, NTIS No. PB 80-120132, Municipal Environmental Research Laboratory, U.S. Environmental Protection Agency, Cincinnati, OH.

USEPA. (1981). *Process Design Manual for Land Treatment of Municipal Wastewaters*, EPA 625/1-77-008, U.S. Environmental Protection Agency, Cincinnati, OH.

USEPA. (1983). *Design Manual: Municipal Wastewater Stabilization Ponds*, EPA 625/1-83-015, Center for Environmental Research Information, U.S. Environmental Protection Agency, Cincinnati, OH.

van Vuuren, L.R.J. and van Duuren, F.A. (1965). Removal of algae from wastewater maturation pond effluent, *J. Water Pollut. Control Fed.*, 37, 1256.

van Vuuren, L.R.J., Meiring, P.G.J., Henzen M.R., and Kolbe, F.F. (1965). The flotation of algae in water reclamation, *Int. J. Air Water Pollut.*, 9(12), 823.

Water Environment Federation. (2001). *Natural Systems for Wastewater Treatment*, 2nd ed., Manual of Practice FD-16, Water Pollution Control Federation, Alexandria, VA.

WHO. (1987). *Wastewater Stabilization Ponds: Principles of Planning and Practice*, WHO Tech. Publ. 10, World Health Organization, Regional Office for the Eastern Mediterranean, Alexandria.

Zirschsky, J. and Thomas, R.E. (1987). State of the art hydrograph controlled release (HCR) lagoons, *J. Water Pollut. Control Fed.*, 59(7), 695–698.

6 Free Water Surface Constructed Wetlands

Wetlands are defined for this book as ecosystems where the water surface is at or near the ground surface for long enough each year to maintain saturated soil conditions and related vegetation. The major wetland types with potential for water quality improvement are swamps that are dominated by trees, bogs that are characterized by mosses and peat, and marshes that contain grasses and emergent macrophytes. The majority of wetlands used for wastewater treatment are in the marsh category, but a few examples of the other two types also exist. The capability of these ecosystems to improve water quality has been recognized for at least 30 years. The use of engineered wetland systems for wastewater treatment has emerged during this period at an accelerating pace. The engineering involved may range from installation of simple inlet and outlet structures in a natural wetland to the design and construction of a completely new wetland where one did not exist before. The design goals of these systems may range from an exclusive commitment for treatment functions to systems that provide advanced treatment or polishing combined with enhanced wildlife habitat and public recreational opportunities. The size of these systems ranges from small on-site units designed to treat the septic tank effluent from a single-family dwelling to 40,000-ac (16,200 ha) wetlands in South Florida for the treatment of phosphorus in agricultural stormwater drainage. These wetland systems are land intensive but offer a very effective biological treatment response in a passive manner so that mechanical equipment, energy, and skilled operator attention are minimized. Where suitable land is available at a reasonable cost, wetland systems can be a most cost-effective treatment alternative, while also providing enhanced habitat and recreational values.

6.1 PROCESS DESCRIPTION

For engineering purposes, wetlands have been described in terms of the position of the water surface. The free water surface (FWS) wetland is characterized by a water surface exposed to the atmosphere. Natural marshes and swamps are FWS wetlands, and bogs can be if the water flows on top of the peat. Most constructed FWS wetlands typically consist of one or more vegetated shallow basins or channels with a barrier to prevent seepage, with soil to support the emergent macrophyte vegetation, and with appropriate inlet and outlet structures. The water depth in this type of constructed wetland might range from 0.2 to 2.6

ft (0.05 to 0.8 m). The design flows for operational FWS treatment wetlands range from less than 1000 gpd (4 m^3/d) to over 20 mgd (75,000 m^3/d).

The biological conditions in these wetlands are similar, in some respects, to those occurring in facultative treatment ponds. The water near the bottom of the wetland is in an anoxic/anaerobic state; a shallow zone near the water surface tends to be aerobic, and the source of that oxygen is atmospheric reaeration. Facultative lagoons, as described in Chapter 4, have an additional source of oxygen that is generated by the algae present in the system. In a densely vegetated wetland, this oxygen source is not available because the plant canopy shades the water surface and algae cannot persist. The most significant difference is the presence, in the wetlands, of physical substrate for the development of periphytic attached-growth microorganisms, which are responsible for much of the biological treatment occurring in the system. In FWS wetlands, these substrates are the submerged leaves and stems of the living plants, the standing dead plants, and the benthic litter layer. In subsurface flow (SSF) wetlands (see Chapter 7), the substrate is composed of the submerged media surfaces and the roots and rhizomes of the emergent plants growing in the system. Many of the treatment responses proceed at a higher rate in a wetland than in facultative lagoons because of the presence of the substrate and these periphytic organisms, and the response in SSF wetlands is typically at a higher rate than in FWS wetlands because of the increased availability of substrate in the gravel media.

In addition to a higher rate of treatment than FWS wetlands, the SSF wetland concept offers several other advantages. Because the water surface is below the top of the gravel, mosquitoes are not a problem as the larvae cannot develop. In cold climates, the subsurface position of the water and the litter layer on top of the gravel offer greater thermal protection for the SSF wetland. The greatest advantage is the minimal risk of public exposure or contact with the wastewater because the water surface is not directly, or easily, accessible; however, the major disadvantage for the SSF concept is the cost of the gravel media. The unit costs for the other system components (e.g., excavation, liner, inlets, outlets) are about the same for either SSF or FWS wetlands, but the cost of gravel in the SSF system adds significantly to project costs. For design flow rates larger than about 50,000 gpd (190 m^3/d), the smaller size of the SSF wetland does not usually compensate for the extra cost of the gravel. Because of these costs, the SSF concept is best suited for those smaller applications where public exposure is an issue, including individual homes, groups of homes, parks, schools, and other commercial and public facilities. It will be more economical to utilize the FWS concept for larger municipal and industrial systems and for other potential wetland applications. The FWS concept also offers a greater potential for incorporation of habitat values in a project. An example of a FWS wetland is shown in Figure 6.1.

The treatment processes occurring in both FWS and SSF wetlands are a complex and interrelated sequence of biological, chemical, and physical responses. Because of the shallow water depth and the low flow velocities, particulate matter settles rapidly or is trapped in the submerged matrix of plants or gravel. Algae are also trapped and cannot regenerate because of the shading

FIGURE 6.1 Free water surface (FWS) wetlands at Arcata, California.

effect in the densely vegetated portions of the wetland. These deposited materials then undergo anaerobic decomposition in the benthic layers and release dissolved and gaseous substances to the water. All of the dissolved substances are available for sorption by the soils and the active microbial and plant populations throughout the wetland. Oxygen is available at the water surface and on microsites on the living plant surfaces and root and rhizome surfaces so aerobic reactions are also possible within the system.

6.2 WETLAND COMPONENTS

The major system components that may influence the treatment process in constructed wetlands include the plants, detritus, soils, bacteria, protozoa, and higher animals. Their functions and the system performance are, in turn, influenced by water depth, temperature, pH, redox potential, and dissolved oxygen concentration.

6.2.1 TYPES OF PLANTS

A wide variety of aquatic plants have been used in wetland systems designed for wastewater treatment. The larger trees (e.g., cypress, ash, willow) often preexist on natural bogs, strands, and "domes" used for wastewater treatment in Florida and elsewhere. No attempt has been made to use these species in a constructed wetland nor has their function as a treatment component in the system been defined. The emergent aquatic macrophytes are the most commonly found species in the marsh type of constructed wetlands used for wastewater treatment. The most frequently used are cattails (*Typha*), reeds (*Phragmites communis*), rushes (*Juncus* spp.), bulrushes (*Scirpus*), and sedges (*Carex*). Bulrush and cattails, or a combination of the two, are the dominant species on most of the constructed wetlands in the United States. A few systems in the United States have *Phragmites*, but this species is the dominant type selected for constructed wetlands in Europe. Systems that are specifically designed for habitat values in addition to treatment usually select a greater variety of plants with an emphasis on food and nesting values for birds and other aquatic life. Information on some typical plant species common in the United States and a discussion of advantages and disadvantages for their use in a constructed wetland are provided in the following text. Further details on the characteristics of these plants can be found in a number of references (Hammer, 1992; Lawson, 1985; Mitsch and Gosselink, 2000; Thornhurst, 1993).

6.2.2 Emergent Species

6.2.2.1 Cattail

Typical varieties are *Typha angustifolia* (narrow leaf cattail) and *Typha latifolia* (broad leaf cattail). Distribution is worldwide. Optimum pH is 4 to 10. Salinity tolerance for narrow leaf is 15 to 30 ppt; broad leaf, <1 ppt. Growth is rapid, via rhizomes; the plant spreads laterally to provide dense cover in less than a year with 2-ft (0.6-m) plant spacing. Root penetration is relatively shallow in gravel (approximately 1 ft or 0.3 m). Annual yield is 14 (dw) ton/ac (30 mt/ha). Tissue (dw basis) is 45% C, 14% N, 2% P; 30% solids. Seeds and roots are a food source for water birds, muskrat, nutria, and beaver; cattails also provide nesting cover for birds. Cattails can be permanently inundated at >1 ft (0.3 m) but can also tolerate drought. They are commonly used on many FWS and SSF wetlands in the United States. The relatively shallow root penetration is not desirable for SSF systems without adjusting the design depth of bed.

6.2.2.2 Bulrush

Typical varieties are *Scirpus acutus* (hardstem bulrush), common tule, *Scirpus cypernius* (wool grass), *Scirpus fluviatilis* (river bulrush), *Scirpus robustus* (alkali bulrush), *Scirpus validus* (soft stem bulrush), and *Scirpus lacustris* (bulrush). Bulrush is known as *Scirpus* in the United States but is referred to as *Schoenoplectus* in the rest of the world (Mitsch and Gosselink, 2000). Distribution is worldwide. Optimum pH is 4 to 9. Salinity tolerance for hardstem, wool grass, river, and soft stem bulrushes is 0 to 5 ppt; alkali and Olney's, 25 ppt. Growth of alkali, wool grass, and river bulrush is moderate, with dense cover achieved in 1 yr with 1-ft (0.3-m) plant spacing; growth of all others is moderate to rapid, with dense cover achieved in 1 yr with 1- to 2-ft (0.3- to 0.60-m) plant spacing. Deep root penetration in gravel is approximately 2 ft (0.6m). Annual yield is approximately 9 (dw) ton/ac (20 mt/ha). Tissue (dw basis) is approximately 18% N, 2% P; 30% solids. Bulrush seeds and rhizomes are a food source for many water birds, muskrats, nutria, and fish; they also provide a nesting area for fish when inundated. Bulrushes can be permanently inundated — hardstem up to 3 ft (1 m), most others 0.5 to 1 ft (0.15 to 0.3 m); some can tolerate drought conditions. They are commonly used for many FWS and SSF constructed wetlands in the United States.

6.2.2.3 Reeds

Typical varieties are *Phragmites australis* (common reed) and wild reed. Distribution is worldwide. Optimum pH is 2 to 8. Salinity tolerance is <45 ppt. Growth is very rapid, via rhizomes; lateral spread is approximately 3 ft/yr (1 m/yr), providing very dense cover in 1 yr with plants spaced at 2 ft (0.6 m). Deep root penetration in gravel is approximately 1.5 ft (0.4 m). Annual yield is approximately 18 (dw) ton/ac (40 mt/ha). Tissue (dw basis) is approximately 45% C, 20% N, 2% P; 40% solids. With regard to habitat values, reeds have low food value for most birds and animals and some value as nesting cover for birds and animals. They can be permanently inundated up to about 1 m (3 ft), and are also very drought resistant. They are considered by some to be an invasive pest species in natural wetlands in the United States. They have been very successfully used at constructed wastewater treatment wetlands in the United States. They are the dominant species used for this purpose in Europe. Because of its low food value, this species is not subject to the damage caused by muskrat and nutria which has occurred in constructed wetlands supporting other plant species.

6.2.2.4 Rushes

Typical varieties are *Juncus articulatus* (jointed rush), *Juncus balticus* (Baltic rush), and *Juncus effusus* (soft rush). Distribution is worldwide. Optimum pH is 5 to 7.5. Salinity tolerance is 0 to <25, depending on type. Growth is very slow, via rhizomes; lateral spread is <0.3 ft/yr (0.1 m/yr), providing dense cover in 1 year with plants spaced at 0.5 ft (0.15 m). Annual yield is 45 (dw) ton/ac (50 mt/ha). Tissue (dw basis) is approximately 15% N, 2% P; 50% solids. Rushes provide food for many bird species, and their roots are food for muskrats. Some rushes can tolerate permanent inundation up to <1 ft (0.3 m), but they prefer dry-down periods. Other plants are better suited as the major species for wastewater wetlands; rushes are well suited as a peripheral planting for habitat enhancement.

6.2.2.5 Sedges

Typical varieties are *Carex aquatilis* (water sedge), *Carex lacustris* (lake sedge), and *Carex stricata* (tussock sedge). Distribution is worldwide. Optimum pH is 5 to 7.5. Salinity tolerance is <0.5 ppt. Growth is moderate to slow, via rhizomes; lateral spread is <0.5 ft/yr (0.15 m/yr), providing dense cover in 1 year with plants spaced at 0.5 ft (0.15 m). Annual yield is <4 (dw) ton/yr (5 mt/ha). Tissue (on a dw basis) is approximately 1% N, 0.1% P; 50% solids. With regard to habitat values, sedges are a food source for numerous birds and moose. Some types can sustain permanent inundation; others require a dry-down period. Other plants are better suited as the major species for wastewater wetlands; sedges are well suited as a peripheral planting for habitat enhancement.

6.2.3 SUBMERGED SPECIES

Submerged plant species have been used in deepwater zones of FWS wetlands and are a component in a patented process that has been used to improve water quality in freshwater lakes, ponds, and golf course water hazards. Species that have been used for this purpose include *Ceratophyllum demersum* (coontail, or hornwart), *Elodea* (waterweed), *Potamogeton pectinatus* (sago pond weed), *Potamogeton perfoliatus* (redhead grass), *Ruppia maritima* (widgeongrass), *Vallisneria americana* (wild celery), and *Myriophyllum* spp. (watermilfoil). The distribution of these species is worldwide. Optimum pH is 6 to 10. Salinity tolerance is <5 to 15 ppt for most varieties. Growth is rapid, via rhizomes; lateral spread is >1 ft/yr (0.3 m/yr), providing dense cover in 1 year with plants spaced at 2 ft (0.6 m). Annual yields vary — coontail, 8.9 (dw) ton/ac (10 mt/ha); *Potamogeton*, 2.7 (dw) ton/ac(3 mt/ha); and watermilfoil, 8 (dw) ton/ac (9 mt/ha). Tissue (dw basis) is approximately 2 to 5% N, 0.1 to 1% P; 5 to 10% solids. These species provide food for a wide variety of birds, fish, and animals; sago pond weed is especially valuable for ducks. These species can tolerate continuous inundation, with the depth of acceptable water being a function of water clarity and turbidity as these plants depend on penetration of sunlight through the water column. Some of these plants have been used to enhance the habitat values in FWS constructed wetlands. Coontail, *Elodea*, and other species have been used for nutrient control in freshwater ponds and lakes; regular harvesting removes the plants and the nutrients.

6.2.4 FLOATING SPECIES

Several floating plants have been used in wastewater treatment systems. These floating plants are not typically a design component in constructed wetlands. The species most likely to occur incidentally in FWS wetlands is *Lemna* (duckweed). The presence of duckweed on the water surface of a wetland can be both beneficial and detrimental. The benefit occurs because the growth of algae is suppressed; the detrimental effect is the reduction in transfer of atmospheric oxygen at the water surface because of the duckweed mat. The growth rate of this plant is very rapid, and the annual yield can be 18 (dw) ton/ac (20 mt/hat) or more. The tissue composition (dw basis) is approximately 6% N, 2% P; solids 5%. Salinity tolerance is less than 0.5 ppt. These species serve as a food source for ducks and other water birds, muskrat, and beaver. The presence of duckweed on FWS wetlands cannot be prevented because the plant also tolerates partial shade. Open-water zones in FWS wetlands should be large enough so wind action can periodically break up and move any duckweed mat to permit desirable reaeration. The decomposition of the unplanned duckweed may also impose an unexpected seasonal nitrogen load on the system.

6.2.5 EVAPOTRANSPIRATION LOSSES

The water losses due to evapotranspiration (ET) should be considered for wetland designs in arid climates and can be a factor during the warm summer months in all locations. In the western United States where appropriative laws govern the use of water, it may be necessary to replace the volume of water lost to protect the rights of downstream water users. Evaporative water losses in the summer months decrease the water volume in the system; therefore, the concentration of pollutants remaining in the system tends to increase even though treatment is very effective on a mass removal basis. For design purposes, the evapotranspiration rate can be taken as being equal to 80% of the pan evaporation rate for the area. This in effect is equal to the lake evaporation rate. In the past, some controversy existed regarding the effect of plants on the evaporation rate. It is the current consensus that the shading effect of emergent or floating plants reduces direct evaporation from the water but the plants still transpire. The net effect is roughly the same rate whether plants are present or not. The first edition of this book indicated relatively high ET rates for some emergent plant species (Reed et al., 1988). These data were obtained from relatively small culture tanks and containers and are not representative of full-scale wetland systems.

6.2.6 OXYGEN TRANSFER

Because of the continuous inundation, the soils or the media in a SSF wetland are anaerobic, which is an environment not well suited to support most vegetative species; however, the emergent plant species described previously have all developed the capability of absorbing oxygen and other necessary gasses from the atmosphere through their leaves and above-water stems, and they have large gas vessels, which conduct those gasses to the roots so the roots are sustained aerobically in an otherwise anaerobic environment. It has been estimated that these plants can transfer between 5 and 45 g of oxygen per day per square meter of wetland surface area, depending on plant density and oxygen stress levels in the root zone (Boon, 1985; Lawson, 1985). However, current estimates are that the transfer is more typically 4 g of oxygen per square meter (Brix, 1994; Vymazal et al., 1998).

Most of this oxygen is utilized at the plant roots, and availability is limited for support of external microbial activity; however, some of this oxygen is believed to reach the surfaces of the roots and rhizomes and create aerobic microsites at these points. These aerobic microsites can then support aerobic reactions such as nitrification if other conditions are appropriate. The plant seems to respond with more oxygen as the demand increases at the roots, but the transfer capability is limited. Heavy deposits of raw sludge at the head of some constructed wetlands have apparently overwhelmed the oxygen transfer capability and resulted in plant die-off. This oxygen source is of most benefit in the SSF constructed wetland, where the wastewater flows through the media and comes in direct contact with the roots and rhizomes of the plants. In the FWS wetland, the wastewater flows above the soil layer and the contained roots and does not come into direct contact with this potential oxygen source. The major oxygen source for the FWS wetland is believed to be atmospheric reaeration at the water surface. To maximize the benefit in the SSF case, it is important to encourage root penetration to the full depth of the media so potential contact points exist throughout the profile. As described in Chapter 7, the removal of ammonia in a SSF wetland can be directly correlated with the depth of root penetration and the availability of oxygen (Reed, 1993).

6.2.7 PLANT DIVERSITY

Natural wetlands typically contain a wide diversity of plant life. Attempts to replicate that diversity in constructed wetlands designed for wastewater treatment have in general not been successful. The relatively high nutrient content of most wastewaters tends to favor the growth of cattails, reeds, etc., and these tend to crowd out the other less competitive species over time. Many of these constructed wetlands in the United States and Europe have been planted as a monoculture or at most with two or three plant species, and these have all survived and provided excellent wastewater treatment. The FWS wetland concept has greater potential for beneficial habitat values because the water surface is exposed and accessible to birds and animals. Further enhancement is possible via incorporation of deep open-water zones and the use of selected plantings to provide attractive food sources (e.g., sago pond weed and similar plants). Nesting islands can also be constructed within these deep water zones for further enhancement. These deep-water zones can also provide treatment benefits as they increase the hydraulic retention time (HRT) in the system and serve to redistribute the flow, if properly constructed. The portions of the FWS wetland designed specifically for treatment can be planted with a single species. Cattails and bulrush are often used but are at risk from muskrat and nutria damage; *Phragmites* offers significant advantages in this regard. A number of FWS and SSF wetlands in the southern United States were initially planted with attractive flowering species (e.g., Canna lily, iris) for esthetic reasons. These plants have soft tissues which decompose very quickly when the emergent portion dies back in the fall and after even a mild frost. The rapid decomposition has resulted in a measurable increase in biological oxygen demand (BOD) and nitrogen leaving the wetland system. In some cases, the system managers utilized an annual harvest for removal of these plants prior to the seasonal dieback or frosts. In most cases, the problems have been completely avoided by replacing these plants with the more resistant reeds, rushes, or cattails, which do not require an annual harvest. Use of soft-tissue flowering species is not recommended for future systems, except possibly as a border.

6.2.8 PLANT FUNCTIONS

The terrestrial plants used in land treatment systems described in Chapter 8 of this book provide the major pathway for removal of nutrients in those systems. In those cases, the system design loading is partially matched to the plant uptake capability of the plants and the treatment area is sized accordingly. Harvesting then removes the nutrients from the site. The emergent aquatic plants used in wetlands also take up nutrients and other wastewater constituents. Harvesting is not, however, routinely practiced in these wetland systems due to problems with access and the relatively high labor costs. Studies have shown that harvesting of the plant material from a constructed wetland provides a minor nitrogen removal pathway as compared to biological activity in the wetland. In two cases (Gearheart et al., 1983; Herskowitz, 1986), a single end-of-season harvest accounted for less than 10% of the nitrogen removed by the system. Harvesting on a more frequent schedule would certainly increase that percentage but would also increase the cost and complexity of system management. Biological activity becomes the dominant mechanism in constructed wetlands as compared to land treatment systems, partially due to the significantly longer HRT in the former systems. When water is applied to the soil surface in most land treatment systems, the residence time for water as it passes from the surface through the active root zone is measured in minutes or hours; in contrast, the residence time in most constructed wetlands is usually measured in terms of at least several days.

In some cases, these emergent aquatic plants are known to take up and transform organic compounds, so harvesting is not required for removal of these pollutants. In the case of nutrients, metals, and other conservative substances, harvesting and removal of the plants are necessary if plant uptake is the design pathway for permanent removal. Plant uptake and harvest are not usually a design consideration for constructed wetlands used for domestic, municipal, and most industrial wastewaters.

Even though the system may be designed as a biological reactor and the potential for plant uptake is neglected, the presence of the plants in these wetland systems is still essential. Their root systems are the major source of oxygen in the SSF concept, and the physical presence of the leaves, stems, roots, rhizomes, and detritus regulates water flow and provides numerous contact opportunities between the flowing water and the biological community. These submerged plant parts provide the substrate for development and support of the attached microbial organisms that are responsible for much of the treatment. The stalks and leaves above the water surface in the FWS wetland provide a shading canopy that limits sunlight penetration and controls algae growth. The exposed plant parts die back each fall, but the presence of this material reduces the thermal effects of the wind and convective heat losses during the winter months. The litter layer on top of the SSF bed adds even more thermal protection to that type of system.

6.2.9 SOILS

In natural wetlands, most of the nutrients required for plant growth are obtained from the soil by emergent aquatic plants. Cattails, reeds, and bulrushes will grow in a wide variety of soils and, as shown in the SSF wetland concept, in relatively fine gravels. The void spaces in the media serve as the flow channels in the SSF wetland. Treatment in these cases is provided by microbial organisms attached to the roots, rhizomes, and media surfaces. Because of the relatively light loading in most SSF wetlands, this microbial growth does not produce thick layers of attached material such as typically occur in a trickling filter, so clogging from this source does not appear to be a problem. The major flow path in FWS wetlands is above the soil surface, and the most active microbial activity occurs on the surfaces of the detrital layer and the submerged plant parts.

Soils with some clay content can be very effective for phosphorus removal. As described in Chapters 3 and 8, phosphorus removal in the soil matrix of a land treatment system can be a major pathway for almost complete phosphorus removal for many decades. In FWS wetlands, the only contact opportunities are at the soil surface; during the first year of system operation, phosphorus removal can be excellent due to this soil activity and plant development. These pathways tend to come to equilibrium after the first year or so, and phosphorus removal will drop off significantly. Soils have been tried in Europe for SSF wetlands, primarily for their phosphorus removal potential. This attempt has not been successful in most cases, as the limited hydraulic capacity of soils results in most of the applied flow moving across the top of the bed rather than through the subsurface voids so the anticipated contact opportunities are not realized. The gravels used in most SSF wetlands have a negligible capacity for phosphorus removal. Soils, again with some clay content, or granular media containing some clay minerals also have some ion exchange capacity. This ion exchange capability may contribute, at least temporarily, to removal of ammonium (NH_4) that exists in wastewater in ionic form. This capacity is rapidly exhausted in most SSF and FWS wetlands as the contact surfaces are continuously under water and continuously anaerobic. In vertical-flow SSF beds, described in Chapter 7, aerobic conditions are periodically restored, and the adsorbed ammonium is released via biological nitrification, which then releases the ion exchange sites for further ammonium adsorption.

6.2.10 ORGANISMS

A wide variety of beneficial organisms, ranging from bacteria to protozoa to higher animals, can exist in wetland systems. The range of species present is similar to that found in the pond systems described in Chapter 4. In the case of emergent aquatic vegetation in wetlands, this microbial growth occurs on the submerged portions of the plants, on the litter, and directly on the media in the SSF wetland case. Wetlands and the overland flow (OF) concept described in Chapter 8 are similar in that they are both "attached-growth" biological systems and share many common attributes with the familiar trickling filters. All of these systems require a substrate for the development of the biological growth; their performance is dependent on the detention time in the system and on the contact opportunities provided and is regulated by the availability of oxygen and by the temperature.

6.3 PERFORMANCE EXPECTATIONS

Wetland systems can effectively treat high levels of BOD, total suspended solids (TSS), and nitrogen, as well as significant levels of metals, trace organics, and pathogens. Phosphorus removal is minimal due to the limited contact opportunities with the soil. The basic treatment mechanisms are similar to those described in Chapter 3 and Chapter 4 and include sedimentation, chemical precipitation and adsorption, and microbial interactions with BOD and nitrogen, as well as some uptake by the vegetation. Even if harvesting is not practiced, a fraction of the decomposing vegetation remains as refractory organics and results in the development of peat in wetland systems. The nutrients and other substances associated with this refractory fraction are considered to be permanently removed.

6.3.1 BOD REMOVAL

The removal of settleable organics is very rapid in all wetland systems and is due to the quiescent conditions in FWS systems and to deposition and filtration in SSF systems. Similar results have been observed with the overland flow systems described in Chapter 8, where close to 50% of the applied BOD is removed within the first few meters of the treatment slope. This settled BOD then undergoes aerobic or anaerobic decomposition, depending on the oxygen status at the point of deposition. The remaining BOD, in colloidal and dissolved forms, continues to be removed as the wastewater comes in contact with the attached microbial growth in the system. This biological activity may be aerobic near the water surface in FWS systems and at the aerobic microsites in SSF systems, but anaerobic decomposition would prevail in the remainder of the system. Removals of BOD in FWS constructed wetlands are presented in Table 6.1.

6.3.2 SUSPENDED SOLIDS REMOVAL

The principal removal mechanisms for TSS are flocculation and sedimentation in the bulk liquid and filtration (mechanical straining, chance contact, impaction, and interception) in the interstices of the detritus. Most of the settleable solids are removed within 50 to 100 ft (15 to 30 m) of the inlet. Optimal removal of TSS requires a full stand of vegetation to facilitate sedimentation and filtration and to prevent the regrowth of algae. Algal solids may require 6 to 10 days of detention time for removal. The removal rates of TSS in constructed wetlands are presented in Table 6.2.

6.3.3 NITROGEN REMOVAL

Nitrogen removal in constructed wetlands is accomplished by nitrification and denitrification. Plant uptake accounts for only about 10% of the nitrogen removal. Nitrification and denitrification are microbial reactions that depend on temperature and detention time. Nitrifying organisms require oxygen and an adequate surface area to grow on and, therefore, are not present in significant numbers in either heavily loaded systems (BOD loading > 100 lb/ac·d) or in newly constructed systems with incomplete plant cover. Based on field experience with FWS systems, it has been found that one to two growing seasons may be necessary to develop sufficient vegetation to support microbial nitrification. Denitrification requires adequate organic matter (plant litter or straw) to convert nitrate to nitrogen gas.

TABLE 6.1
Biochemical Oxygen Demand (BOD) Removal in Free Water Surface Constructed Wetlands

Location	BOD Influent (mg/L)	BOD Effluent (mg/L)	Percent Removal (%)	Ref.
Arcata, California	26	12	54	Gearheart et al. (1989)
Benton, Kentucky	25.6	9.7	62	USEPA (1993a)
Cannon Beach, Oregon	26.8	5.4	84	USEPA (1993a)
Cle Elum, Washington	38	8.9	77	Smith et al. (2002)
Ft. Deposit, Alabama	32.8	6.9	79	USEPA (1993a)
Gustine, California	75	19	75	Crites (1996)
Iselin, Pennsylvania	140	17	88	Watson et al. (1989)
Listowel, Ontario, Canada	56.3	9.6	83	Herskowitz et al. (1987)
Ouray, Colorado	63	11	83	Andrews (1996)
West Jackson County, Mississippi	25.9	7.4	71	USEPA (1993a)
Sacramento County, California	24.2	6.5	73	Nolte Associates (1999)

The reducing conditions in mature FWS constructed wetlands resulting from flooding are conducive to denitrification. If nitrified wastewater is applied to a FWS wetland, the nitrate will be denitrified within a few days of detention. Nitrogen removal is limited by the ability of the FWS system to nitrify. When nitrogen is present in the nitrate form, nitrogen removal is generally rapid and complete. The removal of nitrate depends on the concentration of nitrate, the

TABLE 6.2
Total Suspended Solids (TSS) Removal in Free Water Surface Constructed Wetlands

Location	TSS Influent (mg/L)	TSS Effluent (mg/L)	Percent Removal (%)	Ref.
Arcata, California	30	14	53	Gearheart et al. (1989)
Benton, Kentucky	57.4	10.7	81	USEPA (1993a)
Cannon Beach, Oregon	45.2	8.0	82	USEPA (1993a)
Cle Elum, Washington	32	4.8	85	Smith et al. (2002)
Ft. Deposit, Alabama	91.2	12.6	86	USEPA (1993a)
Gustine, California	102	31	70	Crites (1996)
Iselin, Pennsylvania	300	53	86	Watson et al. (1989)
Listowel, Ontario, Canada	111	8	93	Herskowitz et al. (1987)
Ouray, Colorado	86	14	84	Andrews (1996)
West Jackson County, Mississippi	40.4	14.1	65	USEPA (1993a)
Sacramento County, California	9.2	7.1–11.9	23–29[a]	Nolte Associates (1999)

[a] Effluent collection via surface overflow weir from open water zone contributed to floating solids in the effluent.

detention time, and the available organic matter. Because the water column is nearly anoxic in many wetlands treating municipal wastewater, the reduction of nitrate will occur within a few days. Nitrogen and ammonia removal data are presented in Table 6.3.

TABLE 6.3
Ammonia and Total Nitrogen Removal in Free Water Surface Constructed
Wetlands

Location	Type of Wastewater	Ammonia Influent (mg/L)	Ammonia Effluent (mg/L)	Total Nitrogen Influent (mg/L)	Total Nitrogen Effluent (mg/L)
Arcata, California	Oxidation pond	12.8	10	—	11.6
Beaumont, Texas[a]	Secondary	12	2	—	—
Iselin, Pennsylvania	Oxidation pond	30	13	—	—
Jackson Bottoms, Oregon	Secondary	9.9	3.1	—	—
Listowel, Ontario	Primary	8.6	6.1	19.1	8.9
Pembroke, Kentucky	Secondary	13.8	3.35	—	—
Sacramento County, California[b]	Secondary	14.9	9.1	16.9	11.0
Salem, Oregon[c]	Secondary	12.9	4.7	—	—

[a] USEPA (1999).
[b] Nolte Associates (1999).
[c] City of Salem, Oregon (2003)

Source: Adapted from Crites, R.W. and Tchobanoglous, G., *Small and Decentralized Wastewater Management Systems*, McGraw-Hill, New York, 1998.

6.3.4 PHOSPHORUS REMOVAL

The principal removal mechanisms for phosphorus in FWS systems are adsorption, chemical precipitation, and plant uptake. Plant uptake of inorganic phosphorus is rapid; however, as plants die, they release phosphorus so long-term removal is low. Phosphorus removal depends on soil interaction and detention time. In systems with zero discharge or very long detention times, phosphorus will be retained in the soil or root zone. In flow-through wetlands with detention times between 5 and 10 days phosphorus removal will seldom exceed 1 to 3 mg/L. Depending on environmental conditions within the wetland, phosphorus, as well as some other constituents, can be released during certain times of the year, usually in response to changed conditions within the system such as a change in the oxidation–reduction potential (ORP). Phosphorus removal in wetlands depends on the loading rate and the detention time. Because plants take up phosphorus over the growing season and then release some of it during senescence, reported removal data must be

TABLE 6.4
Phosphorus Removal in Free Water Surface Constructed Wetlands

Location	Hydraulic Loading Rate (in./d)	Total Phosphorus Influent (mg/L)	Total Phosphorus Effluent (mg/L)	Percent Removal (%)
Listowel, Ontario	0.95	1.9	0.7	62
Pembroke, Kentucky	0.30	3.0	0.1	96
Sea Pines, South Carolina	7.95	3.9	3.4	14
Benton, Kentucky	1.86	4.5	4.1	10
Leaf River, Mississippi	4.60	5.2	4.0	23
Lakeland, Florida	2.93	6.5	5.7	13
Clermont, Florida	0.54	9.1	0.2	98
Brookhaven, New York	0.59	11.1	2.3	79
Sacramento County, California	2.45	2.38	2.07	13
Salem, Oregon	0.40	2.2	1.0	55
Average	2.26	4.98	2.36	46

examined as to when the system was sampled and how long the system had been in operation. Removal rates of phosphorus for 10 constructed wetlands are presented in Table 6.4.

6.3.5 METALS REMOVAL

Heavy metal removal is expected to be very similar to that of phosphorus removal although limited data are available on actual removal mechanisms. The removal mechanisms include adsorption, sedimentation, chemical precipitation, and plant uptake. One of the processes that assist in metals removal is burial as metal sulfide precipitates. The process is illustrated in Figure 6.2 (USEPA, 1999). One metal of concern is mercury. Under anaerobic conditions, mercuric ions are biomethylated by microorganisms to methyl mercury, which is the more toxic form of mercury (Kadlec and Knight, 1996). A process that may counteract the methylation is precipitation with sulfides, as illustrated in Figure 6.2. At Sacramento County, California, the mercury concentrations were reduced by 64% to 4 ng/L (Crites, et al., 1997). Metals removal depends on detention time, influent metal concentrations, and metal speciation. Removal data for heavy metals in the Sacramento County demonstration wetlands; in Brookhaven, New York; and in Prague are presented in Table 6.5. The removal of aluminum, zinc, copper, and manganese with distance down a Prague wetland is shown in Table 6.6.

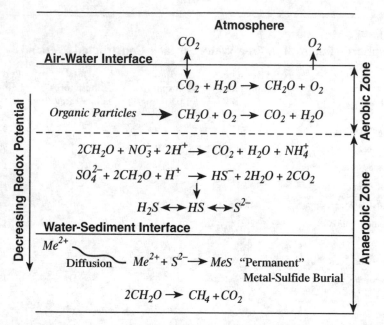

FIGURE 6.2 Metal sulfide burial processes in a wetland. (From USEPA, *Free Water Surface Wetlands for Wastewater Treatment: A Technology Assessment*, Office of Water Management, U.S. Environmental Protection Agency, Washington, D.C., 1999.)

6.3.6 TEMPERATURE REDUCTION

Temperature reduction through free water surface constructed wetlands occurs where the average daily ambient air temperature is lower than the applied wastewater temperature. The expected reduction in temperature through a constructed wetland can be calculated using Equation 6.15 in Section 6.7 later in this chapter. Reductions in temperature achieved at a demonstration constructed wetlands at Sacramento County, California, and at Mt. Angel, Oregon, are presented in Table 6.7.

6.3.7 TRACE ORGANICS REMOVAL

As described in Section 3.3 of Chapter 3 in this book, the removal of trace organic compounds occurs via volatilization or adsorption and biodegradation. The adsorption occurs primarily on the organic matter present in the system. Table 3.6 in Chapter 3 presents the removal of organic chemicals in land treatment systems; removal exceeds 95%, except in a very few cases where >90% was observed. The removal in constructed wetlands is even more effective as the HRT in wetland systems is measured in days as compared to the minutes or hours for land treatment concepts, and significant organic materials for adsorption are almost always present. As a result, the opportunities for volatilization and adsorption/biodegradation are enhanced in the wetland process. Removals observed in

TABLE 6.5
Metals Removal in Free Water Surface Constructed Wetlands

Location	Metal	Influent (μg/L)	Effluent (μg/L)	Percent Removal (%)
Prague	Aluminum	451	<40	91
Sacramento County, California	Antimony	0.43	0.18	58
Sacramento County, California	Arsenic	2.37[a]	2.80	−18
Brookhaven, New York	Cadmium	43	0.6	99
Sacramento County, California	Cadmium	0.08	0.03	63
Brookhaven, New York	Chromium	160	20	88
Sacramento County, California	Chromium	1.43	1.11	23
Brookhaven, New York	Copper	1510	60	96
Sacramento County, California	Copper	7.44	3.17	57
Brookhaven, New York	Iron	6430	2140	67
Sacramento County, California	Lead	1.14	0.23	80
Brookhaven, New York	Lead	1.7	0.4	76
Brookhaven, New York	Manganese	210	120	43
Sacramento County, California	Mercury	0.011	0.004	64
Brookhaven, New York	Nickel	35	10	71
Sacramento County, California	Nickel	5.80	6.84	−18
Sacramento County, California	Silver	0.53	0.09	83
Brookhaven, New York	Zinc	2200	230	90
Sacramento County, California	Zinc	35.82	6.74	81

[a] During the 5 years of monitoring, the influent arsenic dropped from 3.25 to 2.33 μg/L, while the effluent arsenic varied from 2.34 to 3.77 μg/L.

Source: Data from USEPA (1999), Nolte Associates (1999), and Hendry et al. (1979).

pilot-scale constructed wetlands with a 24-hr HRT are presented in Table 6.8. The removals should be even higher and comparable to those in Table 3.6 at the several day HRT commonly used for wetland design.

6.3.8 PATHOGEN REMOVAL

Pathogen removal in wetlands is due to the same factors described in Chapter 3 for pond systems, and Equation 3.25 can be used to estimate pathogen removal in these wetlands. The actual removal should be more effective due to the additional filtration provided by the plants and litter layer in a wetland. Table 3.9 contains performance data for both FWS and SSF systems. The principal removal

TABLE 6.6
Removal of Metals with Length in a Free Water Surface Constructed Wetland at Nucice (Prague)

Metal	0 m	5 m	16 m	32 m	48 m	60 m	62 m
Aluminum	451	126	65	47	46	<40	<40
Copper	11.3	4.1	3.0	<2.0	<2.0	<2.0	<2.0
Manganese	278	47	52	39	41	45	53
Zinc	198	106	12	7.3	3.6	<5.0	<5.0

Source: Vymazal, J. and Krasa, P., Water Sci. Technol., 48(5), 299–305, 2003. With permission.

TABLE 6.7
Reduction of Temperature through Free Water Surface Constructed Wetlands at Sacramento County, California, and Mt. Angel, Oregon

Month	Sacramento County, California			Mt. Angel, Oregon		
	In[a] (°F)	Out (°F)	Reduction (°F)	In[b] (°F)	Out (°F)	Reduction (°F)
January	57.7	48.0	9.7	45.3	44.2	1.1
February	62.4	51.3	11.1	50.2	50.4	-0.2
March	59.0	55.6	3.4	53.5	52.4	1.1
April	64.9	61.1	3.8	63.3	60.9	2.4
May	67.5	59.9	7.6	67.0	62.5	4.5
June	72.1	71.8	0.3	72.8	68.0	4.8
July	74.8	73.6	1.2	73.7	69.1	4.6
August	78.4	72.7	5.7	73.1	66.9	6.2
September	76.1	68.5	7.6	70.3	64.5	5.8
October	64.2	58.6	5.6	59.5	55.9	3.6
November	60.6	57.2	3.4	52.2	50.6	1.6
December	56.3	50.2	6.1	48.4	47.5	0.9
Average	—	—	5.5	—	—	3.0

[a] Five-year average 1994 to 1998 (Nolte Associates, 1999).
[b] Four-year average 1999 to 2002 (City of Mt. Angel, Oregon).

mechanism in SSF wetlands is physical entrapment and filtration. As shown in Table 3.9, the finer textured material used at Iselin, Pennsylvania, was clearly superior to the gravel used at Santee, California. Removals of both bacteria and

TABLE 6.8
Removal of Organic Priority Pollutants
in Constructed Wetlands

Compound	Initial Concentration (μg/L)	Removal in 24 hr (%)
Benzene	721	81
Biphenyl	821	96
Chlorobenzene	531	81
Dimethyl-phthalate	1033	81
Ethylbenzene	430	88
Naphthalene	707	90
p-Nitrotoluene	986	99
Toluene	591	88
p-Xylene	398	82
Bromoform	641	93
Chloroform	838	69
1,2-Dichloroethane	822	49
Tetrachloroethlyene	451	75
1,1,1-Trichloroethane	756	68

Source: Reed, S.C. et al., *Natural Systems for Waste Management and Treatment*, 2nd ed., McGraw-Hill, New York, 1995. With permission.

virus are equally efficient in both SSF and FWS wetlands. The pilot FWS wetlands at Arcata, California, removed about 95% of the fecal coliforms and 92% of the virus with an HRT of about 3.3 d; at the pilot study in Santee, California, the SSF wetland achieved >98% removal of coliforms and >99% virus removal with an HRT of about 6 d.

6.3.9 BACKGROUND CONCENTRATIONS

A successful wetland treatment system is also a successful living ecosystem containing vegetation and related biota. The life and death cycles of this natural biota produce residuals that can then be measured as BOD_5, TSS, nitrogen, phosphorus, and fecal coliforms. It is, therefore, not possible for these wetland systems to produce a zero effluent concentration of these materials; some residual background concentration will always be present. Typical concentrations of these constituents are presented in Table 6.9. These background concentrations are not composed of wastewater constituents, but their concentrations may be indirectly

TABLE 6.9
Background Concentrations of Constituents in Typical Wetlands Effluent

Constituent	Range	Typical
TSS (mg/L)	2–5	3
BOD[a] (mg/L)	2–8	5
Total nitrogen (mg/L)	1–3	2
Nitrate nitrogen (mg/L)	<0.1	<0.1
Ammonia nitrogen (mg/L)	0.2–1.5	1
Organic nitrogen (mg/L)	1–3	<2
Total phosphorus (mg/L)	0.1–0.5	0.3
Fecal coliform (cfu/100 mL)	50–5000	200

[a] A range from 5 to 12 has been reported for fully covered with emergent vegetation.

Note: TSS, total suspended solids; BOD, biochemical oxygen demand.

Source: Data from USEPA (1999, 2000).

related to the system loadings. A wetland system receiving a nutrient rich wastewater is likely to produce a higher background level than a natural wetland receiving clean water. The background concentrations can also vary on a seasonal basis because of the seasonal occurrence of plant decomposition and the variability in bird and wildlife activity.

6.4 POTENTIAL APPLICATIONS

The previous sections of this chapter have provided information on performance expectations, available wetland types, and internal components. This section is intended to provide guidance on the application of constructed wetlands for a variety of purposes. These applications include municipal wastewater, commercial and industrial wastewaters, stormwater runoff, combined sewer overflows (CSO), agricultural runoff, livestock wastewaters, food processing wastewater, landfill leachate, and mine drainage.

6.4.1 MUNICIPAL WASTEWATERS

Examples of FWS constructed wetlands are presented in Table 6.10. The selection of either FWS or SSF constructed wetlands for municipal wastewaters depends on the volume of flow to be treated and on the conditions at the proposed wetland site. As described previously, the SF wetland, because of the higher reaction rates for BOD and nitrogen removal, will require a smaller total surface area than a

TABLE 6.10
Municipal Free Water Surface Constructed Wetlands in the United States

Location	Pretreatment	Flow (mgd)	Area (ac)	Remarks
Arcata, California	Oxidation ponds	2.3	7.5	Early research but now major tourist attraction
Benton, Kentucky	Oxidation ponds	1.0	10	Upgraded with nitrification filter bed (NFB) for ammonia removal
Cle Elum, Washington	Aerated ponds	0.55	5	Alternating vegetated and open water zones
Gustine, California	Aerated ponds	1.0	24	High organic loading
Mt. Angel, Oregon	Oxidation ponds	2.0	10	Seasonal discharge
Ouray, Colorado	Aerated ponds	0.36	2.2	Polishing wetlands
Riverside, California	Secondary	10.0	50	Denitrification wetlands
Sacramento County, California	Secondary	1.0	15	Five-year demonstration project

FWS wetland designed for comparable effluent goals; however, it is not always obvious which concept will be the more cost effective for a particular situation. The final decision will depend on the availability and cost of suitable land and on the cost required for acquisition, transport, and placement of the gravel media used in the SSF bed.

It is likely that economics will favor the FWS concept for very large systems as these are typically located at relatively remote sites and some of the advantages of the SSF concept do not represent a significant benefit. The cost trade-off could occur at design flows less than 0.1 mgd (378 m³/d) and should certainly favor the FWS concept at design flows over 1 mgd (3785 m³/d). In some cases, however, the advantages of the SSF concept outweigh the cost factors. A SSF wetland system has been designed, by the senior author of this book, to treat a portion of the wastewater at Halifax, Nova Scotia, and the thermal advantage of the SSF wetland type justified its selection for that location.

Where nitrogen removal to low levels is a project requirement, the use of *Phragmites* or *Scirpus* in a SSF system is recommended. These species or *Typha* should all be suitable on FWS systems, but *Phragmites* will be less susceptible to damage from animals (see Section 6.2). The use of the nitrifying filter bed (NFB), as described in Section 7.9, should be considered as an alternative when stringent ammonia limits prevail.

Incorporation of deeper water zones in the FWS concept will increase the overall HRT in the wetland and may enhance oxygen transfer from the atmosphere (see Figure 6.3). The individual deep-water zones must be large enough to permit

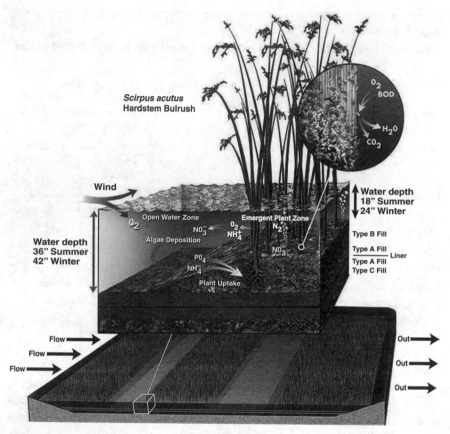

FIGURE 6.3 Open-water sketch for free water surface (FWS) wetlands. (Courtesy of Brown and Caldwell, Walnut Creek, CA.)

movement of the duckweed cover by the wind; a semipermanent layer of duck-weed on the water will prevent any oxygen transfer. The open-water zones, as shown in Figure 6.4 at Cle Elum, Washington, also minimize short-circuiting. If the deep-water zones represent more than 30% of the total system area, the system should be designed as a series of wetlands and ponds using the procedures in this chapter and in Chapter 4. The use of submerged plant species (see Section 6.2) in the deep-water zones will enhance habitat values and may improve water quality. In such cases, the water depth in the zone must be compatible with the sunlight transmission requirements for the plant selected, and the development of a duckweed mat must be avoided.

A careful thermal analysis is necessary for all systems located where sub-freezing temperatures occur during the winter months. This is to ensure adequate performance via the temperature-sensitive nitrogen and BOD removal responses and to determine if restrictive freezing will occur in extremely cold climates. A

FIGURE 6.4 Free water surface (FWS) wetland at Cle Elum showing bulrush and open water.

number of FWS systems designed for northwestern Canada faced the risk of severe winter freezing and therefore have been designed for winter wastewater storage in a lagoon and wetland application during the warm months.

Incorporation of habitat and recreational values is more feasible for the FWS wetland concept because the water surface is exposed and will attract birds and other wildlife. The use of deep water zones with nesting islands will significantly enhance the habitat values of a system, as will the supplemental planting of desirable food source vegetation such as sago pond weed (see Section 6.2).

6.4.2 Commercial and Industrial Wastewaters

Both SSF and FWS wetlands can be suitable for commercial and industrial wastewaters, depending on the same conditions described above for municipal wastewater. Wastewater characterization is especially important for both commercial and industrial wastewaters. Some of these wastewaters are high in strength, low in nutrients, and high or low in pH and contain substances that may be toxic or inhibit biological treatment responses in a wetland. High-strength wastes and high concentrations of priority pollutants are typically subjected to an anaerobic treatment step prior to the wetland component. Constructed wetlands, both SSF and FWS types, are currently in use for wastewater treatment from pulp and paper operations, oil refineries, chemical production, and food processing. In most cases, the wetland component is used as a polishing step after conventional biological treatment. The performance expectations for these wetlands were described in Section 6.3 of this chapter. System design follows the same procedures described in Section 6.5 through Section 6.9. A pilot study may be necessary when unfamiliar toxic substances are present or for design optimization for removal of priority pollutants.

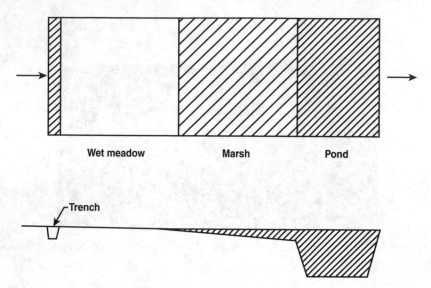

FIGURE 6.5 Stormwater wetlands schematic.

6.4.3 STORMWATER RUNOFF

Sediment removal is typically the major purpose of wetlands designed for treatment of urban stormwater flow from parking lots, streets, and landscapes. In essence, the wetland is a stormwater retention basin with vegetation, and the design uses many of the basic principles of sedimentation basin design. The presence of vegetation fringes, deep and shallow water zones, and marsh segments enhances both the treatment and habitat functions. These wetlands have been shown to provide beneficial responses for BOD, TSS, pH, nitrates, phosphates, and trace metals (Ferlow, 1993).

At a minimum, a stormwater wetland system (SWS) will usually have some combination of deep ponds and shallow marshes. In addition, wet meadows and shrub areas can also be used. Because the flow rate is highly variable and the potential exists for accumulation and clogging with inorganic solids the SSF wetland concept is not practical for this application, so the marsh component in the SWS system will typically be FWS constructed wetlands. These may be configured as shown in Figure 6.5 or in alternative combinations. Key components include an inlet structure, a ditch or basin for initial sedimentation, a spreader swale or weir to distribute the flow laterally if a wet meadow or marsh is the next component, a deep pond, and some type of outlet device that permits overflow conditions during peak storm events and allows slow discharge to the "datum" water level in the system. The "datum" water level is usually established to maintain a shallow water depth in the marsh components. Use of drought-resistant plant species in the marsh components would permit complete dewatering for extended periods.

Typha, *Scirpus*, and *Phragmites* can withstand up to 3 ft (1 m) of temporary inundation, a factor that would establish the maximum water level before overflow in the SWS if these species are used. The maximum storage depth should be about 2 ft (0.6 m), if grassed wet meadows and shrubs are used. The optimum storage capacity of the wetland (the depth between the "datum" and the overflow level) should be a volume equal to 0.5 in. (13 mm) of water on the watershed contributing to the SWS. The minimum storage volume, for effective performance, should be equal to 0.25 in. (6 mm) of water on the contributing water shed. The storage volume for these, or any other depths, can be calculated with Equation 6.1:

$$V = (C)(y)(A_{ws}) \qquad (6.1)$$

where
- V = Storage volume in stormwater wetland (ft^3; m^3).
- C = Coefficient = 3630 for U.S. units; 10 for metric units.
- y = Design depth of water on watershed (mm).
- A_{ws} = Surface area of watershed (ac; ha).

The minimum surface area of the entire SWS, at the overflow elevation, is based on the flow occurring during the 5-year storm event and can be calculated with Equation 6.2:

$$A_{sws} = (C)(Q) \qquad (6.2)$$

where
- A_{sws} = Minimum surface area of SWS at overflow depth (ft^2; m^2).
- C = Coefficient = 180 for U.S. units; 590 for metric units.
- Q = Expected flow from 5-year design storm (ft^3/d; m^3/d).

The aspect ratio of the SWS should be close to 2:1, if possible, and the inlet should be as far as possible from the outlet (or suitable baffles can be used). The spreader swale and inlet zone should be sufficiently wide to reduce the subsequent flow velocity to 1 to 1.5 ft/s (0.3 to 0.5 m/s).

In essence, the SWS performs as a batch reactor. The water is static between storm events, and water quality will continue to improve. When a storm event occurs, the entering flow will displace some or all of the existing volume of treated water before overflow commences. It is possible, using the design models presented in previous sections, to estimate the water quality improvements that will occur under various combinations of storm events. It is necessary to first determine the frequency and intensity of storm events. These data can then be used to calculate the hydraulic retention time during and between storm events; it is then possible to determine the pollutant removal that will occur with the appropriate design model.

6.4.4 COMBINED SEWER OVERFLOW

Management of combined sewer overflow is a significant problem in many urban areas where the older sewerage network carries both stormwater and untreated wastewater. When peak storm events occur, the capacity of the wastewater treatment plant is exceeded; in the past, this condition often led to a temporary bypass and discharge of the untreated CSO to receiving waters. Current regulations now prohibit that practice, and wetlands are being given strong consideration as a treatment alternative for the CSO discharge.

A wetland designed for CSO management faces essentially the same requirements as a stormwater wetland, and the FWS constructed wetland is the preferred concept for the same reasons cited previously. Because the CSO flow always contains some untreated wastewater, the level of pathogens and the mass of pollutants contained in the storm event may be higher than found in normal stormwater flow. The "first flush" with many stormwaters contains the bulk of pollutants, but that may not be the case with CSO discharges because of the wastewater component.

The design of the CSO wetland must commence with an analysis of the frequency and intensity of storm events and the capacity of the existing wastewater treatment facilities. This analysis will be used to determine the volume of excess CSO flow to be contained by the proposed wetland. Containment of the CSO from at least a 5-year or a 10-year storm event is a typical baseline wetland volume. The CSO wetland will act as a batch reactor, and water quality improvements will depend on the intensity and frequency of storm events. Assuming the wetland is sized for the CSO from a 10-year storm event, the flow from any lesser event will be completely contained, and any discharge would be composed of previously contained and treated water.

The hydraulic retention time (HRT) in the wetland must include consideration of precipitation on the wetland, seepage, and evapotranspiration, as well as the input CSO flow. The water quality expectations are usually established by the regulatory authorities. If significant seepage is allowed, then the CSO wetland will perform similarly to the rapid infiltration concept described in Chapter 8 of this book. When the HRT in the wetland has been established for various situations, it is possible to estimate the water quality improvements that will occur by using the design models in this chapter and in Chapter 8 (if seepage is permitted). If the wetland is located adjacent to the ultimate receiving water and the hydrological investigation indicates that the seepage will flow directly to the receiving surface water, then seepage can be very beneficial, particularly with respect to phosphorus removal.

In some cases, trash removal and some form of preliminary treatment are provided separately. If not, these functions should be the initial components in the CSO wetland, with trash racks or similar, and a deep basin for preliminary settling. The wetland component should be designed as a FWS marsh system with a "normal" operating depth of 2 ft (0.6 m). The use of *Phragmites*, *Typha*, or *Scirpus* would permit a temporary inundation of up to 3 ft (1 m) during peak storm events. The use of *Phragmites* should be avoided if the CSO wetland is planned for habitat and recreational benefits in addition to water quality improvement. The wetland component should have at least two parallel trains of two cells each to allow flexibility of management and maintenance.

TABLE 6.11
Water Quality Expectations for a Combined Sewer Overflow (CSO) Wetland at Portland, Oregon

Parameter	Untreated CSO	Preliminary Treatment Effluent[a]	Wetland Seepage	Wetland Overflow
Volume (m³)	31,000	31,000	15,000	3000
BOD (mg/L)	100	85	2	10
TSS (mg/L)	100	70	2	10
TKN (mg/L)	7.0	6.1	3	2
Nitrate nitrogen (mg/L)	0.2	0.2	0.1	0.0
Total phosphorus (mg/L)	0.6	0.45	<0.05	0.17
Fecal coliform (number/100 mL)	110,000	200	<20	10

[a] Disinfection included.

Note: 1000 m³ = 0.26 Mgal; BOD, biochemical oxygen demand; TSS, total suspended solids; TKN, total Kjeldahl nitrogen.

Source: Reed, S.C. et al., *Natural Systems for Waste Management and Treatment*, 2nd ed., McGraw-Hill, New York, 1995. With permission.

Determining the elevation of the bottom of the wetland component is critical for successful performance, particularly in situations where a shallow fluctuating groundwater table exists and where seepage is to be permitted. It is desirable to have bottom soils moist at all times, even during drought conditions, but allowing the groundwater to occupy a significant portion of the containment volume during wet weather should be avoided. *Phragmites* and to a lesser degree *Typha* are drought resistant and would permit location of the wetland bottom in a position that would avoid seasonal groundwater intrusion.

Designing the wetland for inclusion of habitat values complicates this procedure. In this case, the wetland can consist of marsh surfaces above the normal groundwater level and deeper pools that intersect the minimum groundwater level so some water is permanently available for birds and other wildlife.

The results of a feasibility study of a CSO constructed wetland, conducted for the City of Portland, Oregon, are summarized in Table 6.11. The wetland component was designed to contain the 10-year storm event that produced a total CSO flow of about 11.8 Mgal (45,000 m³) from the peak 7-hour flow. Because of land area limitations, it was decided to provide separate facilities for trash removal and preliminary treatment. The potential wetland area contained about 23 ac (9.3 ha), and a 2-ft (0.6-m) water depth in the wetland would contain about

15 Mgal (57,000 m³). The soil beneath and adjacent to the proposed wetland and the ultimate receiving water was a permeable sand. The water quality expectations for this system are given in Table 6.11. The data in Table 6.11 are intended as an example only and cannot be utilized for system design elsewhere. It is necessary to determine the CSO characteristics and site conditions for a wetland for every proposed system because of possibly unique local conditions.

6.4.5 AGRICULTURAL RUNOFF

Nonpoint runoff from cultivated fields adds pollution to receiving water in the form of sediments and nutrients, particularly phosphorus. The Natural Resources Conservation Service (NRCS) has developed a process for treatment and management of these runoff waters. A schematic diagram of the system is shown in Figure 6.5; components include an underdrained wet meadow, a marsh, and a pond in series. An optional final component is a vegetated polishing area. The combined concept is referred to as a Nutrient/Sediment Control System (NSCS) by the NRCS. Several of these systems have been used successfully in northern Maine for treatment of runoff from cultivated fields. The NSCS should not be installed as the sole control system. It should only be used in conjunction with best conservation practices applied for erosion control on the agricultural fields of concern.

Equations 6.3 through 6.7 are used to size the components in the NSCS concept. These are based on an assumed modular width of 100 ft (30.5 m) for the general case. Dimensional modifications are possible to fit the system to specific site constraints as long as the surface area of each NSCS component remains about the same. The design procedure is considered valid for agricultural land including row crops, hay, and pasture with average slopes up to 8%.

Typically, the agricultural runoff will be conveyed to the NSCS in an appropriately sized ditch. The first NSCS component is a trapezoidal sedimentation trench that runs the full width of the system. The bottom width of the trench should be 10 ft (3 m) to facilitate cleaning with a front-end loader. The vegetated side slopes should not be greater than 2:1, and the depth should be at least 4 ft (1.2 m). A ramp is constructed at one end of the trench to allow access for cleaning. The top, downstream edge of the trench includes a level-lip spreader constructed of crushed stone to distribute the water uniformly over the full width of the system. This spreader consists of an 8-ft (2-m)-wide zone of stone, extending the full width of the system and very carefully constructed to ensure a level surface. Within that zone is a trench that is 1 ft (0.3 m deep and 4 ft (1.2 m) wide, also filled with the same stone. The stone size may range from 1 to 3 in. (25 to 76 mm). The necessary surface area of this trench can be calculated with Equation 6.3:

$$\text{Metric units:} \quad \text{AST} = \left[78 + 1.074 W_A + 0.04 W_A^2 \right] \tag{6.3a}$$

$$\text{U.S. units:} \quad \text{AST} = \left[843 + 4.54 W_A + 0.07 W_A^2 \right] \tag{6.3b}$$

where AST is the surface area of sedimentation trench (ft²; m²), and W_A is the area of contributing watershed (ac; ha).

The wet meadow is composed of underdrained, permeable soils planted with cool season grasses (other than Reed Canary grass). This unit must be absolutely level from side to side to promote sheet flow and should slope from 0.5 to 5% in the direction of flow. Underdrain pipe (4 in.; 100 mm) is placed on about 20-ft (6-m) centers perpendicular to the flow direction. These drains are backfilled with a gravel pack, which is covered with an appropriate filter fabric. These drains discharge, below the water surface, in the marsh component. The first drain line should be about 3 m (10 ft) downslope from the level lip spreader. At least 3 in. (76 mm) of topsoil should be spread over the entire wet meadow area prior to grass planting. The surface area of this wet meadow can be calculated with Equation 6.4 and the required slope length in the flow direction with Equation 6.5:

$$\text{Metric units: } AWM = \left[783 + 10.4W_A + 0.37W_A^2\right] \tag{6.4a}$$

$$\text{U.S. units: } AWM = \left[8430 + 45W_A + 0.7W_A^2\right] \tag{6.4b}$$

where AWM is the surface area of wet meadow (ft²; m²), and W_A is the area of contributing watershed (ac; ha).

$$\text{Metric units: } LWM = 22.9 + 0.753W_A \tag{6.5a}$$

$$\text{U.S. units: } LWM = 75 + W_A \tag{6.5b}$$

The wetland or marsh component is the same area as the wet meadow and also extends the full width of the system. Equation 6.4 can be used to determine the surface area of this component. The marsh should be level from side to side of the system and range from zero depth at the interface with the wet meadow to 1.5 ft (0.46 m) deep at the interface with the deep pond. *Typha* is the recommended plant species. The habitat values of the system will be enhanced by planting sago pond weed where the water depth in the marsh will exceed 1.2 ft (0.4 m).

The deep pond (DP) provides a limnetic biological filter for nutrient and fine sediment removal. The area of the pond can be determined with Equation 6.6:

$$\text{Metric units: } ADP = 372 + 55W_A \tag{6.6a}$$

$$\text{U.S. units: } ADP = 4000 + 240W_A \tag{6.6b}$$

The pond should be stocked with indigenous fish that feed on plankton and other microorganisms. Common or golden shiners are often used. The stocking rate should be 250 to 500 fish per 5000 ft² (465 m²) of pond area. The fish may be periodically harvested and sold as bait fish. Freshwater mussels are also stocked at a rate of 100 per 3000 ft² (900 m²). The pond should be between 8 ft (2.4 m) and 12 ft (3.7 m) deep. The principal discharge structure from the pond should be designed to maintained the desired water level and accommodate the expected

TABLE 6.12
Performance of Agricultural Runoff Constructed Wetland

Season	Inflow (m³)	Outflow (m³)	TSS In (kg)	TSS Out (kg)	VSS In (kg)	VSS Out (kg)	TP In (kg)	TP Out (kg)
1990								
Spring	648	1768	7	8	3	7	0.06	0.13
Summer	292	0	1144	0	113	0	3.06	0
Fall	7296	12,295	3884	144	546	35	4.63	1.26
Total	8236	14,062	5036	152	663	42	7.76	1.38
1991								
Spring	1387	7685	54	107	7	26	0.30	0.76
Summer	2023	743	3505	11	393	4	12.4	0.11
Fall	1526	3102	644	34	84	10	3.9	0.70
Total	4936	11,530	4203	152	484	40	16.6	1.57

Note: TSS, total suspended solids; VSS; volatile suspended solids; TP, total phosphorus.

Source: Higgens, M.J. et al., in *Constructed Wetlands for Water Quality Improvement*, Moshiri, G. et al., Eds., Lewis Publishers, Chelsea, MI, 1993, 359–367. With permission.

flow from up to a 5-year storm. A grass-covered emergency spillway is sized and located to accommodate flows in excess of the 5-year storm.

The final optional component is a grassed polishing area that receives the discharge from the deep pond. If practical, another ditch and level lip spreader are desirable to ensure uniform flow in this polishing area. This area can be determined using Equation 6.7:

$$\text{Metric units: } A_p = 232 + 11.5 W_A \qquad (6.7a)$$

$$\text{U.S. units: } A_p = 2{,}500 + 50 W_A \qquad (6.7b)$$

The performance of a NSCS system in northern Maine, over two operational seasons, is summarized in Table 6.12. This system collected the runoff from a 17.3-ac (7-ha) cultivated watershed growing potatoes (Higgens et al., 1993). This system, over the 2 years, achieved an average sediment removal of 96% and total phosphorus removal of 87%.

6.4.6 LIVESTOCK WASTEWATERS

These wastewaters from feed lots, dairy barns, swine barns, poultry operations, and similar activities tend to have high strength, high solids, and high ammonia and organic nitrogen concentrations. It is necessary to reduce the concentration of these materials in a preliminary treatment step, and an anaerobic pond is typically the most cost-effective choice. Procedures in Chapter 4 of this book can be used for design of that system component. In most cases, the FWS wetland will be the cost-effective choice for treatment of these wastewaters, as the smaller land area and other potential advantages of the SSF concept are not usually essential in an agricultural setting. The SSF concept may be at a disadvantage if spills occur in the preliminary treatment step and high solids concentrations are allowed to enter the wetland. The SSF concept may still be desirable for year-round operations in cold climates due to the enhanced thermal protection provided by this system.

Design of a wetland component for this application should follow the same procedures described in Section 6.5 to Section 6.9 of this chapter. A summary of performance data from a two-cell FWS wetland system treating wastewater from swine barns is presented in Table 6.13. An anaerobic lagoon was used as the preliminary treatment step, and that effluent was mixed with periodic discharge from a stormwater retention pond prior to introduction to the wetland component. Because flow rates were not measured, it is not possible to determine the HRT in this system. The volume of flow from the stormwater pond was about 1.5 times the volume from the anaerobic lagoon.

The 500-animal swine operation is estimated to produce 90 kg BOD d^{-1} which is reduced to 36 kg/d in the diluted wetland influent. The organic loading rate on the 3600 m^2 of wetland surface area is 89 lb/ac·d (100 kg/ha·d), and this is identical to the value recommended in Section 6.6 of this chapter.

6.4.7 FOOD PROCESSING WASTEWATER

Several existing FWS systems treat food-processing wastewater (O'Brien et al., 2002). The City of Gustine, California, has a FWS system that receives over 90% of its waste load from food-processing facilities (Crites, 1996). American Crystal Sugar uses primary clarification and anaerobic digestion prior to their 158-acre constructed wetland of sugar beet refinery wastewater in Hillsboro, North Dakota, and another 160-acre wetland at Drayton, North Dakota. At Connell, Washington, a three-stage wetland system is used to treat potato processing wastewater prior to land application (O'Brien et al., 2002). The wetland system consists of a 24-acre FWS wetlands, a 10-acre SSF wetland that nitrifies, and a 5-acre FWS wetland that denitrifies. The 1.4-mgd system produces a 67% removal of total nitrogen from 134 mg/L down to 44 mg/L (O'Brien et al., 2002).

6.4.8 LANDFILL LEACHATES

Both FWS and SSF wetlands have been used for the treatment of landfill leachate. A combination system utilizing a vertical-flow wetland bed (see Chapter 7) followed by a FWS wetland has been proposed for treating landfill leachate in Indiana (Bouldin et al., 1994; Martin et al., 1993; Peverly et al., 1994). In some cases, the leachate is applied directly to the wetland, in others the leachate flows

TABLE 6.13
Performance of Constructed Wetlands Treating Swine Waste

Location	BOD (mg/L)	TSS (mg/L)	TKN (mg/L)	Ammonia Nitrogen (mg/L)	Total Phosphorus (mg/L)	Fecal Coliform (number/100 mL)	Fecal *Streptococcus* (number/100 mL)
Anaerobic lagoon	111	346	116	84	49	817,500	118,750
Stormwater pond	32	51	4	1	3	1022	679
Wetland influent	64	105	26	55	26	175,164	76,727
Wetland effluent cell 1	14	25	18	13	11	2733	3927
Wetland effluent cell 2	10	31	9	5	7	2732	1523

Note: BOD, biological oxygen demand; TSS, total suspended solids; TKN, total Kjeldahl nitrogen.

Source: Hammer, D.A. et al., in *Constructed Wetlands for Water Quality Improvement*, Moshiri, G. et al., Eds., Lewis Publishers, Chelsea, MI, 1993, 343–348. With permission.

to an equalization pond from which it is transferred to the wetland unit. The pond at the Escambia County landfill in Florida is aerated, because septage is also added to the pond (Martin et al., 1993).

Characterization of the leachate is essential for proper wetland design as it can contain high concentrations of BOD, ammonia, and metals, can have a high or low pH, and can possibly include priority pollutants of concern. In addition, the nutrient balance in the leachate may not be adequate to support vigorous plant growth in the wetland, and supplemental potassium, phosphorus, and other micronutrients may be necessary. Because leachate composition will depend on the type and quantity of materials placed in the landfill and on time, a generic definition of characteristics is not possible and data must be collected for each system design.

Examples of leachate water quality from several landfill operations in the Midwest are presented in Table 6.14. These data confirm the earlier statement that BOD, chemical oxygen demand (COD), ammonia, and iron can exist in relatively high concentrations. Some of the volatile organic compounds such as acetone, methyl isobutyl ketone, and phenols can also be present in significant concentrations.

The design of the wetland for leachate treatment will follow the same procedures described in Sections 6.5 to 6.9 of this chapter. The removal of metals and priority pollutants will be as described in Section 6.3. Typically, the wetland will be sized to achieve a specific level of ammonia or total nitrogen in the final effluent. This can be achieved with only a wetland bed or with a wetland bed combined with either a nitrification filter bed (see Section 7.9) or a vertical-flow cell (see Section 7.11). The atmospheric exposure and relatively long HRT provided by any of these options will result in very effective removal of the volatile priority pollutants. If the leachate BOD is consistently above 500 mg/L, then the use of a preliminary anaerobic pond or cell should be considered. Many of the advantages of the SSF wetland concept are not necessary at most landfill locations, so a FWS wetland may be the more cost effective choice even though more land will be required. The exception may be in cold climates where the thermal protection provided by the SSF concept is an operational advantage. The performance of a FWS constructed wetland is shown in Table 6.15.

The nutrient and micronutrient requirements for biological oxidation are presented in Table 6.16. Landfill leachates, industrial and commercial wastewaters, and similar unique discharges should be tested for these components prior to design of a wetland system. If nutrients or micronutrients are deficient in these landfill leachates, the rate constants for BOD and nitrogen removal may be an order of magnitude less than those given in Section 6.6 and Section 6.8.

6.4.9 MINE DRAINAGE

A few hundred FWS wetland systems in the United States are intended for treatment of acid mine drainage. In some cases, the sizing and configurations of these systems were not rationally based. In most cases, however, the systems are

TABLE 6.14
Examples of Landfill Leachate Characteristics

Parameter	Southern Illinois	Berrien County, Michigan	Elkhart County, Indiana	Forest Lawn, Michigan
BOD (mg/L)	2130	—	—	—
COD (mg/L)	4420	2430	—	802
TDS (mg/L)	5210	—	—	—
Sulfate (mg/L)	56	12	<5	—
Oil and grease (mg/L)	15	—	—	—
pH	6.9	—	—	6.3
Ammonia (mg/L)	132	14	160	—
Nitrate (mg/L)	0.6	3	—	—
Chloride (mg/L)	835	275	420	—
Cyanide (mg/L)	0.2	—	<0.005	—
Fluoride (mg/L)	2.9	—	—	—
Aluminum (mg/L)	72	0.3	—	4
Arsenic (mg/L)	0.6	<0.003	<0.005	<0.01
Barium (mg/L)	0.3	—	—	0.32
Boron (mg/L)	3.3	—	—	1.3
Cadmium (mg/L)	<0.02	<0.0002	—	<0.005
Calcium (mg/L)	652	332	—	235
Chromium (mg/L)	0.1	0.003	—	0.014
Cobalt (mg/L)	0.1	—	—	—
Copper (mg/L)	0.1	0.03	—	<0.03
Iron (mg/L)	283	120	14	14
Lead (mg/L)	0.2	<0.001	—	0.015
Magnesium (mg/L)	336	179	—	138
Manganese (mg/L)	9.8	—	0.2	1.34
Mercury (mg/L)	<0.001	<0.0004	0.0002	<0.0002
Nickel (mg/L)	0.2	<0.02	—	0.06
Potassium (mg/L)	157	42	—	378
Phosphorus (mg/L)	—	—	1	—
Selenium (mg/L)	<0.1	—	—	<0.005
Silver (mg/L)	<0.02	—	—	<0.01
Sodium (mg/L)	791	133	—	672
Thallium (mg/L)	<0.1	—	—	—

TABLE 6.14 (cont.)
Examples of Landfill Leachate Characteristics

Parameter	Southern Illinois	Berrien County, Michigan	Elkhart County, Indiana	Forest Lawn, Michigan
Tin (mg/L)	0.1	—	—	<0.03
Zinc (mg/L)	3.5	—	—	0.22
Acetone (ppb)	23,000	—	—	690
Benzene (ppb)	11	20	10	17
Chloroethane (ppb)	53	—	62	19
Diethyl ether (ppb)	840	—	—	94
Ethyl benzene (mg/L)	25	20	400	68
Methylene chloride (mg/L)	58	33	17	290
Methyl ethyl ketone (ppb)	44,300	—	—	2200
Methyl isobutyl ketone (ppb)	220	—	—	58
Tetrahydrofuran (ppb)	2260	—	—	407
Toluene (ppb)	780	150	300	370
m- and *p*-Xylenes (ppb)	13	—	—	155
Di-*n*-butylphthalate (ppb)	24	—	10	—
Phenol (ppb)	555	—	15	—
Atrazine (ppb)	12	—	—	—
2,4-D (ppb)	9	—	—	—

Note: BOD, biochemical oxygen demand; COD, chemical oxygen demand; TDS, total dissolved solids.

Source: Reed, S.C. et al., *Natural Systems for Waste Management and Treatment*, 2nd ed., McGraw-Hill, New York, 1995. With permission.

providing the desired treatment benefits. The major issues of concern are removal of iron and manganese and moderation of the liquid pH. The FWS wetland has been preferred for this service because of the greater potential for aerobic conditions in the system and because the precipitated iron and manganese could result in clogging of a SSF wetland bed. The acidic condition of mine drainage is often caused by oxidation of iron pyrite:

$$2FeS_2 + 2H_2O = 2Fe^{2+} + 4H^+ + 4SO_4^{2-}$$

The ferrous iron produced by the previous reaction undergoes further oxidation in a wetland system:

TABLE 6.15
Removal Efficiency of Free Water Surface Constructed
Wetlands Treating Landfill Leachate

Constituent	Influent	Effluent	Percent Removal (%)
pH	6.32	6.86	—
TSS (mg/L)	1008	30	97
TDS (mg/L)	1078	396	63
COD (mg/L)	456	45	90
TOC (mg/L)	129	17	87
Copper (mg/L)	0.05	0.024	52
Lead (mg/L)	0.078	0.004	94
Mercury (mg/L)	0.0019	0.0019	0
Nickel (mg/L)	0.082	0.01	88
Zinc (mg/L)	0.08	0.03	62

Note: TSS, total suspended solids; TDS, total dissolved solids; COD, chemical oxygen demand; TOC, total organic carbon.

Source: Johnson, K.D. et al., in *Constructed Wetlands for the Treatment of Landfill Leachates*, Mulamoottil, G. et al., Eds., Lewis Publishers, Boca Raton, FL. 1998. With permission.

$$4Fe^{2+} + O_2 + 4H^+ = 4Fe^{3+} + 2H_2O$$

If sufficient alkalinity is not present to provide a buffering capacity, the hydrolysis of the ferric iron (Fe^{3+}) will further decrease the pH in the wetland effluent:

$$Fe^{3+} + 3H_2O = Fe(OH)_3 + 3H^+$$

Several wetland systems described by Brodie et al. (1993) are effective in the removal of iron and manganese but the pH decreases from 6 to about 3 because of the reaction defined above. Previous attempts utilizing exposed limestone filter beds and the addition of buffering agents have been either ineffective or too expensive. Oxides of iron and aluminum would precipitate on the exposed limestone surfaces under aerobic conditions and that surface coating would prevent further calcium dissolution and eliminate any further buffering capacity. To correct this problem, the Tennessee Valley Authority (TVA) has developed an anoxic limestone drain (ALD). Crushed high-calcium-content limestone aggregate (20- to 40-mm size) is placed in a trench 10 to 16 ft (3 to 5 m) wide and to a depth ranging from 2 to 5 ft (0.6 to 1.5 m). The bed cross-section must be large enough

TABLE 6.16
Nutrients and Microorganisms
Required for Biological Oxidation

Parameter	Minimum Required Quantity (kg/kg BOD)
Nitrogen	0.043
Phosphorus	0.006
Manganese	0.0001
Copper	0.00146
Zinc	0.00016
Molybdenum	0.00043
Selenium	14×10^{-10}
Magnesium	0.0030
Cobalt	0.00013
Calcium	0.0062
Sodium	0.00005
Potassium	0.0045
Iron	0.012
Carbonate	0.0027

to pass the maximum expected flow as defined by Darcy's law (see Section 6.5 in this chapter). The exposed portion of the trench is backfilled with compacted clay to seal the bed and ensure anoxic conditions in the limestone. The interface between the clay and the limestone is usually protected with a plastic geotextile. The upstream end of the trench or bed is located to intercept the source of the acid mine drainage.

Brodie et al. (1993) suggested specific guidelines for utilization of the ALD component:

- Existing alkalinity >80 mg/L, Fe <20 mg/L — Only the wetlands system is required.
- Existing alkalinity >80 mg/L, Fe >20 mg/L — A wetlands system without an ALD is probably adequate, although the ALD would be beneficial.
- Existing alkalinity <80 mg/L, Fe >20 mg/L — An ALD is recommended.
- Existing alkalinity <80 mg/L, Fe <20 mg/L — The ALD is not essential but is still recommended.

- Existing alkalinity = 0 mg/L, Fe <20 mg/L — The ALD will be necessary as the Fe concentration approaches 20 mg/L.
- Dissolved oxygen in liquid >2 mg/L or pH >6 and eH >100 mV — These conditions will result in oxide coatings and negate the benefits of an ALD.

A sedimentation pond is recommended as a treatment component prior to a wetland whether or not an ALD component is used in the system. This allows precipitation of a large fraction of the dissolved iron in a basin that can be dredged more easily than the wetland component.

The current practice for design of the wetland component is based on empirical evaluation of the performance of successfully operating systems. The TVA recommends a hydraulic loading from 0.37 to 1.0 gal/ft·d for iron removal depending on the pH, alkalinity, and iron concentration in the inflow. Others recommend a hydraulic loading rate of up to 3.5 gal/ft·d (0.14 m/d) for the same purpose. The treatment cells are designed for the base flow and then sufficient freeboard is provided to accommodate the design storm event. Multiple cells with a water depth in treatment zones of less than 1.5 ft (0.5 m) are recommended. Deep-water zones can also be provided if supplemental habitat values are a project goal. Recommended flow velocities in the wetland cells range from 0.1 to 1.0 ft/s (0.03 to 0.3 m/s). A separate wetland cell should be constructed for each 50 mg/L of iron content in the inflow because of the need for reaeration after oxidation of this amount of iron. If topography permits, a cascade spillway is recommended between these wetland cells.

6.5 PLANNING AND DESIGN

The planning and design of wetland treatment systems involves all of the same factors considered for other natural as well as conventional wastewater treatment systems as described in Chapter 2 of this book. The unique aspect for wetland systems is taking into consideration habitat issues and recreational potential. The functions of a wetland system can range from an exclusive commitment to wastewater treatment to a multipurpose project incorporating environmental enhancement and public recreational benefits. The intended functions of a wetland system must be defined clearly at a very early stage in project development to permit evaluation of feasibility and to ensure cost-effective implementation. All wetland systems, including the gravel-bed SSF type, will attract birds and other wildlife. In a wetland system dedicated for treatment, these habitat values will be incidental and minimal by design. Special features can be introduced to attract specific wildlife and to ensure pleasurable public recreation. Efforts are then required to ensure that toxic or hazardous conditions are not imposed on the attracted wildlife or the public. A desirable combination is to incorporate both approaches and use dedicated treatment wetland units in the early stages of the system followed by wetland units with increasing habitat and recreational values as the water quality in the wetland improves.

6.5.1 SITE EVALUATION

Site evaluation criteria for wetlands and other natural systems are given in Chapter 2 of this manual. The ideal site for a wetland would be within a reasonable distance from the wastewater sources and at an elevation permitting gravity flow to the wetland, between the wetland cells, and to the final discharge point. The site would be available at a reasonable cost, would not require extensive clearing or earthwork for construction, would have a deep nonsensitive groundwater table, and would contain subsoils that, when compacted, would provide a suitable liner. Any divergence from these ideal characteristics will result in increased project costs. The possible future expansion of the system should also be given consideration during the planning and site evaluation effort. The 56-ac (23-ha) FWS wetland system in West Jackson County, Mississippi, was constructed in 1990/1991 with a design capacity of 1.6 mgd. Because of rapid community growth it was necessary to expand the system capacity to 4 mgd with 50 ac (20.2 ha) of new wetland construction in 1997/1998. This expansion was possible because sufficient land was available adjacent to the original wetland system.

6.5.2 PREAPPLICATION TREATMENT

All wetland treatment systems in the United States are preceded by some form of preliminary treatment, ranging from the equivalent of primary to tertiary levels from advanced wastewater treatment systems. The level of preapplication treatment required depends on the functional intent of the wetland component, on the level of public exposure expected, and on the need to protect habitat values. The minimal preliminary treatment for municipal wastewaters would be the equivalent of primary, accomplished with septic tanks or Imhoff tanks for small systems or a pond unit with a deep zone for sludge accumulation for larger systems. It is considered prudent to provide the equivalent of secondary treatment prior to allowing public access to the wetland components or developing specific habitats to encourage birds and other wildlife. This level of treatment could be accomplished in a first-stage wetland unit where public access is restricted and habitat values are minimized. Tertiary treatment with nutrient removal may be necessary prior to discharge to natural wetlands where preservation of the existing habitat and ecosystem is desired. Common preliminary features in stormwater wetlands are a trash rack and a forebay to allow the settling and removal of large objects carried with the stormwater runoff. Wetlands designed for mine drainage treatment may require a preliminary unit for pH or alkalinity adjustment (Brodie et al, 1993).

6.5.3 GENERAL DESIGN PROCEDURES

All constructed wetland systems can be considered to be attached-growth biological reactors, and their performance can be estimated with first-order plug-flow kinetics for BOD and nitrogen removal. Design models are presented in this chapter for removal of BOD, TSS, ammonia nitrogen, nitrate, total nitrogen, and phosphorus, for both FWS and SSF wetlands. In some cases, an alternative model from other sources is also presented for comparison purposes because a universal consensus does not exist on the "best" design approach. The basic relationship for plug-flow reactors is given by Equation 6.8:

$$C_e/C_0 = \exp[-K_T t] \tag{6.8}$$

where

C_e = Effluent constituent concentration (mg/L).
C_0 = Influent constituent concentration (mg/L).
K_T = Temperature-dependent, first-order reaction rate constant (d^{-1}).
t = Hydraulic residence time (d).

The hydraulic residence time in the wetland can be calculated with Equation 6.9:

$$t = LWyn/Q \tag{6.9}$$

where

L = Length of the wetland cell (ft; m).
W = Width of the wetland cell (ft; m).
y = Depth of water in the wetland cell (ft; m).
n = Porosity, or the space available for water to flow through the wetland. Vegetation and litter occupy some space in the FWS wetland, and the media, roots, and other solids do the same in the SSF case. Porosity is a percent (expressed as a decimal).
Q = The average flow through the wetland (ft^3/d; m^3/d):

$$Q = (Q_{in} + Q_{out})/2 \tag{6.10}$$

It is necessary to determine the average flow with Equation 6.10 to compensate for water losses or gains via seepage or precipitation as the wastewater flows through the wetland. A conservative design might assume no seepage and adopt reasonable estimates for evapotranspiration losses and rainfall gains from local records for each month of concern. This requires a preliminary assumption regarding the surface area of the wetland so the volume of water lost or added can be calculated. It is usually reasonable for a preliminary design estimate to assume that Q_{out} equals Q_{in}.

It is then possible to determine the surface area of the wetland by combining Equation 6.8 and Equation 6.9:

$$A_s = (LW)$$

$$= \frac{Q\ln(C_0/C_e)}{K_T yn} \tag{6.11}$$

where A_s is the surface area of wetland (ft^2; m^2). The value used for K_T in Equation 6.1 or Equation 6.4 depends on the pollutant that must be removed and on the temperature; these aspects are presented in later sections of this chapter.

Because the biological reactions involved in treatment are temperature dependent it is necessary, for a proper design, to estimate the water temperature in the wetland. The performance and basic feasibility of FWS wetlands in very cold climates are also influenced by ice formation on the system. In the extreme case, a relatively shallow wetland might freeze to the bottom and effective treatment would cease. This chapter contains calculation procedures for estimating water temperatures in the wetland and for estimating the thickness of ice that will form.

The hydraulic design of the wetland is just as important as the models that determine pollutant removal because those models are based on the critical plug-flow assumption, with uniform flow across the wetland cross-section and minimal short-circuiting. Many of the early designs of both SSF and FWS wetlands did not give sufficient consideration to the hydraulic requirements, and the result was often unexpected flow conditions including short-circuiting and adverse impacts on expected performance. These problems can be avoided using the hydraulic design procedures in this chapter.

A valid design requires consideration of hydraulics and the thermal aspects, as well as removal kinetics. The procedure is usually iterative in that it is necessary to assume a water depth and temperature to solve the kinetic equations. These will predict the wetland area required to remove the pollutant of concern. The pollutant requiring the largest area for removal is the limiting design parameter (LDP), and it controls the size of the wetland. When the wetland area is known, the thermal equations can be used to determine the theoretical water temperature in the wetland. If the original assumed water temperature and this calculated temperature do not agree, further iterations of the calculations are required until the two temperature values converge. The last step is to use the appropriate hydraulic calculations to determine the final aspect ratio (length-to-width) and flow velocity in the wetland. If these final values differ significantly from those assumed for the thermal calculations, further iterations may be necessary.

6.6 HYDRAULIC DESIGN PROCEDURES

The hydraulic design of constructed wetland systems is critical to their successful performance. All of the design models in current use assume uniform flow conditions and unrestricted opportunities for contact between the wastewater constituents and the organisms responsible for treatment. In the SSF wetland concept it is also necessary to ensure that subsurface flow conditions are maintained under normal circumstances for the design life of the system. These assumptions and goals can only be realized through careful attention to the hydraulic design and to proper construction methods. Flow through wetland systems must overcome the frictional resistance in the system which is imposed by the vegetation and litter layer in the FWS type and the media, plant roots, and accumulated solids in the SSF type. The energy to overcome this resistance is provided by the head differential between the inlet and the outlet of the wetland. Some of this differential can be provided by constructing the wetland with a sloping bottom; however, it is neither cost effective nor prudent to depend on just a sloping bottom for the head differential required, as the resistance to flow may increase with time but the bottom slope is fixed for the life of the system. The preferred approach is to construct the bottom with sufficient slope to ensure complete drainage when necessary and to provide an outlet that permits adjustment of the water level at the end of the wetland. This adjustment can then be used to set whatever water surface slope is required and in the lowest position used to drain the wetland. Details on these adjustable outlets can be found in a later section of this chapter.

The aspect ratio (length-to-width) selected for the wetland strongly influences the hydraulic regime and the resistance to flow in the system. In the design of some early FWS systems it was thought that a very high aspect ratio was necessary to ensure plug-flow conditions in the wetland and to avoid short-circuiting, and aspect ratios of at least 10:1 were recommended. A major problem with this approach is that the resistance to flow increases as the length of the flow path increases. A FWS system constructed in California with an aspect ratio of about 20:1 experienced overflow at the head of the wetland after a few years because of the increasing flow resistance from the accumulating vegetative litter. Aspect ratios from less than 1:1 up to about 3:1 or 4:1 are acceptable. Short-circuiting can be minimized by careful construction and maintenance of the wetland bottom, by the use of multiple cells, and by providing intermediate open-water zones for flow redistribution. These techniques are discussed in greater detail in later sections of this chapter.

In essence, a treatment wetland is a shallow body of moving water with a relatively large surface area. The hydraulic design is complicated by the fact that significant frictional resistance to flow develops because of the plants and litter in the FWS case and because of the gravel media in the SSF type. In design it is assumed that the water will move uniformly, at a predictable rate, over the entire surface area. This assumption is hydrologically complicated by the fact that precipitation, evaporation, evapotranspiration, and seepage affect the volume of water present in the wetland, the concentration of pollutants, and the HRT.

Manning's equation is generally accepted as a model for the flow of water through FWS wetland systems. The flow velocity, as described by Equation 6.12, is dependent on the depth of water, the hydraulic gradient (i.e., slope of the water surface), and the resistance to flow:

$$v = (1/n)(y^{2/3})(s^{1/2}) \qquad\qquad (6.12)$$

where

v =   Flow velocity (ft/s; m/s).

n =   Manning's coefficient (s/ft$^{1/3}$; s/m$^{1/3}$).

y =   Water depth (ft; m).

s =   Hydraulic gradient (ft/ft; m/m).

In most applications of Manning's equation, the resistance to flow occurs only on the bottom and the submerged sides of an open channel, and published values of n coefficients for various conditions are widely available in the technical literature. However, in FWS wetlands, the resistance to flow extends through the entire depth of water due to the presence of the emergent vegetation and litter. The relationship between the Manning number (n) and the resistance factor (a) is defined by Equation 6.13:

$$n = a/y^{1/2} \tag{6.13}$$

where a is the resistance factor ($s \cdot ft^{1/6}$; $s \cdot m^{1/6}$).

Reed et al. (1995) presented the following values for resistance factor a in FWS wetlands:

- Sparse, low-standing vegetation — $y > 1.2$ ft (0.4 m), $a = 0.487$ $s \cdot ft^{1/6}$, (0.4 $s \cdot m^{1/6}$)
- Moderately dense vegetation — $y \geq 1.0$ ft (0.3 m), $a = 1.949$ $s \cdot ft^{1/6}$, (1.6 $s \cdot m^{1/6}$)
- Very dense vegetation and litter — $y < 1.0$ ft (0.3 m), $a = 7.795$ $s \cdot ft^{1/6}$, (6.4 $s \cdot m^{1/6}$)

This range of values was experimentally confirmed by Dombeck et al. (1997). The energy required to overcome this resistance is provided by the head differential between the water surface at the inlet and outlet of the wetland. Some of this differential can be provided by constructing the wetland with a sloping bottom. The preferred approach is to construct the bottom with minimal slope that still allows complete drainage when needed and to provide outlet structures that allow adjustment of the water level to compensate for the resistance, which may increase with time. The aspect ratio (length-to-width) selected for a FWS wetland also strongly influences the hydraulic regime because the resistance to flow increases as the length increases. Reed et al. (1995) developed a model that can be used to estimate the maximum desirable length of a FWS wetland channel:

$$L = [(A_s)(y^{2.667})(m^{0.5})(86,400)/(a)(Q_A)]^{0.667} \tag{6.14}$$

where
L = Maximum length of wetland cell (m).
A_s = Design surface area of wetland (m^2).
y = Depth of water in the wetland (m).
m = Portion of available hydraulic gradient used to provide the necessary head (% as a decimal).
a = Resistance factor ($s \cdot m^{1/6}$).
Q_A = Average flow through the wetland (m^3/d) = ($Q_{IN} + Q_{OUT}$)/2.

An initial m value between 10 to 20% is suggested for design to ensure a future reserve as a safety factor. In the general case, this model will produce an aspect ratio of 3:1 or less. The use of the average flow (Q_A) in Equation 6.14 compensates for the influence of precipitation, evapotranspiration, and seepage on the flow through the wetland. The design surface area (A_s) in Equation 6.14 is the bottom area of the wetland as determined by the pollutant removal models presented later in this chapter.

6.7 THERMAL ASPECTS

The temperature conditions in a wetland affect both the physical and biological activities in the system. In the extreme case, sustained low-temperature conditions and the resulting ice formation could result in physical failure of the wetland. The biological reactions responsible for BOD removal, nitrification, and denitrification are known to be temperature dependent (Benefield and Randall, 1980; Gearheart et al., 1989); however, in many cases the BOD removal performance of existing wetland systems in cold climates has not demonstrated clear temperature dependence. This is because the long hydraulic residence time provided by these systems tends to compensate for the lower reaction rates during the winter months. Several systems in Canada and the United States do demonstrate a decrease in nitrogen removal capability during the winter months. This is believed to be caused by a combination of temperature influence on the biological reactions and to a lack of oxygen when an ice cover forms on the water surface.

Temperature-dependent rate constants for the BOD and nitrogen removal models are presented elsewhere in this chapter. It is necessary here, then, to provide a reliable method for estimating the water temperature in the wetland for the proper and effective use of the biological design models. This section presents calculation techniques for the determination of the water temperature in SSF and FWS wetlands and for predicting the thickness of ice that might form on the FWS wetland.

Because the water surface is exposed to the atmosphere in a FWS wetland, some ice formation, at least on a temporary basis, is likely in northern locations that experience periods of subfreezing air temperatures. The presence of some ice can be a benefit in that the ice layer acts as a thermal barrier and slows the cooling rate of the water beneath. In ponds, lakes, and most rivers, the ice layer floats freely and can increase in thickness without significantly reducing the volume available for flow beneath the ice cover. In the case of the FWS wetland, the ice may be held in place by the numerous stems and leaves of the vegetation so the volume available for flow can be significantly reduced as the ice layer thickens. In the extreme case, the ice layer may thicken to the point where flow is constricted, the resulting stresses induced cause cracks in the ice, and flow may commence on top of the ice layer. Freezing of that surface flow will occur, and the wetland is then in a failure mode until warm weather returns. The biological treatment activity in the wetland will also cease at that point. This situation must be prevented or avoided if a constructed wetland is to be considered. In some locations that experience very long periods of very low air temperatures ($<-20°C$; $<0°F$), the solution may be to utilize a seasonal wetland component with wastewater stored in a lagoon during the extreme winter months. A number of systems in South Dakota and northwestern Canada operate in this mode (Bull, 1994; Dornbush, 1993). FWS constructed wetlands have, on the other hand, performed successfully throughout the winter months in Ontario, Canada, and in several communities in Iowa where extreme winter temperatures are also experienced. It is essential for each project in northern climates to conduct a thermal analysis, as described in this section, to ensure that the wetland will be physically stable during the winter months and can sustain water temperatures that allow the biological reactions to proceed.

The calculation procedure presented in this section was derived from Ashton (1986) with the assistance of Darryl Calkins (USA CRREL; Hanover, NH). The procedure has three parts:

1. Calculate water temperatures in the wetland until conditions that allow ice formation (3°C water temperature) commence. Separate calculations are required for densely vegetated wetland segments and for large area open-water zones.
2. Water temperature calculations are then continued for the ice-covered case.
3. An estimate is made of the total depth of ice that may form over the period of concern.

The temperatures determined during steps 1 and 2 are also used to determine the basic feasibility of a FWS wetland in the location under consideration and to verify the temperature assumptions made when sizing the wetland with either the BOD or nitrogen removal models. These BOD and nitrogen models are the first step in the design process because their results are necessary for determining the wetland size, HRT, and flow velocity to be used in the subsequent thermal calculations. The total depth of ice estimated in the third step above also provides an indication of the feasibility of a wetland in the location under consideration and is used to determine the necessary operating water depth during the winter months.

6.7.1 CASE 1. FREE WATER SURFACE WETLAND PRIOR TO ICE FORMATION

Equation 6.25 is used to calculate the water temperature at the point of interest in the wetland. Experience has shown that ice formation commences when the bulk temperature in the liquid approaches 3°C (37°F) because of density differences and convection losses at the water surface (Ashton, 1986; Calkins, 1995). Equation 6.15 is therefore repeated until a temperature of 3°C is reached or until the end of the wetland cell is reached, whichever comes first. If a temperature of 3°C is reached prior to the end of the wetland, then Equation 6.17 is used to calculate the temperatures under an ice cover. If the wetland is composed of vegetated zones interspersed with deeper open water zones, Equation 6.15 must be used sequentially, with the appropriate heat-transfer coefficient (U_s) to calculate the water temperatures:

$$T_w = T_{air} + (T_0 - T_{air})\exp\left[-U_s(x - x_0)/\delta yvc_p\right] \qquad (6.15)$$

where

T_w = Water temperature (°C; °F) at distance x (m; ft).

T_{air} = Average air temperature during period of interest (°C; °F).

T_0 = Water temperature (°C; °F) at distance x_0 (m; ft), the entry point for the wetland segment of interest.

U_s = Heat-transfer coefficient at the wetland surface (W/m²·°C; Btu/ft²·hr·°F).

= 1.5 W/m²·°C (0.264 Btu/ft²·hr·°F) for dense marsh vegetation.

= 10–25 W/m²·°C (1.761–4.403 Btu/ft²·hr·°F) for open water; high value used for windy conditions with no snow cover.

c_p = Specific heat = 1.007 Btu/lb·°F (4215 J/kg·°C).

If the first iteration shows a temperature of less than 37°F (3°C) in the final effluent from the wetland, Equation 6.15 can be rearranged and solved for distance x at which the temperature becomes 37°F:

$$x - x_0 = -[\delta yvc_p/U_s][\ln(37 - T_{air})/(T_0 - T_{air})] \qquad (6.16)$$

Example 6.1

Calculate the water temperature in a three-stage FWS constructed wetland:

Stage 1: Length 300 ft, depth 1 ft, densely vegetated, flow velocity 0.2 ft/hr
Stage 2: Deep open-water zone, length 100 ft, depth 4 ft, flow velocity 0.1 ft/hr
Stage 3: Same as stage 1
Air temperature, 49°F; influent wastewater temperature, 70°F

Solution

1. Use Equation 6.15 to calculate the temperature at the end of stage 1:
   T_w = 49 + (70 − 49) exp[−0.264(300)/(62.4)(1)(11.7)(1.007)]
   T_w = 49 + (21)(0.9) = 67.9°F
2. Calculate the water temperature at the end of stage 2:
   T_w = 49 + (67.9 − 49) exp[−0.264(100)/(62.4)(4)(5.85)(1.007)]
   T_w = 49 + (18.9)(0.836) = 64.8°F
3. Calculate the water temperature at the end of the wetland:
   T_w = 49 + (64.8 − 49) exp[−0.264(300)/(62.4)(1)(11.7)(1.007)]
   T_w = 49 + (15.8)(0.9) = 63.2°F

6.7.2 CASE 2. FLOW UNDER AN ICE COVER

When an ice cover forms, the heat transfer from the underlying water to the ice proceeds at a constant rate that is not influenced by the air temperature or the presence or absence of a snow cover on top of the ice. This is because the ice surface, at the interface with the water, remains at 32°F (0°C) until all of the water is frozen. The rate of ice formation is influenced by the air temperature and the presence or absence of snow, but the cooling rate of the underlying water is not. The wetland water temperature under an ice cover can be estimated with Equation 6.17. This is identical in form to Equation 6.25, except for changes in two of the terms (T_m and U_i) to reflect the presence of the ice cover:

$$T_w = T_m + (T_0 - T_m) \exp\left[-U_i (x - x_0)/\delta y v c_p\right] \qquad (6.17)$$

where

T_m = Ice melting point (32°F; 0°C).

T_0 = Water temperature at distance x_0, assuming 37.4°F (3°C) where an ice cover commences.

U_i = Heat-transfer coefficient at ice/water interface (Btu/ft²·hr·°F; W/m²·°C).

Other terms are as defined previously.

The U_i value in Equation 6.17 depends on the depth of water beneath the ice and the flow velocity:

$$U_i = \Phi\left[v^{0.8}/y^{0.2}\right] \qquad (6.18)$$

where

U_i = Heat-transfer coefficient at ice/water interface (Btu/ft²·hr·°F; W/m²·°C).

Φ = Proportionality coefficient = 0.0022 Btu/ft²·⁶·hr⁰·²·°F (1622 J/m²·⁶·s⁰·²·°C).

v = Flow velocity (ft/hr; m/s), assuming no ice conditions.

y = Depth of water (ft; m).

6.7.3 CASE 3. FREE WATER SURFACE WETLAND AND THICKNESS OF ICE FORMATION

Ice will commence to form on the surface of the FWS wetland when the bulk water temperature reaches 37.4°F (3°C) and will continue as long as the temperature remains at or below 32°F (0°C). In northern climates where extremely low air temperatures can persist for very long periods, the FWS wetland may not be a feasible year-round treatment because extensive ice formation can result in physical failure of the system. The thickness or depth of ice that will form over a 1-d period can be estimated with Equation 6.19:

$$y = \left[(t)(\tau)/(\delta)(\Omega)\right]\left[(T_m - T_{air})/(y_s/k_s + y_i/k_i + 1/U_s) - U_i(T_w - T_m)\right] \qquad (6.19)$$

where

y = Thickness of ice formation per day (ft/d; m/d).

t = Time period of concern (d).

TABLE 6.17
Thermal Conductivities for Wetland Components

Material	k (Btu/ft²·hr·°F)	K (W/m²·°C)
Air (no convection)	0.014	0.024
Snow (new, loose)	0.046	0.08
Snow (long-term)	0.133	0.23
Ice (at 32°F)	1.277	2.21
Water (at 32°F)	0.335	0.58
Wetland litter layer (dry)	0.029	0.05
Dry gravel (25% moisture)	0.867	1.5
Saturated gravel	1.156	2.0
Dry soil[a]	0.462	0.8

[a] This is native soil underlying the wetland bed. Heat transfer is into the wetland bed during the winter and from the wetland bed during the summer. Assume a 3-ft (1-m) depth for this soil layer.

Source: Reed, S.C. et al., *Natural Systems for Waste Management and Treatment*, 2nd ed., McGraw-Hill, New York, 1995. With permission.

τ = Time conversion factor (24 hr/d; 86,400 s/d).

δ = Density of ice (57.2 lb/ft³; 917 kg/m³).

Ω = Latent heat (144 Btu/lb; 334,944 J/kg).

T_m = Melting point of ice (32°F; 0°C).

T_{air} = Average air temperature during time period of concern (°F; °C).

y_s = Depth of snow cover (ft; m).

k_s = Conductivity of snow (from Table 6.17).

y_i = Depth of daily ice formation (ft; m).

k_i = Conductivity of ice (from Table 6.17).

U_s = Heat transfer coefficient at the wetland surface, W/m²·°C (Btu/ft²·hr·°F).

= 1.5 W/m²·°C (0.264 Btu/ft²·hr·°F) for dense marsh vegetation.

= 10–25 W/m²·°C (1.761–4.403 Btu/ft²·hr·°F) for open water; high value used for windy conditions with no snow cover.

U_i = Heat transfer coefficient water to ice (from Equation 6.18).

T_w = Average water temperature during period of concern (from Equation 6.17).

It is necessary to repeat the calculation for each day of interest with appropriate adjustments in the depth of ice and snow in Equation 6.19. The time period of concern for the previous FWS thermal models is equal to the design HRT for the wetland; in this case, the time period of concern may be the entire winter season if significant periods of subfreezing temperatures persist. A reasonable first approximation of potential ice formation can be achieved by using the average monthly air temperatures (in the coldest winter of record) during the period of concern. This model was also derived from Ashton (1986) with the assistance of Darryl Calkins (USA CRREL; Hanover, NH).

The rate of ice formation will be the highest on the first day of freezing, when neither an ice cover nor a snow layer is present to retard heat losses. In addition, the final term in Equation 6.19 is usually small and can be neglected for estimation purposes. As a result, Equation 6.19 reduces to the Stefan formulation (Stefan, 1891):

$$y = m[T_m - T_{air})(t)]^{0.5} \qquad (6.20)$$

where

y = Depth of ice that will form over time period t (ft; m).

m = Proportionality coefficient (ft/°F$^{0.5}$·d$^{0.5}$; m/°C$^{0.5}$·d$^{0.5}$).

 = 0.066 ft/°F$^{0.5}$·d$^{0.5}$ (0.027 m/°C$^{0.5}$·d$^{0.5}$) for open-water zones with no snow.

 = 0.044 ft/°F$^{0.5}$·d$^{0.5}$ (0.018 m/°C$^{0.5}$·d$^{0.5}$) for open-water zones with snow.

 = 0.024 ft/°F$^{0.5}$·d$^{0.5}$ (0.010 m/°C$^{0.5}$·d$^{0.5}$) for wetland with dense vegetation and litter.

T_m = Freezing point of ice (32°F; 0°C).

T_{air} = Average air temperature during time period t (°F; °C).

t = Number of days in the period of interest (d).

Equation 6.20 can be used to estimate total ice formation on FWS wetlands over the entire winter season or for shorter time periods if desired. This equation can be used to determine the feasibility of winter operations for a wetland in locations with very low winter temperatures. For example, a site with persistent air temperatures at −13°F (−25°C) would result in a wetland that is 1.5 ft (0.45 m) deep freezing to the bottom in about 84 days.

The term $(T_m - T_{air})(t)$ is the freezing index and is an environmental characteristic for a particular location (values can be found in published references). Equation 6.20 is also used in Chapter 9 of this book to determine the depth of sludge that can be frozen for dewatering purposes.

6.7.4 SUMMARY

If the thermal models for FWS wetlands predict sustained internal water temperature of less than 33.8°F (1°C), a wetland may not be physically capable of winter operations at the site under consideration at the design hydraulic residence time (HRT). Nitrogen removal is likely to be negligible at those temperatures. Similarly, if Equation 6.20 predicts a seasonal ice thickness greater than about 75%

of the design depth of a FWS wetland, the use of a wetland during the winter months may be questionable. It may be possible to increase the operating depth in these cases as long as the desired treatment results can still be achieved at <37.4°F (<3°C) and beneath an ice cover that will further impede oxygen transfer for nitrogen removal. Constructed wetlands can operate successfully during the winter in most of the northern temperate zone. The thermal models presented in this section should be used to verify the temperature assumptions made when the wetland is sized with the biological models for BOD or nitrogen removal. Several iterations of the calculation procedure may be necessary for the assumed and calculated temperatures to converge.

6.8 DESIGN MODELS AND EFFLUENT QUALITY PREDICTION

Constituent removal design procedures have developed rapidly in recent years. Three design models were compared in the Water Environment Federation's Manual of Practice (WEF, 2001). All three models are based on analysis of wetland input/output data or mass balance relationships, and they all take the general form of a first-order plug-flow model; however, they do not directly account for the complex reactions and interactions that occur in wetlands but instead use a lumped apparent rate constant to account for the change in concentration or mass between the input and output (Tchobanoglous et al., 2003). Such an approach is the best that can be done with the currently available database and understanding of wetland processes. The models are fundamentally equivalent and should be expected to produce similar results, but unfortunately that is not the case. This is partly due to the fact that the models were not developed from the same sets of data and also partly due to differences in the structure and content of the models. The models can be divided into two types: (1) the volumetric models as developed by Reed et al. (1995) and Crites and Tchobanoglous (1998) and (2) the areal loading models developed by Kadlec and Knight (1996). Some of the major advantages and limitations of these two design approaches are listed below.

6.8.1 VOLUMETRIC MODEL

6.8.1.1 Advantages

- The design is based on average flow through the system. This allows compensation for water losses and gains due to precipitation and evapotranspiration.
- The safety factors and irreducible background concentrations are treated as external boundary conditions and have no limiting impact on the mathematical results of design models.

6.8.1.2 Limitations

- The procedure requires knowledge of the water depth in the system. This may be difficult to control during construction of large systems and is likely to change over the long term.
- The porosity of the vegetation and accumulated litter must be known. The assumed design values are based on a limited database, and the value is likely to change over the long term.
- The removal of BOD is assumed to be temperature dependent based on experience with other wastewater treatment processes; however, data from many operating wetland systems do not demonstrate temperature dependence.

6.8.2 AREAL LOADING MODEL

6.8.2.1 Advantages

- The models are based on the mass loading on the wetland surface area; therefore, water depth, which may be difficult to determine for large systems, is not a factor in the design calculations.
- These models are more flexible mathematically. It is possible to produce a better fit of existing data with this two-variable (K, C^*) model as compared to the single-variable (K) volumetric models.

6.8.2.2 Limitations

- The models deal only with input wastewater volume (Q). This does not allow for compensation for water gains and losses in the design calculations.
- The FWS database used for development of these models includes a large number of lightly loaded polishing wetland systems. Use of these data may produce low-valued rate constants (K) that might in turn result in unnecessarily large wetland system designs.
- The internal position of the background concentration (C^*) and safety factor (z) terms in the models for determining wetland area may result in excessive wetland sizes to achieve low concentrations.

6.8.3 EFFLUENT QUALITY PREDICTION

The land area required and the effluent quality can be predicted using the volumetric/detention time models in Table 6.18. Design equations based on the areal loading rate approach are presented in Table 6.19.

TABLE 6.18
Volumetric Process Design Model
Basic Models:

$$C_e/C_0 = \exp(-K_T t) \tag{6.21}$$

$$K_T = K_{20}(\theta)^{(T_w - 20)} \tag{6.22}$$

Treatment area:

$$A_s = Q_A[\ln(C_0/C_e)/K_T(y)(n)] \tag{6.23}$$

where

C_e = Wetland effluent concentration (mg/L).
C_0 = Wetland influent concentration (mg/L).
K_T = Rate constant at temperature T (d^{-1}).
θ = Temperature coefficient at 20°C.
T_w = Average water temperature in wetland during period of concern (°C).
A_s = Treatment area (bottom area) of wetland (m^2).
Q_A = Average flow in the wetland (m^3/d) = $(Q_{IN} + Q_{OUT})/2$.
y = Average depth of water in the wetland (m).
n = Porosity of the wetland (% as a decimal).

Note: (1) The effluent concentration (C_e) cannot be less than the background concentrations listed below. (2) The average flow (Q_A) accounts for water gains and losses from precipitation, evapotranspiration, seepage, etc.

Parameter	FWS Wetland	SSF
Porosity (n) (% as a decimal)	0.70–0.90	—
Depth (y) (m)	0.3–0.6	—

BOD$_5$ Removal:
K_{20} (d^{-1}) = 0.678
θ = 1.06
Background concentration (mg/L) = 6

TSS Removal:

$$C_e/C_0 = [0.1139 + 0.00213(\text{HLR})] \tag{6.24}$$

where HLR = hydraulic loading rate (mm/d × 0.1), and TSS removal is not dependent on temperature. Background concentration (mg/L) = 6

Ammonia Removal:
At 0°C, K_T (d^{-1}) = 0.
At 1°C+, K_{20} = 0.2187.
θ = 1.048.
K_{NH} is a rate constant at 20°C for FWS wetlands (d^{-1}).
Background concentration (mg/L) = 0.2

Note: It is prudent to assume that all TKN (from municipal wastewater) entering the wetland can appear as ammonia, so assume C_0 for ammonia is equal to influent TKN.

Nitrate Removal:
At 0°C, K_T (d^{-1}) = 0.
At 1°C+, K_{20} = 1.000.
θ = 1.15.
Background concentration (mg/L) = 0.2.

Note: It is conservative to assume that all ammonia removed in the previous step can appear as nitrate, so C_0 for nitrate removal design equals C_e from ammonia removal plus any nitrate present in the influent.

Total Nitrogen Removal:

Effluent TN $= C_{e(NO_3)} + \left(C_{e(NH_4)} - C_{e(NO_3)} \right)$

Background concentration (mg/L) = 0.4.

Note: A specific model for total nitrogen removal is not available in this set. The effluent total nitrogen (TN) can be estimated as the sum of residual ammonia and remaining nitrate $(C_0 - C_e)$.

Total Phosphorus Removal:

$$C_e/C_0 = \exp(-K_P/\text{HLR}) \qquad (6.25)$$

where HLR is the average hydraulic loading rate (cm/d), and total phosphorus removal is not dependent on temperature.

K_P (mm/d $\times$ 0.1 = 2.73.
Background concentration (mg/L) = 0.05.

Fecal Coliform Removal:

$$C_e/C_0 \text{ (MPN / 100 mL)} = \left[1/\left(1 + K_T(t/d) \right) \right]^x \qquad (6.26)$$

where t, d = HRT in the system, and x = number of wetland cells in series.

K_{20} (d^{-1}) = 2.6.
θ = 1.19.
Background concentration (cfu/100 mL) = 2000.

Note: This model was developed for facultative ponds and is believed to give a conservative estimate for fecal coliform removal in both FWS and SF wetlands.

Background Concentration:
The background concentration is given for each of the parameters listed above. These values represent an external boundary condition on the design models in this set. None of the models should ever be solved for a concentration less than these background levels.

Wetland Sizing:
The parameter (BOD$_5$, etc.) that requires the largest treatment area for removal is the limiting design factor, and that area should be selected for the intended project. The wetland should then provide acceptable treatment for all other parameters of concern.

Safety Factor:
It is typical in all engineering design projects to apply a safety factor. In most cases, the final safety factor is applied after the preliminary calculations are completed. In this case, the safety factor is applied after the wetland size has been determined and ranges from 15 to 25% depending on the uncertainty of available data and on the stringency of performance expectations. The selection of a safety factor is an engineering judgment and represents a comparable increase in the calculated treatment area.

Source: Adapted from Reed, S.C. et al., *Natural Systems for Waste Management and Treatment*, 2nd ed., McGraw-Hill, New York, 1995. With permission.

TABLE 6.19
Areal-Based Process Design Models

Basic Models:

$$(C_e - C^*)/(C_0 - C^*) = \exp(-K_T/\text{HLR}_A) \qquad (6.27)$$

$$K_T = K_{20}(\theta)^{(T-20)}$$

Treatment area:

$$A_s = (-Q_0/K_T)\ln[(C_e(z) - C^*)/(C_0 - C^*)] \qquad (6.28)$$

where

C_e	=	Wetland effluent concentration (mg/L).
C_0	=	Wetland influent concentration (mg/L).
C^*	=	Background concentration (mg/L).
HLR_A	=	Annual hydraulic loading rate (m/yr).
K_T	=	Rate constant at temperature T (m/yr).
K_{20}	=	Rate constant at 20°C (m/yr).
θ	=	Temperature coefficient.
A_s	=	Treatment area of wetland (m²).
Q_0	=	Annual influent wastewater flow rate (m³/yr).
z	=	Safety factor.

Note: (1) These are areal-based models written in terms of hydraulic loading per unit area as compared to a detention time base for the model in Table 6.16. Detention time (HRT) and hydraulic loading (HLR) are directly related for a specific set of wetland conditions. All of the models should, therefore, produce similar results, but they do not. The difference is due to their derivation from different data sets and to the internal position of C^* and a safety factor (z) in the models in this table. The other models treat the background concentration and a safety factor as external boundary conditions. (2) The Q_0 and the HLR as used in the above models are the influent flow rate only and do not include adjustment for water gains or losses through precipitation, evapotranspiration, or seepage. (3) The porosity (n), water depth (y), and detention time (t) in the wetland are not considered in these areal-based design models. (4) The safety factor (z) is the ratio of the annual average concentration to the maximum monthly concentration for the pollutant of concern as derived from the database used by Kadlec and Knight (1996). (5) Because of the internal position of the C^* and z in this treatment area model, it is not possible to design a system large enough to achieve an effluent with background concentrations. As the required effluent concentration (C_e) approaches background (C^*), the required wetland area approaches infinity.

BOD₅ Removal:

K_{20} (m/yr) = 34.
$\theta = 1.00$.
C^* (mg/L) = $3.5 + 0.053(C_0)$.
θ_z (for C^*) = 1.00.
$z = 0.59$.

Note: The treatment area model cannot be solved for BOD₅ effluent values approaching background levels. For example, if an effluent of 7 mg/L is desired, the influent cannot exceed 12 mg/L. An effluent of 6 mg/L would be impossible to achieve.

TSS Removal:

K_{20} (m/yr) = 1000.
$\theta = 1.00$.
C^* (mg/L) = $5.1 + 0.16(C_0)$.
θ_z (for C^*) = $[C_T^* = C_{20}^* (\theta)^{(T-20)}] = 1.065$.
$z = 0.526$.

Note: The treatment area model cannot be solved for low effluent (C_e) values. For example, if an effluent TSS concentration of 15 mg/L is required, the influent (C_0) cannot exceed 17 mg/L if the area model is to be solved.

Organic Nitrogen Removal:

K_{20} (m/yr) = 17.
θ = 1.05.
C^* (mg/L) 1.5.
z = 0.555.

Ammonia Removal:

K_{20} (m/yr) = 18.
θ = 1.04.
C^* (mg/L) = 0.
z = 0.4.

Nitrate Removal:

K_{20} (m/yr) = 35.
θ = 1.09.
C^* (mg/L) = 0.00.
z = 0.400

Total Nitrogen Removal:

K_{20} (m/yr) = 22.
θ = 1.09.
C^* (mg/L) = 1.5.
z = 0.625.

Total Phosphorus Removal:

K_{20} (m/yr) = 12.
θ = 1.00.
C^* (mg/L) = 0.02.
z = 0.555.

Fecal Coliform Removal:

K_{20} (m/yr) = 75.
θ = 1.00.
C^* (mg/L) = 300.
z = 0.333.

Background Concentration:

The background concentration (C^*) is included internally in each design model.

Wetland Sizing:

The parameter (BOD_5, etc.) that requires the largest treatment area for removal is the limiting design factor, and that area should be selected for the intended project. The wetland should then provide acceptable treatment for all other parameters of concern.

Safety Factor:

The safety factor (z) is included internally in the logarithmic portion of the design model for treatment area.

Source: Kadlec, R.H. and Knight, R., *Treatment Wetlands*, Lewis Publishers, Boca Raton, FL, 1996. With permission.

Example 6.2

A FWS wetland is to be designed to reduce the BOD from 100 mg/L to 15 mg/L. The flow is 0.9 mgd and the wastewater temperature is 68°F. Compare the land areas needed using the volumetric/detention time approach and the areal loading rate approach.

Solution

1. Using Equation 6.23, solve for the land area for the volumetric/detention time approach. Use a depth of 2 ft and a porosity of 0.8. Use $K_T = 0.68$:

   $A_s = Q_A[\ln(C_0/C_e)/K_T(y)(n)]$
   $A_s = (0.9)(3.069 \text{ ac·ft/mil gal})[\ln100/15]/(0.68)(2)(0.8)$
   $A_s = (2.76)(1.897)/(1.088)$
   $A_s = 4.81$ ac
   Use a safety factor of 20%, $A = 5.77$ ac.

2. Calculate the land area using the hydraulic loading rate method. Using Equation 6.25, calculate the land area. Convert flow into m³/yr; use $K_T = 34$; and set $z = 1$, $C^* = 8.8$:

   $A_s = (-Q_0/K_T)\ln[(C_e(z) - C^*)/(C_0 - C^*)]$
   $A_s = (1,243,372 \text{ m}^3/\text{yr})/(34)[(15 - 8.8)/(100 - 8.8)]$
   $A_s = 98,300 \text{ m}^2$
   $A_s = 9.83$ ha $= 24.3$ ac

Comment

The major differences between these models are the C^* values and the internal position of the safety factor (z) in the logarithmic portion of this area model; for example, if safety factor z is 0.59 instead of 1, the calculated area in step 2 would be 67 acres.

6.8.4 DESIGN CRITERIA

The major design parameters are depth, detention time, loading rate, and aspect ratio. The typical design criteria are presented in Table 6.20.

6.9 PHYSICAL DESIGN AND CONSTRUCTION

The basic civil engineering aspects of wetland design and construction are similar to those employed for shallow lagoons. These typically include earthen berms for lateral water containment and some type of seepage control for the bottom of the wetland cell. Unique features of a wetland system are the vegetation and inlet and outlet structures that promote uniform flow across the wetland.

6.9.1 EARTHWORK

Berms for wetland cells are typically built with 3:1 interior side slopes and with a minimum of 2 ft of freeboard above the average water surface in a FWS wetland. The external berms for municipal wetlands should be at least 10 ft (3 m) wide

TABLE 6.20
Typical Design Criteria and Expected Effluent Quality for
Free Water Surface Constructed Wetlands

Item	Unit	Value
Design parameter:		
Detention time	d	2–5 (BOD); 7–14 (N)
BOD loading rate	lb/ac·d	<100
Hydraulic loading rate	in/d	1–5
Water depth	ft	0.2–1.5
Minimum size	ac/mgd	5–10
Aspect ratio	—	2:1 to 4:1
Mosquito control	—	Required
Harvesting interval	yr	3–5
Expected effluent quality:[a]		
Biochemical oxygen demand (BOD$_5$)	mg/L	<20
Total suspended solids (TSS)	mg/L	<20
Total nitrogen (TN)	mg/L	<10
Total phosphorus (TP)	mg/L	<5

[a] Expected effluent quality based on a BOD loading equal to or less than 100 lb/ac·d and typical settled municipal wastewater.

Source: Adapted from Crites, R.W. and Tchobanoglous, G., *Small and Decentralized Wastewater Management Systems*, McGraw-Hill, New York, 1998.

at the top to permit access by service vehicles. Each wetland cell should contain an access ramp for maintenance equipment. If possible, it is desirable to balance the cut and fill on the site to avoid the need for remote borrow pits or spoil disposal. If agronomic-quality topsoil exists on the site it should be stripped and stockpiled. In the case of a FWS wetland, this topsoil can be utilized as the rooting medium for the emergent vegetation and for revegetation of the berm surfaces.

If the wetland system is to meet its performance expectations it is critically important for the water to flow uniformly over the entire surface area provided for treatment. Severe short-circuiting of flow can result from improper grading or nonuniform subgrade compaction. Tolerances for grading will be given in the construction plans and specifications and in general will depend on the size of the system. Very large FWS systems incorporating several thousand acres cannot afford the effort to fine grade to very close tolerances and is not cost effective so the design will typically incorporate a safety factor to compensate. It is usually cost effective, for smaller wetland systems of a few hundred acres or less, to utilize close grading tolerances for construction. Uniform compaction of this subgrade is also important as subsequent construction activity (e.g., liner placement, soil placement for FWS systems) might create ruts and low spots in the subgrade which then result in short-circuiting of flow.

Fine grading and compaction of the native subgrade soils also depend on the liner requirements for the project. If the native soils are sufficiently impermeable (e.g., high clay content) and a liner is not required, then the soil surface should be graded to the specified tolerances and uniformly compacted to the same levels typically used for the subgrade soils in road subgrades. The same procedures should be followed if a membrane liner is used. If a clay liner is used, the native soils should be excavated to the specified depth and any new clay material placed and then compacted and graded to the specified elevations. Generally, all wetland cells are graded level from side to side and either level or with a slight slope in the flow direction. FWS wetlands are often constructed with a small bottom slope ($\leq 0.2\%$) in the flow direction to assist in drainage when cell maintenance is required. Construction activities with either native clay soils or with installed clay liners should only occur in dry weather when the soil moisture content is on the dry side of optimum.

6.9.2 LINERS

All of the conventional materials used to line lagoons and ponds have been used successfully for constructed wetlands, and the basic construction and installation procedures are the same. Membrane liners have been used for both FWS and SSF wetland systems. Both 30-mil polyvinyl chloride (PVC) and high-density polyethylene (HDP) have been successful, as well as 45-mil ethylene propylene diene monomer (EPDM) for smaller systems. Larger systems construct the liner in place but again use conventional procedures for assembly, joint bonding, and anchoring. Puncture of the liner must be prevented during placement and subsequent construction activity. Some currently used membrane liners require protection from ultraviolet solar radiation. Conventional procedures can again be used for this purpose (WEF, 2001). Clay liners have included locally available clay soils and commercially available products such as bentonite. Bentonite is typically mixed with the *in situ* native soils and then graded and compacted. Bentonite liners in the form of pads or blankets are also available; these are laid on a prepared surface and covered with a shallow layer of soil or sand. Clay liners typically have to be a foot or more in depth to provide the necessary hydraulic barrier. In the case of FWS wetlands, the surface of this clay layer should be well compacted to discourage root penetration by the emergent vegetation over the long term.

6.9.3 INLET AND OUTLET STRUCTURES

Uniform influent distribution and effluent collection over the full width of each wetland cell in the system are absolutely essential. Uniform distribution is typically accomplished with perforated manifold pipes for both inlets and outlets.

The size of the manifold and the orifice diameter and spacing are a function of the intended flow rate. For example, cell 1 at the FWS wetland in West Jackson County, Mississippi, is designed for an average flow of 0.6 mgd (2271 m³/d) and utilizes a PVC manifold 12 in. (300 mm) in diameter and extending the full 250-ft (76-m) width of the cell. This manifold is drilled with 2-in. (50-mm)-diameter orifices on 10-ft (3-m) centers. This pipe rests on a concrete footing to ensure stability and discharges to a 6-in. (150-mm) layer of 2-in. (50-mm) coarse aggregate; the coarse aggregate is underlain with a geotextile membrane to protect the underlying soil and prevent weed growth. A single manifold pipe with one central inlet would not be suitable for a very wide wetland cell as it would be difficult to ensure uniform flow from all of the outlets. Multiple manifold pipes (in pairs) could be used for this purpose. Sequential sets of splitter boxes could be used to uniformly divide the flow from the main influent line to whatever number of manifold sets is required.

In northern climates where extended periods of freezing weather are possible, it is necessary to protect these manifold pipes. These manifolds are placed at the bottom of the bed, below the design water surface. In these cases, the water level can be raised at the onset of winter to allow for ice formation at the water surface. Operational adjustment of these submerged manifolds is not possible, so great care must be taken during construction to ensure that the manifold pipe will remain level for the life of the system. At a minimum, some extra efforts at compaction and careful grading in the inlet and outlet zones will be required. In some cases, with potentially unstable soils (e.g., clays) it may be necessary to support the manifold on concrete footings. A clean-out on each end of these submerged manifolds is also recommended to allow flushing if clogging should occur over the long term.

In warm climates it is possible to install the inlet manifold in an exposed position to allow access for maintenance and adjustment. Alternatives to the simple drilled orifice holes allow the operator greater control over flow distribution. Gated aluminum irrigation pipe has been used but is susceptible to clogging, depending on the influent water quality. The TVA has used a nonclogging alternative originally developed in Europe. In this case, the manifold contains a series of pipe "Tees" of the same diameter. These "Tees" are connected to the manifold with O-rings on each side, with the open end of the "Tee" discharging to the wetland bed. Because of the O-rings, the open end of these "Tees" can be rotated vertically. The operator can then adjust each "Tee" as required to maintain a uniform flow distribution along the entire length of the manifold. The perforated effluent manifold in these cases is still placed at the bottom of the bed. Where the local climate permits, the use of an exposed, accessible inlet manifold is recommended for both SSF and FWS wetlands, except in cases where public exposure is an issue.

When submerged inlet and outlet manifolds are used for FWS wetlands, encroachment of the adjacent emergent vegetation must be considered. If the manifold is placed at the same grade as the main wetland bed, the vegetation may encroach on the inlet and outlet zones, and the plant litter and detritus could clog the orifices in the manifold. Several techniques are available to eliminate this problem.

A deep-water zone (approximately 2 ft deeper than the bottom of the main bed) can be incorporated at both the inlet and outlet to prevent growth of the emergent plants. The manifolds can also be placed on top of large rip-rap under-lain by a geotextile membrane, or the manifolds can be enclosed in a berm composed of coarse rip-rap (3 to 6 in. in size); the large-sized stone will not support the growth of emergent plants or weeds. The open water configurations do allow easier access to the manifold but also allow algae growth.

Submerged effluent manifolds must then be connected to an outlet structure containing a device for controlling the water level in the wetland bed. This device could be an adjustable weir or gate, a set of stop logs, or a swiveling elbow, in which case the elbow is attached to the effluent pipe with an O-ring to permit rotation, and a riser in the open end of the elbow sets the maximum water level in the bed. Because the elbow can be rotated at least 90°, the operator can set the water level at any position desired or can drain the bed if necessary.

An alternative to manifolds for inlet and outlet structures is the use of multiple weir or drop boxes. These are usually constructed of concrete, either cast in place or prefabricated. Several boxes along the width of the cell must be used to ensure uniform distribution. Spacing might range from 15 to 50 ft (5 to 15 m) center to center, depending on the width of the cell. In the FWS system at West Jackson County, these boxes are used to transfer water from cell to cell and for final effluent discharge. Box spacing at this site ranges from 70 to 90 ft (21 to 27 m), depending on the width of the cell. These boxes have an advantage as a discharge structure, as the contained weir or gate can be used to adjust water levels and a separate structure is not required for this purpose. They do require an adjacent deep-water zone to prevent vegetation encroachment and in northern climates are at greater risk of freezing as compared to a submerged manifold.

6.9.4 VEGETATION

The presence of the vegetation and litter in the wetland system is absolutely critical for successful performance, but establishing this vegetation is probably the least familiar aspect of wetland construction for most contractors. In recent years, a number of specialty firms have emerged with the necessary expertise for selecting and planting the vegetation on these systems. The use of such a firm is recommended for large-scale systems if the construction contractor does not have prior wetland experience.

Wetland plants can be established from seeds, root and rhizome material (tubers), seedlings (sprigs), and locally obtained clumps. The use of seeds is a low-cost but high-risk endeavor. Hydroseeding has been attempted for FWS wetland systems with marginal success.

Clumps of existing wetland species can sometimes be harvested from local drainage ditches or other acceptable sources. In these cases, most of the stem (to about 1 ft) and leaves are stripped off and the material is planted in clumps of at least a few shoots. Root and rhizome material can be obtained in the same way. If the lead time is sufficient, it is also possible to establish an on-site nursery so seedlings or clumps are available on schedule for planting. A number of commercial nurseries have been established in recent years that can furnish and plant a variety of species in a variety of forms (e.g., seeds, root/rhizomes, seedlings). When selecting such a nursery, it is desirable to utilize a source with a climatic zone similar to the intended site. Commercial seedlings have been used successfully on a number of projects. On larger projects, the use of seedlings allows the use of existing mechanical agricultural equipment for planting. A mechanical tomato planter, for example can easily be adapted for planting wetland seedlings.

Planting in the spring will provide the most successful results for seedlings, root/rhizome stock, or clumps. The planting density can be as close as 1.5-ft (0.45-m) centers or as much as 3 ft (1 m). The higher the density the more rapid will be the development of a mature and completely functional wetland system; however, high-density plantings can significantly increase construction costs. If planted on 3-ft (1-m) centers, a wetland system in a cold climate will typically take two full growing seasons to achieve expected performance objectives.

Planting seedlings or clumps is the simplest approach, as the green part goes up. Some experience with rhizomes is necessary to identify the node that will be the future shoot. The soil should be maintained in a moist condition after planting seeds or any of these other materials. The water level can be increased slowly as new shoots develop and grow. The water level must never be higher than the tips of the green shoots; otherwise, the plants will die.

Providing the necessary water for initial growth can be more complicated for large FWS wetland cells, as it may not be possible to plant the entire surface in a cell at one time. In this case, with a flat-bottomed cell, planting should occur in bands perpendicular to the flow path and a temporary shallow ridge of earth should be created on the upstream side of the band (if the bottom is sloped toward the effluent end, planting should begin at that end and proceed toward the inlets). Sprinklers or shallow flooding can then be used to keep the previously planted areas wet. If mechanical equipment is used for planting, it is important to keep the unplanted areas relatively dry until planting is complete. If hand planting is used, the entire area can be flooded with a few inches of water. The water depth can be increased gradually as the plant shoots grow until the design level is reached. If the FWS wetland is designed to treat primary effluent, the use of a cleaner water source is recommended for this initial planting and growth period. If the intended influent is close to secondary quality, it can be used immediately. If acceptable agronomic soil has been selected and used as the rooting media for the wetland, it should contain sufficient nutrients, and a preliminary application of commercial fertilizers should not be necessary.

It may not be economical to plant very large FWS wetlands on 3-ft (1-m) centers if the total system area comprises at least 1000 ac (400 ha). In this case, the amount of planting that is cost effective should be done in separate bands extending the full width of the wetland cells, with at least some plantings in each cell near the discharge end. These bands can then serve as source material for the future spread of the vegetation; it may require many years for such a wetland to reach complete plant coverage. In some cases, it may be necessary to protect the newly emergent vegetation from the birds and animals that are drawn to the wetland. These new plant shoots are a succulent and an attractive food source for both birds and animals during the early growth stages. The natural emergence of native plant species can also eventually vegetate large wetland areas if a suitable seed bank is available in the local soils.

If the FWS wetland system has been designed only for treatment functions, the plant species to be used are likely to be either bulrush (*Scirpus*) or cattails (*Typha*). Both are hardy plants that can survive some abuse during construction and still be successful. If a portion of the wetland has been designed to provide habitat values it is likely that a larger variety of plants will be specified. Some of these may require special conditions, and a knowledgeable person should be retained for their planting.

6.10 OPERATION AND MAINTENANCE

6.10.1 VEGETATION ESTABLISHMENT

Under ideal conditions, startup of a constructed wetland system would not commence until at least 6 weeks after planting of the vegetation. This time period is required to allow for the new plants to acclimate and grow. In actual practice, startup has sometimes occurred the day after planting was completed at a number of projects. Such emergency responses risk damage to the new plants and may delay achievement of expected performance. Startup procedures are quite simple and involve opening the inlet gates or valves and setting the desired wetland water level at the adjustable outlet or weirs. If the plants have not grown to a height that significantly exceeds the design water level, then the water depth must be increased gradually as the plants grow taller. In northern climates, it is also typical to increase the water depth at the onset of freezing weather to compensate for the expected formation of ice. If the plants are so short that this procedure cannot be implemented, then startup of the system should be deferred until the following spring. Under these circumstances it might also be better to defer planting the vegetation until the following spring.

If startup is defined as that initial period before the system reaches optimum performance, then, in cool temperate climates, the startup period for a wetland system might require in the best case 2 years and in the worst case 3 to 4 years. The system will not attain optimum performance levels until the vegetation and litter are developed fully and at equilibrium. The time required for that to occur is a function of planting density and season. A high-density planting in the spring of the year is likely to be fully developed by the end of the second growing season in cool temperate climates. A low-density planting in late fall just before the first frost in a northern location may require 3 years to reach equilibrium. Start-up time may be more rapid in continuously warm climates. Fortunately, most systems at startup are not faced with the ultimate long-term design flow, and the increased detention time may help compensate for the incomplete vegetative cover. Under these conditions, a spring planting at moderate density (≤3 ft) will usually produce a reasonably dense vegetative cover by the end of the first summer, and the system is likely to meet discharge requirements from that point on. A fall planting, in cold climates, would probably not achieve a comparable vegetative cover until midsummer of the following year.

During the startup period, the operator should inspect the site at least several times per week to observe plant growth and health, the integrity of berms and dikes, and the emergence of mosquitoes (from FWS systems only) and to adjust water levels as required. The experience gained during this initial period will then suggest the inspection frequency required over the long term. During the first spring season after planting and startup, the plant coverage in all wetland cells should be inspected carefully. Any large unvegetated areas should be replanted to avoid the risk of short-circuiting the flow.

The water quality performance during this startup period will not be representative of long-term expectations. In some cases, the performance may be better than the long-term expectations. Phosphorus or nitrogen removal in FWS wetlands is an example. The new wetland has freshly exposed soil surfaces (presumably with some clay content) and rapidly growing vegetation. Both conditions provide a rapid but short-term removal pathway for phosphorus and ammonia nitrogen. These systems may not reach equilibrium for these two parameters until the end of the first or second full growing season, and the equilibrium effluent concentrations are then likely to be higher than the results during startup. The opposite results may be expected for BOD_5 and TSS. A new wetland system has minimal vegetation and minimal substrate for attached growth organisms. As a result, removal of BOD_5 may be marginal and TSS removal poor if algae develop in any exposed open water. The removal of these two parameters can be expected to improve as the plant canopy develops and increases in density.

The operation of a wetland system is very simple and very similar to the requirements for operation of a facultative lagoon. Much of the effort involved is visual observation of conditions and then correction of any problems that develop. The major issues of concern are:

- Water level maintenance
- Uniformity of flow distribution and collection
- Berm and dike integrity
- Health and growth of designed system vegetation
- Control of nuisance pests and insects
- Removal of undesirable vegetation

The key hydraulic requirement is maintenance of uniform flow conditions. The operator must routinely observe and adjust inlet and outlet structures as required, including flow splitter boxes at the inlet end and water level controls at the effluent end. Some temporary surface flow may be observed after surcharge by intense rain storms. If persistent and large-scale surface flow is observed, the operator must then lower the water level an appropriate amount. Surface flow on these systems negates the goal of eliminating the risk of public exposure, and it may also result in the emergence of mosquitoes.

Most municipal FWS wetland systems will have at least two cells in parallel to allow better system control and temporary shut down of one side for maintenance, if required. Seasonal water level adjustments for these systems may be suggested in the operations and maintenance (O&M) manual, even in warm climates. In general, this may require maximum water levels during the winter months and minimal water depths during the warm summer months. The latter is intended to encourage new plant growth and allow maximum dissolved oxygen in the shallow water. The water depths involved might range from 1 to 1.5 ft during the winter and 0.5 to 0.75 ft during the summer. If the O&M manual does not contain such guidance, it is suggested that the operator develop a plan and systematically try similar conditions and observe results. A lag time of several weeks will typically occur before the effects of such a change are observed.

To satisfy their National Pollutant Discharge Elimination System (NPDES) permit requirements, most municipalities only have to measure the specified pollutants in the untreated wastewater and in the final system effluent. Because most systems will have some form of preliminary treatment (e.g., primary, secondary, lagoon, septic tank), these data provide an insufficient basis for the operator to determine if the wetland component is performing to expectations or requires adjustment. It is recommended that the influent to the wetland be sampled and tested periodically for the constituents of concern so the operator can build a record of performance for the wetland component. If problems then develop, these data will be of great assistance in determining the necessary adjustments. Data of this type can also assist the operator in developing a plan for optimized operation (e.g., seasonal water depths, flow splitting to different cells).

A well-designed and properly operated wetland system will not require routine harvest and removal of plant material and litter to achieve water quality goals or sustain expected hydraulic conditions. Harvesting or burning on a few FWS systems has been used to relieve the hydraulic resistance developed in poorly designed systems with very high aspect ratios (length-to-width) and in an attempt to control the habitat for mosquitoes.

The plant litter in these wetlands decomposes over time but leaves a sediment residue that does accumulate (≤ 0.04 in. or 1 mm/yr) over time. When this accumulated sediment and the accumulated refractory solids from the wastewater TSS begin to interfere with the design treatment volume or hydraulics in the wetland, then removal will be necessary. The problem will be most acute near the head of the system, as most of the influent TSS will be removed in the first 20% of the cell length. The access ramp to the cell should, therefore, be located near the head end of the cell. Such maintenance activities have not yet been performed on any operational wetland system in the United States. It is estimated that the need might arise every 50 to 75 years, depending on the wastewater characteristics, the types of plants used in the system, and the local climate.

Maintenance requirements for constructed wetlands are also simple and are similar to those required for facultative lagoons. These include maintenance of berms and dikes (e.g., mowing, erosion control), maintenance of watertight integrity (threatened, for example, by animal burrows, tree growth on berms), and control of nuisance pests and vectors (e.g., muskrats, nutria, mosquitoes). When the wetland is designed to operate at a shallow depth, a special requirement may be the periodic removal of tree seedlings from the wetland bed. If the trees are allowed to reach maturity, they will shade out the emergent vegetation and not provide the necessary substrate for attached-growth organisms. Inlet and outlet structures and water-level control devices must be periodically cleaned and adjusted, including debris removal and cleaning any weir surfaces to remove bacterial growth and other clogging substances. Submerged inlet and outlet manifolds should be flushed periodically and cleaned with a high-pressure hose.

6.10.2 NUISANCE ANIMALS

Nutria and muskrats are of concern in FWS wetlands because they can burrow through berms and eat wetlands vegetation. It may be possible to control animal burrows in the berms by temporarily raising the water level; if necessary, large-diameter rip-rap (4 to 6 in. or 100 to 150 mm) can be applied locally. Tree seedlings should also be removed from the berms and any grass cover routinely cut. An infestation of animals such as muskrat or nutria can eliminate all vegetation in a system, as these animals use cattails and bulrush as both a food source and nesting material. If such damage is noted, a control program should be instituted immediately. Live trapping and release may be successful, but in most cases it has been necessary to kill the animals to solve the problem. Fencing the wetland to exclude such animals has been tried but is not always successful.

6.10.3 MOSQUITO CONTROL

Mosquitoes are common inhabitants of natural wetlands, and their presence at approximately the same density in FWS constructed wetlands is to be expected. Mosquitoes should not be a concern for SSF wetlands as long as the system is properly operated with the water level maintained below the top of the bed. Mosquito control is more difficult in polluted waters with a high organic content, as might exist near the inlet end of a FWS wetland. Insecticide doses might have to be at least double the normal amount in this portion of the FWS wetland. *Gambusia* fish provide effective control during warm weather conditions. An annual restocking of these fish may be necessary in cold climates with low winter temperatures.

The FWS wetland in Arcata, California, successfully used both *Gambusia* fish and a pupaecide (Altosid®) for mosquito control (Gearheart, et al., 1983). Bacterial insecticides (*Bacillus thuringiensis israelensis* and *Bacillus sphaericus*) have been used successfully on a number of wetland systems. The use of *Bacillus thuringiensis israelensis* was recommended (Tennessen, 1993) for use after trials with several insecticides at wetland systems in Kentucky. The side slopes of the containing dikes should be as steep as possible and any vegetation on these surfaces controlled. The presence of duckweed may also contribute to mosquito control by covering the water surface, but this will also interfere with oxygen transfer from the atmosphere.

At the Sacramento County demonstration constructed wetlands, a comprehensive management plan was developed (Williams et al., 1996). Effective mosquito control was managed over the 5-year project using:

- Vegetation control to reduce stagnant areas of excessive plant growth or lodging of tall plants
- Mosquitofish stocking (2 lb/cell) and deep zones to allow overwintering of the fish
- Regular sampling and analysis of larvae production
- Applications of *Bacillus sphaericus* or *Bacillus thuringiensis israelensis* whenever the larvae count exceeded 0.1 larvae per dip

No outbreak of adult mosquitoes was detected when this management plan was followed. These strategies have been verified recently (Knight et al., 2003).

6.10.4 MONITORING

Monitoring needs can include flow, surface water quality, and groundwater quality. Variable-height weirs can be used to monitor flow out of the wetland and to provide a convenient sampling point. Surface water sampling points should be located at catwalks or boardwalks to allow sampling without disturbing the flow.

6.11 COSTS

The cost data in this section were obtained, in part, from site visits to four operational FWS constructed wetland systems, sponsored by USEPA's Center for Environmental Research Information (CERI) in 1997 and from related published sources. The four systems were: Arcata, California; Gustine, California; Ouray, Colorado; and West Jackson County, Mississippi. To provide a common base for cost comparisons all costs have been adjusted, with the appropriate Engineering News Record (ENR) Construction Cost Index (CCI) factor to August 1997 (ENR–CCI = 5854). The major items included in the capital costs of FWS constructed wetlands are:

- Land costs
- Site investigation
- Clearing and grubbing
- Excavation and earthwork
- Liners
- Plants
- Inlet and outlet structures
- Distribution systems

TABLE 6.21
Construction Costs of Free Water Wetlands, <1 mgd

Location	Design Flow (mgd)	Area (ac)	Construction Cost ($/ac)[a]
Armour, South Dakota	0.1	4	31,091
Baltic, South Dakota	0.1	4	34,227
Cannon Beach, Oregon	0.68	16	49,089
Eureka, South Dakota	0.28	40	14,500
Ft. Washakie, Wyoming	0.18	1.6	61,827
Ft. Deposit, Alabama	0.24	14.8	32,906
Mays Chapel, Maryland	0.04	0.6	64,340
Mcintosh, Maryland	0.06	9.2	73,650
Ouray, Colorado	0.36	2.2	53,077
Tabor, South Dakota	0.065	2	31,768
Tripp, South Dakota	0.075	4	29,262
Vermontville, Michigan	0.1	11.4	116,860
Wakonda, South Dakota	0.05	2	26,024
Average			47,586
Median			26,024

[a] Costs are in 1998 dollars (Engineering News Record Construction Cost Index [ENR–CCI] = 5895).

Source: Crites, R.W. and Ogden, M., Costs of constructed wetlands systems, in *Proceedings of WEFTEC 98*, Water Environment Federation, Orlando, FL, October 3–7, 1998. With permission.

- Fencing
- Engineering, legal, contingencies, and contractor's overhead and profit

Summaries of technical and cost data for FWS constructed wetland systems are provided in Table 6.21, Table 6.22, and Table 6.23 (Crites and Ogden, 1998). The median construction cost for systems ≥1 mgd (3785 m³/d) was $14,465/ac, while the median cost for smaller systems was $34,227/ac.

6.11.1 GEOTECHNICAL INVESTIGATIONS

Geotechnical investigations may be used to establish the soil and groundwater conditions at the site. At Mandeville, Louisiana, about $15,000 was allocated in 1989 for site surveys and soil borings in the wetland area. The approximate updated cost for surveying and soil borings at the Mandeville wetland site would be about $1100/ac (1997 dollars).

TABLE 6.22
Construction Costs of Free Water Surface Wetlands, >1 mgd

Location	Design Flow (mgd)	Area (ac)	Construction Cost ($/ac)[a]
Show Low, Arizona	1.4	201	1996
Lakeside, Arizona	1.0	127	4425
Hayward, California	9.7	172	5828
Lakeland, Florida	14.8	1400	6970
Mandan, North Dakota	1.5	41	8155
West Jackson County, Mississippi	2.4	50	14,037
Carolina Bay, South Carolina	2.5	702	14,465
Incline Village, Nevada	3.0	428	16,604
Minot, North Dakota	5.5	34	17,635
Arcata, California	2.3	38.5	18,830
American Crystal Sugar, North Dakota	1.5	81	24,443
Ironbridge, Florida	20	1220	25,165
Mt. Angel, Oregon	2.0	10	39,572
Gustine, California	1.0	24	51,032
Average			17,797
Median			14,465

[a] Costs are in 1998 dollars (Engineering News Record Construction Cost Index [ENR–CCI] = 5895).

Source: Crites, R.W. and Ogden, M., Costs of constructed wetlands systems, in *Proceedings of WEFTEC 98*, Water Environment Federation, Orlando, Florida, October 3–7, 1998. With permission.

6.11.2 CLEARING AND GRUBBING

A wetlands site may require clearing vegetation from the site prior to earthwork operations. Costs ranged from $4871/ac at Ouray to $2000/ac at West Jackson County (WEF, 2001). Costs for clearing and grubbing (on relatively level land) can range from $2000/ac for brush and some small trees to $5000/ac for a tree-covered site.

6.11.3 EARTHWORK

Excavation and earthwork typically include bringing the wetland site to finished grade, constructing berms and access ramps, and, in the case of FWS wetlands, reserving and replacing topsoil in the bed to serve as the vegetation growth medium. Costs ranged from $11,523/ac at Gustine to $8622/ac at West Jackson County.

TABLE 6.23
Annual Operation and Maintenance Costs
for Free Water Surface Wetlands

Location	Design Flow (mgd)	Area (ac)	Annual Cost ($/ac)
Cannon Beach, Oregon	0.68	16	4500
Gustine, California	1.0	24	819
Mt. Angel, Oregon	2.0	10	1780
Ouray, Colorado	0.36	2.2	1364

Source: Data from Crites and Lesley (1998) and WEF (2001).

6.11.4 LINERS

The purpose of a liner is to retain the wastewater in the wetland bed for treatment and to protect the underlying groundwater. A variety of materials, including the *in situ* native soils, can be used as liner material depending on the requirements of the regulatory agencies. The 30-mil HDP liner at Ouray cost $21,332/ac in 1997 dollars. The liner for a 5-acre wetland at Cle Elum, Washington, cost $0.40/ft^2 in 2000.

6.11.5 VEGETATION ESTABLISHMENT

Plant materials can sometimes be obtained locally by cleaning drainage ditches. It is also possible to develop an onsite nursery at the wetland construction site, if the lead time is sufficient, to grow plant sprigs or seedlings from seed and transplant these to the wetland cells. A large and expanding number of commercial nurseries can supply a large variety of plant species for these wetland systems. The majority of the systems surveyed were planted with commercial nursery stock. Small systems are typically planted by hand, but large systems can use mechanical planters, and nursery-grown sprigs or plant seedlings are advantageous for this purpose. Hydroseeding has been attempted but has not been successful to date, although spreading of soil that was laden with bulrush and cattail seed was successful at Sacramento County, California. Cost data for plants and planting from the 1997 USEPA survey sites ranged from $1860/ac for Gustine to $1800/ac for West Jackson County (WEF, 2001).

6.11.6 INLET AND OUTLET STRUCTURES

The inlet and outlet structures for most small- to moderate-sized wetland systems are typically some variation of a perforated manifold pipe. Large wetland systems typically use multiple drop or weir boxes for both inlets and outlets. Weir-type structures costs ranged from $1500 each at Gustine to $2500 each at West Jackson County.

6.11.7 Piping, Equipment, and Fencing

These items would include the piping to get the wastewater to the wetland, the piping from the wetland to final discharge, and any pumps required for this purpose. Fencing is typically utilized around all municipal wastewater treatment systems but is not usually required around the smaller SF wetland beds due to the low risk of public contact and exposure to the wastewater. None of these features is unique to wetland systems, and costs for these items were not available at the sites included in the 1997 EPA survey. The only time fencing might be a unique requirement for a wetland system would be its use to exclude muskrats and nutria. Both of these animals can seriously damage both the wetland vegetation and the berms in the system.

6.11.8 Miscellaneous

These items would cover engineering design and legal fees, construction contingencies, and profit and overhead for the construction contractor. These, again, are not unique to wetland systems and are usually expressed as a percentage of the total construction costs when preparing an estimate. Additional items usually included directly in the construction costs include mobilization and bonding. Typical values for these items are:

- Mobilization — 5% of direct costs
- Bonds — 3% of direct costs
- Engineering design services — 10% of capital costs
- Construction services and start-up — 10% of capital costs
- Contractor's overhead and profit — 15% of capital costs
- Contingencies — 15% of capital costs

6.12 TROUBLESHOOTING

The three most common issues with existing FWS constructed wetlands are (1) short-circuiting, (2) lack of ammonia reduction, and (3) mosquitoes. Short-circuiting can be evaluated using dye or tracer testing or by aerial photography. An example of a lithium chloride tracer study at Sacramento County is shown in Figure 6.6. Baffles, earthen berms, and open-water zones placed perpendicular to the flow path can be used to overcome identified short-circuiting conditions (Crites and Tchobanoglous, 1998). Ammonia removal improvements usually can be attained by adding a nitrification filter bed (NFB), as described in Chapter 7. See Section 6.10 for a description of mosquito control techniques (Andrews, 1996).

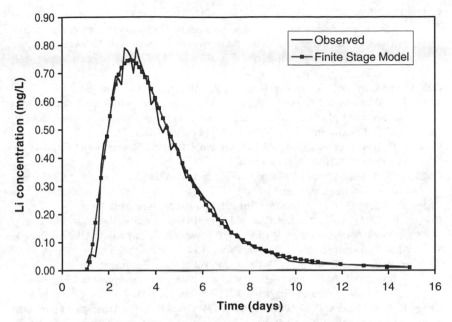

FIGURE 6.6 Tracer study at free water surface (FWS) constructed wetlands in Sacramento County, California.

REFERENCES

Andrews, T. (pers. comm., 1996). Ouray, Colorado, FWS wetlands performance data.

Ashton, G., Ed.(1986). *River and Lake Ice Engineering*, Water Resources Publications, Littleton, CO, 1986.

Benefield, L.D. and Randall, C.W. (1980). *Biological Process Design for Wastewater Treatment*, Prentice–Hall, Englewood Cliffs, NJ.

Black & Veatch. (1992). *Treatment of Combined Sewer Overflows through Constructed Wetlands*, Bureau of Environmental Services, City of Portland, OR.

Boon, A.G. (1985). *Report of a Visit by Members and Staff of WRC to Germany To Investigate the Root Zone Method for Treatment of Wastewaters*, Water Research Centre, Stevenage, England, August 1985, 52 pp.

Bouldin, D.R., Bernard, J.M., and Grunder, D.J. (1994). *Leachate Treatment System Using Constructed Wetlands, Town of Fenton Sanitary Landfill, Broome County, New York*, Report. 94-3, New York State Energy Research and Development Authority, Albany, NY.

Brix, H. (1992). An overview of the use of wetlands for water pollution control in Europe, in *Proceedings: IAWQ Wetlands Systems Conference*, Sydney, Australia, December.

Brix, H. (1994). Constructed wetlands for municipal wastewater treatment in Europe, in *Global Wetlands: Old World and New*, Mitsch, W.J., Ed., Elsevier Science, Amsterdam, 325–333.

Brodie, G.A., Britt, C.R., Tomaszewski, T.M., and Taylor, H.N. (1993). Anoxic limestone drains to enhance performance of aerobic acid drainage treatment wetlands: experiences of the Tennessee Valley Authority, in *Constructed Wetlands for Water Quality Improvement*, Moshiri, G. et al., Eds., Lewis Publishers, Chelsea, MI, 129–138.

Bull, G. (pers. comm., 1994). NovaTec, Inc., Vancouver, British Columbia.

Burka, U. and Lawrence P.C. (1990). A new community approach to waste treatment with higher water plants, in *Constructed Wetlands in Water Pollution Control*, Cooper, P.F. and Findlater, B.C., Eds., IAWPCR, London, 359–371.

Calkins, D. (pers. comm., 1995). U.S. Army Cold Regions Research and Engineering Laboratory (CRREL), Hanover, NH.

CSCE. (1986). *Cold Climate Utilities Manual*, 2nd ed., Canadian Society of Civil Engineers, Montreal, Quebec.

Chapman, A.J. (1974). *Heat Transfer*, 3rd ed., Macmillan, New York.

Cooper, P. (2003). Sizing vertical flow and hybrid constructed wetlands, in *The Use of Aquatic Macrophytes for Wastewater Treatment in Constructed Wetlands*, Dias, V. and Vymazal, J., Eds., ICN and ANAG, Lisbon, Portugal.

Crites, R.W. (1996). Constructed Wetlands for Wastewater Treatment and Reuse, paper presented at the Engineering Foundation Conference, Environmental Engineering in the Food Processing Industry, XXVI, Santa Fe, NM.

Crites, R.W. and Lesley, D. (1998). Constructed Wetlands Remove Algae, paper presented at the Annual HWEA Conference, Hawaii Water Environment Association, Honolulu, HI.

Crites, R.W. and Ogden, M. (1998). Costs of constructed wetlands systems, in *Proceedings of WEFTEC 98*, Water Environment Federation, Orlando, FL, October 3–7, 1998.

Crites, R.W. and Tchobanoglous, G. (1998). *Small and Decentralized Wastewater Management Systems*, McGraw-Hill, New York.

Crites, R.W., Dombeck, G.D., and Williams, C.R. (1996). Two birds with one wetland: constructed wetlands for effluent ammonia removal and reuse benefits, in *Proceedings of Water Environment Federation WEFTEC 96*, Dallas, TX, October 5–9, 1996.

Crites, R.W., Dombeck, G.D., Watson, R.C., and Williams, C.R. (1997). Removal of metals and ammonia in constructed wetlands, *Water Environ. Res.*, 69(2), 132–135.

Dombeck, G., Williams, C., and Crites, R. (1997). Hydraulics in constructed wetlands, in *Proceedings of the 70th Annual Water Environmental Federation*, Chicago, IL.

Dornbush, J.N. (1993). Constructed wastewater wetlands, in *Constructed Wetlands for Water Quality Improvement*, Moshiri, G. et al., Eds., Lewis Publishers, Chelsea, MI, 569–576.

Eckenfelder, W.W. (1989). *Industrial Water Pollution Control*, 2nd ed., McGraw-Hill, New York.

Emond, H., Madison, M., Whitaker, D., Russell, J., and Kessler, F. (2002). A Multifaceted Natural Reclamation System for Wastewater Treatment and Reuse, in *Proceedings of Water Environment Federation WEFTEC 2002*, Chicago, IL, September 28–October, 2, 2002.

Ferlow, D.L. (1993). Stormwater runoff retention and renovation: a back lot function or integral part of the landscape, in *Constructed Wetlands for Water Quality Improvement*, Moshiri, G. et al., Eds., Lewis Publishers, Chelsea, MI.

Gearheart, R.A. (1993). *Construction, Construction Monitoring, and Ancillary Benefits*, Environmental Engineering Department, Humboldt State University, Arcata, CA.

Gearheart, R.J., Wilbur, S., Williams, J., Hull, D., Finney, B., and Sundberg, S. (1983). *Final Report, City of Arcata Marsh Pilot Project Effluent Quality Results: System Design and Management*, Project Report C-06-2270, Department of Public Works, City of Arcata, CA, 150 pp.

Gearhart, R.A., Finney, B.A., Wilbur, S., Williams, J., and Hull, D. (1985). The use of wetland treatment processes in water reuse, in *Future of Water Reuse*, Vol. 2, AWWA Research Foundation, Denver, CO, 617–638.

Gearhart, R.A., Klopp, A.F., and Allen, G. (1989). Constructed free surface wetlands to treat and receive wastewater: pilot project to full scale, in *Constructed Wetlands for Water Quality Improvement*, Moshiri, G. et al., Eds., Lewis Publishers, Chelsea, MI, 121–137,

Godfrey, P.J., Kaynor, E.R., Pelczarski, S., and Benforado, J. (1985). *Ecological Considerations in Wetlands Treatment of Municipal Wastewaters*, Van Nostrand–Reinhold, New York.

Hammer, D.A. (1992). *Creating Freshwater Wetlands*, Lewis Publishers, Chelsea, MI.

Hammer, D.A. and Knight R.L. (1992). Designing constructed wetlands for nitrogen removal, in *Wetland Systems in Water Pollution Control*, IAWQ, Sydney, Australia, 3.1–3.37.

Hammer, D.A., Pullen, B.P., McCaskey, T.A., Eason, J., and Payne, V.W.E. (1993). Treating livestock wastewaters with constructed wetlands, in *Constructed Wetlands for Water Quality Improvement*, Moshiri, G. et al., Eds., Lewis Publishers, Chelsea, MI, 343–348.

Hendry, G.R., Clinton, J., Blumer, K., and Lewin, K. (1979). *Lowland Recharge Project Operations: Physical, Chemical and Biological Changes 1975–1978*, Final Report to the Town of Brookhaven, Brookhaven National Laboratory, Brookhaven, NY.

Herskowitz, J. (1986). *Town of Listowel Artificial Marsh Project Final Report*, Project No. 128RR, Ontario Ministry of the Environment, Toronto, Ontario.

Herskowitz, J. et al. (1987). Artificial marsh treatment project, in *Aquatic Plants for Water Treatment and Resource Recovery*, Reddy, K.R. and Smith, W.H., Eds., Magnolia Publishing, Orlando, FL, 247–254.

Higgens, M.J., Rock, C.A., Bouchard, R., and Wengrezynek, B. (1993). Controlling agricultural runoff by use of constructed wetlands, in *Constructed Wetlands for Water Quality Improvement*, Moshiri, G. et al., Eds., Lewis Publishers, Chelsea, MI, 359–367.

Johnson, K.D., Martin, C.D., Moshiri, G.A., and McCrory, W.C. (1998). Performance of a constructed wetland leachate treatment system at the Chunchula Landfill, Mobile County Alabama, in *Constructed Wetlands for the Treatment of Landfill Leachates*, Mulamoottil, G., McBean, E.A., and Rovers, F., Eds., Lewis Publishers, Boca Raton, FL.

Kadlec, R.H. and Knight, R.L. (1996). *Treatment Wetlands*, CRC Press, Boca Raton, FL.

Knight, R.L., Walton, W.E., O'Meara, G.F., Reisen, W.K., and Wass, R. (2003). Strategies for effective mosquito control in constructed treatment wetlands, *Ecol. Eng.*, 21, 211–232.

Lawson, G.J. (1985). *Cultivating Reeds (Phragmites australis) for Root Zone Treatment of Sewage*, Project Report 965, Institute of Terrestrial Ecology, Cumbria, England.

Martin, C.D., Moshiri, G.A., and Miller, C.C. (1993). Mitigation of landfill leachate incorporating in-series constructed wetlands of a closed-loop design, in *Constructed Wetlands for Water Quality Improvement*, Moshiri, G. et al., Eds., Lewis Publishers, Chelsea, MI, 473–476.

Mitsch, W.J., Ed. (1994). *Global Wetlands: Old World and New*, Elsevier, Amsterdam.

Mitsch, W.J. and Gosselink, J.G. (2000). *Wetlands*, 3rd ed., John Wiley & Sons, New York.

Nolte Associates (1999). *Five-Year Summary Report, 1994–1998, Sacramento Constructed Wetlands Demonstration Project*, Sacramento Regional County Sanitation District, Sacramento, CA.

O'Brien, E., Hetrick, M., and Dusault, A. (2002). *Wastewater to Wetlands: Opportunities for California Agriculture*, prepared by Sustainable Conservation, San Francisco, CA.

Peverly, J., Sanford, W.E., Steenhuis, T.S., and Surface, J.M. (1994). *Constructed Wetlands for Municipal Solid Waste Landfill Leachate Treatment*, Report 94–1, New York State Energy Research and Development Authority, Albany, NY.

Reddy, K.R. and D'Angelo, E.M. (1994). Soil processes regulating water quality in wetlands, in *Global Wetlands: Old World and New*, Mitsch, W., Ed., Elsevier, Amsterdam.

Reed, S.C., Ed. (1990). *Natural Systems for Wastewater Treatment*, MOP FD–16, Water Environment Federation, Alexandria, VA.

Reed, S.C. (1993). *Subsurface Flow Constructed Wetlands for Wastewater Treatment: A Technology Assessment*, EPA 832-R-93-008, U.S. Environmental Protection Agency, Washington, D.C.

Reed, S.C., and Bastian, R.K., Eds. (1980). *Aquaculture Systems for Wastewater Treatment: An Engineering Assessment*, EPA 430/9-80-007, NTIS PB 81156689, U.S. Environmental Protection Agency, Washington, D.C.

Reed, S.C. and Bastian, R.K. (1985). Wetlands for wastewater treatment: an engineering perspective, in *Ecological Considerations in Wetlands Treatment of Municipal Wastewaters*, Godfrey, P.J., Kaynor, E.R., Pelczarski, S., and Benforado, J., Eds., Van Nostrand–Reinhold, New York, 444–450.

Reed, S.C. and Calkins, D. (1996). The thermal regime in a constructed wetland, in *Fifth International Conference on Wetland Systems for Water Pollution Control* (preprint), International Association on Water Quality, London, England, I3–I7

Reed, S.C. and Hines, M. (1995). Constructed wetlands for industrial wastewaters, in *Proceedings 1993 Purdue Industrial Waste Conference*, Lewis Publishers, Chelsea, MI.

Reed, S.C., Middlebrooks, E.J., and Crites, R.W. (1988). *Natural Systems for Waste Management and Treatment*, McGraw-Hill, New York.

Reed, S.C., Crites, R.W., and Middlebrooks, E.J. (1995). *Natural Systems for Waste Management and Treatment*, 2nd ed., McGraw-Hill, New York.

Seidel, K., (1996). Reinigung von Gerwassern durch hohere Pflanzen, *Deutsche Naturwissenschaft*, 12, 297–298.

Skousen, J., Sexstone, A., Garbutt, K., and Sencindiver, J. (1996). Wetlands for treating acid mine drainage, in *Acid Mine Drainage Control and Treatment*, 2nd ed., Skousen, J.G. and Ziemkiewicz, P.F., Eds., National Mine Land Reclamation Center and West Virginia University, Morgantown, VA.

Smith, J.W, Crites R.W., and Leonhard, J. (2002). Performance of constructed wetlands at Cle Elum, Washington, in *Proceedings of Water Environment Federation WEFTEC 2002*, Chicago, IL, September 28–October, 2, 2002.

Stefan, J. (1891). Theory of ice formation, especially in the Arctic Ocean [in German], *Akad. Wiss Wien Sitzunsber Ser. A*, 42(2), 965–983.

Steiner, G.R. and Watson, J.T. (1993). *General Design, Construction, and Operational Guidelines: Constructed Wetlands Wastewater Treatment Systems for Small Users Including Individual Residences*, 2nd ed., TVA/WM–93/10, Tennessee Valley Authority, Chattanooga, TN.

Tchobanoglous, G., Crites, R., Gearheart, R., and Reed, S.C. (2003). A review of treatment kinetics for constructed wetlands, in *The Use of Aquatic Macrophytes for Wastewater Treatment in Constructed Wetlands*, Dias, V. and Vymazal, J., Eds., ICN and ANAG, Lisbon, Portugal.

Tennessen, K.J. (1993). Production and suppression of mosquitoes in constructed wetlands, in *Constructed Wetlands for Water Quality Improvement*, Moshiri, G. et al., Eds., Lewis Publishers, Chelsea, MI, 591–601.

Thornhurst, G.A. (1993). *Wetland Planting Guide for the Northeastern United States*, Environmental Concern, Inc., St. Michaels, MD.

UN Developmental Program. (1993). *Global Environmental Facility: Egyptian Engineered Wetlands*, Lake Manzala, July 1993.

USDA-SCS. (1993). *Nutrient and Sediment Control System*, Technical Note N4, U.S. Department of Agriculture, Soil Conservation Service, Washington, D.C.

USEPA. (1988). *Design Manual: Constructed Wetlands and Aquatic Plant Systems for Municipal Wastewater Treatment*, EPA-625/1-81-013, U.S. Environmental Protection Agency, Cincinnati, OH, 1988.

USEPA. (1993a). *Constructed Wetlands for Wastewater Treatment and Wildlife Habitat*, EPA832-R-93-005, U.S. Environmental Protection Agency, Washington, D.C.

USEPA. (1993b). *Design Manual: Nitrogen Control*, EPA/625/R-93/010, U.S. Environmental Protection Agency, Cincinnati, OH.

USEPA. (1999) *Free Water Surface Wetlands for Wastewater Treatment: A Technology Assessment*, Office of Water Management, U.S. Environmental Protection Agency, Washington, D.C.

USEPA. (2000). *Constructed Wetlands Treatment of Municipal Wastewater*, EPA/625/R-99/010, Office of Research and Development, U.S. Environmental Protection Agency, Cincinnati, OH.

Vidales, J.A., Gerba, C.P., and Karpiscak, M.M. (2003). Virus removal from wastewater in a multispecies subsurface flow constructed wetland, *Water Environ. Res.*, 75(3), 238–245.

Vymazal, J. and Krasa, P. (2003). Distribution of Mn, Al, Cu, and Zn in a constructed wetland receiving municipal sewage, *Water Sci. Technol.*, 48(5), 299–305.

Vymazal, J., Brix, H., Cooper, P.F., Green, M.B., and Haberl, R. (1998). *Constructed Wetlands for Wastewater Treatment in Europe*, Backhuys Publishers, Leiden, The Netherlands.

Watson, J.T., Reed, S.C., Kadlec, R.H., Knight, R.L., and Whitehouse, A.E. (1989). Performance expectations and loading rates for constructed wetlands, in *Constructed Wetlands for Wastewater Treatment*, Hammer, D.A., Ed., Lewis Publishers, Chelsea, MI, 319–351.

WEF. (2001). *Natural Systems for Wastewater Treatment*, 2nd ed., Manual of Practice FD-16, Water Environment Federation, Alexandria, VA.

Williams, C.R., Jones, R.D., and Wright, S.A. (1996). Mosquito control in a constructed wetlands, in *Proceedings of Water Environment Federation WEFTEC 96*, Dallas, TX, October 5–9, 1996.

Witthar, S.R. (1993). Wetland water treatment systems, in *Constructed Wetlands for Water Quality Improvement*, Moshiri, G. et al., Eds., Lewis Publishers, Chelsea, MI, 147–156.

WRc. (1996). *Reed Beds and Constructed Wetlands for Wastewater Treatment*, Water Research Center (WRc), Swindon, England.

7 Subsurface and Vertical Flow Constructed Wetlands

Subsurface flow (SSF) wetlands consist of shallow basins or channels with a seepage barrier and inlet and outlet structures. The bed is filled with porous media and vegetation is planted in the media. The water flow is horizontal in the SSF wetland and is designed to be maintained below the upper surface of the media, hence the title *subsurface flow*. In the United States, the most common medium is gravel, but sand and soil have been used in Europe. The media depth and the water depth in these wetlands have ranged from 1 ft (0.3 m) to 3 ft (0.9 m) in operational systems in the United States. The design flow for most of these systems in the United States is less than 50,000 gpd (189 m³/d). The largest system in the United States (Crowley, LA) has a design flow of 3.5 mgd (13,000 m³/d) (Reed et al., 1995). A schematic of a typical SSF wetland is shown in Figure 7.1.

7.1 HYDRAULICS OF SUBSURFACE FLOW WETLANDS

Darcy's law, as defined by Equation 7.1, describes the flow regime in a porous media and is generally accepted for the design of SSF wetlands using soils and gravels as the bed media. A higher level of turbulent flow may occur in beds using very coarse rock, in which case Ergun's equation is more appropriate. Darcy's law is not strictly applicable to subsurface flow wetlands because of physical limitations in the actual system. It assumes laminar flow conditions, but turbulent flow may occur in very coarse gravels when the design utilizes a high hydraulic gradient. Darcy's law also assumes that the flow in the system is constant and uniform, but in reality the flow may vary due to precipitation, evaporation, and seepage, and local short-circuiting of flow may occur due to unequal porosity or poor construction. If small- to moderate-sized gravel is used as the media, if the system is properly constructed to minimize short-circuiting, if the system is designed to depend on a minimal hydraulic gradient, and if the gains and losses of water are recognized, then Darcy's law can provide a reasonable approximation of the hydraulic conditions in a SSF wetland:

$$v = k_s s$$

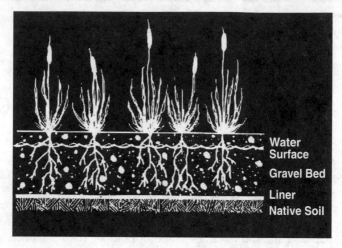

FIGURE 7.1 Schematic of a subsurface flow constructed wetland.

Because

$$v = Q/Wy$$

then

$$Q = k_s A_c s \tag{7.1}$$

where

v = Darcy's velocity, the apparent flow velocity through the entire cross-sectional area of the bed (ft/d; m/d).

k_s = Hydraulic conductivity of a unit area of the wetland perpendicular to the flow direction (ft³/ft²·d; m³/m²·d).

s = Hydraulic gradient, or slope, of the water surface in the flow system (ft/ft; m/m).

Q = Average flow through the wetland (ft³/d; m³/d) = $[Q_{in} + Q_{out}]/2$.

W = Width of the SSF wetland cell (ft; m).

y = Average depth of water in the wetland (ft; m).

A_c = Total cross-sectional area perpendicular to the flow (ft²; m²).

The resistance to flow in the SSF wetland is caused primarily by the gravel media. Over the longer term, the spread of plant roots in the bed and the accumulation of nondegradable residues in the gravel pore spaces will also add resistance. The energy required to overcome this resistance is provided by the head differential between the water surface at the inlet and the outlet of the wetland. Some of this differential can be provided by constructing the wetland with a sloping bottom. The preferred approach is to construct the bottom with sufficient slope to allow complete drainage when needed and to provide outlet structures that allow adjustment of the water level to compensate for the resistance that may increase with time. The aspect ratio (length-to-width) selected for a SSF

TABLE 7.1
Typical Media Characteristics for Subsurface Flow Wetlands

Media Type	Effective Size (D_{10}) (mm)	Porosity (n) (%)	Hydraulic Conductivity (k_s) (ft/d)
Coarse sand	2	28–32	328–3280
Gravelly sand	8	30–35	1640–16,400
Fine gravel	16	35–38	3280–32,800
Medium gravel	32	36–40	32,800–164,000
Coarse rock	128	38–45	164,000–820,000

Note: ft/d × 0.305 = m/d.

wetland also strongly influences the hydraulic regime as the resistance to flow increases as the length increases. Reed et al. (1995) developed a model that can be used to estimate the minimum acceptable width of a SSF wetland channel. It is possible by substitution and rearrangement of terms to develop an equation for determining the acceptable minimum width of the SSF wetland cell that is compatible with the hydraulic gradient selected for design:

$$W = (1/y)[(Q_A)(A_s)/(m)(k_s)]^{0.5} \qquad (7.2)$$

where
W = Width of the SSF wetland cell (ft; m).
y = Average depth of water in the wetland (ft; m).
Q_A = Average flow through the wetland (ft³/d; m³/d).
A_s = Design surface area of the wetland (ft²; m²).
m = Portion of available hydraulic gradient used to provide the necessary head, as a decimal.
k_s = Hydraulic conductivity of the media used (ft³/ft²/d; m³/m²/d).

The m value in Equation 7.2 typically ranges from 5 to 20% of the potential head available. When using Equation 7.2 for design it is recommended that not more than one third of the effective hydraulic conductivity (k_s) be used in the calculation and that the m value not exceed 20% to provide a large safety factor against potential clogging and other contingencies not defined at the time of design. Typical characteristics for media (medium gravel is most commonly used in the United States) with the potential for use in SSF wetlands are given in Table 7.1.

For large projects, it is recommended that the hydraulic conductivity (k_s) be directly measured with a sample of the media to be used in the field or laboratory prior to final design. A permeameter is the standard laboratory device, but it is not well suited to the coarser gravels and rocks often used in these systems. A

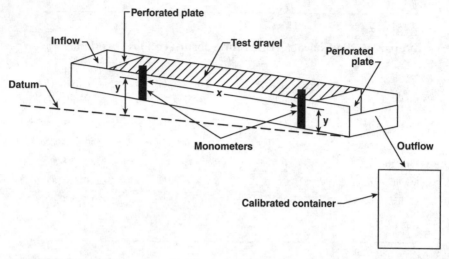

FIGURE 7.2 Permeameter trough for measuring hydraulic conductivity of subsurface flow media.

permeameter trough that has been used successfully to measure the effective hydraulic conductivity of a range of gravel sizes is shown in Figure 7.2.

The total length of the trough is about 16.4 ft (5 m), with perforated plates located about 1.5 ft (0.5 m) from each end. The space between the perforated plates is filled with the media to be tested. The manometers are used to observe the water level inside the permeameter, and they are spaced about 9 ft (3 m) apart. Jacks or wedges are used to slightly raise the head end of the trough above the datum. Water flow into the trough is adjusted until the gravel media is flooded but without free water on the surface. The discharge (Q) is measured in a calibrated container and timed with a stopwatch. The cross-sectional flow area (A_c) is estimated by noting the depth of the water as it leaves the perforated plate at the end of the trough and multiplying that value by the width of the trough. The hydraulic gradient (s) for each test is $(y_1 - y_2)/x$ (dimensions are shown on Figure 7.2). It is then possible to calculate the hydraulic conductivity because the other parameters in Equation 7.2 have all been measured. The Reynolds number should also be calculated for each test to ensure that the assumption of laminar flow was valid.

The porosity (n) of the media to be used in the SSF wetland should also be measured prior to final system design. This can be measured in the laboratory using a standard American Society for Testing and Materials (ASTM) procedure. An estimate is possible in the field by using a large container with a known volume. The container is filled with the media to be tested, and construction activity is simulated by some compaction or lifting and dropping the container. The container is then filled to a specified mark with a measured volume of water. The volume of water added defines the volume of voids (V_v). Because the total volume (V_t) is known, it is possible to calculate the porosity (n):

$$n = V_v/V_t(100) \qquad (7.3)$$

Many existing SSF wetlands were designed with a high aspect ratio (length-to-width ratio of 10:1 or more) to ensure plug flow in the system. Such high aspect ratios are unnecessary and have induced surface flow on these systems because the available hydraulic gradient is inadequate to maintain the intended subsurface flow. Some surface flow will occur on all SSF wetlands in response to major storm events, but the pollutant concentrations are proportionally reduced and treatment efficiency is not usually affected. The system should be initially designed for the average design flow and the impact of peak flows and storm events evaluated.

The previous recommendation that the design hydraulic gradient be limited to not more than 10% of the potential head has the practical effect of limiting the feasible aspect ratio for the system to relatively low values (<3:1 for beds 2 ft deep; 0.75:1 for beds 1 ft deep). SSF systems in Europe with soil instead of gravel have been constructed with up to 8% slopes to provide an adequate hydraulic gradient, and they have still experienced continuous surface flow due to an inadequate safety factor.

7.2 THERMAL ASPECTS

The actual thermal status of a SSF wetland bed can be a very complex situation. Heat gains or losses can occur in the underlying soil, the wastewater flowing through the system, and the atmosphere. Basic thermal mechanisms involved include conduction to or from the ground, conduction to or from the wastewater, conduction and convection to or from the atmosphere, and radiation to or from the atmosphere. It can be shown that energy gains or losses to the ground are a minor component and can therefore be neglected. It is conservative to ignore any energy gains from solar radiation but is appropriate at northern sites where the temperature conditions are most critical. In the southwest, where solar radiation can be very significant on a year-round basis, this factor might be included in the calculations. Convection losses can be significant due to wind action on an open water surface, but this should not be the case for most SSF wetlands where a dense stand of vegetation, a litter layer, and a layer of relatively dry gravel are typically present. These damp out the wind effects on the underlying water in the wetland, and, as a result, convection losses will be relatively minor and can be ignored in the thermal model. The simplified model developed below is therefore based only on conduction losses to the atmosphere and is conservative. This procedure was developed from basic heat-transfer relationships (Chapman, 1974) with the assistance of experts on the topic (Calkins, 1995; Ogden, 1994).

The temperature at any point in the SSF wetland can be predicted by comparing the estimated heat losses to the energy available in the system. The losses are assumed to occur via conduction to the atmosphere, and the only energy source available is assumed to be the water flowing through the wetland. As water is cooled, it releases energy, and this energy is defined as the specific heat. The specific heat of water is the amount of energy that is either stored or released as the temperature is either increased or decreased. The specific heat is dependent on pressure and to a minor degree on temperature. Because atmospheric pressure will prevail at the water surface in the systems discussed in this book, and because the temperature influence is minor, the specific heat is assumed to be a constant for practical purposes. For the calculations in this book, the specific heat is taken as 1.007 BTU/lb·°F (4215 J/kg·°C). The specific heat relationship applies down to the freezing point of water (32°F; 0°C). Water at 32°F will still not freeze until the available latent heat is lost. The latent heat is also assumed to be a constant and equal to 144 BTU/lb (334,944 J/kg). The latent heat is, in effect, the final safety factor, protecting the system against freezing; however, when the temperature drops to 32°F (0°C), freezing is imminent and the system is on the verge of physical failure. To ensure a conservative design, the latent heat is only included as a factor in these calculations when a determination of potential ice depth is made.

The available energy in the water flowing through the wetlands is defined by Equation 7.4:

$$q_G = c_p(\delta)(A_s)(y)(n) \tag{7.4}$$

where

q_G = Energy gain from water (Btu/°F; J/°C).
c_p = Specific heat capacity of water (1.007 Btu/lb·°F; 4215 J/kg·°C).
δ = Density of water (62.4 lb/ft³; 1000 kg/m³).
A_s = Surface area of wetland (ft²; m²).
y = Depth of water in wetland (ft; m).
n = Porosity of wetland media (percent).

If it is desired to calculate the daily temperature change of the water as it flows through the wetland, the term A_s/t is substituted for A_s in Equation 7.4:

$$q_G = (c_p)(\delta)(A_s/t)(y)(n) \tag{7.5}$$

where q_G is the energy gain during 1 d of flow (Btu/d·°F; J/d·°C), t is the hydraulic residence time in the system (d), and the other terms are as defined previously.

The heat losses from the entire SSF wetland can be defined by Equation 7.6:

$$q_L = (T_0 - T_{air})(U)(\sigma)(A_s)(t) \tag{7.6}$$

where

q_L = Energy lost via conduction at the atmosphere (Btu; J).
T_0 = Water temperature entering wetland (°F; °C).
T_{air} = Average air temperature during period of concern (°F; °C).
U = Heat-transfer coefficient at the surface of the wetland bed (Btu/ft²·hr·°F; W/m²·°C).
σ = Time conversion (24 hr/d; 86,400 s/d).
A_s = Surface area of wetland (ft²; m²).
t = Hydraulic residence time in the wetland (d).

TABLE 7.2
Thermal Conductivity of Subsurface Flow Wetland Components

Material	k (Btu/ft^2·hr·°F)	k (W/m·°C)
Air (no convection)	0.014	0.024
Snow (new, loose)	0.046	0.08
Snow (long-term)	0.133	0.23
Ice (at 0°C)	1.277	2.21
Water (at 0°C)	0.335	0.58
Wetland litter layer	0.029	0.05
Dry (25% moisture) gravel	0.867	1.5
Saturated gravel	1.156	2.0
Dry soil	0.462	0.8

The T_{air} values in Equation 7.6 can be obtained from local weather records or from the closest weather station to the proposed wetland site. The year with the lowest winter temperatures during the past 20 or 30 years of record is selected as the "design year" for calculation purposes. It is desirable to use an average air temperature over a time period equal to the design hydraulic residence time (HRT) in the wetland for these thermal calculations. If monthly average temperatures for the "design year" are all that is available, they will usually give an acceptable first approximation for calculation purposes. If the results of the thermal calculations suggest that marginally acceptable conditions will prevail then further refinements are necessary for a final system design.

The conductance (U) value in Equation 7.6 is the heat-conducting capacity of the wetland profile. It is a combination of the thermal conductivity of each of the major components divided by its thickness as shown in Equation 7.7:

$$U = 1/[(y_1/k_1) + (y_2/k_2) + (y_3/k_3) + (y_n/k_n)] \tag{7.7}$$

where

U = Conductance (Btu/ft^2·hr·°F; W/m^2·°C).

$k_{(1-n)}$ = Conductivity of layers 1 to n (Btu/ft^2·hr·°F; W/m·°C).

$y_{(1-n)}$ = Thickness of layers 1 to n (ft; m).

Values of conductivity for materials that are typically present in SSF wetlands are presented in Table 7.2. The conductivity values of the materials, except the wetland litter layer, are well established and can be found in numerous literature sources. The conductivity for a SSF wetland litter layer is believed to be conservative but is less well established than the other values in Table 7.2.

Example 7.1

Determine the conductance of a SSF wetland bed with the following characteristics: 8-in. litter layer, 6 in. of dry gravel, and 18 in. of saturated gravel. Compare the value to the conductance with a 12-in. layer of snow.

Solution

1. Calculate the U value without snow using Equation 7.7:

$$U = 1/[(0.67/0.029) + (0.5/0.867) + (1.5/1.156)]$$
$$= 0.040 \text{ Btu/ft}^2\cdot\text{hr}\cdot°\text{F}$$

2. Calculate the U value with snow:

$$U = 1/[(1/0.133) + (0.67/0.029) + (0.5/0.867) + (1.5/1.156)]$$
$$= 0.031 \text{ Btu/ft}^2\cdot\text{hr}\cdot°\text{F}$$

Comment

The presence of the snow reduces the heat losses by 23%. Although snow cover is often present in colder climates, it is prudent for design purposes to assume that the snow is not present.

The change in temperature due to the heat losses and gains defined by Equation 7.5 and Equation 7.6 can be found by combining the two equations:

$$T_c = q_L/q_G = (T_0 - T_{air})(U)(\sigma)(A_s)(t)/(c_p)(\delta)(A_s)(y)(n) \tag{7.8}$$

where T_c is the temperature change in the wetland (°F; °C), and the other terms are as defined previously.

The effluent temperature (T_e) from the wetland is:

$$T_e = T_0 - T_c \tag{7.9}$$

or

$$T = T_0 - (T_0 - T_{air})[(U)(\sigma)(t)/(c_p)(\delta)(y)(n)] \tag{7.10}$$

The calculation must be performed on a daily basis. The T_0 value is the temperature of the water entering the wetland that day, T_e is the temperature of the effluent from the wetland segment, and T_{air} is the average daily air temperature during the time period.

The average water temperature (T_w) in the SSF wetland is, then:

$$T_w = (T_0 - T_c)/2 \tag{7.11}$$

This average temperature is compared to the temperature value assumed when the size and the HRT of the wetland were determined with either the biochemical oxygen demand (BOD) or nitrogen removal models. If the two temperatures do not closely correspond, then further iterations of these calculations are necessary until the assumed and calculated temperatures converge.

Further refinement of this procedure is possible by including energy gains and losses from solar radiation and conduction to or from the ground. During the winter months, conduction from the ground is likely to represent a small net gain of energy because the soil temperature is likely to be higher than the water temperature in the wetland. The energy input from the ground can be calculated with Equation 7.6; a reasonable U value would be 0.056 Btu/ft^2·hr·°F (0.32 W/m^2·°C), and a reasonable ground temperature might be 50°F (10°C).

The solar gain can be estimated by determining the net daily solar gain for the location of interest from appropriate records. Equation 7.12 can then be used to estimate the heat input from this source. The results from Equation 7.12 should be used with caution. It is possible that much of this solar energy may not actually reach the water in the SSF wetland because the radiation first impacts on the vegetation and litter layer and a possible reflective snow cover, so an adjustment is necessary in Equation 7.12. As indicated previously, it is conservative to neglect any heat input to the wetland from these sources:

$$q_{solar} = (\Phi)(A_s)(t)(s) \tag{7.12}$$

where

q_{solar} = Energy gain from solar radiation (Btu; J).

Φ = Solar radiation for site (Btu/ft^2·d; J/m^2·d).

A_s = Surface area of wetland (ft; m).

t = HRT for the wetland (d).

s – Fraction of solar radiation energy that reaches the water in the SSF wetland, typically 0.05 or less.

If these additional heat gains are calculated, they should be added to the results from Equation 7.4 or Equation 7.5 and this total used in the denominator of Equation 7.10 to determine the temperature change in the system.

If the thermal models for SSF wetlands predict sustained internal water temperature of less than 33.8°F (1°C), a wetland may not be physically capable of winter operations at the site under consideration at the design HRT. Nitrogen removal is likely to be negligible at those temperatures.

Constructed wetlands can operate successfully during the winter in most of the northern temperate zone. The thermal models presented in this section should be used to verify the temperature assumptions made when the wetland is sized with the biological models for BOD or nitrogen removal. Several iterations of the calculation procedure may be necessary for the assumed and calculated temperatures to converge.

7.3 PERFORMANCE EXPECTATIONS

The performance expectations for SSF constructed wetlands are considered in the following discussion. As with the free water system (FWS; see Chapter 6), process performance depends on design criteria, wastewater characteristics, and operations. Removal mechanisms are described in Chapter 3.

TABLE 7.3
Total BOD Removal Observed in Subsurface Flow Wetlands

Location	Pretreatment	Influent	Effluent	Removal (%)	Nominal Detention Time (d)
Benton, Kentucky[a]	Oxidation pond	23	8	65	5
Mesquite, Nevada[b]	Oxidation pond	78	25	68	3.3
Santee, California[c]	Primary	118	1.7	88	6
Sydney, Australia[d]	Secondary	33	4.6	86	7

[a] Full-scale operation from March 1988 to November 1988, operated at 80 mm/d (Watson et al., 1989).

[b] Full-scale operation, January 1994 to January 1995.

[c] Pilot-scale operation in 1984, operated at 50 mm/d (Gersberg et al., 1985).

[d] Pilot-scale operation at Richmond, New South Wales, near Sydney, Australia, operated at 40 mm/d from December 1985 to February 1986 (Bavor et al., 1986).

7.3.1 BOD REMOVAL

Performance data for BOD removal are presented in Table 7.3. The removal of BOD appears to be faster and somewhat more reliable with SSF wetlands than for FWS wetlands, partly because the decaying plants are not in the water column, thereby producing slightly less organic matter in the final effluent.

7.3.2 TSS REMOVAL

Subsurface flow wetlands are efficient in the removal of suspended solids, with effluent total suspended solids (TSS) levels typically below 10 mg/L. Removal rates are similar to FWS wetlands.

7.3.3 NITROGEN REMOVAL

Although the SSF system at Santee, California, was able to remove 86% of the nitrogen from primary effluent, other SSF systems have reported removals of from 20 to 70%. When detention times exceed 6 to 7 d, an effluent total nitrogen concentration of about 10 mg/L can be expected, assuming a 20- to 25-mg/L influent nitrogen concentration. If the applied wastewater has been nitrified (using extended aeration, overland flow, or recirculating sand filters), the removal of nitrate through denitrification can be accomplished with detention times of 2 to 4 d.

TABLE 7.4
Removal for Metals through Subsurface Flow Constructed Wetlands at Hardin, Kentucky

Metal	Influent Range (μg/L)	Influent Average (μg/L)	Effluent Range (μg/L)	Effluent Average (μg/L)	Removal (%)
Aluminum	380–3800	1696	<50–100	<50	>97
Copper	20–190	77	<10	<10	>87
Iron	310–2400	1111	370–2900	1234	–10
Manganese	44–480	258	64–590	288	–10
Zinc	20–120	64	<10	<10	>84

Source: Hines, M., Southeast Environmental Engineering, Concord, Tennessee (personal communication).

7.3.4 PHOSPHORUS REMOVAL

Phosphorus removal in SSF wetlands is largely ineffective because of limited contact between adsorption sites and the applied wastewater. Depending on the loading rate, detention time, and media characteristics, removals may range from 10 to 40% for input phosphorus in the range from 7 to 10 mg/L. Crop uptake is generally less than 10% (about 0.5 lb/ac·d or 0.55 kg/ha·d).

7.3.5 METALS REMOVAL

Limited data are available on metals removal using municipal wastewater in SSF systems. In acid mine drainage systems, removal of iron and manganese is significant. Total iron has been shown to be reduced from 14.3 to 0.8 mg/L and total manganese from 4.8 to 1.1 mg/L (Brodie et al., 1989). At Santee, California, removal of copper, zinc, and cadmium was 99%, 97%, and 99%, respectively, during a 5.5-d detention time (Gersberg et al., 1984). The removal of metals at the Hardin, Kentucky, SSF system is presented in Table 7.4. The Hardin system has an activated sludge system for pretreatment, an HRT of 3.3 d, and two parallel cells, one planted to *Phragmites* and one planted to *Scirpus*.

7.3.6 PATHOGEN REMOVAL

A removal of 99% (2 log) of total coliform was found when primary effluent was applied at 2 in./d (detention time of 6 d) at Santee, California (Gersberg et al., 1989).

Natural Wastewater Treatment Systems

7.4 DESIGN OF SSF WETLANDS

Subsurface flow wetlands are designed based on hydraulic detention time and average design flow. The shortest detention times are usually necessary for BOD, nitrate nitrogen, and TSS removal from municipal wastewater, while ammonia and metals removal usually requires longer detention times.

7.4.1 BOD Removal

The recommended approach to design for BOD removal in SSF wetlands is the volume-based detention time model, as expressed in Equation 7.13:

$$A_s = Q(\ln C_0 - \ln C_e)/K_T(y)(n) \qquad (7.13)$$

where

A_s = Wetland surface area (ac; m²).
Q = Average design flow (ac-ft/d; m³/d).
C_0 = Influent BOD concentration (mg/L).
C_e = Effluent BOD concentration (mg/L).
K_T = Rate constant = 1.1 d⁻¹ at 20°C.
y = Design depth (ft; m).
n = Porosity of media (see Table 7.1).

The temperature of the wastewater will affect the rate constant according to Equation 7.14:

$$K_T = K_{20}(1.06)^{(T-20)} \qquad (7.14)$$

where

K_T = Rate constant at temperature T.
K_{20} = 1.1 d⁻¹.
T = Wastewater temperature (°C).

Most operational SSF wetlands in the United States have a treatment zone and operating water depth of 2 ft (0.6 m). A few, in warm climates where freezing is not a significant risk, operate with a bed depth of 1 ft (0.3 m). The shallow depth enhances the oxygen transfer potential but requires a greater surface area and the system is at greater risk of freezing in cold climates. A bed 2 ft (0.6 m) deep also requires special operation to induce desirable root penetration to the bottom of the bed.

Subsurface flow wetlands in the United States utilize at least the equivalent of primary treatment as the preliminary treatment prior to the wetland component. This can be obtained with septic tanks, Imhoff tanks, ponds, conventional primary treatment, or similar systems. The purpose of the preliminary treatment is to reduce the concentration of easily degraded organic solids that otherwise would accumulate in the entry zone of the wetland system and result in clogging, possible odors, and adverse impacts on the plants in the entry zone.

7.4.2 TSS Removal

The removal of total suspended solids in SSF wetlands is due to physical processes and is only influenced by temperature through the viscosity effects on the flow of water. Because the settling distance for particulate matter is relatively small and the residence time in the wetland is very long, the viscosity effects can be neglected. The removal of TSS in these wetlands is not likely to be the limiting design parameter for sizing the wetland, because TSS removal is very rapid as compared to either BOD or nitrogen.

Most of the solids in domestic, municipal, and many industrial wastewaters are organic in nature and will decompose in time, leaving minimal residues. The equivalent of primary treatment, as with BOD, will provide an acceptable level of preliminary treatment prior to the wetland component for these types of wastewaters. The subsequent decomposition of the remaining solids in the wetland should leave minimal residues and result in minimal clogging. Wetland systems designed for stormwater, combined sewer overflows, and some industrial wastewaters that have high concentrations of inorganic solids may not require primary treatment but should consider use of a settling pond or cell as the first unit in a wetland system to avoid a rapid accumulation of inorganic solids in the wetland.

The removal of TSS in SSF wetlands has been correlated to the hydraulic loading rate (HLR) as shown in Equation 7.15:

$$C_e = C_0[0.1058 + 0.0011(\text{HLR})] \tag{7.15}$$

where
  C_e = Effluent TSS (mg/L).
  C_0 = Influent TSS (mg/L).
  HLR = Hydraulic loading rate (cm/d).

The hydraulic loading rate is the flow rate divided by the surface area. Equation 7.15 is valid for HLR values between 0.4 and 75 cm/d. To use Equation 7.15, calculate the HLR by dividing the flow in ac-ft by the area in acres. Then convert the HLR in in./d to cm/d by dividing by 2.54 cm/in.

7.4.3 Nitrogen Removal

Because the water level is maintained below the media surface in SSF wetlands, the rate of atmospheric reaeration is likely to be significantly less than the FWS wetland type; however, as described previously, the roots and rhizomes of the vegetation are believed to have aerobic microsites on their surfaces, and the wastewater as it flows through the bed has repeated opportunities for contact with these aerobic sites in an otherwise anaerobic environment. As a result, conditions for nitrification and denitrification are present in the same reactor. Both of these biological nitrification and denitrification reactions are temperature dependent, and the rate of oxygen transfer to the plant roots may vary somewhat with the season.

TABLE 7.5
**Performance Comparison for Vegetated and Unvegetated Cells at
Subsurface Flow Wetlands in Santee, California**

Bed Condition	Root Penetration (in.)	Effluent BOD (mg/L)	Effluent TSS (mg/L)	Effluent NH_3 (mg/L)
Scirpus	30	5.3	3.7	1.5
Phragmites	>24	22.3	7.9	5.4
Typha	12	30.4	5.5	17.7
No vegetation	0	36.4	5.6	22.1

Note: HRT = 6 d; primary effluent applied: BOD = 118 mg/L, TSS = 57 mg/L, NH_3 = 25 mg/L; depth = 2.5 ft.

Source: Gersberg, R.M. et al., *Water Res.*, 20, 363–367, 1985. With permission.

The major carbon sources supporting denitrification are the dead and decaying roots and rhizomes, the other organic detritus, and the residual wastewater BOD. These carbon sources are probably more limited for SSF wetlands, during initial operations, as compared to the FWS case because most of the plant litter collects on top of the bed. After a few years of litter build-up and decay, both types of wetlands may have comparable carbon sources for support of denitrification.

Because a major source of oxygen in the SSF case is the plant roots, it is absolutely essential to ensure that the root system penetrates to the full design depth of the bed. Any water that flows beneath the root zone is in a completely anaerobic environment, and nitrification will not occur except by diffusion into the upper layers. This response is illustrated by the data in Table 7.5, where removal of ammonia can be directly correlated with the depth of penetration by the plant roots. The beds containing *Typha* (root penetration about 40% of the bed depth) achieved only 32% ammonia removal as compared to the *Scirpus* beds, which achieved 94% removal and had complete root penetration.

Many existing SSF systems in the United States were designed with the assumption that regardless of the plant species selected the roots would somehow automatically grow to the bottom of the bed and supply all of the necessary oxygen. This has not occurred, and many of these systems cannot meet their discharge limits for ammonia. This problem can be avoided in the future if proper care is taken during design and operation of the system. The root depths listed in Table 7.5 for Santee, California, probably represent the maximum potential depth for the plant species listed because Santee has a warm climate with a continuous growing season and the applied wastewater contains sufficient nutrients. This suggests that the design depth of the bed should not be greater than the potential root depth of the plant intended for use, if oxygen is required for ammonia removal.

TABLE 7.6
Potential Oxygen from Emergent Wetland Vegetation

Plant Type	Root Depth (ft)	Available Oxygen (g/m³·d)[a]	Available Oxygen (g/m²·d)[b]
Scirpus	2.5	7.7	5.7
Phragmites	2.0	8.0	4.8
Typha	1.0	7.0	2.1
Average	—	7.5	—

[a] Available oxygen per unit volume of measured root zone.
[b] Available oxygen per unit surface area of a 2.5-ft-deep bed.

Operational methods for actually achieving the maximum potential root penetration will still be necessary because the plants can obtain all of the necessary moisture and nutrients with the roots in a relatively shallow position. In some European systems, the water level is lowered gradually in the fall of each year to induce deep root penetration. It is claimed that three growing seasons are required to achieve full penetration by *Phragmites* using this method. Another approach, in cool climates where winter treatment requirements typically require a larger area, is to construct the bed with three parallel cells and only operate two for a month at a time during the warm periods. The roots in the dormant cell should penetrate as the nutrients in the water are consumed. In warm climates, where freezing is not a risk, it is possible to limit the bed depth to 1 ft (0.3 m), which should allow rapid and complete root penetration. The volume of gravel required will be constant regardless of the bed depth, but the surface area required to achieve the same level of treatment will increase as the depth decreases.

7.4.3.1 Nitrification

No consensus has been reached with regard to how much oxygen can be furnished to the root zone in SSF wetlands or regarding the oxygen transfer efficiency of various plant species. It is generally agreed that these emergent plants transmit enough oxygen to their roots to stay alive under normal stress levels, but disagreement arises (as discussed in Chapter 6) over how much oxygen is available at the root surfaces to support biological activity. The oxygen demand from the wastewater BOD and other naturally present organics may utilize most of this available oxygen, but based on the ammonia removals observed at Santee (Table 7.5) there must still be significant oxygen in the root zone to support nitrification.

If the ammonia removals observed at Santee are assumed to be due to biological nitrification, it is possible to calculate the amount of oxygen that should have been available for that purpose, as it requires about 5 g of oxygen to nitrify 1 g of ammonia. The results of these calculations are shown in Table 7.6.

The oxygen available for nitrification per unit of wetland surface area ranged from 2.1 to 5.7 g/m²·d because the depth of root penetration varied with each plant species. These oxygen values are in the published range (4 to 5 g O_2 per m²·d); however, the available oxygen, when expressed in terms of the actual root zone of the various plants, is about the same, regardless of the species (average 7.5 g O_2 per m³·d). This suggests that, at least for these three species, the oxygen available for nitrification will be about the same so the rate of nitrification is therefore dependent on the depth of the root zone present in the SSF bed. Equation 7.16 defines this relationship:

$$K_{NH} = 0.01854 + 0.3922(r_z)^{2.6077} \qquad (7.16)$$

where K_{NH} is the nitrification rate constant at 20°C (d⁻¹) and r_z is the fraction of SSF bed depth occupied by the root zone (decimal).

The K_{NH} value would be 0.4107 with a fully developed root zone and 0.01854 if there were no vegetation on the bed. These values are consistent with performance results observed at several SSF sites evaluated in the United States (Reed, 1993). Independent confirmation of this rate constant is provided by the design model published by Bavor et al. (1986). Bavor's model takes the same form as Equation 7.17 with a rate constant at 20°C of 0.107 d⁻¹ in a gravel bed system where the plant root zone occupied between 50 and 60% of the bed depth.

Having defined the basic rate constant K_{NH}, it is possible to determine the ammonia removal, via nitrification, in a SSF wetland with Equation 7.17 and Equation 7.18:

$$C_e/C_0 = \exp(-K_T t) \qquad (7.17)$$

$$A_s = Q(\ln C_0 - \ln C_e)/K_T(y)(n) \qquad (7.18)$$

where

C_e	=	Effluent ammonia concentration (mg/L).
C_0	=	Influent ammonia concentration (mg/L).
K_T	=	Temperature-dependent rate constant (d⁻¹).
t	=	Hydraulic residence time (d).
A_s	=	Surface area of wetland (ac; m²).
Q	=	Average flow through the wetland (ac-ft/d; m³/d).
y	=	Depth of water in the wetland (ft; m).
n	=	Porosity of the wetland (see Table 7.1).

The temperature dependence of the rate constant K_T is given by:

$$\text{At 0°C: } k_0 = 0 \text{ d}^{-1} \qquad (7.19)$$

$$\text{At 1°C: } K_T = K_{NH}(0.4103) \text{ d}^{-1} \qquad (7.20)$$

$$\text{At 1°C+: } K_T = K_{NH}(1.048)^{(T-20)} \text{ d}^{-1} \qquad (7.21)$$

For temperatures below 10°C, it is necessary to solve Equation 7.16 to determine the K_{NH} value. Interpolation can be used for temperatures between 0 and 1°C.

It is unacceptable to assume that the root zone will automatically occupy the entire bed volume, except for relatively shallow (1 ft or 0.3 m) systems using small-sized gravel (20 mm). Deep beds (2 ft or 0.6 m) require the special measures discussed previously to induce and maintain full root penetration. If these special measures are not utilized it would be conservative to assume that the root zone occupies not more than 50% of the bed depth unless measurements show otherwise. It is also unlikely, based on observations at numerous operational systems, that the plant roots will penetrate deeply in the large void spaces occurring when large-size rock (>2 in. or >50 mm) is selected as the bed media.

Equation 7.19 will typically require an HRT of between 6 to 8 d to meet stringent ammonia limits under summer conditions with a fully developed root zone and an even longer period at low winter temperatures. A cost-effective alternative to a large SSF wetland designed for ammonia removal may be the use of a nitrification filter bed (NFB). In that case, the SSF wetland can be designed for BOD removal only, and the relatively compact NFB can be used for ammonia removal. The combination of the SSF wetland and the NFB bed should require less than one half of the total area that would be necessary for a SSF wetland designed for ammonia removal. The NFB bed can also be used to retrofit existing wetland systems. Design details for the NFB concept are presented in a later section of this chapter.

7.4.3.2 Denitrification

Equation 7.16 to Equation 7.21 only account for conversion of ammonia to nitrate and predict the area required for a given level of conversion. When actual removal of nitrogen is a project requirement, it is necessary to consider the denitrification requirements and size the wetland accordingly. In the general case, most of the nitrate produced in a SSF wetland will be denitrified and removed within the area provided for nitrification and without supplemental carbon sources. FWS wetlands can be more effective for nitrate removal than the SSF type because of the greater availability of carbon from the plant detritus, at least during the first few years of operation. Even though the SSF wetland has more surface area for biological responses, it is likely that the availability of carbon in the system limits the denitrification rate so that SSF and FWS wetlands perform in a comparable manner. The recommended design model for estimating nitrate removal via denitrification is provided by Equation 7.22 and Equation 7.23:

$$C_e/C_0 = \exp(-K_T t) \tag{7.22}$$

$$A_s = Q\ln(C_e/C_0)/K_T yn \tag{7.23}$$

where

A_s = Surface area of wetland (ac; m^2).

C_e = Effluent nitrate-nitrogen concentration (mg/L).

C_0 = Influent nitrate-nitrogen concentration (mg/L).

K_T = Temperature-dependent rate constant (d^{-1}) = 0 d^{-1} at 0°C, and $1.00(1.15)^{(T-20)}$ d^{-1} at 1°C+.

n = Porosity of the wetland (see Table 7.1 for typical values).

t = Hydraulic residence time (d).

y = Depth of water in the wetland (ft; m).

Q = Average flow through the wetland (ac-ft/d; m^3/d).

The influent nitrate concentration (C_0) used in Equation 7.22 or Equation 7.23 is the amount of ammonia oxidized, as calculated in Equation 7.17. Because Equation 7.17 determines the ammonia remaining after nitrification in the SSF wetland, it can be conservatively assumed that the difference $(C_0 - C_e)$ is available as nitrate nitrogen. The rate of denitrification between 0°C and 1°C can be determined by interpolation. For practical purposes, denitrification is insignificant at these temperatures. It must be remembered that Equation 7.22 and Equation 7.23 are only applicable for nitrate nitrogen that is present in the wetland system.

Because the SSF wetland is generally anoxic but also has aerobic sites on the surfaces of the roots and rhizomes, it is possible to obtain both nitrification and denitrification in the same reactor volume. Equation 7.23 gives the wetland surface area required for denitrification. This denitrification area is not in addition to the area required for nitrification as determined with Equation 7.18; it is usually less than or equal to the results from Equation 7.18, depending on the input level of nitrate in the untreated wastewater and the water temperature.

7.4.3.3 Total Nitrogen

When denitrification is required, a discharge limit on total nitrogen (TN) usually exists. The TN in the SSF wetland effluent is the sum of the results from Equation 7.17 and Equation 7.22. The determination of the area required to produce a specific effluent TN value is an iterative procedure using Equation 7.17 and Equation 7.22:

1. Assume a value for residual ammonia (C_e) and solve Equation 7.18 for the area required for nitrification. Determine the HRT for that system.
2. Assume that $(C_0 - C_e)$ is the nitrate produced by Equation 7.17 and use this value as the influent (C_0) in Equation 7.23. Determine effluent nitrate using Equation 7.22.
3. The effluent TN is the sum of the C_e values from Equation 7.17 and Equation 7.22. If that TN value does not match the required TN, another iteration of the calculations is necessary.

7.4.4 ASPECT RATIO

The aspect ratio is the ratio of the length-to-width of the normally rectangular SSF beds. The early SSF systems had large aspect ratios and influent clogging, and surfacing of water occurred when little attention was paid to the hydraulics (Reed et al., 1995; USEPA, 1993). At Mesquite, Nevada, a SSF wetlands was successfully designed with an aspect ratio of 0.25:1 (Lekven et al., 1993). Current thinking is that the aspect ratio should be between 0.25:1 and 4:1.

7.5 DESIGN ELEMENTS OF SUBSURFACE FLOW WETLANDS

The design elements for SSF wetlands include pretreatment, media, vegetation, and inlet and outlet structures.

7.5.1 PRETREATMENT

Both FWS and SSF wetlands in the United States utilize at least the equivalent of primary treatment as the preliminary treatment prior to the wetland component. This might be obtained with septic tanks, Imhoff tanks, ponds, conventional primary treatment, or similar systems. The purpose of the preliminary treatment is to reduce the concentration of easily degraded organic solids that otherwise would accumulate in the entry zone of the wetland system and result in clogging, possible odors, and adverse impacts on the plants in that entry zone. A system designed for step feed of untreated wastewater might overcome these problems. A preliminary anaerobic reactor would be useful to reduce the organic and solids content of high-strength industrial wastewaters. Many of the SSF wetland systems in Europe apply screened and degritted wastewater to a wetland bed. This approach results in sludge accumulation, odors, and clogging but is acceptable in remote locations. In some cases, an inlet trench is used for solids deposition and the trench is cleaned periodically.

7.5.2 MEDIA

The SSF wetland bed typically contains up to 2 ft (0.6 m) of the selected media. This is sometimes overlain with a layer of fine gravel that is 3 to 6 in. (76 mm to 150 mm) deep. The fine gravel serves as an initial rooting medium for the vegetation and is maintained in a dry condition during normal operations. If relatively small gravel (<20 mm) is selected for the main treatment layer, a finer top layer is probably not necessary, but the total depth should be slightly increased to ensure a dry zone at the top of the bed. Most operational SSF wetlands in the United States have a treatment zone and operating water depth of 2 ft (0.6 m). A few systems, in warm climates where freezing is not a significant risk, operate with a bed depth of 1 ft (0.3 m). The shallow depth enhances the oxygen transfer potential but requires a greater surface area, and the system is at greater risk of freezing in cold climates. The deep (2 ft or 0.6 m) bed also requires special operation to induce desirable root penetration to the bottom of the bed.

7.5.3 VEGETATION

Vegetation for SSF wetlands should be perennial emergent plants such as bulrush, reeds, and cattails. The SSF wetland concept has significantly less potential

FIGURE 7.3 Inlet manifold for subsurface flow wetlands at Hardin, Kentucky.

habitat value as compared to the FWS wetland because the water is below the surface of the SSF media and not directly accessible to birds and animals. The presence of open-water zones within a SSF system negates many of the advantages of the concept and such zones are not normally included in the system plan. Enhancement of habitat values or esthetics is possible via selected plantings around the perimeter of the SSF bed. Because optimum wastewater treatment is the basic purpose of the SSF concept, it is acceptable to plan for a single plant species; based on successful experience in both the United States and Europe, *Phragmites* offers a number of advantages. A number of SSF wetlands in the southern states were initially planted with attractive flowering species (e.g., Canna lily, iris) for esthetic reasons. These plants have soft tissues that decompose very quickly when the emergent portion dies back in the fall and after even a mild frost. The rapid decomposition has resulted in a measurable increase in BOD and nitrogen leaving the wetland system. In some cases, the system managers have utilized an annual harvest for removal of these plants prior to the seasonal dieback or frosts. In most cases, the problems have been completely avoided by replacing these plants with the more resistant reeds, rushes, or cattails, which do not require an annual harvest. Use of these soft-tissue flowering species is not recommended on future systems, except possibly as a border.

7.5.4 INLET DISTRIBUTION

Inlet devices have ranged from open trenches to single-point weir boxes to perforated pipe manifolds. A surface manifold developed by the Tennessee Valley Authority (TVA) uses multiple, adjustable outlet ports (Steiner and Freeman, 1989; Watson et al., 1989). Having the manifold on the surface allows for operational adjustments if differential settlement occurs. An example of a surface manifold at Hardin, Kentucky, is presented in Figure 7.3. Subsurface manifolds

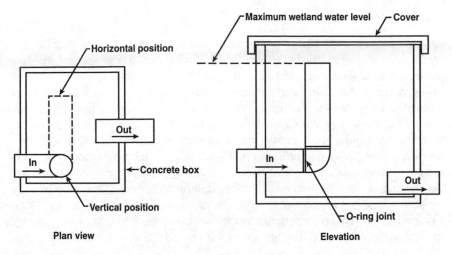

FIGURE 7.4 Adjustable outlet for subsurface flow wetlands.

encased in coarse gravel have also been used successfully. The disadvantage of this type of manifold is the potential for differential settlement and clogging from nuisance animals or solids. The advantage of a subsurface manifold is that the growth of algae on the outlets is avoided and thermal protection is provided.

7.5.5 OUTLET COLLECTION

Outlet collection should incorporate a manifold to avoid short-circuiting to a single outlet. A subsurface manifold is recommended to ensure the flow path is through the media. An adjustable outlet weir or swivel elbow allows control of the hydraulic gradient, as shown in Figure 7.4.

7.6 ALTERNATIVE APPLICATION STRATEGIES

Most SSF wetlands have been designed for continuous-flow applications. The lack of oxygen transfer, noted by Reed et al. (1995; USEPA, 1993) as the principal limitation of nitrification in SSF wetlands, led to researchers trying batch flow, rapid drainage of SSF beds, and reciprocating wetlands to get more oxygen into the wastewater.

7.6.1 BATCH FLOW

A number of modes of batch flow have been attempted. The case study of SSF wetlands at Minoa, New York (Section 7.8) illustrates one approach. Other approaches are described under the section on vertical flow wetlands (Section 7.11).

7.6.2 Reciprocating (Alternating) Dosing (TVA)

Researchers at the TVA developed and patented a "reciprocating" dosing of SSF wetlands in which the wastewater is quickly drained from one wetland cell and pumped into a second parallel cell (Behrends et al., 1996). The draining and filling occur within 2 hr, and then the process is reversed; the second cell is drained quickly and the first cell is refilled. The reciprocating flow process is repeated continuously, with a small amount of influent continually being added to the first cell and a fraction of the wastewater continually being withdrawn from the second cell as system effluent (USEPA, 2000). The reciprocating two-cell system was compared to a conventional two-cell system for 6 months in side-by-side testing in late 1995 and early 1996 at Benton, Tennessee. Operation of both two-cell pairs in the reciprocating mode has continued since May of 1996. Comparing conventional operation to the reciprocating mode, the reciprocating mode has produced significantly lower effluent BOD and ammonia nitrogen (USEPA, 2000).

7.7 POTENTIAL APPLICATIONS

The applications for SSF wetlands are many and expanding. Municipal wastewater examples are numerous, onsite wetlands are widely used, and a variety of industrial wastewaters have been treated. Some examples are presented here.

7.7.1 Domestic Wastewater

In the majority of cases, the utilization of SSF wetlands is preferred over the FWS type for on-site systems treating domestic wastewaters. This is because of the advantages of the SSF approach, which excludes mosquitoes and other insect vectors and eliminates risks of personal contact or exposure with the wastewater being treated. In northern climates, the additional thermal protection provided by the SSF concept is also an advantage. The design of these systems should follow the recommendations given in Section 7.6, supplemented as required. If nitrogen removal is a project requirement, the use of either *Phragmites* or *Scirpus* as the system vegetation is recommended. If stringent nitrogen limits prevail, the use of a compact recirculating NFB bed with plastic media should be considered to minimize the total area of the wetland (see Section 6.8 for details). In locations with relatively warm winter conditions, a 1-ft deep bed with *Typha* would also be suitable, but such a bed would require twice the surface area as compared to a 2-ft deep *Phragmites* or *Scirpus* bed. If nitrogen removal is not required, then the use of ornamental plants or shrubs is acceptable. In these cases, a layer of suitable mulch on the bed surface will enhance plant growth. The use of at least two parallel wetland cells is recommended, except for the smallest applications at single-family dwellings.

7.7.2 LANDFILL LEACHATE

The HRT in the "coarse" gravel cell at Tompkins County, New York, was estimated to be about 15 d. The total HRT in the two SF cells at Broome County, New York, at the estimated leachate flow of about 260 gal/d (1 m³/d) is calculated to be about 22 d. At these long detention times, the expected removal of BOD and ammonia should have been much greater than indicated by the results in Table 6.14. The poor performance observed for BOD removal at both of these systems is believed to be due to insufficient phosphorus in the untreated leachate to support the necessary biological reactions. The phosphorus concentration was only 0.15 mg/L at Tompkins County and was not measured at the Broome County site. This very low phosphorus level is insufficient to effectively remove the BOD loading applied, regardless of the detention time provided in the system. There appears to be sufficient quantities of nitrogen and other essential micronutrients to support BOD and ammonia removal. Treatment optimization at these two landfills and possibly at many others would require regular additions of at least supplemental phosphorus.

7.7.3 CHEESE PROCESSING WASTEWATER

A subsurface flow wetlands with supplemental aeration has been constructed for Eichten Cheese near Center City, Minnesota. The treatment system consists of a septic tank, a SSF wetland, and an infiltration bed. The forced-air aeration system improved the SSF wetland BOD reduction performance from 17 to 94%. The aeration system consisted of a blower and a perforated aeration tubing system (Wallace, 2001).

7.7.4 AIRPORT DEICING FLUIDS TREATMENT

Glycol is used at airports to deice the wings of airplanes. Runoff of stormwater with glycol in it is an environmental problem that SSF wetlands can help to solve. SSF wetlands are appropriate because open water is not acceptable near airport runways and close-growing vegetation can be used. SSF wetlands have been used to treat deicing fluids at Edmonton and Toronto, Canada; at Airborne Express Airport in Wilmington, Ohio; and at Heathrow Airport in London (Richter et al., 2003; Karrh et al., 2001). A SSF wetlands was designed for glycol treatment at Westover Air Reserve Base in western Massachusetts, and the design criteria are presented in Table 7.7. The expected BOD removal for the system was 90% (Karrh et al., 2001).

7.8 CASE STUDY: MINOA, NEW YORK

The Village of Minoa, New York, near Syracuse, has a three-cell SSF constructed wetland. The conceptual design was prepared by Sherwood C. Reed in 1994. The treatment capacity of the 1.1-ac (0.45-ha) wetland as constructed was

TABLE 7.7
Design Criteria for Subsurface Flow Wetlands Treating Deicing Runoff

Parameter	Value	Notes
Design flow (mgd)	0.1	Flow constrained by limited bed area
Peak wetlands flow (mgd)	0.4	Flow constrained by limited bed area
Hydraulic loading rate (in./d)	5.7	—
System residence time (d)	2.2	—
Wetlands residence time (d)	1.85	Volume/design flow
Bed area (ac)	0.6	Constrained by site
Bed length (ft)	212	Distance perpendicular to the flow
Bed width (ft)	110	Distance in the direction of flow
Length-to-width ratio	1.9	—
Bed depth (ft)	2.0	—
Bed bottom slope	0.0001	Allows for bed drainage
Bed media (D_{50}) (in.)	1.2	Material should have <1% fines
Media porosity	0.47	High for most gravels
Inlet/outlet rock (D_{50}) (in.)	7.0	Rip-rap-sized material
Inlet/outlet width (ft)	10	Distribution and collection trenches

Source: Karrh, J.D. et al., in *Wetlands and Remediation II: Proceedings of the Second International Conference on Wetlands and Remediation*, Nehring, K.W. and Brauning, S.E., Eds., Battelle, Columbus, OH, 2001. With permission.

130,000 gal/d (454 m^3/d) with a hydraulic residence time of 2.4 d (Reed and Giarrusso, 1999). BOD reduction of the primary effluent to 30 mg/L was the design objective. The three cells were constructed at different elevations, and piping was provided to allow either series or parallel operation. Each cell was provided with water level controls, drainage, and internal sampling wells. The slope of each bed was 1%, but the media surface was level so the depth of water varied from 1.6 ft (0.5 m) at the inlet to 3 ft (0.9 m) at the outlet. The top 4 in. (100 mm) of the bed was 0.25-in. (0.6-mm) pea gravel and served as the rooting medium for the plants. The treatment zone in the bed used 1.5-in. (40-mm) washed and screened coarse gravel, obtained as crushed stone from a local source. The cell bottoms were lined with a 60-mil high-density polyethylene (HDPE) liner. Each cell was also divided by a longitudinal barrier so the influence on performance of two different plant species (*Phragmites* and *Scirpus*) could be evaluated.

Start-up occurred in June 1995 using secondary effluent from the trickling filters. In January 1996, the operation was switched to the primary effluent. The effluent BOD value during the spring and summer of 1996 was only 84 mg/L,

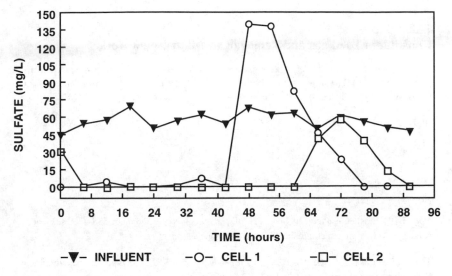

FIGURE 7.5 Sulfate reduction at Minoa, New York, subsurface flow wetlands.

which was unacceptable. In addition, objectionable sulfide odors were noted at the outlet structures of the cell, and a black "sludge-like" substance was observed accumulating in the gravel void spaces in each of the wetland cells. Suggested remedial actions, including reduction in the loading rate, dilution of the influent with secondary effluent, and chemical oxidants, were tried without success. The operator, Steve Giarrusso, began to operate the cells by sequentially draining and refilling the cells on a regular basis. BOD removal, which had averaged 44% in 1996, increased to 95% in 1997.

In 1997, the typical sequence consisted of opening the drain for cell 1 on a Tuesday morning while the full flow continued to enter cell 1. After 24 hr, the drain to cell 1 was closed and the drain to cell 2 was opened. The next day, the drain for cell 2 was closed and the drain for cell 3 was opened. On Friday, the drain for cell 3 was closed, the drain for cell 1 was opened, and the cycle was repeated. Cells 1 and 2 took 4 to 5 hr to drain to their lowest levels and 24 hr to refill, so for about 20 hr the media was exposed to aerobic conditions.

The wastewater contained about 50 mg/L of sulfate, and, after a few hours of conventional loading, the water in the cell became anaerobic and the sulfates were reduced to sulfides. During a 90-hr test conducted by Clarkson University, the sulfates were found to be reduced to near zero until the draining and reaeration restored the aerobic conditions and stopped the reduction of sulfates. This phenomenon is shown in Figure 7.5. The improvement in treatment between 1996, when continuous flow was practiced, and 1997, when the sequential fill/drain operation was initiated, is shown in Table 7.8 (Reed and Giarrusso, 1999).

TABLE 7.8
Constituent Loadings and Removals at Minoa, New York

Constituent	Loading (lb/ac·d)	Removal in 1996 (lb/ac·d)	Removal in 1997 (lb/ac·d)	Improvement from 1996 to 1997 (%)
BOD$_5$	160	65	107	64
COD	305	121	273	125
TSS	75	59	73	24
NH$_3$–N	16	1.1	1.9	75
TP	3.9	1.7	1.8	5

Note: BOD, biochemical oxygen demand; COD, chemical oxygen demand; TSS, total suspended solids; NH$_3$–N, nitrogen ammonia; TP, total phosphorus.

Source: Reed, S.C. and Giarrusso, S., in *Proceedings of WEFTEC 1999*, Water Environment Federation, New Orleans, LA, October 9–13, 1999.

7.9 NITRIFICATION FILTER BED

The nitrification filter bed (NFB) concept was developed by Sherwood C. Reed as a retrofit for existing wetland systems having difficulty meeting their ammonia discharge limits. It has been used successfully for both FWS and SSF wetland systems. As shown in Figure 7.6, it consists of a vertical-flow gravel filter bed on top of the existing SSF or FWS wetland bed. In the latter case, the fine-gravel NFB is supported by a layer of coarse gravel to maintain aerobic conditions in the NFB.

The NFB unit can be located at the head of the wetland channel or near the end. In either case, the wetland effluent is pumped to the top of the NFB and uniformly distributed. The inlet location has advantages in that the nitrified percolate will mix with the influent wastewater. The resulting denitrification will remove nitrogen from the system, further reduce the BOD, and recover some of the alkalinity consumed during the nitrification step. Locating the NFB near the end of the wetland cell will produce the desired level of nitrification but there is insufficient time for significant denitrification so most of the nitrate produced will pass out of the system with the effluent. Pumping capacity and power costs will be higher for the inlet location, particularly for retrofit of long, narrow wetland channels. A U-shaped wetland channel with the inlet adjacent to the outlet would retain the advantages of denitrification and minimize the pumping requirements.

The NFB is similar in concept to the familiar recirculating sand filter (see Chapter 10), which has been used successfully for many years to polish and nitrify septic tank effluent (Crites and Tchobanoglous, 1998). These recirculating

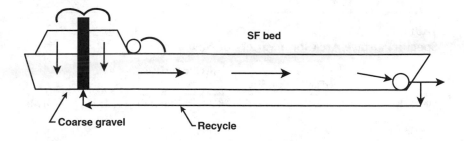

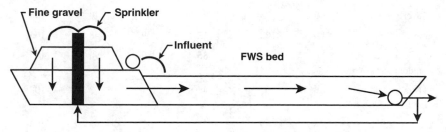

FIGURE 7.6 Schematic diagram of nitrification filter bed.

sand filters normally operate with a hydraulic loading of less than 5 gal/ft²·d (0.2 m/d). Gravel is used in the NFB to increase the hydraulic conductivity of the media and permit much higher hydraulic loading rates on the system. The hydraulic loading (with a 3:1 recycle ratio) is about 100 gal/ft²·d (4 m/d) on one of the operational NFB systems at Benton, Kentucky.

The design procedure for the NFB is based on nitrification experience with trickling filter and RBC attached growth concepts where the removal capability is related to the specific surface area available for development of the attached growth nitrifying organisms (USEPA, 1993). Several conditions are required for successful nitrification performance:

- The BOD level must be low (BOD/TKN < 1).
- Exposure to the atmosphere or to an oxygen source must be sufficient to maintain aerobic conditions in the attached film of nitrifying organisms.
- The surface must be moist at all times to sustain organism activity at optimum rates.
- Alkalinity must be sufficient to support the nitrification reactions (≈10 g alkalinity per 1 g ammonia).

Equation 7.24 can be used to determine the specific surface area (A_v) required to achieve a particular effluent ammonia (C_e) at the bottom of the NFB:

$$A_v = [2713 - 1115(C_e) + 204(C_e)2 - 12(C_e)^3]/K_T \qquad (7.24)$$

TABLE 7.9
Specific Surface Area for a Variety of Media Types

Media Type	Median Particle Size		Specific Surface Area		Void Ratio	k_s (m/d)[a]
	(in.)	(mm)	(ft²/ft³)	(m²/m³)		
Medium sand	0.12	3	270	886	40	1
Pea gravel	0.57	14.5	85	280	28	104
Gravel	1.0	25	21	69	40	105
Gravel	4.0	102	12	39	48	106
Plastic media, random pack	1.0	25	85	280	90	107
Plastic media, random pack	2.0	50	48	157	93	108
Plastic media, random pack	3.5	89	38	125	95	108

[a] Maximum potential hydraulic conductivity; NFB design should utilize a small fraction of this value to ensure unsaturated flow.

where

A_v = Specific surface area (m²/kg NH₄·d; 1 m²/kg NH₄/d × 4.882 = ft²/lb·d).

C_e = Desired NFB effluent ammonia concentration (mg/L).

K_T = Temperature-dependent coefficient: $[1(1.048)^{(T-20)}]$ at 10°C+, and $[0.626(1.15)^{(T-10)}]$ at 1–10°C.

Equation 7.24 is based on curve fitting of performance data from attached growth nitrification reactors, and the units involved are not dimensionally compatible. It will still, however, give a reasonably accurate estimate of the specific surface area required to achieve effluent ammonia levels in the range of 0 to 6 mg/L. Equation 7.24 has been verified in a recent full-scale application at Mandeville, Louisiana (Reed et al., 2003).

Information on the specific surface area per unit volume for a number of potential media types is presented in Table 7.9. The specific surface area available for the natural sand and gravel media types tends to increase as the potential hydraulic conductivity decreases. The plastic media listed in Table 7.9 are spherical in shape with a variety of internal members to increase the available surface area per unit. These have a very high specific surface and an equally high potential hydraulic conductivity. Rigid corrugated plastic media and flexible hanging plastic sheets are also available. These plastic media are commonly used in trickling filter units designed for nitrification and could be used in the same capacity as a nitrification component for a wetland system. A relatively small container filled with this plastic media has been proposed as a nitrification component for small-scale wetland systems.

The natural sand and fine gravel media do not drain quickly, and it is usually necessary to design for intermittent wet and dry cycles to allow a portion of the bed to drain and restore aerobic conditions. The coarse gravel and plastic media can be exposed to continuous hydraulic loading (at a reasonable rate) and still maintain aerobic conditions in the media. It is also necessary to keep the media surfaces completely wet at all times to ensure optimum responses from the nitrifying organisms. The minimum hydraulic loading, for this purpose, on the plastic media is in the range of 590 to 1757 gpd/ft^2 (24 to 72 m^3/d·m^2) of bed surface area. The typical hydraulic loading on an intermittent sand filter bed is 0.75 to 15 gpd/ft^2 (0.03 to 0.06 m/d). Recycle may not be necessary as long as wetting of the media is complete and sufficient oxygen is present in the profile. In the case of sand and fine gravel systems, a larger bed area, divided into cells, is provided to allow for intermittent hydraulic loading and drainage periods. Assuming one half of the system is draining at any one time, the pumping rate would have to be $2Q$ as compared to $1Q$ for a continuously operated bed.

Typically, the effluent from the wetland cell is applied to the NFB to ensure a low BOD concentration in the liquid. Nitrification in the NFB can be expected when the applied water has a BOD/TKN ratio of less than 1.0 and the soluble BOD concentration is less than 12 mg/L (USEPA, 1993). The ratio of soluble to total BOD in typical wetland effluents is about 0.6 to 0.8 (Reed, 1991, 1993).

Equation 7.15 is used to determine the specific surface area required to achieve the necessary effluent ammonia level. The characteristics of an appropriate media are selected from Table 7.9 to determine the volume of media required. Usually, the NFB bed will be 1 to 2 ft (0.3 to 0.6 m) deep and extend the full width of the wetland cell to ensure complete mixing with the wastewater flowing through the wetland. The use of sprinklers, for distribution on top of the NFB, is recommended to provide proper distribution and maximum aeration. In cold climates with extended periods of subfreezing temperatures, an exposed bed with sprinklers, as shown in Figure 7.6, may not be feasible. In this case, the use of plastic media in a protected tank or similar container should be considered. Such a tank would have to be vented to provide the necessary air flow.

A design for a retrofit NFB at an existing wetland has to conform to the existing wetland configuration and effluent water quality conditions. In many cases, the combination of an NFB and a wetland designed for BOD removal may be more cost effective than the much larger area required for a wetland to remove both BOD and ammonia. In this case the wetland is sized for BOD removal to 5 to 10 mg/L; the ammonia removal expected in this wetland is determined with appropriate models, and then the NFB is designed for the balance of ammonia requiring removal. A cost comparison will then show if the NFB combination is more economical than a larger wetland system.

7.10 DESIGN OF ON-SITE SYSTEMS

On-site systems are defined as relatively small facilities serving a single waste-water source or possibly a cluster of residential units in a development. Usually, the on-site system is at the same location as the wastewater source, but in some cases pumping to a remote site is used if suitable soils for in-ground discharge do not exist at the original wastewater source. Preliminary treatment is typically provided by septic tanks or similar devices, but in some cases packaged secondary treatment plants have been used. In most cases, the advantages inherent in the SSF wetland concept (e.g., no insect vectors, subsurface flow so no risk of public contact with the untreated wastewater) favor its use for these on-site systems. The disposal of the final effluent from a wetland is still a project requirement, even in arid climates where evaporation and seepage (if allowed) may account for a large fraction of the wastewater.

Surface discharge and in-ground disposal are the only two alternatives available. In-ground disposal methods are described in Chapter 10 of this book. Surface discharges must meet the applicable state and local discharge requirements; many states and local governments will not permit surface discharges from small on-site systems so this alternative must be explored with the appropriate agencies prior to any design. The site investigation requirements for on-site in-ground disposal are discussed in Chapter 10. The simple percolation test may be marginally adequate for very small systems at single-family dwellings but is not adequate for larger facilities and flows. In these cases, it is necessary to determine the actual hydraulic conductivity of the *in situ* soils and to determine the groundwater position and gradient to ensure that mounding and system failure will not occur.

Most current criteria for in-ground disposal systems via leach fields, beds, mounds, etc. specify a hydraulic loading rate (gpd/ft) based on the results of the site investigation as modified by prior performance experience. These hydraulic loading rates are based in part on the hydraulic characteristics of the soil and in part on the clogging potential of typical septic tank effluent, because a clogging layer accumulates at the soil/disposal bed interface.

Because the use of a wetland system prior to the disposal step can produce the equivalent of tertiary effluent, the potential for clogging is significantly reduced, and it should be possible to reduce the surface area of the disposal bed or trenches significantly. A disposal bed or trench, after a wetland system can typically be at least one third to one half the "normal" infiltration area because of the improved water quality. It is still essential to measure (or estimate for very small systems) the actual hydraulic conductivity of the receiving soils to validate the size reduction. Heavy clay soils, for example, have limited permeability regardless of the quality of the water applied. In some cases, in-ground disposal on coarse, highly permeable soils is also prevented because the applied wastewater does not have enough time and contact for adequate treatment. The use of a on-site wetland prior to in-ground disposal should alleviate the problem and allow development on such soils.

Several approaches are available for designing on-site wetland systems. One of the most prominent is to utilize guidelines issued by the TVA. In an evaluation published by the USEPA (Steiner and Watson, 1993), it was concluded that these TVA guidelines are probably adequate for the design of small-scale systems at single-family dwellings but are deficient for larger flows and for surface discharging systems. The deficiencies relate to the lack of soils investigations for the larger disposal fields, the lack of design criteria for nitrogen removal, and the lack of any temperature dependence, which will affect winter water quality in colder climates. The USEPA evaluation recommended that the design of on-site systems should follow the same procedures used for large-scale systems because the design principles and thermal constraints are the same. As a result, the design procedures found in earlier sections of this chapter should also be used for on-site wetlands.

The USEPA document recommended several simplifying assumptions for the design of wetlands for smaller on-site systems:

- Determine the design flow; 60 gpd (0.23 m/d) is a reasonable assumption for per-capita flow for residential systems. State or local criteria will govern.
- Use a multicompartment septic tank. Use one tank for single-family dwellings; use two or more tanks in series for larger scale (>10,000 gpd) projects. The total volume of the tanks should be at least twice the design daily flow.
- Assume that the BOD_5 leaving the septic tanks is a conservative 100 mg/L. Assume that the wetland effluent BOD will not exceed 10 mg/L.
- Use clean, washed gravel as the treatment media in the bed with a size range of 0.5–1 in. (1.25–2.5 cm), with a total depth of 2 ft (0.6 m). For design, assume the "effective" water depth in the bed is 1.8 ft (0.55 m). Reasonable estimates include: hydraulic conductivity (k_s) = 5000 ft^3/ft^2/d (1500 m^3/m^2/d); porosity = 0.38. If a large number of systems is to be installed using the same materials, field or laboratory testing for hydraulic conductivity (k_s) and porosity (n) is recommended.
- Use reeds (*Phragmites*) as the preferred plant species.
- Estimate the summer and winter water temperatures to be expected in the bed. In the summer and in year-round warm climates, 2°C is reasonable. In cold winter climates, a winter water temperature of 1°C is a reasonable assumption.
- Determine the bed surface area with:

$$A_s = (L)(W) = Q[\ln(C_0/C_e)]/K_T dn \qquad (7.13)$$

- As a safety factor, use a rate constant K_{20} that is 75% of the base value (1.104 d^{-1}). So, for the design of small on-site systems, K_{20} = 0.828 d^{-1}. At 20°C, and with the other factors defined above, this equation reduces to:

  Metric: $A_s = 13.31(Q) = $ m^2 (Q in m^3/d)
  U.S. units: $A_s = 4.07(Q) = $ ft^2 (Q in ft^3/d)

At 6°C:

Metric: $A_s = 30.1(Q) = $ m^2 (Q in m^3/d)
U.S. units: $A_s = 9.2(Q) = $ ft^2 (Q in ft^3/d)

- Adjustments for other temperatures, other media types, etc. should use the basic design equations. Adopt an aspect ratio (length-to-width) of 2:1; calculate bed length (L) and width (W) because the surface area was determined above. In the general case, an aspect ratio of 2:1, or less, with a bed depth of 2 ft (0.6 m) will satisfy the Darcy's law constraints on hydraulic design of the bed, so hydraulic calculations are not required. If site conditions will not permit the use of a length-to-width radio of 2:1 for the bed and a 2-ft (0.6-m) bed depth, then hydraulic calculations as described previously will be necessary. This approach will give an HRT of about 2.8 d (at 20°C) in the bed which is more than adequate for BOD removal to 10 mg/L. If nitrogen removal to 10 mg/L is required, the size of the system should be doubled to produce an HRT of about 6 days. Nitrogen removal during the winter months in cold climates may require an HRT of about 10 d. In these cases, heat-loss calculations should be performed to be sure the bed is adequately protected against freezing.

- Construct the bed as a single cell for single-family dwellings. Use multiple cells (at least two) in parallel for larger sized systems. Use clay or a synthetic liner to prevent seepage from the bed.

- Construct the bed with a flat bottom and a perforated effluent manifold at the bottom of the bed. A perforated inlet manifold a few inches above the bottom of the bed is adequate for most small systems. These inlet and outlet zones should use 1- to 2-in. (2.5- to 5-cm) washed rock for a length of about 3 ft (1 m) and for the full depth of the bed.

- The effluent manifold should connect to either a swiveling standpipe or a flexible hose for discharge to allow control of the water level in the bed. The inlet and effluent manifolds should have accessible cleanouts at the surface of the bed.

The system described here should produce an effluent with BOD of <10 mg/L, TSS of <10 mg/L, and TN of <10 mg/L and should therefore be suitable for either surface or in-ground discharge. The excellent water quality should permit a significant reduction in the area required for the disposal field. For example, a typical conventional on-site system for a family of four (300 gpd, 1 m^3/d) might include a 1000-gal (4-m^3) septic tank and a 500-ft^2 (46-m^2) infiltration area in a sandy loam soil. Addition of a wetland component with a 6-d HRT would require about 300 ft^2 (28 m^2) of area. If appropriate credit for the higher level of treatment is allowed, the total area for the wetland cell and the infiltration bed could be less than 500 ft^2 (<46 m^2).

7.11 VERTICAL-FLOW WETLAND BEDS

In the vertical-flow wetland concept, the wastewater is uniformly applied to the top of the bed, and the effluent is withdrawn via perforated pipes on the bottom, parallel to the long axis of the bed. The concept is based on the work of Seidel (1966) and is in use at several locations in Europe. A system typically consists of two groups, or stages, of vertical-flow cells in series followed by one or more horizontal-flow polishing cells. Each stage of vertical-flow units consists of several individual wetland cells in parallel because wastewater is applied intermittently in rotation. The operational systems in Europe apply either primary effluent (typically from a septic tank) or in some cases untreated raw wastewater.

Typically, the beds are dosed for up to 2 d and then rested for 4 to 8 d. A 2-d wet and 4-d dry cycle (2/4) would require a minimum of three sets of stage I cells; a 2/8 cycle would require at least five cells. The number of stage II cells is one half that of the stage I components, and these are also loaded in rotation.

The main advantage of the concept is the restoration of aerobic conditions during the periodic resting and drying period. This allows removal of BOD and ammonia nitrogen at higher rates than can be achieved in the continuously saturated and generally anaerobic horizontal flow SSF wetland bed. As a result, the vertical-flow beds can be somewhat smaller in area than a comparable SSF wetland designed for the same performance level.

During the dosing period, hydraulic loading on the stage I beds is typically 7.4 gal/ft·d (0.3 m/d) for primary effluent, and double that value for the stage II cells. Such a two-stage system can typically achieve better than 90% BOD and TSS removal. The bed profile contains several layers of various sized granular materials. A typical profile, from the top of the bed, would include:

10 in. (25 cm) freeboard
3 in. (8 cm) coarse sand, planted with *Phragmites*
6 in. (15 cm) pea gravel (6 mm size)
4 in. (10 cm) washed medium gravel (12 mm size)
6 in. (15 cm) washed coarse gravel (40 mm size)

Perforated underdrain pipes are laid on the bottom of the cell on about 3-ft (1-m) centers. The upstream end of these pipes extends up to and above the bed surface to create a "chimney" effect and encourage oxygen transfer to the profile. The upper portion of this perforated pipe is contained within a solid pipe jacket to prevent short-circuiting of percolate flow. Additional vertical "chimney" pipes are placed at 6-ft (2-m) centers in the rows between the perforated effluent piping. These vertical pipes are perforated in the bottom layer of gravel and solid from there to the above-surface end.

Insufficient performance data are available for this concept to permit development of a rational design model. The equations below are based on the performance of a system in the United Kingdom with a 2-d wet and 4-d dry cycle.

They can be used with extreme caution (because of the limited database) to estimate the performance of similar systems.

BOD removal, per stage:

$$C_e/C_0 = \exp(-K_T/\text{HLR}) \tag{7.25}$$

where

C_e = Effluent BOD (mg/L)

C_0 = Influent BOD (mg/L)

K_T = Temperature-dependent rate constant (d^{-1}) = $0.317(1.06)^{(T-20)}$ d^{-1}.

HLR = Average daily hydraulic loading rate during the dosing cycle (m/d).

Ammonia removal, per stage:

$$C_e/C_0 = \exp(-K_T/\text{HLR}) \tag{7.26}$$

where

C_e = Effluent ammonia (mg/L).

C_0 = Influent ammonia (mg/L).

K_T = Temperature dependent rate constant (d^{-1}) = $0.1423(1.06)^{(T-20)}$ d^{-1}.

HLR = Average daily hydraulic loading rate, during the dosing cycle (m/d).

Ordinarily, a higher rate of ammonia removal should be expected in the second stage of a two-stage system; however, in this two-stage system, the rate of ammonia removal per stage is about equal because the BOD loading on the second stage is still higher than desired for optimum nitrification, as discussed previously for the nitrification filter bed (in Section 7.9). This response suggests that further improvements and optimization of the vertical flow concept as used in Europe are desirable. The first stage should be large enough to produce an effluent BOD in the range of 10 to 15 mg/L. The second stage could then be optimized for ammonia removal, and the principal role of SSF wetland used as the third component would be denitrification and final polishing.

7.11.1 Municipal Systems

In a vertical-flow wetland system, the wetland is divided into a number of distinct beds or cells that operate in parallel. The cells contain about 3 ft (0.9 m) of granular media, which are typically planted with bulrush to maintain the porosity of the bed. The wastewater is uniformly applied to the top of the cell and flows vertically downward through the bed. The effluent is withdrawn via perforated pipes on the bottom, parallel to the long axis of the bed. The beds are intermittently dosed, allowing time for the beds to rest between dosing cycles. Hydraulic loading rates, during dosing, are typically 30 cm per day (7.4 gal/ft^{2}·d) for primary effluent and double that for secondary cells.

The operational vertical-flow wetland systems in Europe apply either primary effluent (typically from a septic tank) or in some cases untreated raw wastewater.

TABLE 7.10
Vertical Flow Wetlands Performance
for Salem, Oregon[a]

Constituent	Influent (mg/L)	Effluent (mg/L)
BOD_5	10.7	3
TSS	8.4	3
Ammonia nitrogen	13.3	1.3
Nitrate nitrogen	1.0	10.4

[a] Average for year 2003.

Note: BOD_5, biochemical oxygen demand; TSS, total suspended solids.

Municipal wastewater applications in North America, in addition to the demonstration project at Salem, Oregon, include Pelee Island, Ontario; Niagara-on-the-Lake, Ontario; and Vineland, Ontario. The experience at the Salem, Oregon, demonstration natural reclamation system with vertical-flow wetlands has been very positive. The system was constructed in 2002. Secondary effluent is flooded onto beds at 4 cycles per day with 1 hour on and 5 hours off. The beds are 3 ft (0.9 m) deep and graded from sand to gravel from top to bottom. The bottom material is gravel for the collection of the underdrainage. The 2003 year results when the beds were loaded at 15.3 in. (39 cm) are shown in Table 7.10. The VF wetlands are shown in Figure 7.7.

7.11.2 TIDAL VERTICAL-FLOW WETLANDS

Tidal flow wetlands involve cyclic flooding and draining of a media bed (Sun et al., 1999). In 1901, a patent application was made for a tidal flow wetland (Monjeau, 1901). Recent research by David Austin of Living Machines, in Taos, New Mexico, has concentrated on media characteristics, such as ammonium adsorption. The results of the pilot testing of the Living Machines tidal vertical-flow wetlands are presented in Table 7.11. The influent flow was 450 gpd (1.7 m³/d), the wetland area was 96 ft² (8.9 m²), and the average hydraulic loading rate was 4.7 gal/ft²·d (19 cm/d) (Austin et al., 2003).

7.11.3 WINERY WASTEWATER

Winery wastewater at the EastDell Estates Winery in Ontario, Canada, is pretreated with a vertical-flow wetland for BOD reduction. The wastewater BOD was reduced by 65% through a septic tank followed by a 96% reduction through the vertical-flow wetlands (Rozema, 2004).

(a)

(b)

FIGURE 7.7 Salem, Oregon, vertical flow wetlands: (a) recently planted bed showing distribution system, (b) bulrush growth, which matures in 1 year.

7.12 CONSTRUCTION CONSIDERATIONS

Both types of wetlands typically require an impermeable barrier to ensure containment of wastewater and to prevent contamination of groundwater. In some cases, such a barrier may be provided if clay is naturally present or if *in situ* soils can be compacted to a nearly impermeable state. Chemical treatments, a bentonite layer, and asphalt or membrane liners are also possibilities. In the case of a

TABLE 7.11
Tidal Vertical Flow Wetlands Design and Performance

Design Factor	Units	Value
Influent flow	gal/d	450
Recycle ratio	—	3:1 to 14:1
Flow regime	—	Downflow flood and drain
Hydraulic residence time	H	24
Area of five cells	ft^2	96
Media depth	ft	2
Media type	mm × mm	9.5 × 2.4 expanded shale

Performance	Influent (mg/L)	Effluent (mg/L)
BOD$_5$	428	5.2
Total nitrogen	48	8.3
Nitrate nitrogen	3.0	7.0
TKN	45	1.3
TSS	<50	3.5

Note: BOD$_5$, biochemical oxygen demand; TKN, total Kjeldahl nitrogen; TSS, total suspended solids.

Source: Austin, D. et al., in *Proceedings of WEFTEC 2003*, Water Environment Federation, Los Angeles, CA, October 11–15, 2003.

wetland treating landfill leachate, a double liner with leak detection may be required by some regulatory agencies.

The bottom surface must be level from side to side for the entire length of the wetland bed. Both types of wetlands may have a slight uniform slope to ensure drainage, but as described previously the bottom slope should not be designed to provide the necessary hydraulic conditions for flow in the system. The necessary hydraulic gradient and water level control in each wetland cell are provided by an adjustable outlet device. The bottom of the wetland, during the final grading operations, should be compacted to a degree similar to that used for highway subgrades. The purpose is to maintain the design surface during subsequent construction activities. Several constructed wetland systems, both SSF and FWS types, have been found with significant flow short-circuiting due to inadequate grade control during system construction. A particular concern for the SSF type is trucks delivering the gravel media. The ruts from just a few of these vehicles can induce permanent short-circuiting in the completed system. Construction traffic should not be permitted on the cell bottom during wet weather conditions.

The membrane liner, if used, is placed directly on the completed cell bottom. The SSF media can be placed directly on heavy-duty liner materials. In the case of FWS wetlands, a layer of reserved topsoil is placed on top of the liner to serve as the rooting medium for the vegetation.

The selection of SSF media type is critical to the successful performance of the system. Unwashed crushed stone has been used in a large number of existing projects. Truck delivery of such material during construction can lead to problems due to segregation of fines in the truck during transit and then deposition of all of the fine material in a single spot when the load is dumped. This can result in a number of small blockages in the flow path and internal short-circuiting in the system. Washed stone or gravel is preferred. Coarse aggregates for concrete construction are commonly available throughout the United States and would be suitable for construction of SSF wetland systems.

The dikes and berms for the wetland cells are constructed in the same manner as those for lagoons and similar water impoundments. For large-scale systems, the top of the berm should be wide enough for small trucks and maintenance equipment. Each cell in the system must have a ramp into the cell to permit access for maintenance vehicles.

7.12.1 VEGETATION ESTABLISHMENT

Establishing vegetation at an appropriate density is a critical requirement for construction of both types of wetland systems. Local plants are already adapted to the regional environment and are preferred, if available. Several commercial nurseries are also capable of providing the plant stock for large projects. Planting densities are discussed in Section 6.2; the closer the initial spacing, the sooner the system will be at full density. Most of the species will propagate from seed, and aerial seeding might be considered for large-scale projects. Plant development from seed takes significant time and requires very careful water control, and seed consumption by birds can be a problem. The quickest and most reliable approach is to transplant rhizomes of the vegetation of choice in the prepared treatment bed.

Each rhizome cutting should have at least one bud or preferably a growing shoot and is planted with one end about 2 in. (5 cm) below the surface of the medium with the bud or shoot exposed to the atmosphere, above the saturated media. Planting of seeds or rhizomes can occur in the spring after the last frost; rhizome material can also be planted in the fall. The bed is flooded and the water level maintained at the soil or media surface for at least 6 weeks or until significant new growth has developed and emerged. At this stage, the wetland can be placed in full operation as long as the water level is not above the tops of the new plant growth. If freshwater is used during the incubation period, the use of some supplemental fertilizer is desirable to accelerate plant growth.

The design of very large systems might consider planting the vegetation in parallel bands, with the long axis of the band perpendicular to the flow direction.

TABLE 7.12
Construction Costs for Subsurface Flow Constructed Wetlands

Location	Design Flow (gpd)	Area (ac)	Construction Cost ($/ac)
La Siesta, Hobbs, New Mexico	5000	0.11	198,900
Howe, Indiana	6000	0.14	221,700
McNeil, Arkansas	15,000	0.39	263,300
Santa Fe Opera, New Mexico	17,000	0.15	374,000
Phillips H.S. Bear Creek, Alabama	20,000	0.50	94,600
Carville, Louisiana	100,000	0.57	234,700
Benton, Louisiana	310,000	1.19	294,100
Mesquite, Nevada	400,000	4.8	130,800
Carlisle, Arkansas	860,000	1.09	379,500

Note: Costs updated to June 1998.

Source: Crites, R.W. and Ogden, M., in *Proceedings of WEFTEC 1998*, Water Environment Federation, Orlando, FL, October 3–7, 1998.

Each band would commence operation with relatively dense vegetation, and the spaces between bands can be filled in over the long term. If cost constraints are an issue, it is advantageous to put about 75% of the vegetation stock in the last half of the cell and 25% in the first half.

7.13 OPERATION AND MAINTENANCE

Constructed wetlands provide passive treatment and therefore require minimal operating labor. The issues requiring operational attention in SSF and vertical-flow wetlands are maintenance of inlet and outlet manifolds and monitoring of water quality. The vegetation, once established, requires very little attention, unless it is attacked by predatory animals. Water level control usually is maintained by the outlet device and may be modified seasonally.

7.14 COSTS

The cost elements for constructed wetlands are described in Chapter 6. Additionally, for SSF wetlands and vertical-flow wetlands bed media are a significant cost item. Construction costs for SSF wetlands are summarized in Table 7.12. The cost for media depends on local gravel costs and the cost for hauling. Three SSF systems with detailed media costs are presented Table 7.13.

TABLE 7.13
Costs of Media for Subsurface Flow Constructed Wetlands

Location and Gravel Size	Gravel Depth (ft)	Gravel Quantity (yd³/ac)	Cost ($/yd³)	1997 cost ($/ac)
Mesquite, Nevada				
3/8 to 1 in.	2.67	4308	8.40	43,800
Carville, Louisiana				
3/4-in. top layer	0.5	806	20.75	18,103
1/2- to 3-in. bed	2.0	3226	15.45	53,952
Ten Stones, Vermont				
3/8-in. top layer	0.5	806	19.17	15,451
3/4- to 1-in. bed	2.0	3226	9.18	29,615

Source: Adapted from WEF, *Natural Systems for Wastewater Treatment*, 2nd ed., Manual of Practice FD-16, Water Environment Federation, Alexandria, VA, 2001.

7.15 TROUBLESHOOTING

Troubleshooting in SSF and VF wetlands may be necessary to address:

- Hydraulic problems due to clogging
- Water quality problems due to metal sulfide precipitation

Historical problems with surfacing of water in SSF wetlands can usually be traced to inadequate design of head loss through the media and outlet devices being placed too high and without the ability to be lowered. These systems often have an excessively high aspect ratio, which exacerbates the problem by concentrating the applied solids in the first 10% of the bed length. If organic solids are the problem, either resting or drying of the bed or applications of hydrogen peroxide can be used. Hydrogen peroxide seemed to overcome organic clogging at Mesquite, New Mexico (Hanson et al., 2001). Metal sulfide problems, such as caused by anaerobic conditions at Minoa, New York, can be overcome by batch dosing and draining of the beds. Other water quality problems, such as inadequate ammonia removal, can be overcome by using nitrifying filter beds (Section 7.9) or converting to vertical-flow wetlands (Section 7.11).

REFERENCES

Askew, G.L., Hines, M.W., and Reed S.C. (1994). Constructed wetland and recirculating gravel filter system, full scale demonstration and testing, in *Proceedings of the Seventh International Symposium on Individual and Small Community Sewage Systems, Atlanta, GA*, American Society of Agricultural Engineers, St Joseph, MI, 84–91.

Austin, D., Lohan, E., and Verson, E. (2003). Nitrification and denitrification in a tidal vertical flow wetland pilot, in *Proceedings of WEFTEC 2003*, Water Environment Federation, Los Angeles, CA, October 11–15, 2003.

Bavor, H.J., Roser, D.J., Fisher, P.J., and Smalls, I.C. (1986). *Joint Study on Sewage Treatment Using Shallow Lagoon–Aquatic Plant Systems*, Water Research Laboratory, Hawkesbury Agricultural College, Richmond, NSW, Australia.

Baybrook, T. et al. (1998). Cold Climate Performance of Sub-Surface Flow Constructed Wetlands for Wastewater Treatment, paper presented at the Annual Technology Symposium, Water Environment Association of Ontario, Toronto.

Behrends, L.L., Sikora, F.J. Coonrod, H.S. Bailey, E., and McDonald, C. (1996). Reciprocating subsurface flow constructed wetlands for removing ammonia, nitrate, and chemical oxygen demand: potential for treating domestic, industrial and agricultural wastewater, *Proc. WEFTEC 1996*, 5, 251–263.

Behrends, L., Houke, L., Bailey, E., Jansen, P., and Brown D. (2000). Reciprocating constructed wetlands for treating industrial, municipal, and agricultural wastewater, in *Proceedings of the International Conference on Wetland Systems for Water Pollution Control*, Lake Buena Vista, FL, November 11–16, 2000.

Belmont, M.A. and Metcalfe, C.D. (2003). Feasibility of using ornamental plants in subsurface flow constructed wetlands to remove nitrogen, chemical oxygen demand and nonylphenol ethoxylate surfactants: a laboratory-scale study, *Ecol. Eng.*, 21, 233–247.

Brodie, G.A., Hammer, D.A., and Tomljanovich, D.A. (1989). Treatment of acid drainage with a constructed wetlands at the Tennessee Valley Authority 950 coal mine, in *Constructed Wetlands for Wastewater Treatment*, Hammer, D.A., Ed., Lew Publishers, Chelsea, MI, 201–209

Calkins, D. (pers. comm., 1995). U.S. Army Cold Regions Research and Engineering Laboratory (CRREL), Hanover, NH.

CSCE. (1986). *Cold Climate Utilities Manual*, 2nd ed., Canadian Society of Civil Engineers, Montreal, Quebec, Canada.

Chapman, A.J. (1974). *Heat Transfer*, 3rd ed., Macmillan, New York, 1974.

Crites, R.W. and Ogden, M. (1998). Costs of constructed wetlands systems, in *Proceedings of WEFTEC 1998*, Water Environment Federation, Orlando, FL, October 3–7, 1998.

Crites, R.W. and Tchobanoglous, G. (1998). *Small and Decentralized Wastewater Management Systems*, McGraw-Hill, New York.

Gersberg, R.M., Lyons, S.R., Elkins, B.V., and Goldman, C.R. (1984). The removal of heavy metals by artificial wetlands, in *Proceedings of Water Reuse Symposium III*, Vol. 2, American Water Works Association Research Foundation, Denver, CO, 639–648.

Gersberg, R.M., Elkins, B.V., Lyons, S.R., and Goldman, C.R. (1985). Role of aquatic plants in wastewater treatment by artificial wetlands, *Water Res.*, 20, 363–367.

Gersberg, R.M., Gearhart, R.A., and Ives, M. (1989). Pathogen removal in constructed wetlands, in *Constructed Wetlands for Wastewater Treatment: Municipal, Industrial and Agricultural*, Hammer, D.A., Ed., Lewis Publishers, Chelsea, MI, 431–445.

Hanson, A., Mimbela, L.E., Polka, R., and Zachritz, W. (2001). Unplugging the bed of a subsurface-flow wetlands using hydrogen peroxide, in *Wetlands and Remediation II: Proceedings of the Second International Conference on Wetlands and Remediation*, Nehring, K.W. and Brauning, S.E., Eds., Battelle, Columbus, OH.

He, Q. and Mankin, K.R. (2002). Performance variations of COD and nitrogen removal by vegetated submerged bed wetlands, *J. Am. Water Res. Assoc.*, 38(6), 1679–1689.

Hines, M. (pers. comm., 2004). Southeast Environmental Engineering, Concord, TN.

Karrh, J.D., Moriarty, J., Kornuc, J.J., and Knight, R.L. (2001). Sustainable management of aircraft anti/de-icing process effluents using a subsurface-flow treatment wetland, in *Wetlands and Remediation II: Proceedings of the Second International Conference on Wetlands and Remediation*, Nehring, K.W. and Brauning, S.E., Eds., Battelle, Columbus, OH.

Kucuk, O.S., Sengul, F., and Kapdan, I.K. (2003). Removal of ammonia from tannery effluents in a reed bed constructed wetland, *Water Sci. Technol.*, 48(11–12), 179–186.

Lawson, G.J. (1985). *Cultivating Reeds* (Phragmites australis) *for Root Zone Treatment of Sewage*, Project Report 965, Institute of Terrestrial Ecology, Cumbria, England, 64 pp.

Lekven, C.C., Crites, R.W., and Beggs, R.A. (1993). Subsurface flow wetlands at Mesquite, Nevada, in *Constructed Wetlands for Water Quality Improvement*, Moshiri, G.A., Ed., Lewis Publishers, Boca Raton, FL.

Monjeau, C. (1901). U.S. Patent Number 681,884, September 3, 1901.

Ogden, M., (pers. comm., 1994). Natural Systems, Inc. (NSI), Santa Fe, NM.

Reed, S.C. (1991). *Constructed Wetlands Characterization: Carville and Mandeville, Louisiana*, Risk Reduction Engineering Laboratory (RREL), U.S. Environmental Protection Agency, Cincinnati, OH.

Reed, S.C. (1993). *Constructed Wetlands Characterization: Hammond and Greenleaves, Louisiana*, Risk Reduction Engineering Laboratory (RREL), U.S. Environmental Protection Agency, Cincinnati, OH.

Reed, S.C. and Calkins, D. (1996). The thermal regime in a constructed wetland, in *Proceedings of the Fifth International Conference on Wetland Systems for Water Pollution Control* (preprint), International Association on Water Quality, London, I3–I7,

Reed, S.C. and Giarrusso, S. (1999). Sequencing operation provides aerobic conditions in a constructed wetland, in *Proceedings of WEFTEC 1999*, Water Environment Federation, New Orleans, LA, October 9–13, 1999.

Reed, S.C., Crites, R.W., and Middlebrooks, E.J. (1995). *Natural Systems for Waste Management and Treatment*, 2nd ed., McGraw-Hill, New York.

Reed, S.C., Hines, M., and Ogden, M. (2003). Improving ammonia removal in municipal constructed wetlands, in *Proceedings of WEFTEC 2003*, Water Environment Foundation, Los Angeles, CA.

Richter, K.M., Margetts, J.R., Saul, A.J., Guymer, I., and Worrall, P. (2003). Baseline hydraulic performance of the Heathrow constructed wetlands subsurface flow system, *Water Sci. Technol.*, 47(7–8), 177–181.

Rousseau, D.P.L., Vanrolleghem, P.A., and De Pauw, N. (2004). Model-based design of subsurface flow constructed treatment wetlands: a review, *Water Res.*, 38, 1484–1493.

Rozema, L. (pers. comm., 2004). Aqua Technologies, Ontario, Canada.

Seidel, K. (1996). Reinigung von Gerwassern durch hohere Pflanzen [German], *Deutsche Naturwiss.*, 12, 297–298.

Steiner, G.R. and Freeman, R.J. (1989). Configuration and substrate design considerations for constructed wetlands wastewater treatment, in *Constructed Wetlands for Wastewater Treatment: Municipal, Industrial and Agricultural*, Hammer, D., Ed., Lewis Publishers, Chelsea., MI.

Steiner, G.R. and Watson, J.T. (1993). *General Design, Construction, and Operational Guidelines Constructed Wetlands Wastewater Treatment Systems for Small Users Including Individual Residences*, 2nd ed., TVA/WM–93/10, Tennessee Valley Authority, Chattanooga, TN.

Sun, G., Gray, K.R., Biddlestone, A.J., and Cooper, D.J. (1999). Treatment of agricultural wastewater in a combined tidal flow–downflow reed bed system, *Water Sci. Technol.*, 40(3), 139–146.

USEPA. (1993). *Subsurface Flow Constructed Wetlands for Wastewater Treatment: A Technology Assessment*, EPA 832-R-93-008, Office of Wastewater Management, U.S. Environmental Protection Agency, Washington, D.C.

USEPA. (2003). *Constructed Wetlands Treatment of Municipal Wastewaters*, EPA/625/R-99/010, Office of Research and Development, U.S. Environmental Protection Agency, Cincinnati, OH.

Wallace, S.D. (2001). Treatment of cheese-processing waste using subsurface-flow wetlands, in *Wetlands and Remediation II: Proceedings of the Second International Conference on Wetlands and Remediation*, Nehring, K.W. and Brauning, S.E., Eds., Battelle, Columbus, OH.

Watson, J.T., Reed, S.C., Kadlec, R.H., Knight, R.L., and Whitehouse, A.E. (1989). Performance expectations and loading rates for constructed wetlands, in *Constructed Wetlands for Wastewater Treatment: Municipal, Industrial and Agricultural*, Hammer, D., Ed., Lewis Publishers, Chelsea MI, 319 351.

WEF. (2001). *Natural Systems for Wastewater Treatment*, 2nd ed., Manual of Practice FD-16, Water Environment Federation, Alexandria, VA.

Worrall, P., Revitt, D.M., Prickett, G., and Brewer, D. (2001). Constructed wetlands for airport runoff: the London Heathrow experience, in *Wetlands and Remediation II: Proceedings of the Second International Conference on Wetlands and Remediation*, Nehring, K.W. and Brauning, S.E., Eds., Battelle, Columbus, OH.

8 Land Treatment Systems

Land treatment systems include slow rate (SR), overland flow (OF), and soil aquifer treatment (SAT) or rapid infiltration (RI). In addition, the on-site soil absorption systems discussed in Chapter 10 utilize soil treatment mechanisms.

8.1 TYPES OF LAND TREATMENT SYSTEMS

The process of land treatment is the controlled application of wastewater to soil to achieve treatment of constituents in the wastewater. All three processes use the natural physical, chemical, and biological mechanisms within the soil–plant–water matrix. The SR and SAT processes use the soil matrix for treatment after infiltration of the wastewater, the major difference between the processes being the rate at which the wastewater is loaded onto the site. The OF process uses the soil surface and vegetation for treatment, with limited percolation, and the treated effluent is collected as surface runoff at the bottom of the slope. The characteristics of these systems are compared in Table 8.1 and the treatment performance expectations were summarized in Table 1.3 in Chapter 1.

8.1.1 SLOW-RATE SYSTEMS

The slow rate process is the oldest and most widely used land treatment technology. The process evolved from "sewage farming" in Europe in the sixteenth century to a recognized wastewater treatment system in England in the 1860s (Jewell and Seabrook, 1979). By the 1880s, the United States had a number of slow-rate systems. In a survey of 143 wastewater facilities in 1899, slow rate land treatments systems were the most frequently used form of treatment (Rafter, 1899). Slow rate land treatment was rediscovered at Penn State in the mid-1960s (Sopper and Kardos, 1973). By the 1970s, both the U.S. Environmental Protection Agency (USEPA) and the U.S. Corps of Engineers had invested in land treatment research and development (Pound and Crites, 1973; Reed, 1972). By the late 1970s, a number of long-term effects studies on slow-rate systems had been conducted (Reed and Crites, 1984). A list of selected municipal slow-rate systems is presented in Table 8.2. A large SR system at Dalton, Georgia, occupies 4605 acres of sprinkler irrigated forest, as shown in Figure 8.1 (Crites et al., 2001).

8.1.2 OVERLAND FLOW SYSTEMS

The overland flow process was developed to take advantage of slowly permeable soils such as clays. Treatment occurs in OF systems as wastewater flows down vegetated, graded-smooth, gentle slopes that range from 2 to 8% in grade. A

TABLE 8.1
Characteristics of Land Treatment Systems

Characteristic	Slow Rate (SR)	Overland Flow (OF)	Soil Aquifer Treatment (RI)
Application method	Sprinkler or surface	Sprinkler or surface	Usually surface
Preapplication treatment	Ponds or secondary	Fine screening or primary	Ponds or secondary
Annual loading (ft/yr)	2–18	10–70	18–360
Field area (ac/mgd)	60–560	16–112	3–60
Use of vegetation	Nutrient uptake and crop revenue	Erosion control and habitat for microorganisms	Usually not used
Disposition of applied wastewater	Evapotranspiration and percolation	Surface runoff, evapotranspiration, some percolation	Percolation, some evaporation

TABLE 8.2
Selected Municipal Slow-Rate Land Treatment Systems

Location	Flow (mgd)	System Area (ac)	Application Method
Bakersfield, California	19.4	5088	Surface irrigation
Clayton County, Georgia	20.0	2370	Solid-set sprinklers
Dalton, Georgia	33.0	4605	Solid-set sprinklers
Lubbock, Texas	16.5	4940	Center-pivot sprinklers
Mitchell, South Dakota	2.45	1284	Center-pivot sprinklers
Muskegon County, Michigan	29.2	5335	Center-pivot sprinklers
Petaluma, California	5.3	555	Hand-move, solid-set sprinklers
Santa Rosa, California	20.0	6362	Solid-set sprinklers

schematic showing both surface application and sprinkler application is presented in Figure 8.2. The treated runoff is collected at the bottom of the slope. The process was pioneered in the United States by the Campbell Soup Company, first at Napoleon, Ohio, in 1954 and subsequently at Paris, Texas (Gilde et al., 1971). Research was conducted on the OF process using municipal wastewater at Ada, Oklahoma (Thomas et al., 1974) and at Utica, Mississippi (Carlson et al., 1974). As a result of this and other research (Martel, 1982; Smith and Schroeder, 1985), over 50 municipal OF systems have been constructed for municipal wastewater treatment. A list of selected municipal overland flow systems is presented in Table 8.3.

FIGURE 8.1 Typical sprinkler irrigation system at the forested slow rate site at Dalton, Georgia.

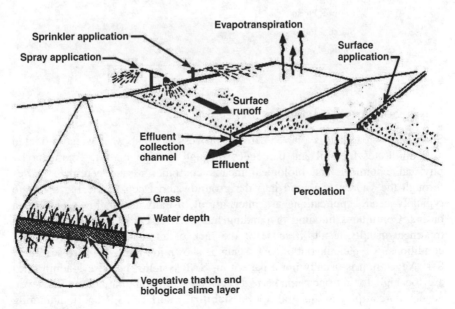

FIGURE 8.2 Overland flow process.

TABLE 8.3
Municipal and Industrial Overland Flow Systems in the United States

Municipal Systems	Industrial Systems
Alma, Arkansas	Chestertown, Maryland
Alum Creek Lake, Ohio	El Paso, Texas
Beltsville, Maryland	Middlebury, Indiana
Carbondale, Illinois	Napoleon, Ohio
Cleveland, Michigan	Paris, Texas
Corsicana, Texas	Rosenberg, Texas
Davis, California	Woodbury, Georgia
Falkner, Michigan	
Gretna, Virginia	
Heavener, Oklahoma	
Kenbridge, Virginia	
Lamar, Arkansas	
Minden-Gardnerville, Nevada	
Mt. Olive, New Jersey	
Newman, California	
Norwalk, Iowa	
Raiford, Florida	
Starke, Florida	
Vinton, Louisiana	

8.1.3 SOIL AQUIFER TREATMENT SYSTEMS

Soil aquifer treatment is a land treatment process in which wastewater is treated as it infiltrates the soil and percolates through the soil matrix. Treatment by physical, chemical, and biological means continues as the percolate passes through the vadose zone and into the groundwater. Deep permeable soils are typically used. Applications are intermittent, usually to shallow percolation basins. Continuous flooding or ponding has been practiced, but less complete treatment usually results because of the lack of alternate oxidation/reduction conditions. A typical layout of SAT basins is shown in Figure 8.3 (also see Table 8.4). Vegetation is usually not a part of an SAT systems, because loading rates are too high for nitrogen uptake to be effective. In some situations, however, vegetation can play an integral role in stabilizing surface soils and maintaining high infiltration rates (Reed et al., 1985).

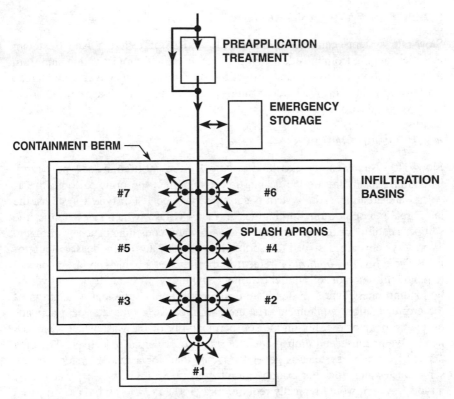

FIGURE 8.3 Typical layout of soil aquifer treatment basins.

TABLE 8.4
Selected Soil Aquifer Treatment Systems

Location	Hydraulic Loading(ft/yr)
Brookings, South Dakota	40
Calumet, Michigan	115
Darlington, South Carolina	92
Fresno, California	44
Hollister, California	50
Lake George, New York	190
Los Angeles County Sanitary District, California	330
Orange County, Florida	390
Tucson, Arizona	331
West Yellowstone, Montana	550

8.2 SLOW-RATE LAND TREATMENT

Slow-rate systems can encompass a wide variety of different land treatment facilities ranging from hillside spray irrigation to agricultural irrigation, and from forest irrigation to golf course irrigation. The design objectives can include wastewater treatment, water reuse, nutrient recycling, open space preservation, and crop production.

8.2.1 DESIGN OBJECTIVES

Slow-rate systems can be classified as type 1 (slow infiltration) or type 2 (crop irrigation), depending on the design objective. When the principal objective is wastewater treatment, the system is classified as type 1. For type 1 systems, the land area is based on the limiting design factor (LDF), which can be either the soil permeability or the loading rate of a wastewater constituent such as nitrogen. Type 1 systems are designed to use the most wastewater on the least amount of land. The term *slow infiltration* refers to type 1 systems being similar in concept to rapid infiltration or soil aquifer treatment but having substantially lower hydraulic loading rates. Type 2 systems are designed to apply sufficient water to meet the crop irrigation requirement. The area required for a type 2 system depends on the crop water use, not on the soil permeability or the wastewater treatment needs. Water reuse and crop production are the principal objectives. The area needed for type 2 systems is generally larger than for a type 1 system for the same wastewater flow. For example, for 1 mgd (3785 m³/d) of wastewater flow, a type 1 system would typically require 60 to 150 ac (24 to 60 ha) as compared to the 200 to 500 ac (80 to 200 ha) for a type 2 system.

8.2.1.1 Management Alternatives

Unlike SAT and overland flow, slow-rate systems can be managed in several different ways. The other two land treatment systems require that the land be purchased and the system managed by the wastewater agency. For slow-rate systems, the three major options are (1) purchase and management of the site by the wastewater agency, (2) purchase of the land and leasing it back to a farmer, and (3) contracts between the wastewater agency and farmers for use of private land for the slow rate process. The latter two options allow farmers to manage the slow rate process and harvest the crop. A representative list of small SR systems that use each of the different management alternatives is presented in Table 8.5.

8.2.2 PREAPPLICATION TREATMENT

Preliminary treatment for an SR system can be provided for a variety of reasons including public health protection, nuisance control, distribution system protection, or soil and crop considerations. For type 1 systems, preliminary treatment, except for solids removal, is de-emphasized because the SR process can usually

TABLE 8.5
Management Alternatives Used in Selected Slow-Rate Systems

Purchase and Management by Agency	Flow (mgd)	Agency Purchase and Lease to Farmer	Flow (mgd)	Farmer Contract	Flow (mgd)
Dinuba, California	1.5	Coleman, Texas	0.4	Camarillo, California	3.8
Fremont, Michigan	0.3	Kerman, California	0.5	Dickinson, North Dakota	1.5
Kennett Square, Pennsylvania	0.05	Lakeport, California	0.5	Mitchell, South Dakota	2.4
Lake of the Pines, California	0.6	Modesto, California	20.0	Quincy, California	0.75
Oakhurst, California	0.25	Perris, California	0.8	Petaluma, California	4.2
West Dover, Vermont	1.6	Winter, Texas	0.5	Sonoma Valley, California	2.7
Wolfeboro, New Hampshire	0.3	Santa Rosa, California	15.0	Sonora, California	1.2

Source: Adapted from Crites, R.W. and Tchobanoglous, G., Small and Decentralized Wastewater Management Systems, McGraw-Hill, New York, 1998.

TABLE 8.6
Pretreatment Guidelines for Slow-Rate Systems

Level of Pretreatment	Acceptable Conditions
Primary treatment	Acceptable for isolated locations with restricted public access
Biological treatment by lagoons or in-plant processes, plus control of fecal coliform count to less than 1000 MPN per 100 mL	Acceptable for controlled agricultural irrigation, except for human food crops to be eaten raw
Biological treatment by lagoons or in-plant processes, with additional BOD or SS control as needed for aesthetics, plus disinfection to log mean of 200 MPN per 100 mL (USEPA fecal coliform criteria for bathing waters)	Acceptable for application in public access areas such as parks and golf courses

Note: MPN, most probable number; BOD, biological oxygen demand; SS, suspended solids.

Source: USEPA, *Process Design Manual for Land Treatment of Municipal Wastewater*, EPA 625/1-81-013, U.S. Environmental Protection Agency, Cincinnati, OH, 1981.

achieve final water quality objectives with minimal pretreatment. Public health and nuisance control guidelines for type 1 SR systems have been issued by the EPA (USEPA, 1981) and are given in Table 8.6. Type 2 systems are designed to emphasize reuse potential and require greater flexibility in the handling of wastewater. To achieve this flexibility, preliminary treatment levels are usually higher. In many cases, type 2 systems are designed for regulatory compliance following preliminary treatment so irrigation can be accomplished by other parties such as private farmers.

8.2.2.1 Distribution System Constraints

Preliminary treatment is generally required to prevent problems of capacity reduction, plugging, and localized generation of odors in the distribution system. For this reason, a minimum primary treatment (or its equivalent) is recommended for all SR systems to remove settleable solids and oil and grease. For sprinkler systems, it is further recommended that the size of the largest particle in the applied wastewater be less than one third the diameter of the sprinkler nozzle to avoid plugging.

8.2.2.2 Water Quality Considerations

The total dissolved solids (TDS) in the applied wastewater can affect plant growth, soil characteristics, and groundwater quality. Guidelines for interpretation of water quality for salinity and other specific constituents for SR systems are presented in Table 8.7. The term "restriction on use" does not indicate that the

TABLE 8.7
Guidelines for Interpretation of Water Quality

Problem and Related Constituent	No Restriction	Slight to Moderate Restriction	Severe Restriction	Crops Affected
Salinity as TDS (mg/L)	<450	450–2000	>2000	Crops in arid areas affected by high TDS; impacts vary
Permeability:				
SAR = 0–3	TDS >450	130–450	<130	All crops
SAR = 3–6	TDS >770	200–770	<200	
SAR = 6–12	TDS >1200	320–1200	<320	
SAR = 12–20	TDS >1860	800–1860	<800	
SAR = 20–40	TDS >3200	1860–3200	<1860	
Specific ion toxicity:				
Sodium (mg/L)	<70	>70	>70	Tree crops and woody
Chloride (mg/L)	<140	140–350	>350	ornamentals; fruit
Boron (mg/L)	<0.7	0.7–3.0	>3.0	trees and some field
Residual chlorine (mg/L)	<1.0	1.0–5.0	>5.0	crops; ornamental, only if overhead sprinklers are used

Note: TDS, total dissolved solids; SAR, sodium adsorption ratio.

Source: Ayers, R.S. and Westcot, D.W., *Water Quality for Agriculture*, FAO Irrigation and Drainage Paper 29, Revision 1, Food and Agriculture Organization of the United Nations, Rome, 1985.

effluent is unsuitable for use; rather, it means there may be a limitation on the choice of crop or need for special management. Sodium can adversely affect the permeability of soil by causing clay particles to disperse. The potential impact is measured by the sodium adsorption ratio (SAR) which is a ratio of sodium concentration to the combination of calcium and magnesium. The SAR is defined in Equation 8.1.

$$SAR = \frac{Na}{\sqrt{\left(\frac{Ca + Mg}{2}\right)}} \tag{8.1}$$

where
SAR　=　Sodium adsorption ratio (unitless).
Na　　=　Sodium concentration (mEq/L; mg/L divided by 23).
Ca　　=　Calcium concentration (mEq/L; mg/L divided by 20).
Mg　　=　Magnesium concentration (mEq/L; mg/L divided by 12.15).

In type 2 SR systems the leaching requirement must be determined based on the salinity of the applied water and the tolerance of the crop to soil salinity. Leaching requirements range from 10 to 40% with typical values being 15 to 25%. Specific crop requirements for soil–water salinity must be used to determine the required leaching requirement (Reed and Crites, 1984; Reed et al., 1995).

8.2.2.3 Groundwater Protection

Most SR systems with secondary preapplication treatment are protective of the receiving groundwater. The concern over emerging chemical constituents, such as endocrine disruptors and pharmaceutical chemicals, has led to research on the ability of the soil profile to remove these trace organic compounds (Muirhead et al., 2003).

8.2.3 DESIGN PROCEDURE

A flowchart of the design procedure for slow-rate systems is presented in Figure 8.4. The procedure is divided into a preliminary and final design phase. Determinations made during the preliminary design phase include: (1) crop selection, (2) preliminary treatment, (3) distribution system, (4) hydraulic loading rate, (5) field area, (6) storage needs, and (7) total land requirement. When the preliminary design phase is completed, economic comparisons can be made with other wastewater management alternations. The text will focus on preliminary or process design with references to detailed design procedures (Hart, 1975; Pair, 1983; USDA, 1983; USEPA, 1981).

8.2.4 CROP SELECTION

The selection of the type of crop in a slow-rate system can affect the level of preliminary treatment, the selection of the type of distribution system, and the hydraulic loading rate. The designer should consider economics, growing season, soil and slope characteristics, and wastewater characteristics in selecting the type of crop. Forage crops or tree crops are usually selected for type 1 systems, and higher value crops or landscape vegetation are often used in type 2 systems.

8.2.4.1 Type 1 System Crops

In type 1 SR systems, the crop must be compatible with high hydraulic loading rates, have a high nutrient uptake capacity, a high consumptive use of water, and a high tolerance to moist soil conditions. Other characteristics of value are tolerance to wastewater constituents (such as TDS, chloride, boron) and limited requirements for crop management. The nitrogen uptake rate is a major design variable for design of a type 1 system. Typical nitrogen uptake rates for forage, field, and tree crops are presented in Table 8.8. The largest nitrogen removal can be achieved with perennial grasses and legumes. Legumes, such as alfalfa, can

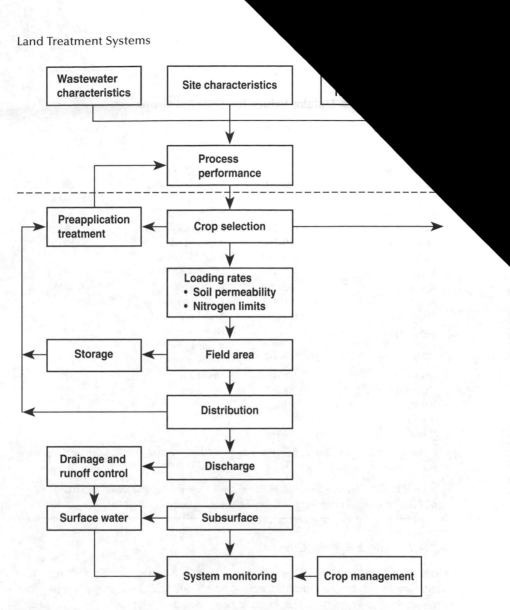

FIGURE 8.4 Flowchart of the design procedure for slow rate land treatment.

fix nitrogen from the air; however, they will preferentially take nitrate from the soil solution if it is provided. The use of legumes (clovers, alfalfa, vetch) in type 1 systems should be limited to well-draining soils because legumes generally do not tolerate high soil moisture conditions. The most common tree crops for type 1 systems are mixed hardwoods and pines (Nutter et al., 1986). Tree crops provide revenue potential as firewood, pulp, or biomass fuel. Tree species with high growth response such as eucalyptus and hybrid poplars will maximize nitrogen uptake.

Water quality requirements

...elected Crops

Crop	Nitrogen Uptake (lb/ac·yr)	Crop	Nitrogen Uptake (lb/ac·yr)
		Eastern forest	
		Mixed hardwoods	200
		Red pine	100
		White spruce	200
		Pioneer succession	200
		Aspen sprouts	100
		Southern forest	
		Mixed hardwoods	250
Ryegrass	160–250	Loblolly pine	200–250
Sweet clover[a]	155	*Lake states forest*	
Tall fescue	130–290	Mixed hardwoods	100
Field crops		Hybrid poplar	140
Barley	110	*Western forests*	
Corn	155–180	Hybrid poplar	270
Cotton	65–100	Douglas fir	200
Grain sorghum	120		
Potatoes	200		
Soybeans[a]	220		
Wheat	140		

[a] Legume crops can fix nitrogen from the air but will take up most of their nitrogen from applied wastewater nitrogen.

Source: USEPA, *Process Design Manual for Land Treatment of Municipal Wastewater*, EPA 625/1-81-013, U.S. Environmental Protection Agency, Cincinnati, OH, 1981.

8.2.4.2 Type 2 System Crops

Crop irrigation or water reuse systems can use a broad variety of crops and landscape vegetation including trees, grass, field, and food crops. Field crops often include corn, cotton, sorghum, barley, oats, and wheat.

8.2.5 HYDRAULIC LOADING RATES

Hydraulic loading rates for SR systems are expressed in units of in./wk (mm/wk) or ft/yr (m/yr). The basis of determination varies from type 1 to type 2.

8.2.5.1 Hydraulic Loading for Type 1 Slow-Rate Systems

The hydraulic loading rate for a type 1 system is determined by using the water balance equation:

$$L_w = ET - Pr + P \tag{8.2}$$

where

L_w = Wastewater hydraulic loading rate (in./mo; mm/mo).

ET = Evapotranspiration rate (in./mo; mm/mo).

Pr = Precipitation rate (in./mo; mm/mo).

P = Percolation rate (in./mo; mm/mo).

The water balance is generally used on a monthly basis. The design values for precipitation and evapotranspiration are generally chosen for the wettest year in 10, to be conservative. For slow-rate systems, the surface runoff (tailwater) is usually captured and reapplied. An exception is the forested type 1 system, where surface and subsurface seepage is allowed by the regulatory agency. Seepage (the surfacing of groundwater) may occur on or off the site without causing water quality problems.

The design percolation rate is based on the permeability of the limiting layer in the soil profile. For type 1 systems, the permeability is often measured in the field using cylinder infiltrometers, sprinkler infiltrometers, or the basin flooding technique. The range of soil permeability is usually contained in the detailed soil survey from Natural Resources Conservation Service (NRCS). Although the given range is often wide (0.2 to 0.6 in./hr; 5 to 15 mm/hr), the lower value is often used in preliminary planning. The design percolation rate is calculated from the soil permeability taking into account the variability of the soil conditions and the overall cycle of wetting (application) and drying (resting) of the site:

$$P \text{ (daily)} = K(0.04 \text{ to } 0.10)(24 \text{ hr/d}) \qquad (8.3)$$

where

P = Design percolation rate (in./d; mm/d).

K = Permeability of limiting soil layer (in./hr; mm/hr).

0.04 to 0.10 = Adjustment factor to account for the resting period between applications and the variability of the soil conditions.

Using either NRCS permeability data or field test results, it is recommended that the daily design percolation rate should range from 4 to 10% of the total rate. Selection of the adjustment factor depends on the site and the degree of conservativeness desired. For most SR systems, the wetting period is 5 to 15% of a given month. If the soil is only wet for 5% of the time, then only that percent of the time (in a given month) should be used as percolation time. The 4% factor should be used when the soil type variation is large, when the wet/dry ratio is small (5% or less), and the soil permeability is less than 0.2 in./hr (5 mm/hr). The high percentages, up to 10%, can be used where soil permeabilities are higher, the soil permeability is more uniform, and the wet/dry ratio is higher than 7%.

8.2.5.2 Hydraulic Loading for Type 2 Slow-Rate Systems

For crop irrigation systems, the hydraulic loading rate is based on the crop irrigation requirements. The loading rate can be calculated using Equation 8.4:

$$L_w = \left(\frac{ET - Pr}{1 + LR} \right) \left(\frac{100}{E} \right) \tag{8.4}$$

where

L_w = Wastewater hydraulic loading rate (in./yr; mm/yr).
ET = Crop evapotranspiration rate (in./yr; mm/yr).
Pr = Precipitation rate (in./yr; mm/yr).
LR = Leaching requirement (fraction).
E = Irrigation efficiency (percent).

The leaching requirement depends on the crop, the total dissolved solids (TDS) of the wastewater, and the amount of precipitation. The leaching requirement is typically 0.10 to 0.15 for low TDS wastewater and a tolerant crop such as grass. For higher TDS wastewater (750 mg/L or more), the leaching requirement can range from 0.20 to 0.30. The irrigation efficiency is the fraction of the applied wastewater that corresponds to the crop evapotranspiration. The higher the efficiency, the less water that percolates through the root zone. Sprinkler systems usually have efficiencies of 70 to 80%, while surface irrigation systems usually have efficiencies of 65 to 75%.

8.2.6 DESIGN CONSIDERATIONS

Design considerations for both types of SR systems are described in the following text. Considerations for nitrogen loading, organic loading, land requirements, storage requirements, distribution systems, application cycles, surface runoff control, and underdrainage are presented.

8.2.6.1 Nitrogen Loading Rate

The limiting design factor (LDF) for many SR systems is the nitrogen loading rate. The total nitrogen loading (nitrate nitrogen, ammonia nitrogen, and organic nitrogen) is important because the soil microorganisms will convert organic nitrogen to the plant-available inorganic forms. Limitations on the total nitrogen loading rate are based on meeting a maximum nitrate nitrogen concentration of 10 mg/L in the receiving groundwater at the boundary of the project (usually 20 to 100 ft or 6 to 30 m downgradient of the wetted field area). To make certain that the groundwater nitrate nitrogen concentration limit is met, the usual practice is to set the percolate nitrate nitrogen concentration at 10 mg/L prior to commingling of the percolate with the receiving groundwater.

The nitrogen loading rate must be balanced against crop uptake of nitrogen, denitrification, and the leakage of nitrogen with the percolate. The nitrogen balance is given in Equation 8.5:

TABLE 8.9
Denitrification Loss Factor for Slow-Rate Systems

Type of Wastewater	Carbon/Nitrogen Ratio	Warm Climate f Factor	Cold Climate f Factor
High-strength wastewater	>20	0.8	0.5
Moderate-strength industrial wastewater	8–20	0.5	0.4
Primary effluent	3–5	0.4	0.25
Secondary effluent	1–1.5	0.25	0.2
Tertiary effluent	<1	0.15	0.1

Source: Adapted from Crites, R.W. and Tchobanoglous, G., *Small and Decentralized Wastewater Management Systems*, McGraw-Hill, New York, 1998.

$$L_n = U + fL_n + AC_p P \qquad (8.5)$$

where

L_n = Nitrogen loading rate (lb/ac·yr; kg/ha·yr).

U = Crop uptake of nitrogen (lb/ac·yr; kg/ha·yr).

f = Fraction of applied nitrogen lost to nitrification/denitrification, volatilization, and soil storage (see Table 8.9).

A = Conversion factor (0.23; 10.0).

C_p = Concentration of nitrogen in percolate (mg/L).

P = Percolate flow (in./yr; m/yr).

By combining the nitrogen balance and water balance equation, the hydraulic loading rate that will meet the nitrogen limits can be calculated using Equation 8.6:

$$L_{wn} = \frac{C_p(Pr - ET) + 4.4U}{C_w(1 - f) - C_p} \qquad (8.6)$$

where L_{wn} is the hydraulic loading rate controlled by nitrogen (in./yr; m/yr), C_w is the concentration of nitrogen in the applied wastewater (mg/L), and the other terms are as defined previously.

Crop uptake of nitrogen can be estimated from Table 8.8. The fraction of applied nitrogen that is lost to denitrification, volatilization, and soil storage depends on the wastewater characteristics and the temperature. The fraction will be highest for warm climates and high-strength wastewaters with carbon-to-nitrogen ratios of 20 or more (see Table 8.9).

TABLE 8.10
BOD Loading Rates at Industrial Slow-Rate Systems

Location	Industry	BOD Loading Rate, Cycle Average (lb/ac·d)
Almaden Winery; McFarland, California	Winery stillage	420
Anheuser-Busch; Houston, Texas	Brewery	360
Bronco Wine; Ceres, California	Winery	128
Citrus Hill; Frostproof, Florida	Citrus	448
Contadina; Hanford, California	Tomato processing	84–92
Frito-Lay; Bakersfield, California	Potato processing	84
Harter Packing; Yuba City, California	Tomato and peach processing	150–351
Hilmar Cheese; Hilmar, California	Cheese processing	420
Ore-Ida Foods; Plover, Wisconsin	Potato processing	190
SK Foods; Lemoore, California	Tomato processing	210
TRI Valley Growers; Modesto, California	Tomato processing	200

Source: Data from Crites and Tchobanoglous (1998) and Smith and Murray (2003).

8.2.6.2   Organic Loading Rate

Organic loading rates do not limit municipal SR systems but may be important for industrial SR systems. Loading rates for biological oxygen demand (BOD) often exceed 100 lb/ac·d (110 kg/ha·d) and occasionally exceed 300 lb/ac·d (330 kg/ha·d) for SR systems applying screened food processing and other high-strength wastewater. A list of industrial SR systems with organic loading rates in the above range is presented in Table 8.10. Odor problems have been avoided in these systems by providing adequate drying times between wastewater applications. Organic loading rates beyond 450 lb/ac·d (500 kg/ha·d) of BOD should generally be avoided unless special management practices are used (Reed et al., 1995). Procedures for managing organic loadings from high-strength industrial wastewater are presented in Section 8.6.

8.2.6.3   Land Requirements

The land requirements for a slow-rate system include the field area for application, plus land for roads, buffer zones, storage ponds, and preapplication treatment. The area can be calculated using Equation 8.7:

$$A = \frac{Q}{0.027L_w} \qquad (8.7)$$

where

A = Field area (ac; ha).

Q = Annual flow (Mgal/yr; m³/yr).

0.027 = Conversion constant.

L_w = Design hydraulic loading rate (in./yr; mm/yr).

The design hydraulic loading rate can be based on soil permeability, crop irrigation requirements, or nitrogen loading rate. Modification to the land requirement based on storage is discussed in the section on storage.

Example 8.1. Land Area for a Slow-Rate System

Calculate the land requirements for a type 1 slow-rate system for a community of 1000 persons. The climate is moderately warm, and the design wastewater flow rate is 65,000 gal/d. A partially mixed aerated lagoon produces an effluent with 50 mg/L BOD and 30 mg/L total nitrogen. A site has been located that has relatively uniform soil with a limiting soil permeability (K) of 0.2 in./hr. The selected mix of forage grasses will take up 300 lb/ac·yr of nitrogen. The water balance of evapotranspiration and precipitation shows a net evapotranspiration of 18 in./yr.

Solution

1. Calculate the design percolation rate using Equation 8.3. Use a 7% factor to account for relatively uniform soil and moderate permeability:

   P (annual) $= K(0.07)(24 \text{ hr/d})(365 \text{ d/yr})$

   $P = 613(0.2)$

   $P = 123 \text{ in./yr}$

2. Calculate the wastewater loading rate L_w using Equation 8.2.

   $L_w = (ET - Pr) + P$

   $L_w = (\text{Net } ET) + P$

   $L_w = 18 \text{ in.} + 123 \text{ in.}$

   $L_w = 141 \text{ in./yr}$

3. Calculate the field area based on soil permeability limits using Equation 8.7:

   $$Q = \frac{65,000 \text{ gal}}{d} \times \frac{365 \text{ d}}{yr} \times \frac{1 \text{ Mgal}}{10^6 \text{ gal}} = \frac{23.7 \text{ Mgal}}{yr}$$

   $$A = \frac{Q}{0.027 \times L} = \frac{23.7 \text{ Mgal / yr}}{0.027 \times 141 \text{ in./ yr}} = 6.2 \text{ ac}$$

4. Calculate the hydraulic loading rate controlled by nitrogen, using a percolate nitrate nitrogen limit of 10 mg/L (Equation 8.6). Use a denitrification percentage of 25%:

$$L_{wn} = \frac{C_p(Pr - ET) + 4.4U}{C_w(1-f) - C_p} = \frac{10(-18) + 4.4(300)}{30(1-0.25) - 10} = 91.2 \text{ in./yr}$$

5. Calculate the field area based on nitrogen limits using Equation 8.7:

$$A = \frac{Q}{0.027 \times L_{wn}} = \frac{23.7 \text{ Mgal/yr}}{0.027 \times 91.2 \text{ in./yr}} = 9.6 \text{ ac}$$

6. Calculate the organic loading rate, assuming that 9.6 ac based on the nitrogen limits will be the required field area:

$$\text{BOD in wastewater} = \frac{0.065 \text{ Mgal}}{d} \times \frac{50 \text{ mg}}{L} \times 8.34 = 27.1 \text{ lb/d}$$

$$\text{BOD loading} = \frac{27.1 \text{ lb/d}}{9.6 \text{ ac}} = 2.8 \text{ lb/ac·d}$$

Therefore, the BOD loading is not limiting because it is much less than 450 lb/ac·d.

7. Determine the field area required. Because the area for nitrogen limits (9.6 ac) is larger than the area required for soil permeability (6.2 ac), the required field area is 9.6 ac.

Comment

Nitrogen is the limiting design factor for this example.

8.2.6.4 Storage Requirements

Wastewater is usually stored during periods when it is too wet or too cold to apply to the fields. Except for forested sites, where year-round application is possible, most systems also store wastewater during crop harvesting, planting, or cultivation. Storage for cold weather is generally required by most state regulatory agencies unless it can be shown that groundwater quality standards will not be violated by winter applications and that surface runoff will not occur as a result of wastewater application. The conservative estimate of the storage period is to equate it to the nongrowing season for the crop selected. A more exact site-specific method is to use the water balance as shown in Example 8.2.

Example 8.2. Storage Requirements for a Slow-Rate System

Estimate the storage requirements for the SR system from Example 8.1 using the water balance approach. The monthly precipitation and evapotranspiration data are presented in the following table. The temperatures are too cold in January for wastewater application. The maximum percolation rate is 10.3 in./month.

Month (1)	Crop Evapotranspiration[a] (2)	10-Year Rainfall[b] (3)
January	1.1	7.2
February	2.0	7.0
March	2.7	4.5
April	3.9	3.0
May	5.6	0.4
June	7.0	0.1
July	8.6	0.1
August	7.4	0.2
September	5.9	0.6
October	3.7	1.2
November	2.0	4.0
December	1.2	4.8
Total	51.1	33.1

[a] Forage crop evapotranspiration.
[b] Average distribution of rainfall for the wettest year in 10.

Solution

1. Determine the available wastewater each month:

$$\text{Available wastewater (in./mo)} = \frac{\text{Monthly flow}}{\text{Area}}$$

$$= \left(\frac{65{,}000 \text{ gal}}{d}\right)\left(\frac{365 \text{ d}}{yr}\right)\left(\frac{yr}{12 \text{ mo}}\right)$$

$$\times \left(\frac{1}{9.6 \text{ ac}}\right)\left(\frac{ac}{43{,}560 \text{ ft}^2}\right)$$

$$\times \left(\frac{ft^3}{7.48 \text{ gal}}\right)\left(\frac{12 \text{ in.}}{ft}\right)$$

$$= 7.6 \text{ in./mo}$$

2. Complete the water balance table as shown on the next page:

Month (1)	Crop Evapotranspiration[a] (2)	10-Year Rainfall[b] (3)	Design Percolation[c] (4)	Wastewater Loading[d] (5)	Available Wastewater[e] (6)	Change in Storage[f] (7)	Cumulative Storage (8)
January	1.1	7.2	6.1	0.0	7.6	+7.6	8.5
February	2.0	7.0	10.3	5.3	7.6	+2.3	10.8[g]
March	2.7	4.5	10.3	8.5	7.6	-0.9	9.9
April	3.9	3.0	8.1	9.0	7.6	-1.4	8.5
May	5.6	0.4	3.7	8.9	7.6	-1.3	7.2
June	7.0	0.1	2.0	8.9	7.6	-1.3	5.9
July	8.6	0.1	0.4	8.9	7.6	-1.3	4.6
August	7.4	0.2	1.7	8.9	7.6	-1.3	3.3
September	5.9	0.6	3.6	8.9	7.6	-1.3	2.0
October	3.7	1.2	6.4	8.9	7.6	-1.3	0.7
November	2.0	4.0	10.3	8.3	7.6	-0.7	0.0
December	1.2	4.8	10.3	6.7	7.6	+0.9	0.9
Total	51.1	33.1	73.2	91.2	91.2	—	—

[a] Forage crop evapotranspiration.
[b] Average distribution of rainfall for the wettest year in 10.
[c] Maximum percolation rate is 10.3 in./mo.
[d] Loading rate is limited by percolation rate from November through March (January has zero loading due to cold weather); loading rate for April through October is limited by the annual nitrogen loading.
[e] Based on 65,000 gal/d and a field area of 9.6 ac.
[f] Available wastewater minus the wastewater loading.
[g] February is the maximum month.

3. The design percolation rate (column 4 data) is 10.3 in./mo when that much rainfall or wastewater is applied. From April to October, the wastewater loading is limited by the nitrogen loading, and the design percolation rate is the difference between the wastewater loading (column 5) and the net evapotranspiration (evapotranspiration – precipitation) (column 2 minus column 3).

4. The wastewater loading is limited by the nitrogen balance from April through October; by the precipitation and percolation rates for November, December, February, and March; and by the cold weather in January.

5. Determine the change in storage by subtracting the wastewater loading (column 5) from the available wastewater (column 6). Enter the amount in column 7.

6. The cumulative storage calculations (column 8) begin with the first positive month for storage in the fall/winter (December). The maximum month for storage is February, with a value of 10.8 in. This depth is converted to million gallons as follows:

$$\text{Storage volume} = (10.8 \text{ in.})(9.6 \text{ ac})\left(\frac{\text{ft}}{12 \text{ in.}}\right)\left(\frac{43,560 \text{ ft}^2}{\text{ac}}\right)$$

$$\times \left(\frac{7.48 \text{ gal}}{\text{ft}^3}\right)\left(\frac{\text{Mgal}}{10^6 \text{ gal}}\right)$$

$$= 2.82 \text{ Mgal}$$

7. Convert the required storage volume into equivalent days of flow:

$$\text{Days of storage} = \frac{\text{Volume (Mgal)}}{\text{Flow (mgd)}} = \frac{2.82 \text{ Mgal}}{0.065 \text{ mgd}} = 43.4 \text{ d}$$

Comment

The estimated storage volume from the above procedure can be adjusted during final design to account for the net gain or loss in volume from precipitation, evaporation, and seepage. In the wettest year in 10, the storage volume should be reduced to zero at one point in time during the year. To estimate the area needed for the storage pond, divide the required volume in ac·ft by a typical depth, such as 10 ft. The net precipitation falling on the surface area can then be added to the storage volume. Typical seepage rates that are allowed by state regulations range from 0.062 to 0.25 in./d. These state standards for pond seepage are becoming more stringent, and compaction or lining requirements are becoming more common; therefore, a conservative approach would be to assume zero seepage.

TABLE 8.11
Comparison of Suitability Factors for Distribution Systems

Distribution System	Suitable Crops	Minimum Infiltration Rate (in./hr)	Maximum Slope (%)
Sprinkler systems:			
Portable hand move	Pasture, grain, alfalfa, orchards, vineyards, vegetable and field crops	0.10	20
Wheel-line (side-roll)	All crops less than 3 ft high	0.10	10–15
Solid set	No restriction	0.05	No restriction
Center pivot or linear	All crops except trees	0.20	15
Traveling gun	Pasture, grain, alfalfa, field crops, vegetables	0.30	15
Surface systems:			
Graded borders, narrow (border strip) 15 ft wide	Pasture, grain, alfalfa, vineyards	0.30	7
Graded borders, wide, up to 100 ft	Pasture, grain, alfalfa, orchards	0.30	0.5–1.0
Straight furrows	Vegetables, row crops, orchards, vineyards	0.10	3
Drip systems:			
Drip tube or microjets	Orchards, landscape, vineyards, vegetables	0.02	No restriction

8.2.6.5 Distribution Techniques

The three principal techniques used for effluent distribution are sprinkler, surface, and drip application. Sprinkler distribution is often used in the newer SR systems (see Figure 8.1), in most industrial wastewater (high solids content), and in all forested systems. Surface application includes border strip, ridge-and-furrow, and contour flooding. Drip irrigation should only be attempted with high-quality filtered effluent. A comparison of suitability factors for distribution systems is presented in Table 8.11. Selection of the distribution technique depends on the soil, crop type, topography, and economics. Of the sprinkler systems, the portable hand move and solid set are most common for small systems because of the relatively high flow rates required for the other systems. Continuous-move systems usually require 300 to 500 gal/min (1135 to 1890 L/min) to operate.

8.2.6.6 Application Cycles

Sprinkler systems operate between once every 3 days and once every 10 days or more. Surface application systems operate once every 2 to 3 weeks. For all systems, the total field area is divided into subsections or sets which are irrigated sequentially over the application cycle. For type 1 systems, the application schedule depends on the climate, crop, and soil permeability. For type 2 (crop irrigation) systems, the schedule depends on the crop, climate, and soil moisture depletion.

8.2.6.7 Surface Runoff Control

The surface runoff of applied wastewater from SR systems is known as *tailwater* and must be contained on-site. Collection of tailwater and its return to the distribution system or storage pond are integral parts of the design of surface application systems. Sprinkler systems on steep slopes or on slowly permeable soil may also use tailwater collection and recycle. A typical tailwater return system consists of a perimeter collection channel, a sump or pond, a pump, and a return forcemain to the storage or distribution system. Tailwater volumes range from 15% of applied flows for slowly permeable soils to 25 to 35% for moderately permeable soils (Hart, 1975). Storm-induced runoff does not need to be retained on-site; however, stormwater runoff should be considered in site selection and site design. Erosion caused by stormwater runoff can be minimized by terracing steep slopes, contour plowing, no-till farming, and grass border strips. If effluent application is stopped before the storm, the stormwater can be allowed to drain off the site.

8.2.6.8 Underdrainage

In some instances, subsurface drainage is necessary for SR systems to lower the water table and prevent water-logging of the surface soils. The existence of a water table within 5 ft indicates the possibility of poor subsurface drainage and should lead to an examination of the underdrains. For small SR sites (less than 10 ac or 4 ha) the need for underdrains may make the site uneconomical to develop. Underdrains usually consist of 4 to 6 in. (100 to 150 mm) of perforated plastic pipe buried 6 to 8 ft (1.8 to 2.4 m) deep. In sandy soils, drain spacings are 300 to 400 ft (91 to 122 m) apart in a parallel pattern. In clayey soils, the spacings are much closer, typically 50 to 100 ft (15 to 30 m) apart. Procedures for designing underdrains are described in Van Schilfgaarde (1974), USDA (1972), and USDoI (1978).

8.2.7 Construction Considerations

In most instances, the slow rate site can be developed according to local agricultural practices (Crites, 1997). Local extension services, NRCS representatives, or agricultural engineering experts should be consulted. One of the key concerns is to pay attention to the soil infiltration rates. Earthworking operations should be conducted to minimize soil compaction, and soil moisture should generally be substantially below optimum during these operations. High-flotation tires are recommended for all vehicles, particularly for soils with high percentages of fines. Deep ripping may be necessary to break up hardpan layers, which may be present below normal cultivation depths.

8.2.8 OPERATION AND MAINTENANCE

Proper operation of an SR system requires management of the applied wastewater, crop, and soil profile. Applied wastewater must be rotated around the site through the application cycle to allow time for drying maintenance, cultivation, and crop harvest. The soil profile must also be managed to maintain infiltration rates, avoid soil compaction, and maintain soil chemical balance. Compaction and surface sealing can reduce the soil infiltration or runoff. The causes can include (WEF, 2001):

1. Compaction of the surface soil by harvesting or cultivating equipment.
2. Compaction from grazing animals when the soil is too wet (wait 2 to 3 d after irrigation to allow grazing by animals).
3. A clay or silt crust can develop on the surface as the result of precipitation or wastewater application.
4. Surface clogging as a result of suspended solids application.

The compaction, solids accumulation, and crusting of surface soils may be broken up by cultivating, plowing, or disking when the soil surface is dry. At sites where clay pans (hard, slowly permeable soil layers) have formed, it may be necessary to plow to a depth of 2 to 6 ft (0.6 to 1.8 m) to mix the impermeable soil layers with more permeable surface soils. A check of the soil chemical balance is required periodically to determine if the soil pH and percent exchangeable sodium are in the acceptable range. Soil pH can be adjusted by adding lime (to increase pH) or gypsum (to decrease pH). Exchangeable sodium can be reduced by adding sulfur or gypsum followed by leaching to remove the displaced sodium.

8.3   OVERLAND FLOW SYSTEMS

Overland flow is a fixed-film biological treatment system in which the grass and vegetative litter serve as the matrix for biological growth. Process design objectives, system performance design criteria and procedures, and land and storage requirements are described in this section.

8.3.1   DESIGN OBJECTIVES

Overland flow (OF) can be used as a pretreatment step to a water reuse system or can be used to achieve secondary treatment, advanced secondary treatment, or nitrogen removal, depending on discharge requirements. Because OF produces a surface water effluent, a discharge permit is required (unless the water is reused).

In most cases, the discharge permit will limit the discharge concentrations of BOD and total suspended solids (TSS), and that is the basis of the design approach in this chapter.

8.3.2 SITE SELECTION

Overland flow is best suited to sites with slowly permeable soil and sloping terrain. Sites with moderately permeable topsoil and impermeable or slowly permeable subsoils can also be used. In addition, moderately permeable soils can be compacted to restrict deep percolation and ensure a sheet flow down the graded slope. Overland flow may be used at sites with existing grades of 0 to 12%. Slopes can be constructed from level terrain (usually the minimum of a 2% slope is constructed). Steep terrain can be terraced to a finished slope of 8 to 10%. At the wastewater application rates in current use, the site grade is not critical to performance when it is within the range of 2 to 8% (Smith and Schroeder, 1982). Site grades of less than 2% will require special attention to avoid low spots that will lead to ponding. Grades above 8% have an increased risk of short-circuiting, channeling, and erosion.

8.3.3 TREATMENT PERFORMANCE

Overland flow systems are effective in removing BOD, TSS, nitrogen, and trace organics. They are less effective in removing phosphorus, heavy metals, and pathogens. Performance data and expectations are described in this section.

8.3.3.1 BOD Loading and Removal

In municipal systems, the BOD loading rate typically ranges from 5 to 20 lb/ac·d. Biological oxidation accounts for the 90 to 95% removal of BOD normally found in OF systems. Based on experience with food processing wastewater, the BOD loading rate can be increased to 100 lb/ac·d (110 kg/ha·d) for most wastewater without affecting BOD removal. The industrial wastewater system at Paris, Texas, continues to remove 92% of applied BOD (Tedaldi and Loehr, 1991). BOD removals from four overland flow systems are presented in Table 8.12 along with the application rate and slope length. A typical BOD concentration in the treated runoff water is about 10 mg/L.

8.3.3.2 Suspended Solids Removal

Overland flow is effective in removing biological and most suspended solids, with effluent TSS levels commonly being 10 to 15 mg/L. Algae are not removed effectively in most OF systems because many algal types are buoyant and resist removal by filtration or sedimentation (Peters et al., 1981). If effluent TSS limits are 30 mg/L or less, the use of facultative or stabilization ponds that generate high algae concentrations is not recommended prior to overland flow. If OF is otherwise best suited to a site with an existing pond system, design and operational

TABLE 8.12
BOD Removal for Overland Flow Systems

Location	Wastewater Type	Application Rate (gal/ft·min)	Slope Length (ft)	Influent BOD (mg/L)	Effluent BOD (mg/L)
Ada, Oklahoma	Raw wastewater	0.10	120	150	8
	Primary effluent	0.13	120	70	8
	Secondary effluent	0.27	120	18	5
Easley, South Carolina	Raw wastewater	0.29	180	200	23
	Pond effluent	0.31	150	28	15
Hanover, New Hampshire	Primary effluent	0.17	100	72	9
	Secondary effluent	0.10	100	45	5
Melbourne, Australia	Primary effluent	0.32	820	507	12

TABLE 8.13
Nitrogen Removal for Overland Flow Systems

Parameter	Ada, Oklahoma	Hanover, New Hampshire	Utica, Mississippi
Type of wastewater	Screened raw wastewater	Primary effluent	Pond effluent
Application rate (gal/ft·min)	0.10	0.17	0.087
BOD/N ratio	6.3	2.3	1.1
Total nitrogen (lb/ac·yr):			
Applied	1070	850	590
Removed	980	790	445–535
Crop uptake	100	190	220
Nitrification/denitrification	880	600	225–325
Removal, mass balance (%)	92	94	75–90
Denitrification (% of total removal)	90	76	50–60
Total nitrogen (mg/L):			
Applied	23.6	36.6	20.5
Runoff	2.2	5.4	4.3–7.5
Nitrogen removal, concentration basis (%)	91	85	63–79

Source: USEPA, *Process Design Manual for Land Treatment of Municipal Wastewater*, EPA 625/1-81-013, U.S. Environmental Protection Agency, Cincinnati, OH, 1981.

procedures are available to overcome the algae removal issue. The application rate should not exceed 0.12 gal/min·ft (0.10 m³/m·hr) for such systems, and a nondischarge mode of operation can be used during algae blooms. In the non-discharge mode, short application periods (15 to 30 min) are followed by 1- to 2-hr rest periods. The OF systems at Heavener, Oklahoma, and Sumrall, Michigan, operate in this manner during algae blooms (WEF, 2001).

8.3.3.3 Nitrogen Removal

The removal of nitrogen by OF systems depends on nitrification/denitrification and crop uptake of nitrogen. The removal of nitrogen in several OF systems is presented in Table 8.13, which shows that denitrification can account for 60 to 90% of the nitrogen removed with denitrification rates of 800 lb/ac·yr or more. Up to 90% removal of ammonia was reported at 0.13 gal/min·ft (0.10 m³/hr·m) at the OF system at the City of Davis, California, where oxidation lagoon effluent was applied (Kruzic and Schroeder, 1990). Further research at the Davis site proved that the wet/dry ratio was also very important (Johnston and Smith, 1988). The effect of the wet/dry ratio in ammonia removal is illustrated in Figure 8.5.

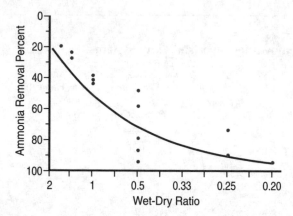

FIGURE 8.5 Effect of wet/dry ratio on the removal of ammonia by overland flow. (From Johnston, J. and Smith, R., Operating Schedule Effects on Nitrogen Removal in Overland Flow Treatment Systems, paper presented at the 61st Annual Conference of the Water Pollution Control Federation, Dallas, TX, 1988.)

To obtain effective nitrification, the wet/dry ratio must be 0.5 or less. At Sacramento County, California, secondary effluent was nitrified at an application rate of 0.70 gal/min·ft (0.54 m³/hr·m). Ammonia concentrations were reduced from 14 to 0.5 mg/L (Nolte Associates, 1997). At Garland, Texas, nitrification studies were conducted with secondary effluent to determine whether a 2-mg/L summer limit for ammonia and a 5-mg/L winter limit could be attained. Application rates ranged from 0.43 to 0.74 gal/min·ft (0.33 to 0.57 m³/hr·m). Winter values for effluent ammonia ranged from 0.03 to 2.7 mg/L and met the effluent requirements. The recommended application rate for Garland was 0.56 gal/min·ft (0.43 m³/hr·m) for an operating period of 10 hr/d and a slope length of 200 ft (61 m) with sprinkler application (Zirschky et al., 1989).

8.3.3.4 Phosphorus and Heavy Metal Removal

Phosphorus removal in OF is limited to about 40 to 50% because of the lack of soil–wastewater contact. If needed, phosphorus removal can be enhanced by the addition of chemicals such as alum or ferric chloride. Heavy metals are removed using the same general mechanisms as with phosphorus: absorption and chemical precipitation. Heavy metal removal will vary with the constituent metal from about 50 to about 80% (WEF, 2001).

8.3.3.5 Trace Organics

Trace organics are removed in OF systems by a combination of volatilization, absorption, photodecomposition, and biological degradation. If removal of trace organics is a major concern, Reed et al. (1995) and Jenkins et al. (1980) should be reviewed.

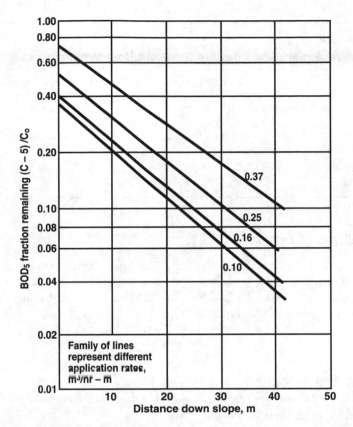

FIGURE 8.6 Relationship between application rate and slope length for BOD removal in the design of overland flow systems.

8.3.3.6 Pathogens

Overland flow is not very effective in removing microorganisms. Fecal coliforms will be reduced by about 90% when raw or primary effluent is applied; however, minimal removal occurs when secondary effluent (with much lower coliform levels than primary) is applied (USEPA, 1981). Enteric virus removals up to 85% have been observed with overland flow.

8.3.4 PREAPPLICATION TREATMENT

The usual treatment prior to OF application is primary settling. For small systems, Imhoff tanks or 1- to 2-d detention aerated ponds are recommended. Static or rotating fine screens have also been used successfully at Davis, California, and Hall's Summit, Louisiana (WEF, 2001).

TABLE 8.14
Suggested Application Rates for Overland Flow Systems

Preapplication Treatment	Stringent Requirements (BOD = 10–15 mg/L) (gal/min·ft)	Less Stringent Requirements (BOD = 30 mg/L) (gal/min·ft)
Screening, septic tank, or short-term aerated cell	0.1–0.15	0.25–0.33
Sand filter, trickling filter, secondary	0.2–0.25	0.3–0.45

8.3.5 DESIGN CRITERIA

The principal design criteria for the OF process are application rate and slope length. Other design criteria include hydraulic loading rate, application period, and organic loading rate. The relationship between hydraulic loading rate and application rate is shown in Equation 8.8:

$$L_w = qPF/Z \qquad (8.8)$$

where

L_w = Hydraulic loading rate (in./d; mm/d).
q = Application rate per unit width of slope (gal/ft·min; m³/m·min).
P = Application period (hr/d).
F = Conversion factor (96.3).
Z = Slope length (ft; m).

8.3.5.1 Application Rate

The application rate used for design of municipal OF systems depends on the limiting design factor (usually BOD), the preapplication treatment, limitations, and the climate. A range of suggested application rates is presented in Table 8.14. The lower end of the range shown in Table 8.14 should be used for cold climates and the upper end for warm climates. The relationship between application rate and slope length is shown in Figure 8.6. As mentioned previously, facultative ponds are not recommended as preapplication treatment for OF. If OF is used in conjunction with facultative ponds, however, the application rate should not exceed 0.12 gal/min·ft (0.10 m³/hr·m).

8.3.5.2 Slope Length

Slope lengths in OF practice have typically ranged from 100 to 200 ft (30 to 60 m). The longer the slope has been, the greater has been the removal of BOD, TSS, and nitrogen. The recommended slope length depends on the method of application. For gated pipe or spray heads that apply wastewater at the top of the slope, a slope length of 120 to 150 ft (36 to 45 m) is recommended. For sprinkler application, the slope length should be between 150 and 200 ft (45 to 61 m). The minimum slope length for sprinkler application (usually positioned one third the distance down the slope) should be the wetted diameter of the sprinkler plus about 65 to 70 ft (20 to 21 m).

8.3.5.3 Hydraulic Loading Rate

The hydraulic loading rate, expressed in in./d or in./wk, was the principal design parameter in the USEPA's *Design Manual* (USEPA, 1981). Selecting the application rate, however, and calculating the resultant hydraulic loading rate have a more rational basis. Using the application rate approach allows the designer to consider varying the application rate and application period to accomplish a reduction or increase in hydraulic loading.

8.3.5.4 Application Period

A range of application periods has been used successfully, with 6 to 12 hr/d being most common. A typical application period is 8 hr/d. With an 8-hr/d application period, the total field area is divided into three sections. The application period can then be increased to 12 hr/d for grass harvest or system maintenance. The application period can be increased to 24 hr/d for a short time (3 to 5 d) without adverse impacts on BOD or TSS performance; however, ammonia and trace metals may be released.

8.3.6 DESIGN PROCEDURE

The procedure for design of OF systems is to establish the limiting design parameter; select the application rate, application period, and slope length; calculate the hydraulic loading rate; and calculate the field. The storage volume, if any, must also be determined, and the field area increased to account for stored volume.

8.3.6.1 Municipal Wastewater, Secondary Treatment

A relationship between BOD removal and application rates has been developed (Reed et al., 1995; Smith and Schroeder, 1982; USEPA, 1981). If a system is to be designed for BOD removal only, the use of the "rational" model (Figure 8.6) will predict a higher application rate than listed in Table 8.14. For small systems, it is recommended that the application rate be selected from Table 8.14.

8.3.6.2 Industrial Wastewater, Secondary Treatment

For industrial wastewater with BOD concentrations of 400 to 2000 mg/L or more, the organic loading rate is often limiting. The procedure for process design is as follows:

1. Calculate the BOD load from the concentration and flow:

$$\text{BOD load} = 8.34QC \tag{8.9}$$

where BOD load is the daily BOD load (lb/d; kg/d); 8.34 is the conversion factor; Q is the flow (mgd; m³/d); and C is the BOD concentration (mg/L).

2. Calculate the land area from Equation 8.10:

$$A = (\text{BOD load})/100 \tag{8.10}$$

where A is the field area (ac), and 100 is the limiting loading of BOD (lb/ac·d; kg/ha·d).

When the BOD of the applied wastewater exceeds approximately 800 mg/L, the oxygen transfer from the atmosphere through the fixed film becomes limiting. The BOD removal rate will decline unless effluent recycling is practiced. In some industrial applications, effluent recycle to dilute the BOD in the raw wastewater has been used. For example, the BOD from a food processing wastewater was reduced from 1800 down to 500 mg/L with an effluent recycle ratio of 3:1 (Perry et al., 1981).

8.3.7 DESIGN CONSIDERATIONS

Design considerations for OF include land area requirements, storage requirements, vegetation selection, distribution systems, and runoff collection.

8.3.7.1 Land Requirements

The field area required for OF depends on the flow, the required storage, and the loading rate. For an OF system that operates without storage, the field area is calculated using Equation 8.11:

$$A = QF/L_w \tag{8.11}$$

where
A = Field area (ac; ha).
Q = Design wastewater flow (mgd; m³/d).
F = Conversion factor (36.8 ac·in./Mgal).
L_w = Hydraulic loading rate (in./d; mm/d).

If wastewater must be stored because of cold weather, the field area is determined using Equation 8.12:

$$A = (365Q + V_S)(F)/DL_w \tag{8.12}$$

where
A = Field area (ac).
Q = Design wastewater flow (mgd).

V_S = Net gain or loss in storage volume from precipitation, evaporation, and seepage (Mgal).

F = Conversion factor (36.8 ac·in./Mgal).

D = Number of operating days per year.

L_w = Hydraulic loading rate (in./d).

8.3.7.2 Storage Requirements

Storage of wastewater may be required for cold weather, wet weather, or crop harvesting. Cold weather storage is the most common with operations ceasing when temperatures fall below 32°F (0°C). Design storage days should be estimated from climatic records for the site. Storage for wet weather is generally not necessary. Rainfall effects on BOD removal are minimal, and storage is not required during normal rainfall events. If storage is not necessary for cold or wet weather, it is usually advisable to provide a week of storage to accommodate emergencies such as equipment problems, crop harvest, or maintenance. The storage pond should be located offline so it contains pretreated effluent for a minimum time and is drained as soon as application is possible.

Example 8.3. Process Design for Overland Flow

Calculate the field area for an overland flow system to treat 0.5 mgd of septic tank effluent. Preapplication is by a community septic tank, and the discharge BOD limit is 30 mg/L. The climate is moderately warm with 25 days of winter storage required.

Solution

1. Select an application rate appropriate for the degree of preapplication treatment, climate, and BOD removal requirement. Select 0.30 gal/ft·min from Table 8.14.

2. Select a slope length and application period. Select 150 ft for slope length and 8 hr/d for the application period.

3. Calculate the hydraulic loading rate using Equation 8.8:

   $L_w = (q)(P)(96.3)/Z$

   $L_w = (0.30)(8)(96.3)/150$

   $L_w = 1.5$ in./d

4. Calculate the number of operating days per year:

   D = 365 – 25 = 340 d/yr

5. Use a net gain from precipitation on the storage pond of 2 Mgal. Calculate the field area using Equation 8.12:

   $A = [(365)(Q) + V_S]\ 36.8/DL_w$

   $A = [365(0.5) + 2]36.8/(340)(1/5)$

   $A = 13.3$ ac

8.3.7.3 Vegetation Selection

Perennial water-tolerant grass is recommended for OF vegetation. The function of the grass is to provide a support medium for microorganisms, prevent channeling and erosion, ensure thin sheet flow, and allow for filtration and sedimentation of solids in the vegetative layer. Nutrient uptake is another role — less critical in most cases. The grass crop is harvested usually three or more times per season and is sold for hay or green chop. A mixture of grasses is recommended, and the mixture should include warm season and cool season species. Warm season grasses that have been used successfully include common and coastal Bermuda grass, dallis grass, and bahia grass. Cool season grasses include reed canary grass, tall fescue, redtop, Kentucky bluegrass, and orchard grass. Some grasses such as reed canary require a nurse crop (such as rye grass) for a year or two before they become well established. Local agricultural advisers should be consulted for the best mix of grasses for the particular site.

8.3.7.4 Distribution System

Surface application using gated pipe is an economically attractive alternative to impact sprinklers for small systems. For municipal wastewater, the surface application technique offers lower energy demand and avoids spray drift and the attendant setback distances of sprinklers. For industrial wastewater, the standard solid set sprinklers are recommended.

8.3.7.5 Runoff Collection

Treated runoff is collected in grass-lined opened drainage channels at the toe of the slope. These channels are typically V-type channels with side slopes of 4:1 or more. Runoff channels should be sloped at 0.5 to 1% to avoid ponding.

8.3.8 Construction Considerations

The slope or terrace of an OF system must be graded to a uniform smoothness with no low spots or reversals in grade. Finish slopes of 1 to 8% are usually acceptable, although 2% is considered minimum in some states. Cross slopes should not exceed 0.5%, especially when finish slopes are 1 to 2%. Where extensive cut-and-fill operations are necessary, the slope should be watered and allowed to settle after rough grading. Any depressions should then be filled, and the slope should be final graded, disked, and landscaped (USEPA, 1984).

8.3.9 Operation and Maintenance

Operation and maintenance considerations include fine tuning of the application cycle, vegetation harvesting, and maintenance of the slope and runoff collection channels. Pest control must consider mosquitoes and invasions of army worms (WEF, 2001). Periodic mowing of the cover grass is necessary to maintain a healthy stand of grass and reduce bunching. A minimum of four mowings per year is recommended. The slopes should be dried completely before harvesting.

8.4 SOIL AQUIFER TREATMENT SYSTEMS

Soil aquifer treatment (SAT) has the highest hydraulic loading rate of any land treatment system. The site selection criteria for SAT are also more stringent. Design objectives, site selection procedures, treatment performance, design considerations, construction considerations, and operation and maintenance for SAT systems are described in this section.

8.4.1 DESIGN OBJECTIVES

The principal design objective for SAT systems is wastewater treatment. Other design objectives can include recharge of streams by interception of shallow groundwater; recovery of water by walls or underdrains, with subsequent reuse; groundwater recharge; and temporary (seasonal) storage of renovated water in an aquifer. Most SAT systems are designed to treat wastewater and avoid a direct discharge to a surface water course.

8.4.2 SITE SELECTION

Site selection is very important to the success of an SAT project because failure of SAT systems is most often related to improper or insufficient site evaluation (Reed et al., 1985). The important factors in site evaluation and selection are the soil depth, soil permeability, depth to groundwater, and groundwater flow direction.

8.4.3 TREATMENT PERFORMANCE

Soil aquifer treatment is an effective process for BOD, TSS, and pathogen removal. Removal of phosphorus and metals depends on travel distance and soil texture. Nitrogen removal can be significant when systems are managed for that objective.

8.4.3.1 BOD and TSS Removal

Typical values of BOD loadings and BOD removals for SAT systems are presented in Table 8.15. Suspended solids are typically 1 to 2 mg/L in the percolate from SAT systems as a result of filtration through the soil profile. BOD loadings on industrial SAT systems range from 100 to 600 lb/ac·d (112 to 667 kg/ha·d). BOD loadings beyond 300 lb/ac·d (336 kg/ha·d) require careful management to avoid odor production. Suspended solids loadings of 100 to 200 lb/ac·d (112 to 224 kg/ha·d) or more require frequent disking or scarifying of the basin surface to avoid plugging of the soil. For example, a 150-lb/ac·d (168-kg/ha·d) loading of TSS at Hollister, California, required disking after each 3-week application/drying cycle (Pound et al., 1978).

8.4.3.2 Nitrogen Removal

Nitrification/denitrification is the principal mechanism for removal of ammonia and nitrate from the wastewater in SAT systems. Ammonia adsorption also plays

TABLE 8.15
BOD Removal for Soil Aquifer Treatment Systems

Location	Applied Wastewater BOD (lb/ac·d[a])	Applied Wastewater BOD (mg/L)	Percolate Concentration (mg/L)	Removal (%)
Boulder, Colorado	48[b]	131[b]	10[b]	92
Brookings, South Dakota	11	23	1.3	94
Calumet, Michigan	95[b]	228[b]	58[b]	75
Ft. Devens, Massachusetts	77	112	12	89
Hollister, California	156	220	8	96
Lake George, New York	47	38	1.2	97
Milton, Wisconsin	138	28	5.2	81
Phoenix, Arizona	40	15	0–1	93–100
Vineland, New Jersey	43	154	6.5	96

[a] Total lb/ac·yr divided by the number of operating days in the year.
[b] Chemical oxygen demand (COD) basis.

Source: Crites, R.W. and Tchobanoglous, G., *Small and Decentralized Wastewater Management Systems*, McGraw-Hill, New York. 1998. With permission.

an important role in retaining ammonia in the soil long enough for biological conversion. Nitrification and denitrification are affected by low temperatures and proceed slowly at temperatures of 36 to 41°F (3.6 to 5°C). In addition, denitrification requires an adequate carbon source and the absence of available oxygen.

The ANAMMOX (anaerobic ammonia oxidation) process of ammonia and nitrate reduction was found to be occurring in SAT systems at Lake Tahoe–Truckee, California (Woods et al., 1999) and at Mesa, Arizona (Gable and Fox, 2000). The process appears to be continuing through the saturated soil.

Experience with nitrification has been that rates of up to 60 lb/ac·d (67 kg/ha·d) can be achieved under favorable moisture and temperature conditions. Total nitrogen loadings should be checked to verify that they are not in excess of the 50 to 60 lb/ac·d (56 to 67 kg/ha·d) range. Ammonia will be retained in the upper soil profile when temperatures are too low (below 36°F or 2.2°C) for nitrification.

Nitrogen removal is a function of detention time, BOD-to-nitrogen ratio (adequate carbon source), and anoxic conditions. Detention time is related to the hydraulic loading rate through the soil profile. For effective nitrogen removal (80% or more), the loading rate should not exceed 6 in./d (Lance et al., 1976). The BOD-to-nitrogen ratio must be 3:1 or more to ensure adequate carbon to drive the denitrification reaction. Secondary effluent will have a BOD-to-nitrogen

ratio of about 1:1, while primary effluent usually has a BOD-to-nitrogen ratio of
3:1. To overcome the low BOD-to-nitrogen ratio in secondary effluent, a longer
application period (7 to 9 d) is necessary (Bouwer et al., 1980). Typical removals
of total nitrogen and percolate concentration of nitrate nitrogen and total nitrogen
are presented in Table 8.16.

8.4.3.3 Phosphorus Removal

Phosphorus removal in SAT is accomplished by adsorption and chemical precipitation. The adsorption occurs quickly, and the slower occurring chemical precipitation replenishes the adsorption capacity of the soil. Typical phosphorus
removals for SAT are presented in Table 8.17, including travel distances through
the soil. If phosphorus removal is critical, a phosphorus adsorption test using the
specific site soil can be conducted (Reed and Crites, 1984). To conduct an
adsorption test, about 10 g of soil is placed in containers containing known
concentrations of phosphorus in solution. After periodic shaking (for up to 5 d),
the solution is decanted and analyzed for phosphorus. The difference in concentrations is attributed to adsorption onto the soil particles. The detailed procedure
is presented by Enfield and Bledsoe (1975). Actual (long-term) phosphorus retention at an SAT site will be 2 to 5 times greater than the values obtained in the
5-d phosphorus adsorption test (USEPA, 1981). If the travel distance to the critical
point for phosphorus removal is known, the "worst case" phosphorus concentration can be calculated using Equation 8.13:

$$P_x = Pe^{-kt} \tag{8.13}$$

where
  P_x = Total phosphorus at a distance x on the flow path (mg/L).
  P = Total phosphorus in the applied wastewater (mg/L).
  k = 0.048 (d).
  t = Detention time $(d^{-1}) = x(0.40)/K_xG$, where x is distance along the flow
           path (ft), K_x is the hydraulic conductivity in soil in direction x (ft/day),
           and G is the hydraulic gradient ($G = 1$ for vertical flow; H/L for
           horizontal flow).

8.4.3.4 Heavy Metal Removal

Heavy metal removal, using the same mechanisms as described for slow rate,
will range from 50 to 90% for SAT. Metals applied at very low concentrations
(below drinking water standards) may not be affected by passage through sand
(Crites, 1985a).

8.4.3.5 Trace Organics

Trace organics are removed in SAT systems by volatilization, sorption, and
degradation. Removals depend on the constituent, applied concentration, loading

TABLE 8.16
Nitrogen Removal for Soil Aquifer Treatment Systems

Location	Applied Total Nitrogen (lb/ac-d)	Applied Total Nitrogen (mg/L)	Percolate Nitrate Nitrogen (mg/L)	Percolate Total Nitrogen (mg/L)	Total Nitrogen Removal (%)
Calumet, Michigan	20.7	24.4	3.4	7.1	71
Dan Region, Israel	28.9	13.0	6.5	7.2	45
Ft. Devens, Massachusetts	37.0	50.0	13.6	19.6	61
Hollister, California	14.9	40.2	0.9	2.8	93
Lake George, New York	12.5	12.0	7.0	7.5	38
Phoenix, Arizona	40.0	18.0	5.3	5.5	69
W. Yellowstone, Montana	115.6	28.4	4.4	14.1	50

Source: Crites, R.W., in *Artificial Recharge of Groundwater,* Asano, T., Ed., Butterworth Publishers, Stoneham, MA, 1985, 579–608. With permission.

TABLE 8.17
Phosphorus Removal for Soil Aquifer Treatment Systems

Location	Applied Phosphorus[a] (mg/L)	Vertical Travel Distance (ft)	Horizontal Travel Distance (ft)	Percolate Concentration (mg/L)	Removal (%)
Boulder, Colorado	6.2	8–10	0	0.2–4.5	40–97
Brookings, South Dakota	3.0[b]	2.6	0	0.45	85
Calumet, Michigan	3.5	10–30	0–400	0.1–0.4	89–97
	3.5	30[c]	5580[c]	0.03	99
Ft. Devens, Massachusetts	9.0[b]	50	100	0.1	99
Hollister, California	10.5	22	0	7.4	29
Lake George, New York	2.1[b]	10	0	<1	>52
	2.1[b]	10[c]	1970[c]	0.014	99
Phoenix, Arizona	8–11	30	0	2–5	40–80
	7.9	20	100	0.51	94
Vineland, New Jersey	4.8[b]	6.5–60	0	1.54	68
	4.8[b]	13–52	850–1700	0.27	94

[a] Total phosphate measured, except as noted.
[b] Soluble phosphate measured.
[c] Seepage.

Source: USEPA, *Process Design Manual for Land Treatment of Municipal Wastewater*, EPA 625/1-81-013, U.S. Environmental Protection Agency, Cincinnati, OH, 1981.

TABLE 8.18
Recorded Trace Organic Concentrations at Selected Soil Aquifer Treatment Sites (ng/L)

Pesticide	Vineland, New Jersey			Milton, Wisconsin		
	Applied Wastewater	Shallow Groundwater	Control Groundwater	Applied Wastewater	Shallow Groundwater	Downgradient Groundwater
Endrin	<0.03	<0.03	<0.03	<0.03	<0.03	<0.03
Lindane	2830–1227	453–1172	21.3	41	157.6	3.9
Methoxychlor	<0.01	<0.01	<0.01	<0.01	<0.01	<0.01
Toxaphene	<0.1	<0.1	<0.1	<0.1	<0.1	<0.1
2,4-D	9.5–10.5	16.4–13.0	10.4	53.8	92.4	23.6
2,4,5-TP silvex	72	26.8–120	185	16.2	41.2	38.7

Source: Data from Benham-Blair & Affiliates (1979).

rate, and presence of easily degradable organics to serve as a primary substrate (Crites, 1985b). Removals studied at Phoenix, Arizona; Ft. Devens, Massachusetts; and Whittier Narrows, California, have ranged from 10 to 96%. Levels of pesticides in applied effluent and in groundwater at two SAT sites are presented in Table 8.18.

8.4.3.6 Endocrine Disruptors

Soil aquifer treatment systems have been utilized for the removal of endocrine-disrupting chemicals found in municipal wastewaters (Conroy et al., 2001; Quanrad et al., 2002). Endocrine disruptors originate from industrial, agricultural, and domestic sources. These include a combination of natural hormones, pharmaceutical products, and industrial chemicals such as polychlorinated biphenyls, organochlorine pesticides, phenoxyacid herbicides, phthalates, and tirazines. Following conventional secondary treatment, percolation through approximately 120 ft (36 m) of unconsolidated sediments to the local aquifer reduced residual estrogenic activity by >95% (Table 8.19) (Quanrad et al., 2002). The fate of micropollutants

TABLE 8.19
Fractional Attenuation of Estrogenic Activity (Relative to Primary Effluent) During Secondary Treatment and Soil Aquifer Treatment

Sample Location	Fractional Removal
Primary	0.00
Secondary unchlorinated	0.62
Secondary chlorinated	0.65
Secondary dechlorinated	0.65
Storage pond	0.68
0.8 m (2.5 ft)	0.77
3.1 m (10 ft)	0.83
5.2 m (17 ft)	0.83
18.3 m (60 ft)	0.93
36.6 m (120 ft)	0.99

Source: USEPA, *Process Design Manual for Land Treatment of Municipal and Industrial Wastewater,* Center for Environmental Research Information (CERI), U.S. Environmental Protection Agency, Cincinnati, OH, 2004.

originating from pharmaceuticals and active ingredients in personal-care products have been studied at two groundwater recharge facilities in Arizona (Drewes et al., 2001). Preliminary studies indicate that groundwater recharge offers a high potential to remove acidic drugs such as lipid regulators and analgesics. Other compounds such as antiepileptic drugs and x-ray contrast agents showed no clear indication of removal during travel times of more than 6 years.

Additional studies of long-term SAT at field sites in Mesa, Arizona, indicate that substantial removal of effluent organic matter can occur. Identified trace organics were efficiently removed as a function of travel time to very low concentrations or below detection limits. Based on the characterization techniques used, the character of bulk organics present in final SAT water resembled the character of natural organic matter present in drinking water (Drewes et al., 2001).

8.4.3.7 Pathogens

Pathogens are filtered out by the soil and adsorbed onto clay particles and organic matter. Fecal coliform are removed by 2 to 4 orders of magnitude in many SAT systems (USEPA, 1981). At the SAT site in Phoenix, Arizona, 99.99% virus removal was achieved after travel through 30 ft (10 m) of sand at a loading rate of 300 ft/yr (100 m/yr) (Crites, 1985b).

8.4.4 PREAPPLICATION TREATMENT

When the overall treatment objective has been established, the appropriate level of preapplication treatment should be determined. For SAT, the minimum preapplication treatment is primary sedimentation or the equivalent. For small systems a short detention-time pond is recommended. Long-detention-time facultative or aerobic ponds are not recommended because of their propensity to produce high concentrations of algae. The algae produced in stabilization ponds will reduce infiltration rates in SAT systems significantly. If facultative or stabilization ponds are to be used with SAT, it is recommended that an aquatic treatment or constructed wetland system be used between the pond and the SAT basins to reduce TSS levels.

8.4.5 DESIGN PROCEDURE

The process design procedure for a typical SAT system is outlined in Table 8.20. If the hydraulic pathway is toward surface water, the limiting design factor will be related to surface water quality requirements, which could range from BOD and TSS to nutrients and trace organics. Groundwater discharges are more often controlled by pathogen and nitrate requirements. If nitrogen removal is a process design consideration, the following six steps should also be followed after the annual hydraulic loading is calculated (step 6); if necessary, steps 5 and 6 may be repeated for different levels of preapplication treatment to achieve the required level of overall treatment:

TABLE 8.20
Process Design Procedure for Soil Aquifer Treatment

Step	Description
1	Characterize the soil and groundwater conditions with field measurements of lateral and vertical permeability.
2	Predict the hydraulic pathway of the percolate.
3	Select the infiltration rate from the field data.
4	Determine the overall treatment requirements.
5	Select the appropriate level of preapplication treatment.
6	Calculate the annual hydraulic loading rate.
7	Calculate the field (basin) area.
8	Check for groundwater mounding.
9	Select the final hydraulic loading cycle.
10	Determine the number of basins.
11	Determine the monitoring requirements.

1. Calculate the mass of ammonia that can be adsorbed by the soil cation exchange capacity.
2. Calculate the length of the application period that can be used without exceeding the mass of ammonia that can be adsorbed.
3. Compare the nitrogen loading rate to the 50 to 60 lb/ac·d (56 to 67 kg/ha·d) limit for nitrification.
4. Balance the ammonia adsorption with the available oxygen to establish the application and drying periods for nitrification.
5. Balance the nitrate nitrogen produced by nitrification against the applied BOD to ensure an adequate BOD-to-nitrogen ratio. Revise the application/drying cycle if necessary.
6. If necessary, consider reducing the soil infiltration rate to increase the detention time for higher nitrogen removal. Reduction of infiltration can be accomplished by incorporating silt or finer textured topsoil, reduction of the depth of flooding, or compaction of the soil.

8.4.6 Design Considerations

Design considerations for SAT systems include hydraulic loading rates, nitrogen loading rates, organic loading rates, land requirements, hydraulic loading cycle, infiltration system design, and groundwater mounding.

TABLE 8.21
Typical Design Factors Used To Convert Measured Infiltration Rates to Soil Aquifer Treatment Hydraulic Loading Rates

	Percent of Measured Values	
Test Procedure	Conservative Range	Less Conservative Range
Basin flooding test	5–7	8–10
Cylinder infiltrometer or air entry permeameter	1–2	2–4

Source: Crites, R.W. and Tchobanoglous, G., *Small and Decentralized Wastewater Management Systems*, McGraw-Hill, New York, 1998. With permission.

8.4.6.1 Hydraulic Loading Rates

The hydraulic characteristics of the soil and aquifer system usually determine the design hydraulic loading rate of a site. In some instances, the nitrogen or BOD loading may control the area needed; however, the limiting design factor (LDF) for SAT systems is usually the infiltration/percolation rate. The design hydraulic loading rate is the measured clean water infiltration rate multiplied by a design factor. The design factor depends on the procedure used for measuring the infiltration rate, on the variability of the infiltration test results, on the variation in soil characteristics over the site, and on the conservatism of the designer. Design factors account for the cyclical (intermittent loading and drying) nature of SAT applications, the variability of site conditions and test measurements, and a long-term decrease in infiltration rates as a result of wastewater loadings. Typical design factors for different field tests are presented in Table 8.21.

8.4.6.2 Nitrogen Loading Rates

Where nitrogen removal is important, the total nitrogen rate should be kept below 60 lb/ac·d (67 kg/ha·d). To determine the nitrogen loading rate from the hydraulic loading rate, use Equation 8.14:

$$\text{NLR} = (L_w)(0.23)(C)/D \tag{8.14}$$

where
NLR = Nitrogen loading rate (lb/ac·d; kg/ha·d).
L_w = Hydraulic loading rate (in./yr; m/yr).
0.23 = Conversion factor (10.0).
C = Wastewater nitrogen concentration (mg/L).
D = Number of operating days per year.

For a typical municipal wastewater with 40 mg/L of total nitrogen and an SAT system operated 365 d/yr, the 60-lb/ac·d (67-kg/ha·d) nitrogen loading rate corresponds to a 200-ft/yr hydraulic loading rate.

8.4.6.3 Organic Loading Rates

The limit on organic loading rates depends on the climate, the nature of the wastewater, and the remoteness of the site. From experience with food processing and winery wastewater, the BOD loading rate should generally be less than 600 lb/ac·d (667 kg/ha·d). For municipal systems, a limit of about 300 lb/ac·d (336 kg/ha·d) is recommended.

8.4.6.4 Land Requirements

The field area (basin bottoms) for an SAT system can be calculated using Equation 8.15:

$$A = CQ(365 \text{ d/yr})/L_w \tag{8.15}$$

where
A =   Field area (ac; ha).
C =   Conversion factor (3.07 ac·ft/mgd).
Q =   Flow (mgd; m³/d).
L_w =   Hydraulic loading rate(ft/yr; m/yr).

The limiting design factor (LDF) for an SAT system must be determined by calculating the field area using Equation 8.15 and comparing that value to the field area required based on nitrogen or organic loading rates. The field area based on nitrogen or organic loading rates is calculated using Equation 8.16:

$$A = (8.34)(C)(Q)/L \tag{8.16}$$

where
A =   Field area (ac; ha).
8.34 =   Conversion factor.
C =   Concentration of nitrogen or BOD (mg/L).
Q =   Flow (mgd; m³/d).
L =   Limiting loading rate (lb/ac·d; kg/ha·d).

In addition to the field area, land requirements for an SAT system include basin side slopes, berms, access roads, and land for preapplication treatment.

8.4.6.5 Hydraulic Loading Cycle

In SAT systems, the hydraulic loading cycle consists of an application (flooding) period followed by a drying (resting) period. This intermittent cycle is key to the successful performance of an SAT system. Application periods range from 1 to 9 d, while drying periods range from 5 to 20 d. Hydraulic loading cycles can be

TABLE 8.22
Hydraulic Loading Cycles for Soil Aquifer Treatment

Loading Objective	Wastewater Applied	Season	Application Period (d)	Drying/Resting Period (d)
Maximize infiltration rates	Primary	Summer	1–2	6–7
		Winter	1–2	7–12
	Secondary	Summer	2–3	5–7
		Winter	1–3	6–10
Maximize nitrification	Primary	Summer	1–2	6–9
		Winter	1–2	7–13
	Secondary	Summer	2–3	5–6
		Winter	1–3	6–10
Maximize nitrogen removal	Primary	Summer	1–2	10–13
		Winter	1–2	13–20
	Secondary	Summer	6–7	9–12
		Winter	9–12	12–16

Source: Adapted from Crites, R.W. and Tchobanoglous, G., *Small and Decentralized Wastewater Management Systems*, McGraw-Hill, New York, 1998.

selected to maximize infiltration rates, maximize nitrification, or maximize nitrogen removal. Hydraulic loading cycles for these three objectives are presented in Table 8.22. For mild climates, the shorter drying periods and longer application periods are used. For cold or wet climates, the longer drying periods are necessary.

8.4.6.6 Infiltration System Design

Although sprinklers and subsurface perforated pipe may be used for distribution, the most common method is shallow level spreading basins. Gravity distribution is used through pipes or ditches, and the basins are divided by berms to allow periodic drying and scarification of the basin surface. The number of basins depends on the site topography, hydraulic loading cycle, and wastewater flow. For small systems, the basins are generally 0.5 to 2 ac in size. The minimum number of basins is typically 3 to 4 and may be as many as 10 to 15, depending on the loading cycle (Reed and Crites, 1984).

8.4.6.7 Groundwater Mounding

During the application period the applied wastewater percolates through the soil profile and reaches either a slowly permeable layer or the groundwater table. When the water cannot flow vertically, it tends to form a temporary mound at the interface with the groundwater before it can move laterally with the groundwater. If the magnitude of the mound amounts to 1 to 2 ft of overall rise, the mounding is of little consequence. If the mound rises to within 2 ft of the basin surface, however, the treatment performance and rate of infiltration will diminish. Groundwater mounding equations (Reed and Crites, 1984) and nomographs (USEPA, 1981) can be used to determine the impact of groundwater mounding.

Example 8.4. Process Design for Soil Aquifer Treatment

Calculate the field area required for a soil aquifer treatment system that treats 0.75 mgd of municipal wastewater. Preapplication treatment in an aerated pond reduces the BOD to 120 mg/L and the total nitrogen to 25 mg/L. The soils on the site have a moderate variability, and the minimum infiltration rate, using basin infiltration tests, is 2.5 in./hr. Upgradient groundwater has an average nitrate nitrogen concentration of 8 mg/L, which limits the downgradient groundwater quality to 8 mg/L.

Solution

1. Calculate the hydraulic loading rate (using Equation 8.3) and a 7% design factor from Table 8.21:

   $L_w = 0.07(24 \text{ hr/d})(365 \text{ d/yr})(2.5 \text{ in./hr})(1 \text{ ft/12 in.}) = 128 \text{ ft/yr}$

2. Compare the hydraulic loading to the acceptable nitrogen loading rate using Equation 8.14:

   $L_n = L_w CF/D = (128)(25)(2.7)/365 = 23.7 \text{ lb/ac·d}$

   Because the calculated 23.7 lb/ac·d is well below the 60-lb/ac·d limit, nitrogen loading is not limiting for the field area determination.

3. Calculate the BOD loading in lb/d:

   BOD load = (8.34)(0.75 mgd)(120 mg/L) = 750 lb/d

4. Calculate the required field area on the basis of hydraulic loading rate:

   $A = 3.06(0.75)(365)/128 = 6.5 \text{ ac}$

5. Calculate the field area needed to keep the BOD loading below 300 lb/ac-d:

   $A = 750 \text{ lb/d}/300 \text{ lb/ac·d} = 2.5 \text{ ac}$

6. Determine the field area required. Because the area needed for the hydraulic loading is greater than the area required for the BOD loading rate, the limiting field area is 6.5 ac.

8.4.7 CONSTRUCTION CONSIDERATIONS

Construction of infiltration basins must be conducted carefully to avoid compacting the infiltration surface. Basin surfaces should be located in cut sections with excavated material being placed and compacted in the berms. The berms do not have to be higher than 3 to 4 ft (1 to 1.3 m) in most cases. Erosion off the berm slopes should be avoided because erodible material is often fine textured and can blind or seal the infiltration surface.

8.4.8 OPERATION AND MAINTENANCE

The operation and maintenance considerations for SAT systems include maintenance of the infiltration rate, avoiding freezing and ice formation conditions in cold climates, and avoidance of solids clogging.

8.4.8.1 Cold Climate Operation

Storage is generally not provided for SAT even where cold winters would limit operation of SR or OF systems (Reed et al., 1995). Proper thermal protection is needed for pumps, piping, and valves (Reed and Crites, 1984). Wastewater can continue to be land applied in SAT basins throughout subfreezing weather provided the soil profile does not freeze with moisture in it. Approaches that can be used to avoid critical ice formation include:

1. Design of one basin with excess freeboard to accept continuous loading for up to 2 or 3 weeks during extreme conditions; this basin would then be rested for an extended period during warmer weather and the basin surface scarified.
2. Ridge and furrow surface application combined with a floating ice cover; the ice gives thermal protection to the soil and is supported on the ridges as wastewater infiltrates in the flowing furrows.
3. Use of snow fences to retain snow on the basins to insulate the soil.
4. Use of wastewater (perhaps bypassed from the headworks) with minimal preapplication treatment to retain the available heat in the wastewater.

8.4.8.2 System Management

It is essential that SAT systems be operated with an application and a drying period. The drying period is critical to effective treatment, restoration of aerobic conditions, and maintenance of infiltration rates. The length of time required to dry each basin of visible water should be recorded, and any increasing trend in required drying time should be noted. An increase in the necessary drying time can signal the need for basin surface maintenance. Such maintenance can include disking, scarifying (tilling or breaking up the surface), or scraping off surface solids.

8.5 PHYTOREMEDIATION

Phytoremediation is the process by which plants are used to treat or stabilize contaminated soils and groundwater (USEPA, 2000). The technology is complex and is only introduced here (Lasat, 2002). The technology has emerged as a response to the clean-up efforts for sites contaminated with toxic and hazardous wastes. Contaminants that have been successfully remediated with plants include petroleum hydrocarbons, chlorinated solvents, metals, radionuclides, and nutrients such as nitrogen and phosphorus. Glass (1999) estimated that in 1998 at least 200 field remediations or demonstrations had been completed or were in progress around the world; however, the "remediation" technology as currently used is not new but rather draws on the basic ecosystem responses and reactions documented in this and other chapters in this book. The most common applications depend on the plants to draw contaminated soil water to the root zone, where either microbial activity or plant uptake of the contaminants provides the desired removal. Evapotranspiration during the growing season provides for movement and elimination of the contaminated groundwater. Once taken up by the plant, the contaminants are either sequestered in plant biomass or possibly degraded and metabolized to a volatile form and transpired. In some cases, the plant roots can also secrete enzymes, which contribute to degradation of the contaminants in the soil.

Obviously, food crops and similar vegetation, which might become part of the human food chain, are not used on these remediation sites. Grasses and a number of tree species are the most common choices. Hybrid poplar trees have emerged as the most widely used species. These trees grow faster than other northern temperate zone trees, they have high rates of water and nutrient uptake, they are easy to propagate and establish from stem cuttings, and the large number of species varieties permits successful use at a variety of different site conditions. Cottonwood, willow, tulip, eucalyptus, and fir trees have also been used. Wang et al. (1999), for example, have demonstrated the successful removal by hybrid poplar trees (H11-11) of carbon tetrachloride (15 mg/L in solution). The plant degrades and dechlorinates the carbon tetrachloride and releases the chloride ions to the soil and carbon dioxide to the atmosphere. Indian mustard and maize have been studied for the removal of metals from contaminated soils (Lombi et al., 2001). Alfalfa has been used to remediate a fertilizer spill (Russelle et al., 2001).

8.6 INDUSTRIAL WASTEWATER MANAGEMENT

Food processing and other high-strength industrial wastewaters have been treated using land treatment for over 50 years (Bendixen et al., 1969; Brown and Caldwell et al., 2002; Crites et al., 2000). Current issues include: (1) organic loading and oxygen transfer, (2) total acidity loading, and (3) salinity. These issues are discussed in this section.

8.6.1 ORGANIC LOADING RATES AND OXYGEN BALANCE

The soil profile removes biodegradable organics through filtration, adsorption, and biological reduction and oxidation. Most of the biological activity occurs near the surface, where organics are filtered by the soil, and oxygen is present to support biological oxidation; however, biological activity will continue with depth if a food source and nutrients are present. The BOD loading rate is defined as the average BOD applied over the field area in one application cycle. The oxygen demand created by the BOD is balanced by the atmospheric reaeration of the soil profile during the drying period. Excess organic loading can result in (1) odorous anaerobic conditions, (2) reduced soil environments mobilizing oxidized forms of iron and manganese, or (3) increases in percolate hardness and alkalinity via carbon dioxide dissolution. Prevention from excess loading of organics is a function of maintaining an aerobic soil profile, which is managed by organic loading, hydraulic loading, drying time, oxygen flux, and cycle time, not organic loading alone.

Aerobic conditions and carbon dioxide venting can be maintained by balancing the total oxygen demand with oxygen diffusion into the soil. McMichael and McKee (1966) reviewed methods for determining oxygen diffusion in the soil after an application of wastewater. They discussed three principal mechanisms for reaeration: (1) dissolved air carried in the soil by percolating water, (2) the hydrodynamic flow of air resulting from a "piston-like" movement of a slug of water, and (3) diffusion of air through the soil pores. Dissolved oxygen in wastewater has an insignificant impact on high BOD waste streams. The "piston-like" effect may have a substantial impact on the oxygen available immediately after drainage, but quantifying the exact amount is dependent on the difficult-to-model dynamics of draining soils. McMichael and McKee (1966) solved the non-steady-state equation of oxygen diffusion based on Fick's law. They used the equation as a tool for determining the flux of oxygen (mass of O_2 per area) that diffuses in the soil matrix over a given time.

The flux of oxygen across the soil surface does not address the destination of the oxygen, but as long as a gradient exists the oxygen will continue to diffuse into the soil pores. The gradient is based on the oxygen concentration at the soil surface and the initial concentration in the soil. McMichael and McKee (1966) assumed total depletion of oxygen in the soil matrix. Overcash and Pal (1979) assumed a more conservative 140 g/m^3 based on a plant-growth-limiting concentration (Hagen et al., 1967).

The total oxygen demand (TOD) is the sum of the BOD and the nitrogenous oxygen demand (NOD) and plant requirement. The NOD is defined as:

$$NOD = 4.56 \times \text{nitrifiable nitrogen} \tag{8.17}$$

Nitrifiable nitrogen is the ammonium concentration, which is often insignificant when compared to high BOD waste streams. Thus, the TOD is defined as:

$$TOD = BOD + NOD \tag{8.18}$$

From the TOD, the time required to diffuse an equivalent amount of oxygen can be determined. The diffusion equation follows:

$$N_{O2} = 2(C_{O2} - C_p) \times (D_p \cdot t/\pi)1/2 \tag{8.19}$$

where

N_{O2} = Flux of oxygen crossing the soil surface (g/m^2).

C_{O2} = Vapor phase O_2 concentration above the soil surface (310 g/m^3).

C_p = Vapor phase O_2 concentration required in soil to prevent adverse yields or root growth (140 g/m^3).

D_p = Effective diffusion coefficient = $0.6(s)(D_{O2})$, where s is the fraction of air-filled soil pore volume at field capacity, and D_{O2} is the oxygen diffusivity in air $(1.62 \text{ m}^2/\text{d})$.

t = Aeration time = cycle time – infiltration time.

Equation 8.6 can be solved with respect to time:

$$t = D_p \times [N_{O2}/2(C_{O2} - C_p)]^2 \qquad (8.20)$$

Cycle time is a function of required aeration time plus the time for the soil to reach field capacity. The time to reach field capacity is estimated with the infiltration time calculated by dividing the depth applied by the steady-state infiltration rate:

$$t_i = 3600(d/I) \qquad (8.21)$$

where

t_i = Time to infiltrate (hr).

d = Depth (cm).

I = Steady-state infiltration rate (cm/s).

Numerous variables are involved in determining the oxygen balance, all of which must be evaluated on a site-specific basis. An important point to note is that supplemental irrigation water without a significant oxygen demand can increase the required cycle time due to increasing drain time. The time required for the upper zone of the soil to drain is a function of climatic conditions and the depth of the wastewater applied. To achieve the desired loading in surface applications, mixing of supplemental water is often required because of larger applications. Most surface applications cannot apply less than 7.6 cm (3 in.) in a uniform manner.

8.6.2 TOTAL ACIDITY LOADING

Natural biochemical reactions in the soil drive the soil pH to a neutral condition. A range of wastewater pH between 3 and 11 has been applied successfully to land treatment systems. Extended duration of low pH can change the soil fertility and lead to leaching of metals. When the acidity is comprised of mostly organic acids, the water will be neutralized as the organics are oxidized. The acidity of wastewater can be characterized with the total acidity with units of mg $CaCO_3$ per L. The total acidity represents the equivalent mass as $CaCO_3$ required to adjust the pH to a specific pH, commonly defined as 7.0. The soil buffer capacity is reported as mg $CaCO_3$ per kg or tons $CaCO_3$ per ac. The buffer capacity represents the soils ability to neutralize an equivalent amount of acidity. A balance between the total acidity applied in the wastewater and the buffer capacity of the soil can indicate the capacity of the soil to effectively neutralize the acid in the wastewater. The buffer capacity of the soil is restored after organic acids are broken. Most field crops grow well in soils with a pH range of 5.5 to 7.0. Some crops that have a high calcium requirement, such as asparagus or cantaloupes, prefer a soil pH greater than 7.0. If the pH of the soil begins to drop, liming is recommended to return the pH to the desirable range for crop production. Because of the ability of the soil to treat large amounts of organics acids, it is recommended that the pH of wastewater only be adjusted for extreme pH conditions (pH <5.0 or >9). If the mineral (nonorganic) cause of the high or low pH is a threat to crops or groundwater, adjustment may be necessary.

8.6.3 SALINITY

Municipal effluent has an increase of 150 to 380 mg/L of TDS over the source water. In nonoxidized waste streams, approximately 40% of the dissolved solids will consist of volatile dissolved solids that will be removed in the treatment process or will degrade in the soil. The initial dissolved solids plus 40% of the incremental increase are fixed dissolved solids (FDS) or salts. Plant macronutrients, such as nitrogen, phosphorous, and potassium, and minerals, such as calcium and magnesium, are part of the FDS and are partially removed in land application systems that incorporate growing and harvesting of crops. The remaining inorganic dissolved solids are either leached from the soil profile or precipitate out into nonsoluble forms. When inorganic dissolved solids accumulate in the soil, they increase the osmotic stress in plants, resulting in reduced yields or failed germination.

The recommended maximum TDS concentrations for reclaimed water are 500 to 2000 mg/L (USEPA, 2005). At 1000 mg/L of FDS, 32 in. of reclaimed water is equivalent to 7200 lb/ac of salts. A chemical 15–15–15 fertilizer (15% N, 15% P, 15% K) applied at 300 units of N will also apply 2000 pounds of fertilizer salts. It should be recognized that a significant fertilizer salt load is often avoided by reusing water with nutrient value. Salt removal by plants is estimated using the ash content of the harvested crop and can be calculated with similarly as nutrient uptake. Ash content is approximately 10% of the dry weight. Often, salts in excess of crop uptake are applied and leaching of salts is required to limit salt build-up in the root zone.

The leaching requirement is the ratio of the depth of deep percolation to the depth of the applied water. The same ratio exists between the concentration of the conservative salts applied and the concentration of conservative salts in the percolate. The EC of water can reliably indicate the salt concentration when little or no dissolved organics are present. A simple form of this relationship is presented in Equation 8.22; the equation is only valid when weathering and precipitation of salts are insignificant (Hoffman, 1996):

$$LR = \frac{D_d}{D_a} = \frac{C_a}{C_d} \tag{8.22}$$

where
  LR = Leaching fraction (unitless).
  D_d = Drainage depth (m).

D_a = Depth applied (m).

C_a = Concentration of salt applied (dS/m).

C_d = Concentration of salt in drainage (dS/m).

If Equation 8.10 is solved for C_d, the salt concentration of the drainage is equal to the concentration of the salt applied divided by the leaching fraction:

$$C_a = \frac{C_d}{LR} \qquad (8.23)$$

All terms are as described above. The leaching requirement is determined based on the crop sensitivity. The average root zone salts are calculated based on solving the continuity equation for salt throughout the root zone (Hoffman and van Genuchten, 1983):

$$\frac{\overline{C}}{C_a} = \frac{1}{L} + \frac{\delta}{Z \cdot LR} \cdot \ln\left[L + (1-L)e^{-Z/\delta}\right] \qquad (8.24)$$

$\overline{C}$ = Mean root zone salt concentration (dS/m).

C_a = Salt concentration of applied water (dS/m).

δ = Empirical constant = 0.2Z.

Z = Root zone depth (m).

LR = Leaching fraction as defined in Equation 8.10.

To determine the desired EC value of drainage, both the crop sensitivity to salinity and the groundwater quality should be reviewed. The groundwater uses, quality, and flux beneath the site should be reviewed to determine the impact of the leachate of groundwater. High EC values can be offset by small leaching depths resulting in insignificant loading to the groundwater. Also, precipitation of minerals continues to occur below the root zone reducing the loading to groundwater.

REFERENCES

Ayers, R.S. and Westcot, D.W. (1985). *Water Quality for Agriculture*, FAO Irrigation and Drainage Paper 29, Revision 1, Food and Agriculture Organization of the United Nations, Rome.

Bendixen, T.W., Hill, R.D., DuByne, F.T., and Robeck, G.G. (1969). Cannery waste treatment by spray irrigation runoff, *J. Water Pollut. Control Fed.*, 41, 385.

Benham-Blair & Affiliates, Inc., and Engineering Enterprises, Inc. (1979). *Long-Term Effects of Land Application of Domestic Wastewater: Milton, Wisconsin, Rapid Infiltration Site*, EPA-600/2-79-145, U.S. Environmental Protection Agency, Washington, D.C.

Bouwer, H. et al. (1980). Rapid-infiltration research at Flushing Meadows project, Arizona, *J. Water Pollut. Control Fed.*, 52(10), 2457–2470.

Brown and Caldwell et al. (2002). *Manual of Good Practice for Land Application of Food Process/Rinse Water*, California League of Food Processors, Sacramento, CA.

Carlson, C.A., Hunt, P.G., and Delaney, Jr., T.B. (1974). *Overland Flow Treatment of Wastewater*, Misc. Paper Y-74-3, U.S. Waterways Experiment Station, Vicksburg, MS.

Conroy, O., Turney, K.D., Lansey, K.E., and Arnold, R.G. (2001). Endocrine disruption in wastewater and reclaimed water, in *Proceedings of the Tenth Biennial Symposium on Artificial Recharge of Groundwater*, Tucson, AZ, June 7–9, 2001, 171–179.

Crites, R.W. (1985a). Micropollutant removal in rapid infiltration, in *Artificial Recharge of Groundwater*, Asano, T., Ed., Butterworth Publishers, Stoneham, MA, 579–608.

Crites, R.W. (1985b). Nitrogen removal in rapid infiltration systems, *J. Environ. Eng. Div. ASCE*, 111(6), 865–873.

Crites, R.W. (1997). Slow rate land treatment, in *Natural Systems for Wastewater Treatment: Manual of Practice*, Water Environment Federation, Alexandria, VA.

Crites, R.W. and Tchobanoglous, G. (1998). *Small and Decentralized Wastewater Management Systems*, McGraw-Hill, New York.

Crites, R. W., Reed, S.C., and Bastian, R.K. (2000). *Land Treatment Systems for Municipal and Industrial Wastes*, McGraw-Hill. New York.

Crites, R.W., Reed, S.C., and Bastian, R.K. (2001). Applying treated wastewater to land, *BioCycle*, 42, 32–36.

Drewes, J., Heberer, T., and Reddersen, K. (2001). Fate of pharmaceuticals during groundwater recharge, in *Proceedings of the Tenth Biennial Symposium on Artificial Recharge of Groundwater*, June 7–9, 2001, 181–190.

Enfield, C.G. (1978). Evaluation of phosphorus models for prediction of percolate water quality in land treatment, in *Proceedings of the International Symposium on Land Treatment of Wastewater*, Vol. 1, Cold Regions Research and Engineering Laboratory (CRREL), Hanover, NH, 153

Enfield, C.G. and Bledsoe, B.E. (1975). *Kinetic Model from Orthophosphate Reactions in Mineral Soils*, EPA-660/2-75-022, Office of Research and Development, U.S. Environmental Protection Agency, Ada, OK.

Gable, J.E. and Fox, P. (2000). Nitrogen Removal During Soil Aquifer Treatment by Anaerobic Ammonium Oxidation (ANAMMOX), paper presented at Joint WEF/AWWA Conference, San Antonio, TX.

Gilde, L.C., Kester, A.S., Law, J.P., Neeley, C.H., and Parmelee, D.M. (1971). A spray irrigation system for treatment of cannery wastes, *J. Water Pollut. Control Fed.*, 43, 2011.

Glass, D. G. (1999). *International Activities in Phytoremediation: Industry and Market Overview, Phytoremediation and Innovative Strategies for Specialized Remedial Applications*, Battelle Press, Columbus, OH, 95–100.

Hagen, R.M., Haise, H.R., and Edminster, T.W., Eds. (1967). *Irrigation of Agriculture Lands*, Agronomy Ser. No. 11, American Society of Agronomy, Madison, WI.

Hart, W.E. (1975). *Irrigation System Design*, Colorado State University, Department of Agricultural Engineering, Fort Collins, CO.

Hoffman, G.J. (1996). Leaching fraction and root zone salinity control, in *Agricultural Salinity Assessment and Management*, Tanji, K.K., Ed., ASCE No. 71, American Society of Civil Engineers, New York.

Hoffman, G.J. and van Genuchten, M.Th. (1983). Soil properties and efficient water use: water management for salinity control, in *Limitations to Efficient Water Use in Crop Production*, Taylor, H.M., Jordan, W., and Sinclair, T., Eds., American Society of Agronomy, Madison, WI, 73–85.

Jenkins, T.F., Leggett, D.C., and Martel, C.J. (1980). Removal of volatile trace organics from wastewater, *J. Environ. Sci. Health*, A15, 211.

Jewell, W.J. and Seabrook, B.L. (1979). *History of Land Application as a Treatment Alternative*, EPA 430/9-79-012, U.S. Environmental Protection Agency, Washington, D.C.

Johnston, J. and Smith, R. (1988). Operating Schedule Effects on Nitrogen Removal in Overland Flow Treatment Systems, paper presented at the 61st Annual Conference of the Water Pollution Control Federation, Dallas, TX.

Kruzic, A.J. and Schroeder, E.D. (1990). Nitrogen removal in the overland flow wastewater treatment process: removal mechanisms, *Res. J. Water Pollut. Control Fed,*, 62(7), 867–876.

Lance, J.C., Whisler, F.D., and Rice, R.C. (1976). Maximizing denitrification during soil filtration of sewage water, *J. Environ. Qual.*, 5, 102.

Lasat, M.M. (2002). Phytoextraction of toxic metals: a review of biological mechanisms, *J. Environ. Qual.*, 31, 109–120.

Lombi, E., Zhao, F.J., Dunham, S.J., and McGrath, S.P. (2001). Phytoremediation of heavy-metal-contaminated soils: natural hyperaccumulation versus chemically enhanced phytoextraction, *J. Environ. Qual.*, 30, 1919–1926.

Martel, C.J., Jenkins, T.F., Diener, C.J., and Butler P.L. (1982). *Development of a Rational Design Procedure for Overland Flow Systems*, CRREL Report 82-2, Cold Regions Research and Engineering Laboratory (CRREL), Hanover, NH.

McMichael, F.C. and McKee, J.E. (1966). *Wastewater Reclamation at Whittier Narrows*, Publ. No. 33, State Water Quality Control Board, Los Angeles, CA.

Muirhead, W.M., Kawata, E.M., and Crites, R.W. (2003). Assessment of recycled water irrigation in central Oahu, in *Proceedings of WEFTEC 2003*, Water Environment Federation, Los Angeles, CA, October 11–15, 2003.

Nolte Associates. (1997). *Demonstration Wetlands Project: 1996 Annual Report*, Sacramento Regional Wastewater Treatment Plant, Sacramento County, CA.

Nutter, W. L. (1986). Forest land treatment of wastewater in Clayton County, Georgia: a case study, in *The Forest Alternative for Treatment and Utilization of Municipal and Industrial Wastes*, Cole, D.W., Henry, C.L., and Nutter, W.L., University of Washington Press, Seattle, WA.

Overcash, M.R. and Pal, D. (1979). *Design of Land Treatment Systems for Industrial Wastes: Theory and Practice*, Ann Arbor Science, Ann Arbor, MI.

Pair, C. H., Ed. (1983). *Irrigation*, 5th ed., Irrigation Association, Silver Spring, MD.

Perry, L.E., Reap, E.J., and Gilliand, M. (1981). Pilot scale overland flow treatment of high strength snack food processing wastewaters, in *Proceedings National Conference on Environmental Engineering*, American Society of Civil Engineers, Environmental Engineering Division, Atlanta, GA.

Peters, R.E., Lee C.R., and Bates, D.J. (1981). *Field Investigations of Overland Flow Treatment of Municipal Lagoon Effluent*, Tech. Report EL-81-9, U.S. Waterways Experiment Station, Vicksburg, MS.

Pound, C.E. and Crites, R.W. (1973). *Wastewater Treatment and Reuse by Land Application*, EPA 660/2-73-006b, U.S. Environmental Protection Agency, Washington, D.C.

Pound, C.E., Crites, R.W., and Olson, J.V. (1978). *Long-Term Effects of Land Application of Domestic Wastewater: Hollister, California, Rapid Infiltration Site*, EPA-600/2-78-084, U.S. Environmental Protection Agency, Washington, D.C.

Quanrad, D., Conroy, Q., Turney, K., Lansey, K., and Arnold, R. (2002). Fate of estrogenic activity in reclaimed water during soil aquifer treatment, in *Proceedings of the Water Sources Conference*, American Water Works Association, Las Vegas, NV.

Rafter, G.W. (1899). *Sewage Irrigation*. Part II. *Water Supply and Irrigation Papers*, No. 22, U.S. Geological Survey, U.S. Government Printing Office, Alexandria, VA.

Reed, S.C. (1972). *Wastewater Management by Disposal on the Land*, Special Report 171, Cold Regions Research and Engineering Laboratory (CRREL), Hanover, NH.

Reed, S.C. and Crites, R.W. (1984). *Handbook of Land Treatment Systems for Industrial and Municipal Wastes*, Noyes Publications, Park Ridge, NJ.

Reed, S.C., Crites, R.W. and Wallace, A.T. (1985). Problems with rapid infiltration: a post mortem analysis, *J. Water Pollut. Control Fed.*, 57(8), 854–858.

Reed, S.C., Crites, R.W., and Middlebrooks, E.J. (1995). *Natural Systems for Waste Management and Treatment*, 2nd ed., McGraw-Hill, New York.

Russelle, M.P., Lamb, J.F.S., Montgomery, B.R., Elsenheimer, D.W., Miller, B.S., and Vance, C.P. (2001). Alfalfa rapidly remediates excess inorganic nitrogen at a fertilizer spill site, *J. Environ. Qual.*, 30, 30–36.

Smith, J.W and Crites, R.W. (2001). Rational method for the design of organic loading rates in a land application system, in *Proceedings of WEFTEC 2001*, Water Environment Federation, Atlanta, GA.

Smith, J.W and Murray, R. (2003). Vadose zone investigation of land application food processing wastewater, in *Proceedings of WEFTEC 2003*, Water Environment Federation, Los Angeles, CA.

Smith, R.G. and Schroeder, E.D. (1982). *Demonstration of the Overland Flow Process for the Treatment of Municipal Wastewater: Phase II Field Studies*, Report to California State Water Resources Control Board, Department of Civil Engineering, University of California, Davis.

Smith, R.G. and Schroeder, E.D. (1985). Field studies of the overland flow process for the treatment of raw and primary treated municipal wastewater, *J. Water Pollut. Control Fed.*, 57(7), 785–794.

Sopper, W.E. and Kardos, L.T. (1973). *Recycling Treated Municipal Wastewater and Sludge Through Forest and Cropland*, The Pennsylvania State University Press, University Park, PA.

Tedaldi, D.J. and Loehr, R.C. (1991). Performance of an overland flow system treating food-processing wastewater, *Res. J. Water Pollut. Control Fed.*, 63, 266.

Thomas, R.E., Jackson, K., and Penrod, L. (1974). *Feasibility of Overland Flow for Treatment of Raw Domestic Wastewater*, EPA 660/2-74-087, U.S. Environmental Protection Agency, Washington, D.C.

USDA. (1972). *Drainage of Agricultural Land: A Practical Handbook for the Planning, Design, Construction, and Maintenance of Agricultural Drainage Systems*, U.S. Department of Agriculture, Soil Conservation Service, U.S. Government Printing Office, Washington, D.C.

USDA. (1983). Sprinkler irrigation, in *Irrigation SCS National Engineering Handbook*, U.S. Department of Agriculture, Soil Conservation Service, U.S. Government Printing Office, Washington, D.C., chap. 11, sect. 15

USDOI. (1978). *Drainage Manual*, Bureau of Reclamation, U.S. Department of Interior, Washington, D.C.

USEPA. (1981). *Process Design Manual for Land Treatment of Municipal Wastewater*, EPA 625/1-81-013, U.S. Environmental Protection Agency, Cincinnati, OH.

USEPA. (1984). *Process Design Manual for Land Treatment of Municipal Wastewater, Supplement on Rapid Infiltration and Overland Flow*, EPA 625/1-81-0139, Center for Environmental Research Information (CERI), U.S. Environmental Protection Agency, Cincinnati, OH.

USEPA. (2000). *Introduction to Phytoremediation*, EPA/600/R-99/107, U.S. Environmental Protection Agency, Washington, D.C.

USEPA. (2005). *Process Design Manual for Land Treatment of Municipal and Industrial Wastewater*, Center for Environmental Research Information (CERI), U.S. Environmental Protection Agency, Cincinnati, OH.

Van Schifgaarde, J., Ed. (1974). *Drainage for Agriculture*, No. 17, Series on Agronomy, American Society of Agronomy, Madison, WI.

Wang, X., L. E. Newman, M. P. Gordon (1999). Biodegradation of carbon tetrachloride by poplar trees: results from cell culture and field experiments, in *Phytoremediation and Innovative Strategies for Specialized Remedial Applications*, Battelle Press, Columbus, Ohio.

WEF. (2001). *Natural Systems for Wastewater Treatment, Draft, Manual of Practice*, Water Environment Federation, Alexandria, VA.

Woods, C., Bouwer, H., Svetich, R., Smith, S., and Prettyman R. (1999). Study finds biological nitrogen removal in soil aquifer treatment system offers substantial advantages, in *Proceedings of WEFTEC 99*, Water Environment Federation, New Orleans, LA, October 9–13, 1999.

Zirschky, J., Crawford, D., Norton, L., Richards, S., and Deemer, D. (1989). Meeting ammonia limits using overland flow, *J. Water Pollut. Control Fed.*, 61(12), 1225–1232.

9 Sludge Management and Treatment

Approximately 6.9 million ton of biosolids were generated in the United States in 1998, and about 60% of it was used beneficially in land applications, composting, and landfill cover. It is estimated that, by 2010, 8.2 million tons will be generated, and 70% of the biosolids is expected to be used beneficially (USEPA, 1999). Recycling options are described in various documents (Crites and Tchobanoglous, 1998; Crites et al., 2000; USEPA, 1994a, 1995a,c). Sludges are a common by-product from all waste treatment systems, including some of the natural processes described in previous chapters. Sludges are also produced by water treatment operations and by many industrial and commercial activities. The economics and safety of disposal or reuse options are strongly influenced by the water content of the sludge and the degree of stabilization with respect to pathogens, organic content, metals content, and other contaminants. This chapter describes several natural methods for sludge treatment and reuse. In-plant sludge processing methods, such as thickening, digestion, and mechanical methods for conditioning and dewatering, are not included in this text; instead, Grady et al. (1999), ICE (2002), Metcalf & Eddy (2003), Reynolds and Richards (1996), and USEPA (1979, 1982) are recommended for that purpose.

9.1 SLUDGE QUANTITY AND CHARACTERISTICS

The first step in the design of a treatment or disposal process is to determine the amount of sludge that must be managed and its characteristics. Deriving a solids mass balance for the treatment system under consideration can produce a reliable estimate. The solids input and output for every component in the system must be calculated. Typical values for solids concentrations from in-plant operations and processes are reported in Table 9.1. Detailed procedures for conducting mass balance calculations for wastewater treatment systems can be found in Grady et al. (1999), Metcalf & Eddy (2003), Reynolds and Richards (1996), and USEPA (1979). The characteristics of wastewater treatment sludges are strongly dependent on the composition of the untreated wastewater and on the unit operations in the treatment process. The values reported in Table 9.2 and Table 9.3 represent typical conditions only and are not a suitable basis for a specific project design. The sludge characteristics must be either measured or carefully estimated from similar experience elsewhere to provide the data for final designs.

TABLE 9.1
Typical Solids Content from Treatment Operations

Treatment Operation	Percent (%)[a]	Typical Dry Solids (kg/10³ m³)[b]
Primary Settling		
Primary only	5	150
Primary and waste-activated sludge	1.5	45
Primary and trickling-filter sludge	5	150
Secondary Reactors		
Activated sludge:		
Pure oxygen	2.5	130
Extended aeration	1.5	100
Trickling filters	1.5	70
Chemical Plus Primary Sludge		
High lime (>800 mg/L)	10	800
Low lime (<500 mg/L)	4	300
Iron salts	7.5	600
Thickeners		
Gravity type:		
Primary sludge	8	140
Primary and waste-activated sludge	4	70
Primary and trickling filter	5	90
Flotation	4	70
Digestion		
Anaerobic:		
Primary sludge	7	210
Primary and waste-activated sludge	3.5	105
Aerobic:		
Primary and waste-activated sludge	2.5	80

[a] Percent solids in liquid sludge.
[b] kg/10³ m³ = dry solids/1000 m³ liquid sludge.

Source: Metcalf & Eddy, *Wastewater Engineering: Treatment, Disposal, and Reuse*, 3rd ed., McGraw-Hill, New York, 1991. With permission.

TABLE 9.2
Typical Composition of Wastewater Sludges

Component	Untreated Primary	Digested
Total solids (TS; %)	5	10
Volatile solids (% of TS)	65	40
pH	6	7
Alkalinity (mg/L as $CaCO_3$)	600	3000
Cellulose (% of TS)	10	10
Grease and fats (ether soluble; % of TS)	6–30	5–20
Protein (% of TS)	25	18
Silica (SiO_2; % of TS)	15	10

Source: Metcalf & Eddy, *Wastewater Engineering: Treatment, Disposal, and Reuse*, 3rd ed., McGraw-Hill, New York, 1991. With permission.

TABLE 9.3
Nutrients and Metals in Typical Wastewater Sludges

Component	Median	Mean
Total nitrogen (%)	3.3	3.9
NH_4^+ (as N; %)	0.09	0.65
NO_3^- (as N; %)	0.01	0.05
Phosphorus (%)	2.3	2.5
Potassium (%)	0.3	0.4

Component	Mean	Standard Deviation
Copper (mg/kg)	741	962
Zinc (mg/kg)	1200	1554
Nickel (mg/kg)	43	95
Lead (mg/kg)	134	198
Cadmium (mg/kg)	7	12
PCB-1248 (mg/kg)	0.08	1586

Source: Data from USEPA (1983, 1990) and Whiting (1975).

TABLE 9.4
Pond Sludge Accumulation Data Summary

	Facultative Ponds (Utah)		Aerated Ponds (Alaska)	
Parameter	A	B	C	D
Flow (m³/d)	37,850	694	681	284
Surface (m²)	384,188	14,940	13,117	2520
Bottom (m²)	345,000	11,200	8100	1500
Operated since last cleaning (yr)	13	9	5	8
Mean sludge depth (cm)	8.9	7.6	33.5	27.7
Total solids (g/L)	58.6	76.6	85.8	9.8
Volatile solids (g/L)	40.5	61.5	59.5	4.8
Wastewater, suspended solids (mg/L)	62	69	185	170

Source: Schneiter, R.W. et al., *Accumulation, Characterization and Stabilization of Sludges from Cold Regions Lagoons*, CRREL Special Report 84-8, U.S. Army Cold Regions Research and Engineering Laboratory, Hanover, NH, 1984. With permission.

9.1.1 SLUDGES FROM NATURAL TREATMENT SYSTEMS

A significant advantage for the natural wastewater treatment systems described in previous chapters is the minimal sludge production in comparison to mechanical treatment processes. Any major quantities of sludge are typically the result of preliminary treatments and not the natural process itself. The pond systems described in Chapter 4 are an exception in that, depending on the climate, sludge will accumulate at a gradual but significant rate, and its ultimate removal and disposal must be given consideration during design. In colder climates, studies have established that sludge accumulation proceeds at a faster rate, so removal may be required more than once over the design life of the pond. The results of investigations in Alaska and Utah (Schneiter et al., 1984) on sludge accumulation and composition in both facultative and partial-mix aerated lagoons are reported in Table 9.4 and Table 9.5.

A comparison of the values in Table 9.4 and Table 9.5 with those in Table 9.2 and Table 9.3 indicates that the pond sludges are similar to untreated primary sludges. The major difference is that the solids content, both total and volatile, is higher for most pond sludges than for primary sludge, and the fecal coliforms are significantly lower. This is reasonable in light of the very long detention time in ponds as compared with primary clarifiers. The long detention time allows for significant die-off of fecal coliforms and for some consolidation of the sludge solids. All four of the lagoons described in Table 9.4 and Table 9.5 are assumed to be located in cold climates. Pond systems in the southern half of the United States might expect lower accumulation rates than those indicated in Table 9.4.

TABLE 9.5
Composition of Pond Sludges

Parameter	Facultative Ponds (Utah)		Aerated Ponds (Alaska)	
	A	B	C	D
Total solids (%)	5.9	7.7	8.6	0.89
Total solids (mg/L)	586,000	766,600	85,800	9800
Volatile solids (%)	69.1	80.3	69.3	48.9
Total organic carbon (mg/L)	5513	6009	13,315	2651
pH	6.7	6.9	6.4	6.8
Fecal coliforms ([number/100 mL] × 10⁵)	0.7	1	0.4	2.5
Total Kjeldahl nitrogen (mg(L)	1028	1037	1674	336
Total Kjeldahl nitrogen (% of TS)	1.75	1.35	1.95	3.43
Ammonia nitrogen (as N; mg/L)	72.6	68.6	93.2	44.1
Ammonia nitrogen (as N; % of TS)	0.12	0.09	0.11	0.45

Source: Schneiter, R.W. et al., *Accumulation, Characterization and Stabilization of Sludges from Cold Regions Lagoons*, CRREL Special Report 84-8, U.S. Army Cold Regions Research and Engineering Laboratory, Hanover, NH, 1984. With permission.

9.1.2 SLUDGES FROM DRINKING-WATER TREATMENT

Sludges occur in water treatment systems as a result of turbidity removal, softening, and filter backwash. The dry weight of sludge produced per day from softening and turbidity removal operations can be calculated using Equation 9.1 (Lang et al., 1985):

$$S = 84.4Q(2Ca + 2.6Mg + 0.44Al + 1.9Fe + SS + A_x) \tag{9.1}$$

where
S = Sludge solids (kg/d).
Q = Design water treatment flow (m^3/s).
Ca = Calcium hardness removed (as $CaCO_3$; mg/L).
Mg = Magnesium hardness removed (as $CaCO_3$; mg/L).
Al = Alum dose (as 17.1% Al_2O_3; mg/L).
Fe = Iron salts dose (as Fe; mg/L).
SS = Raw-water suspended solids (mg/L).
A_x = Additional chemicals (e.g., polymers, clay, activated carbon) (mg/L).

The major components of most of these sludges are due to the suspended solids (SS) from the raw water and the coagulant and coagulant aids used in treatment.

TABLE 9.6
Characteristics of Water Treatment Sludges

Characteristic	Range of Values
Volume (as percent of water treated)	<1.0
Suspended solids concentration	0.1–1000 mg/L
Solids content	0.1–3.5%
Solids content after long-term settling	10–35%
Composition, alum sludge:	
Hydrated aluminum oxide	15–40%
Other inorganic materials	70–35%
Organic materials	15–25%

Source: Lang, L.E. et al., *Procedures for Evaluating and Improving Water Treatment Plant Processes at Fixed Army Facilities*, Report of the U.S. Army Construction Engineering Research Laboratory, Champaign, IL, 1985.

Sludges resulting from coagulation treatment are the most common and are typically found at all municipal water treatment works. Typical characteristics of these sludges are reported in Table 9.6.

9.2 STABILIZATION AND DEWATERING

Stabilization of wastewater sludges and dewatering of most all types of sludge are necessary for economic, environmental, and health reasons. Transport of sludge from the treatment plant to the point of disposal or reuse is a major factor in the costs of sludge management. Table 9.7 presents the desirable sludge solids content for the major disposal and reuse options. Sludge stabilization controls offensive odors, lessens the possibility for further decomposition, and significantly reduces pathogens. Typical pathogen contents in unstabilized and anaerobically digested sludges are compared in Table 3.10. Research on the use of various fungal strains as a means to stabilize sludges has been conducted with mixed results but may hold promise in some cases (Alam et al., 2004).

9.2.1 METHODS FOR PATHOGEN REDUCTION

The pathogen content of sludge is especially critical when the sludge is to be used in agricultural operations or when public exposure is a concern. Four processes to significantly reduce pathogens and seven processes to further reduce pathogens are recognized by the U.S. Environmental Protection Agency (EPA), as described by Bastian (1993), Crites et al. (2000), and USEPA (2003a).

TABLE 9.7
Solids Content for Sludge Disposal or Reuse

Disposal/Reuse Method	Reason To Dewater	Required Solids (%)
Land application	Reduce transport and other handling costs	>3
Landfill	Regulatory requirements	>10[a]
Incineration	Process requirements to reduce fuel required to evaporate water	>26

[a] Greater than 20% in some states.

Source: USEPA, *Process Design Manual: Land Application of Municipal Sludge*, EPA 625/1-83-016, Center for Environmental Research Information, U.S. Environmental Protection Agency, Cincinnati, OH, 1983.

9.3 SLUDGE FREEZING

Freezing and then thawing a sludge will convert an undrainable jelly-like mass into a granular material that will drain immediately upon thawing. This natural process may offer a cost-effective method for dewatering.

9.3.1 EFFECTS OF FREEZING

Freeze–thawing will have the same effect on any type of sludge but is particularly beneficial with chemical and biochemical sludges containing alum which are extremely slow to drain naturally. Energy costs for artificial freeze–thawing are prohibitive, so the concept must depend on natural freezing to be cost effective.

9.3.2 PROCESS REQUIREMENTS

The design of a freeze dewatering system must be based on worst-case conditions to ensure successful performance at all times. If sludge freezing is to be a reliable expectation every year, the design must be based on the warmest winter during the period of concern (typically 20 years or longer). The second critical factor is the thickness of the sludge layer that will freeze within a reasonable period if freeze–thaw cycles are a normal occurrence during the winter. A common mistake with past attempts at sludge freezing has been to apply sludge in a single deep layer. In many locations, a large single layer may never freeze completely to the bottom, so only the upper portion goes through alternating freezing and thawing cycles. It is absolutely essential that the entire mass of sludge be frozen completely for the benefits to be realized; also, when the sludge has frozen and thawed, the change is irreversible.

9.3.2.1 General Equation

The freezing or thawing of a sludge layer can be described by Equation 9.2:

$$Y = m(\Delta T \times t)^{1/2} \qquad (9.2)$$

where

Y	=	Depth of freezing or thawing (cm; in.).
m	=	Proportionality coefficient (cm (°C·d)$^{-1/2}$) = 2.04 cm (°C·d)$^{-1/2}$ = 0.60 in. (°F·d)$^{-1/2}$.
ΔT	=	Temperature difference between 0°C (32° F) and the average ambient air temperature during the period of interest (°C; °F).
t	=	Time period of concern (d).
$\Delta T \times t$	=	Freezing or thawing index (°C·d; °F·d).

Equation 9.2 has been in general use for many years to predict the depth of ice formation on ponds and streams. The proportionality coefficient m is related to the thermal conductivity, density, and latent heat of fusion for the material being frozen or thawed. A median value of 2.04 was experimentally determined for wastewater sludges in the range of 0 to 7% solids (Reed et al., 1984). The same value is applicable to water treatment and industrial sludges in the same concentration range.

The freezing or thawing index in Equation 9.2 is an environmental characteristic for a particular location. It can be calculated from weather records and can also be found directly in other sources (Whiting, 1975). The factor ΔT in Equation 9.2 is the difference between the average air temperature during the period of concern and 32°F (0°C). Example 9.1 illustrates the basic calculation procedure.

Example 9.1. Determination of Freezing Index

The average daily air temperatures for a 5-d period are listed below. Calculate the freezing index for that period.

Day	Mean Temperature (°C)
1	0
2	–6
3	–9
4	+3
5	–8

Solution

1. The average air temperature during the period is –4°C.
2. The freezing index for the period is ΔT d = [0 – (–4)](5) = 20°C·d.

The rate of freezing decreases with time under steady-state temperatures, because the frozen material acts as an insulating barrier between the cold ambient air and the remaining unfrozen sludge. As a result, it is possible to freeze a greater total depth of sludge in a given time if the sludge is applied in thin layers.

9.3.2.2 Design Sludge Depth

In very cold climates with prolonged winters, the thickness of the sludge layer is not critical; however, in more temperate regions, particularly those that experience alternating freeze–thaw periods, the layer thickness can be very important. Calculations by Equation 9.2 tend to converge on a 3-in. (8-cm) layer as a practical value for almost all locations where freezing conditions occur. At 23°F (–5°C), a 3-in. (8-cm) layer should freeze in about 3 days; at 30°F (–1°C) it would take about 2 weeks. A greater depth should be feasible in colder climates. Duluth, Minnesota, for example, successfully freezes sludges from a water treatment plant in 9-in. (23-cm) layers (Schleppenbach, 1983). It is suggested that a 3-in. (8-cm) depth may be used for feasibility assessment and preliminary designs. A larger increment may then be justified by a detailed evaluation during final design.

9.3.3 Design Procedures

The process design for sludge freezing must be based on the warmest winter of record to ensure reliable performance at all times. The most accurate approach is to examine the weather records for a particular location and determine how many 3-in. (8-cm) layers could be frozen each winter. The winter with the lowest total depth is then the design year. This approach might assume, for example, that the first layer is applied to the bed on November 1 each year. Equation 9.2 is rearranged and used with the weather data to determine the number of days required to freeze the layer:

$$t = \frac{(Y/m)^2}{\Delta T} \tag{9.3}$$

With an 8-cm layer and $m = 2.04$, the equation becomes:

$$t = \frac{15.38}{\Delta T}$$

In U.S. customary units (3-in. layer, $m = 0.6$ in. $[°F·d]^{-1/2}$):

$$t = \frac{25.0}{\Delta T}$$

9.3.3.1 Calculation Methods

The mean daily air temperatures are used to calculate the ΔT value. The calculations take account of thaw periods, and a new sludge application is not made until the previous layer has frozen completely. One day is then allowed for a new sludge application and cooling, and calculations with Equation 9.3 are repeated to again determine the freezing time. The procedure is repeated through the end of the winter season. A tabular summary is recommended for the data and calculation results. This procedure can be easily programmed for rapid calculations with a spreadsheet or desktop calculator.

9.3.3.2 Effect of Thawing

Thawing of previously frozen layers during a warm period is not a major concern, as these solids will retain their transformed characteristics. Mixing of a new deposit of sludge with thawed solids from a previously frozen layer will extend the time required to refreeze the combined layer (solve Equation 9.3 for the combined thickness). If an extended thaw period occurs, removal of the thawed sludge cake is recommended.

9.3.3.3 Preliminary Designs

A rapid method, useful for feasibility assessment and preliminary design, relates the potential depth of frozen sludge to the maximum depth of frost penetration into the soil at a particular location. The depth of frost penetration is also dependent on the freezing index for a particular location; published values can be found in the literature (e.g., Penner, 1962; Whiting, 1975). Equation 9.4 correlates the total depth of sludge that could be frozen if applied in 3-in. (8-cm) increments with the maximum depth of frost penetration:

$$\Sigma Y = 1.76\left(F_p\right) - 101 \quad \text{(metric units)} \tag{9.4a}$$

$$\Sigma Y = 1.76\left(F_p\right) - 40 \quad \text{(U.S. units)} \tag{9.4b}$$

where ΣY is the total depth of sludge that can be frozen in 3-in. (8-cm) layers during the warmest design year, in inches or centimeters, and F_p is the maximum depth of frost penetration, in inches or centimeters. The maximum depths of frost penetration for selected locations in the northern United States and Canada are reported in Table 9.8.

9.3.3.4 Design Limits

It can be demonstrated using Equation 9.4 that sludge freezing will not be feasible unless the maximum depth of frost penetration is at least 22 in. (57 cm) for a particular location. In general, that will begin to occur above the 38th parallel of

TABLE 9.8
**Maximum Depth of Frost Penetration and Potential Depth
of Frozen Sludge**

Location	Maximum Frost Penetration (cm)	Potential Depth of Frozen Sludge (cm)
Bangor, Maine	183	221
Concord, New Hampshire	152	166
Hartford, Connecticut	124	117
Pittsburgh, Pennsylvania	97	70
Chicago, Illinois	122	113
Duluth, Minnesota	206	261
Minneapolis, Minnesota	190	233
Montreal, Quebec	203	256

latitude and will include most of the northern half of the United States, with the exception of the west coast; however, sludge freezing will not be cost effective if only one or two layers can be frozen in the design year. A maximum frost penetration of about 39 in. (100 cm) would allow sludge freezing for a total depth of 30 in. (75 cm). The process should be cost effective at that stage, depending on land and construction costs. The results of calculations using Equation 9.4 are plotted in Figure 9.1, which indicate the potential depth of sludge that could be frozen at all locations in the United States. This figure or Equation 9.3 can be

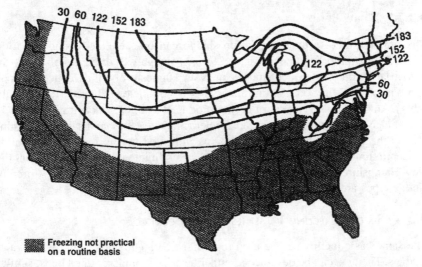

FIGURE 9.1 Potential depth of sludge that could be frozen when applied in 8-cm layers.

TABLE 9.9
Effects of Sludge Freezing

	Percent Solids Content	
Location and Sludge Type	Before Freezing	After Freezing
Cincinnati, Ohio		
Wastewater sludge, with alum	0.7	18
Water treatment, with iron salts	7.6	36
Water treatment, with alum	3.3	27
Ontario, Canada		
Waste-activated sludge	0.6	17
Anaerobically digested	5.1	26
Aerobically digested	2.2	21
Hanover, New Hampshire		
Digested, wastewater sludge, with alum	2–7	25–35
Digested primary	3–8	30–35

Source: Data from Farrell(1970), Reed et al. (1984), Rush and Strickland (1979), and Schleppenbach (1983).

used for preliminary estimates, but the final design should be based on actual weather records for the site and the calculation procedure described earlier.

9.3.3.5 Thaw Period

The time required to thaw the frozen sludge can be calculated using Equation 9.2 and the appropriate thawing index. Frozen sludge will drain quite rapidly. In field trials with wastewater sludges in New Hampshire, solids concentrations approached 25% as soon as the material was completely thawed (Reed et al., 1984). An additional 2 weeks of drying produced a solids concentration of 54%. The sludge particles retain their transformed characteristics, and subsequent rainfall on the bed will drain immediately, as indicated by the fact that the solids concentration was still about 40% 12 hours after an intense rainfall (4 cm) at the New Hampshire field trial (Reed et al., 1984). The effects for a variety of different sludge types are reported in Table 9.9.

9.3.4 SLUDGE FREEZING FACILITIES AND PROCEDURES

The same basic facility can be used for water treatment sludges and wastewater sludges. The area can be designed as either a series of underdrained beds, similar in detail to conventional sand drying beds, or deep, lined, and underdrained trenches. The Duluth, Minnesota, water treatment plant uses the trench concept (Schleppenbach, 1983). The sludges are pumped to the trenches on a routine basis throughout the year. Any supernatant is drawn off just prior to the onset of winter. After an initial ice layer has formed, sludge is pumped up from beneath the ice, spread in repeated layers on the ice surface, and allowed to freeze. The sand bed approach requires sludge storage elsewhere and application to the bed after the freezing season has begun.

9.3.4.1 Effect of Snow

Neither beds nor trenches require a roof or a cover. A light snowfall (less than 4 cm) will not interfere with the freezing, and the contribution of the meltwater to the total mass will be negligible. What must be avoided is application of sludge under a deep snow layer. The snow in this case will act as an insulator and retard freezing of the sludge. Any deep snow layers should be removed prior to a new sludge application.

9.3.4.2 Combined Systems

If freezing is the only method used to dewater wastewater sludges, then storage is required during warm periods. A more cost-effective alternative is to combine winter freezing with polymer-assisted summer dewatering on the same bed. In a typical case, winter sludge application might start in November and continue in layers until about 3 ft (1 m) of frozen material has accumulated. In most locations, this will thaw and drain by early summer. Polymer-assisted dewatering can then continue on the same beds during the summer and early fall. Sludge storage in deep trenches during the warm months is better suited for water treatment operations where putrefaction and odors are not a problem.

9.3.4.3 Sludge Removal

It is recommended that the drained wastewater sludges be removed each year. Inert chemical sludges from water treatment and industrial operations can remain in place for several years. In these cases, a trench 7 to 10 ft (2 to 3 m) deep can be constructed, so the dried solids residue remains on the bottom. In addition to new construction, the sludge freezing concept can allow the use of existing conventional sand beds, which are not now used in the winter months.

Example 9.2

A community near Pittsburgh, Pennsylvania, is considering freezing as the dewatering method for their estimated annual wastewater sludge production of 0.4 million gallons (1500 m³, 7% solids). Maximum frost penetration (from Table 9.8) is 38 in. (97 cm).

Solution

1. Use Equation 9.4 to determine potential design depth of frozen sludge:

$$\Sigma Y = 1.76(F_p) - 101 = 1.76(97) - 101 = 70 \text{ cm}$$

2. Then, determine the bed area required for freezing:

$$\text{Area} = \frac{1500 \text{ m}^3}{0.70 \text{ m}} = 2143 \text{ m}^2$$

   This area could be provided by 16 freezing beds, each 7 m by 20 m.
   Allow 30 cm for freeboard. Constructed depth = 0.70 + 0.30 = 1.0 m.

3. Determine the time required to thaw the 0.70-m sludge layer, if average
   temperatures are 10°C in March, 17°C in April, and 21°C in May. Use
   Equation 9.3 with a sludge depth of 70 cm:

$$\Delta T \cdot t = \left(\frac{Y}{m}\right)^2 = 1177°C \cdot d$$

$$\text{(March)} + \text{(April)} + \text{(May 1–17)}$$

$$\Delta T \cdot t = (31)(10) + (30)(17) + (17)(21) = 1177°C \cdot d$$

Therefore, the sludge layer should be completely thawed by May 18
under the assumed conditions.

9.3.4.4 Sludge Quality

Although the detention time for sludge on the freezing beds may be several
months, the low temperatures involved will preserve the pathogens rather than
destroy them. As a result, the process can be considered only as a conditioning
and dewatering operation, with little additional stabilization provided; however,
wastewater sludges treated in this way may be "cleaner" than sludges that are air
dried on typical sand beds. This is due to the rapid drainage of sludge liquid after
thawing, which carries away a significant portion of the dissolved contaminants.
In contrast, air-dried sludges will still contain most of the metal salts and other
evaporation residues.

9.4 REED BEDS

Reed bed systems are similar in some ways to the vertical flow constructed
wetlands described in Chapter 7. In this case, the bed is composed of selected
media supporting emergent vegetation, and the flow path for liquid is vertical
rather than horizontal. These systems have been used for wastewater treatment,

landfill leachate treatment, and sludge dewatering. This section describes the sludge dewatering use, where the bed is typically underdrained and the percolate is returned to the basic process for further treatment. These beds are similar in concept and function to conventional sand drying beds.

In conventional sand beds, each layer of sludge must be removed when it reaches the desired moisture content, prior to application of the next sludge layer. In the reed bed concept, the sludge layers remain on the bed and accumulate over a period of many years before removal is necessary. The significant cost savings from this infrequent cleaning are the major advantage of reed beds. Frequent sludge removal is necessary on conventional sand beds, as the sludge layer develops a crust and becomes relatively impermeable, with the result that subsequent layers do not drain properly and the new crust prevents complete evaporation. When reeds are used on the bed, the penetration of the stems through the previous layers of sludge maintains adequate drainage pathways and the plant contributes directly to dewatering through evapotranspiration.

This sludge dewatering method is in use in Europe, and approximately 50 operational systems are located in the United States. All of the operational beds have been planted with the common reed *Phragmites*. Experience has shown that it is necessary to apply well-stabilized wastewater sludges to these beds. Aerobically or anaerobically digested sludges are acceptable, but untreated raw sludges with a high organic content will overwhelm the oxygen-transfer capability of the plants and may kill the vegetation. The concept will also work successfully with inorganic water treatment plant sludges and high-pH lime sludges.

The structural facility for a reed bed is similar in construction to an open, underdrained sand drying bed. Typically, either concrete or a heavy membrane liner is used to prevent groundwater contamination. The bottom medium layer is usually 10 in. (25 cm) of washed gravel (20 mm) and contains the underdrain piping for percolate collection. An intermediate layer of pea gravel about 3 in. (8 cm) thick prevents intrusion of sand into the lower gravel. The top layer is 4 in. (10 cm) of filter sand (0.3 to 0.6 mm). The *Phragmites* rhizomes are planted at the interface between the sand and gravel layers. At least 3 ft (1 m) of freeboard is provided for long-term sludge accumulation. The *Phragmites* are planted on about 12-in. (30-cm) centers, and the vegetation is allowed to become well established before the first sludge application (Banks and Davis, 1983b).

9.4.1 FUNCTION OF VEGETATION

The root system of the vegetation absorbs water from the sludge, which is then lost to the atmosphere via evapotranspiration. It is estimated that during the warm growing season this evapotranspiration pathway can account for up to 40% of the liquid applied to the bed. As described in Chapter 7, these plants are capable of transmitting oxygen from the leaf to the roots; thus, aerobic microsites (on the root surfaces) exist in an otherwise anaerobic environment that can assist in sludge stabilization and mineralization.

9.4.2 DESIGN REQUIREMENTS

Sludge application to these reed beds is similar to the freezing process previously described, in that sequential layers of sludge are applied during the operational season. The solids content of the sludge can range up to 4%, but 1.5 to 2% is preferred (Banks and Davis, 1983a). Solids content greater than 4% will not allow uniform distribution of the sludge on the densely vegetated bed. The annual loading rate is a function of the solids content and whether the sludge has been digested anaerobically or aerobically. Aerobically digested sludges impose less stress on the plants and can be applied at slightly higher rates. At 2% solids, anaerobically digested sludges can be applied at a hydraulic loading of about 25 gal/ft^2·yr (1 m^3/m^2·yr) and aerobically digested sludges at 50 gal/ft^2·yr (2 m^3/m^2·yr). The corresponding solids loadings would be 4.2 lb/ft^2·yr (20 kg/m^2·yr) for anaerobic sludges and 8.3 lb/ft^2·yr (40 kg/m^2·yr) for aerobic sludges. For each 1% increase in solids content (up to 4%), the hydraulic loading should be reduced by about 10% (for example, for aerobic sludge at 4% solids, the hydraulic loading is 1.6 m^3/m^2·yr). For comparison, the recommended solids loading on conventional sand beds would be about 16.4 lb/ft^2·yr (80 kg/m^2·yr) for typical activated sludges. This suggests that the total surface area required for these reed beds will be larger than for conventional sand beds.

The typical operational cycle allows a sludge application every 10 d during the warm months and every 20 to 24 d during the winter. This schedule allows 28 sludge applications per year; for 2% solids aerobic sludges, each layer of sludge would be about 4 in. (10.7 cm). It is recommended that during the first year of operation the loadings be limited to one half the design values to limit stress on the developing plants.

An annual harvest of the *Phragmites* plants is typically recommended. This usually occurs during the winter months, after the top of the sludge has frozen. Electrical or gasoline-powered hedge clippers can be used. The plant stems are cut at a point that will still be above the top of the sludge layers expected during the remainder of the winter. This allows the continued transfer of air to the roots and rhizomes. In the spring, the new growth will push up through the accumulated sludge layers without trouble. The harvest produces about 25 ton/ac, dry solids 2.5 ton/ac (56 mt, wet weight per hectare). The major purpose of the harvest is to physically remove this annual plant production and thereby allow the maximum sludge accumulation on the bed. The harvested material can be composted or burned.

Sludge applications on a bed are stopped about 6 months before the time selected for cleaning. This allows additional undisturbed residence time for the pathogen content of the upper layer to be reduced. Typically, sludge application is stopped in early spring, and the bed is cleaned out in late fall. The cleaning operation removes all of the accumulated sludge in addition to the upper portion of the sand layer. New sand is then placed to restore the original depth. New plant growth occurs from the roots and rhizomes that are present in the gravel layer.

The number of separate reed beds at a facility will depend on the frequency of sludge wasting and the volume wasted during each event. Typically, the winter period controls the design because of the less frequent sludge applications (21 to 24 d of resting) permitted. For example, assume that a facility wastes aerobically digested sludge on a daily basis at a rate of 10 m^3/d (2% solids). The minimum total bed area required is (10 m^3/d)(365 d/yr)/(2 m^3/m^2·yr) = 1825 m^2. Try 12 beds, each 152 m^2 in area; assume that each is loaded for 2 d in sequence to produce a 24-d resting cycle during the winter months. The unit loading is then (10 m^3/d)(2 d)/(152 m^2) = 0.13 m = 13 cm. This is close to the recommended 10.7-cm layer depth for a single application; therefore, in this case, a minimum of 12 cells would be acceptable.

9.4.3 PERFORMANCE

It is estimated that 75 to 80% of the volatile solids (VSS) in the sludge will be reduced during the long detention time on the bed. As a result of this reduction and the moisture loss, a 10-ft-deep (3-m) annual application will be reduced to 2.4 to 4 in. (6 to 10 cm) of residual sludge. The useful life of the bed is therefore 6 to 10 yr between cleaning cycles. With one exception, all the reed bed systems in the United States are located where some freezing weather occurs each winter. The exception is the reed bed system at Fort Campbell, Kentucky. Observations at these systems indicate that the volume reduction experienced at Fort Campbell is significantly less than that experienced at systems in colder climates. The reason is believed to be the freezing and thawing of the sludge that occurs in the colder climates, which results in much more effective drainage of water from the accumulated sludge layers. This suggests that reed beds in cold climates should follow the criteria described in a previous section for freezing rather than the arbitrary 21-d cycle for winter sludge applications. This should result in a more effective process and, in colder climates, more frequent sludge application.

The loss of volatile solids during the long detention time on these reed beds raises the concern that the metals concentration of the residual sludges could increase to the point where beneficial uses of the material or normal disposal options are limited. Table 9.10 summarizes data from the reed bed system serving the community of Beverly, New Jersey. The reed bed system in Beverly has been in operation for 7 yr; therefore, the average age of the accumulated sludge was 3.5 yr. The applied sludges sampled from 1990 to 1992 are believed to be representative of the entire period. The tabulated data on accumulated sludge represents a core sample of the entire 7-yr sludge accumulation on the bed. The total volatile solids experienced a 71% reduction, and the total solids demonstrate a 251% increase due to the effective dewatering. All of the metals concentrations show an increase. If beneficial use of the removed sludge is a project goal, it is suggested that the critical metals in the accumulated sludge be measured on an annual basis. These data will provide the basis for following the trend of increasing concentration and can be used to decide when to remove the sludge from the bed prior to developing unacceptable metal concentrations.

TABLE 9.10
Comparison of Applied vs. Accumulated Sludge

Parameter	Applied Sludges[a]	Accumulated Sludge[b]
Total solids (%)	7.1	17.8
Volatile solids (%)	81.14	56
pH	5.3	6
Arsenic (mg/kg)	0.64	1
Cadmium (mg/kg)	6	8.3
Chromium (mg/kg)	16.3	62.3
Copper (mg/kg)	996.5	2120
Lead (mg/kg)	510	1130
Mercury (mg/kg)	10.2	28.3
Nickel (mg/kg)	29.8	45.7
Zinc (mg/kg)	4150	6400

[a] Digested primary sludges applied to the bed from 1990 to 1992.
[b] Accumulated dewatered sludge on the bed March 12, 1992.

Source: Costic & Associates, *Engineers Report: Washington Township Utilities Authority Sludge Treatment Facility*, Costic & Associates, Long Valley, NJ, 1983. With permission.

Another issue of concern in some states is the use of *Phragmites* on these systems. The *Phragmites* plant has little habitat value and has been known to crowd out more beneficial vegetation species in marshes. The risk of seeds or other plant material escaping from the operational reed bed and infesting a natural marsh is negligible; however, when the sludge is cleaned out of the bed, some root and rhizome material may also be removed with the sludge. The final sludge disposal site may have to be considered if regrowth of the *Phragmites* at that site would pose a problem. Disposal in landfills or utilization in normal agricultural applications should not create problems. If it is absolutely necessary, the removed sludges can be screened and the root and rhizome stock separated. It also should be possible to stockpile the removed sludge and cover it with dark plastic for several additional months to kill the rhizome material.

9.4.4 BENEFITS

The major advantage of the reed bed concept is the ease of operation and maintenance and the very high final solids content (suitable for landfill disposal). This significantly reduces the cost for sludge removal and transport. A 6- to 7-yr cleaning cycle for the beds seems to be a reasonable assumption. One disadvantage is the requirement for an annual harvest of the vegetation and disposal of that material; however, over a 7-yr cycle, the total mass of sludge residue and vegetation requiring disposal will be less than the sludge requiring disposal from sand drying beds or other forms of mechanical dewatering.

Example 9.3

A community near Pittsburgh, Pennsylvania (see Example 9.2), produces 3000 m³ of sludge (at 3.5% solids) per year. Compare reed beds for dewatering with a combination reed–freezing bed system.

Solution

Assume a 4-month freezing season, a design loading for reeds of 2.0 m³/m², and a design depth for freezing of 70 cm (satisfactory value; Example 9.2 indicates a maximum potential depth of 70 cm as feasible). Use 12 beds:

1. Calculate bed area if reed dewatering is used alone:

$$\text{Total area} = \frac{3000 \text{ m}^3}{2 \text{ m}^3/\text{m}^2 - \text{yr}} = 1500 \text{ m}^2$$

$$\text{Individual bed} = \frac{1500 \text{ m}^2}{12} = 125 \text{ m}^2$$

The schedule allows 28 sludge applications per year to the reed beds. Then, 3000 m³/12 beds/28 applications/125 m²/bed = 0.07 m/application = 7 cm.

2. 21 warm-weather applications = 21 × 7 cm = 147 cm.
   7 winter applications using reed bed criteria = 7 × 7 = 49 cm.

3. Freeze–thaw criteria allow a total winter application of 70 cm; therefore, an additional 21 cm or three additional applications are allowed, for a total of 10, and an annual total of 31. At 31 annual applications, the allowable loading is 2.17 m³/m²·yr, and the required bed area is 3000 m³/2.17 m³/m²·yr = 1382 m², so each of the individual beds can be reduced in area to 115 m². This savings in area might be very significant in climates colder than in New Jersey.

9.4.5 Sludge Quality

The dewatered material removed from the reed beds will be similar in character to composted sludge with respect to pathogen content and stabilization of organics. The long detention times combined with the final 6-month rest period prior to sludge removal ensure a stable final product for reuse or disposal. If metals are a concern, then a routine monitoring program can track the metals content of the accumulating sludge. In some cases, the metal content may be the basis for sludge removal rather than the volumetric capacity of the bed.

9.5 VERMISTABILIZATION

Vermistabilization (i.e., sludge stabilization and dewatering using earthworms) has been investigated in numerous locations and has been successfully tested full scale on a pilot basis (Donovan, 1981; Eastman et al., 2001). A potential cost advantage for the concept in wastewater treatment systems is the capability for stabilization and dewatering in one step as compared to thickening, digestion, conditioning, and dewatering in a conventional process. Vermistabilization has also been used successfully with dewatered sludges and solid wastes. The concept is feasible only for sludges that contain sufficient organic matter and nutrients to support the worm population.

9.5.1 WORM SPECIES

In most locations, the facilities required for the vermistabilization procedure will be similar to an underdrained sand drying bed enclosed in a heated shelter. Studies at Cornell University evaluated four earthworm species: *Eisenia foetida*, *Eudrilus eugeniae*, *Pheretima hawayana*, and *Perionyx excavatus*. *E. foetida* showed the best growth and reproductive responses, with temperatures in the range of 68 to 77°F (20 to 25°C). Temperatures near the upper end of the range are necessary for optimum growth of the other species. Worms are placed on the bed in a single initial application of about 0.4 lb/ft^2 (2 kg/m^2) (live weight). Sludge loading rates of about 0.2 lb/ft^3/wk (1000 g of sludge volatile solids per m^2 per wk) were recommended for liquid primary and liquid waste-activated sludge (Loehr et al., 1984). Liquid sludges used in the Cornell University tests ranged from 0.6 to 1.3% solids, and the final stabilized solids ranged from 14 to 24% total solids (Loehr et al., 1984). The final stabilized sludge had about the same characteristics regardless of the type of liquid sludge initially applied. Typical values were as follows:

- Total solids (TS) = 14–24%
- Volatile solids = 460–550 g per kg TS
- Chemical oxygen demand = 606–730 g per kg TS
- Organic nitrogen = 27–35 g per kg TS
- pH = 6.6–7.1

Thickened and dewatered sludges have also been used in operations in Texas with essentially the same results (Donovan, 1981). Application of very liquid sludges (<1%) is feasible as long as the liquid drains rapidly so aerobic conditions can be maintained in the unit. Final sludge removal from the unit is required only at long intervals, about 12 months.

9.5.2 LOADING CRITERIA

The recommended loading of 0.2 lb/ft^2·wk (1000 g/m^2·wk) is equivalent, for typical sludges, to a design area requirement of 4.5 ft^2/capita (0.417 m^2/capita).

This is about 2.5 times larger than a conventional sand drying bed. The construction cost difference will be even greater, as the vermistabilization bed must be covered and possibly heated; however, major cost savings are possible for the overall system, because thickening, digestion, and dewatering units may not be required if vermistabilization is used with liquid sludges.

9.5.3 PROCEDURES AND PERFORMANCE

At an operation in Lufkin, Texas, thickened (3.5 to 4% solids) primary and waste-activated sludge are sprayed at a rate of 0.05 lb/ft²·d (0.24 kg/m²·d) dry solids over beds containing worms and sawdust. The latter acts as a bulking agent and absorbs some of the liquid, assisting in maintaining aerobic conditions. An additional layer of sawdust, 1 to 2 in. (2.5 to 5 cm) thick, is added to the bed after about 2 months. The original sawdust depth was about 8 in. (20 cm) when the beds were placed in operation. The mixture of earthworms, castings, and sawdust is removed every 6 to 12 months. A small front-end loader is driven into the bed to move the material into windrows. A food source is spread adjacent to the windrows, and within 2 days essentially all the worms have migrated to the new material. The concentrated worms are collected and used to inoculate a new bed. The castings and sawdust residue are removed, and the bed is prepared for the next cycles.

Human pathogen reduction in a field experiment with vermiculture (vermicomposting) was found to reduce fecal coliforms, *Salmonella* spp., enteric viruses, and helminth ova more effectively than composting (Eastman et al., 2001). The ratio of earthworms (*Eisenia foetida*) to biosolids was 1:1.5 wet weight. After 144 hr, fecal coliforms showed a 6.4-log reduction, while a control experiment showed only a 1.6-log reduction. *Salmonella* spp. reduction was 8.6 log, and the control reduction was 4.9 log. Enteric viruses were reduced by 4.6 log as compared to 1.8 log reduction in the control. Helminth ova reduction was 1.9 log vs. 0.6 log in the control.

Example 9.4

Determine the bed area required to utilize vermistabilization for a municipal wastewater treatment facility serving 10,000 to 15,000 people. Compare the advantages of liquid vs. thickened sludge.

Solution:

1. Assuming an activated sludge system or the equivalent, the daily sludge production will be about 1 mt dry solids per day. If the sludge contains about 65% volatile solids (see Table 9.2), the Cornell loading rate of 1 kg/m²·wk is equal to 1.54 kg/m²·wk of total solids. Assume downtime of 2 wk per year for bed cleaning and general maintenance. The Lufkin, Texas, loading rate for thickened sludge is equal to 1.78 kg/m²·wk of total solids.

2. Calculate the bed area for liquid (1% solids or less) and for thickened (3 to 4% solids) sludges.

For liquid sludge:

Bed area = (1000 kg/d)(365 d/yr)/(1.54 kg/m^2·wk)(50 wk) = 4740 m^2

For thickened sludge:

Bed area = (1000 kg/d)(365 d/yr)/(1.78 kg/(m^2·wk)(50 wk) = 4101 m^2

3. A cost analysis is required to identify the most cost-effective alternative. The smaller bed area for the second case is offset by the added costs required to build and operate a sludge thickener.

9.5.4 SLUDGE QUALITY

The sludge organics pass through the gut of the worm and emerge as dry, virtually odorless castings. These are suitable for use as a soil amendment or low-order fertilizer if metal and organic chemical content are within acceptable limits (see Table 9.16 for metals criteria). Only limited quantitative data are available with regard to removal of pathogens with this process. The Texas Department of Health found no *Salmonella* in either the castings or the earthworms at a vermistabilization operation in Shelbyville, Texas, that received raw sludge (Donovan, 1981). A market may exist for the excess earthworms harvested from the system. The major prospect is as bait for freshwater sport fishing. Use as animal or fish food in commercial operations has also been suggested, but numerous studies have shown that earthworms accumulate very significant quantities of cadmium, copper, and zinc from wastewater sludges and sludge-amended soils; therefore, worms from a sludge operation should not be the major food source for animals or fish in the commercial production of food for human consumption.

9.6 COMPARISON OF BED-TYPE OPERATIONS

The physical plants for freezing systems, reed systems, and vermistabilization systems are similar in appearance and function. In all cases, a bed is required to contain the sand or other support medium, the bed must be underdrained, and a method for uniform distribution of sludge is essential. Vermistabilization beds must be covered and probably heated during the winter months in most of the United States. The other two concepts require neither heat nor covers. Table 9.11 summarizes the criteria and the performance expectations for these three concepts. The annual loading rate for the vermistabilization process is much less than for the other concepts discussed in this section; however, vermistabilization may still be cost effective in small to moderate-sized operations, as thickening, digestion, conditioning, and dewatering can all be eliminated from the basic process design. Freezing sludge does not provide any further stabilization. Digestion or other stabilization of wastewater sludges is strongly recommended prior to application on freezing or reed beds to avoid odor problems.

TABLE 9.11
Comparison of Bed-Type Operations

Sludge Types	Freezing (All)	Reeds (Nontoxic)[a]	Freezing and Reeds (Nontoxic)	Worms (Organic Nontoxic)
Bed enclosure	None	None	None	Yes
Heat required	No	No	No	Yes
Initial solids (%)	1–8	3–4	3–8	1–4
Typical loading rate (kg/m²/yr)[b]	40[c]	60	50	<20
Final solids (%)	20–50[d]	50–90[d]	20–90[d]	15–25
Further stabilization provided	No	Some	Some	Yes
Sludge removal frequency (yr)	1	10[e]	10[e]	1

[a] Assumes year-round operation in a warm climate.
[b] Annual loading in terms of dry solids.
[c] Includes use of bed for conventional drying in summer
[d] Final solids amount depends on length of final drying period.
[e] The vegetation is typically harvested annually.

9.7 COMPOSTING

Composting is a biological process for the concurrent stabilization and dewatering of sludges. If temperature and reaction time satisfy the required criteria, the final product should meet the class A pathogen and vector attraction reduction requirements (see Chapter 3). The three basic types of compost systems are (USEPA, 1981a):

- *Windrow* — The material to be composted is placed in long rows, which are periodically turned and mixed to expose new surfaces to the air.
- *Static pile* — The material to be composted is placed in a pile, and air is either blown or drawn through the pile by mechanical means. Figure 9.2 illustrates the various configurations of static pile systems.
- *Enclosed reactors* — These can range from complete, self-contained reactor units to structures that partially or completely enclose static pile or windrow-type operations. The enclosure in these latter cases is usually for odor and climate control.

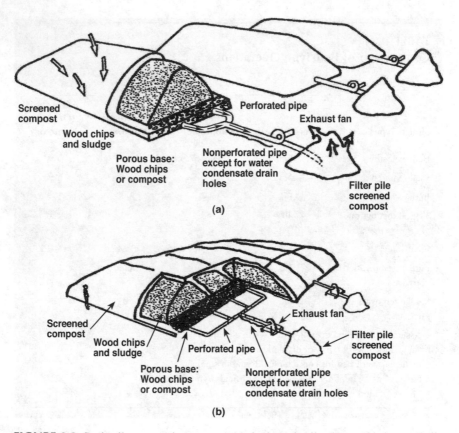

FIGURE 9.2 Static pile composting systems: (a) single static pile; (b) extended aerated pile.

The process does not require digestion or stabilization of sludge prior to composting, although there may be increased odor production issues to deal with when composting raw sludges. Composting projects are frequently designed based on 20% solids, but many operating projects are starting with 12 to 18% solids and as a result end up using more bulking agent to absorb moisture to get to approximately 40% solids in the mix of sludge and bulking agent. The end product is useful as a soil conditioner (and is sold for that purpose in many locations) and has good storage characteristics.

The major process requirements include: oxygen at 10 to 15%, a carbon-to-nitrogen ratio of 26:1 to 30:1, volatile solids over 30%, water content 50 to 60%, and pH 6 to 11. High concentrations of metals, salts, or toxic substances may affect the process as well as the end use of the final product. Ambient site temperatures and precipitation can have a direct influence on the operation. Most municipal sludges are too wet and too dense to be effectively composted alone, so the use of a bulking agent is necessary. Bulking agents that have been used successfully include wood chips, bark, leaves, corncobs, paper, straw, peanut and rice hulls, shredded tires, sawdust, dried sludge, and finished compost. Wood chips have been the most common agent and are often separated from the finished compost mixture and used again. The amount of bulking agent required is a function of sludge moisture content. The mixture of sludge and bulking agent should have a moisture content between 50 and 60% for effective composting. Sludges with 15 to 25% solids might require a ratio of between 2:1 and a 3:1 of wood chips to sludge to attain the desired moisture content in the mixture (USDA/USEPA, 1980).

Mixing of the sludge and the bulking agent can be accomplished with a front-end loader for small operations. Pugmill mixers, rototillers, and special composting machines are more effective and better suited for larger operations (USEPA, 1984). Similar equipment is also used to build, turn, and tear down the piles or windrows. Vibratory-deck, rotary, and trommel screens have all been used when separation and recovery of the bulking agent are process requirements. The pad area for either windrow or aerated pile composting should be paved. Concrete has been the most successful paving material. Asphalt may be suitable, but it may soften at higher composting temperatures and may itself be susceptible to composting reactions.

Outdoor composting operations have been somewhat successful in Maine and in other locations with severe winter conditions. The labor and other operational requirements are more costly for such conditions. Covering the composting pads with a simple shed roof will provide greater control and flexibility and is recommended for sites that will be exposed to subfreezing temperatures and significant precipitation. If odor control is a concern, it may be necessary to add walls to the structure and include odor control devices in the ventilation system.

For static pile systems, the aeration piping shown in Figure 9.2 is typically surrounded by a base of wood chips or unscreened compost about 12 to 18 in. (30 to 45 cm) deep. This base ensures uniform air distribution and also absorbs excess moisture. In some cases, permanent air ducts are cast into the concrete base pad. The mixture of sludge and bulking agent is then placed on the porous base material. Experience has shown that the total pile height should not exceed 13 ft (4 m) to avoid aeration problems. Typically, the height is limited by the capabilities of most front-end loaders. A blanket of screened or unscreened compost is used to cover the pile for thermal insulation and to adsorb odors. About 18 in. (45 cm) of unscreened or about 10 in. (25 cm) of screened compost is used. Where the extended pile configuration is used, an insulating layer only 3 in. (8 cm) thick is applied to the side that will support the next composting addition. Wood chips or other coarse material are not recommended, as the loose structure will promote heat loss and odors.

The configuration shown in Figure 9.2 draws air into the pile and exhausts it through a filter pile of screened compost. This pile should contain about 35 ft^3 (1 m^3) of screened compost for every 3.3 ton (3 mt) of sludge dry solids in the compost pile. To be effective, this filter pile must remain dry; when the moisture content reaches 70%, the pile should be replaced.

Several systems, both experimental and operational, use positive pressure to blow air through the compost pile (Kuter et al., 1985; Miller and Finstein, 1985). The blowers in this case are controlled by heat sensors in the pile. The advantages claimed for this approach include more rapid composting (12 vs. 21 d), a higher level of volatile solids stabilization, and a drier final product. The major concern is odors, as the air is exhausted directly to the atmosphere in an outdoor operation. Positive aeration, if not carefully controlled, can result in desiccation of the lower part of the pile and therefore incomplete pathogen stabilization. The approach seems best suited to larger operations with enclosed facilities, in which the increased control will permit realization of the potential for improved efficiency.

The time and temperature requirements for either pile or windrow composting depend on the desired level of pathogen reduction. If "significant" reduction is acceptable, then the requirement is a minimum of 5 d at 105°F (40°C) with 4 hr at 130°F (55°C) or higher. If "further" reduction is necessary, then 130°F (55°C) for 3 d for the pile method or 130°F (55°C) for 15 d with five turnings for the windrow method is required. In both cases, the minimum composting time is 21 d, and the curing time in a stockpile, after separation of the bulking agent, is another 21 d.

A system design requires a mass balance approach to manage the input and output of solid material (sludge and bulking agent) and to account for the changes in moisture content and volatile solids. A continuing materials balance is also essential for proper operation of the system. The pad area for a composting operation can be determined using Equation 9.5:

$$A = \frac{1.1S(R+1)}{H}$$
(9.5)

where
A = Pad area for active compost piles (m^2; ft^2).
S = Total volume of sludge produced in 4 wk (m^3; ft^3).
R = Ratio of bulking agent volume to sludge volume.
H = Height of pile, not including cover or base material (m; ft).

A design using odor-control filter piles should allow an additional 10% of the area calculated above for that purpose. Equation 9.5 assumes a 21-d composting period but provides an additional 7 d of capacity to allow for low temperature, excessive precipitation, and malfunctions. If enclosed facilities are used or if positive pile aeration is planned, proportional reductions in the design area are possible.

The area calculated using Equation 9.5 assumes that mixing of sludge and bulking material will occur directly on the composting pad. Systems designed for a sludge capacity of more than 15 dry ton per day should provide additional area for a pugmill or drum mixer.

In many locations the finished compost from the suction-type aeration will still be very moist, so spreading and additional drying are typically included. The processing area for this drying and screening procedure to separate the bulking agent is typically equal in size to the composting area for a site in cool, humid climates. This can be reduced in more arid climates and where positive-draft aeration is used.

An area capable of accommodating 30 d of compost production is recommended as the minimum for all final curing locations. Additional storage area may be necessary, depending on the end use of the compost. Winter storage may be required — for example, if the compost is used only during the growing season.

Access roads, turnaround space, and a wash rack for vehicles are all required. If runoff from the site and leachate from the aeration system cannot be returned to the sewage treatment plant, then a runoff collection pond must also be included. Detention time in the pond might be 15 to 20 d, with the effluent applied to the land as described in Chapter 8. Most composting operations also have a buffer zone around the site for odor control and visual esthetics; the size will depend on local conditions and regulatory requirements.

The aeration rate for the suction-type aerated pile is typically 8 ft³/min (14 m³/hr) per ton sludge dry solids. Positive-pressure aeration, at higher rates, is sometimes used during the latter part of the composting period to increase drying (Miller and Finstein, 1985). Kuter et al. (1985) used temperature-controlled positive-pressure aeration at rates ranging from 47 to 200 ft³/min (80 to 340 m³/hr) per ton sludge dry solids and achieved a stable compost in 17 d or less. These high aeration rates result in lower temperatures in the pile (below 113°F [45°C]). The direction of air flow can be reversed during the latter stages to elevate the pile temperature above the required 131°F (55°C). The temperatures in the final curing pile should be high enough to ensure the required pathogen kill so the composting operation can be optimized for stabilization of volatile solids.

Monitoring is essential in any composting operation to ensure efficient operations as well as the quality of the final product. Critical parameters to be determined include:

- *Moisture content* in sludge and bulking material to ensure proper operations
- *Metals and toxics* in sludge to ensure product quality and compost reactions
- *Pathogens* as required by regulations
- *pH* in sludge, particularly if lime or similar chemicals are used
- *Temperature* taken daily until the required number of days above 130°F (55°C) is reached; weekly thereafter at multiple sites to ensure that the entire mass is subjected to appropriate temperatures
- *Oxygen*, initially, to set blower operation

Example 9.5

Determine the area required for a conventional extended-pile composting operation for the wastewater treatment system described in Example 9.3 (1500 m³ sludge production per year at 7% solids). Assume that a site is available next to the treatment plant so runoff and drainage can be returned to the treatment system.

Solution

1. Use wood chips as a bulking agent. At 7% solids, the sludge is still "wet," so a mixing ratio of at least 5 parts of wood chips to 1 part sludge will be needed. Assume top of compost at 2 m. Thus:

   4-wk sludge production = $(1500)(4)/(52) = 115.4$ m^3

2. Use Equation 9.5 to calculate the composting area:

   $A = 1.1$ S $(R + 1)/H$
   $A = [1.1(115.4)(5 + 1)]/2 = 381$ m^2

3. Filter piles for aeration = 10% of $A = 0.1 \times 381 = 38.1$ m^2.
4. Processing and screening area = $A = 381$ m^2.
5. Curing area: Assume 150 m^2.
6. Wood chip and compost storage: Assume 200 m^2.
7. Roads and miscellaneous: Allow 20% of total.

   Total $A = 381 + 38.1 + 381 + 150 + 200 = 1150$ m^2
   Roads = $(0.2)(1150) = 230$ m^2

8. Total area including roads = 1380 m^2.

A buffer zone might also be necessary, depending on site conditions. The area calculated here is significantly less than the area calculated in Example 9.2 for freeze drying beds. This is because composting can continue on a year-round basis, but the freezing beds must be large enough to contain the entire annual sludge production.

9.8 LAND APPLICATION AND SURFACE DISPOSAL OF BIOSOLIDS

Standards (40 CFR Part 503) for the use or disposal of sewage sludge were published in the *Federal Register* on February 19, 1993 (Bastian, 1993; Crites et al., 2000; USEPA, 1994a). The regulation discusses land application, surface disposal, pathogen and vector attraction reduction, and incineration. Land application is defined as beneficial use of the sludge at agronomic rates, while all other placement on the land is considered to be surface disposal. Heavy-metal concentrations are limited by two levels of sludge quality: pollutant ceiling concentrations and pollutant concentrations ("high quality"). Two classes of quality with regard to pathogen densities (class A and class B) are described. Two types of vector attraction reduction are presented: sewage sludge processing or the use of physical barriers.

The USEPA rules were evaluated by the National Research Council (NRC, 2002) and the results were presented in a report issued in July 2002. The NRC concluded that there was no documented scientific evidence that the regulations had failed to protect the public health; however, uncertainty on possible health effects exists. Further research was recommended to address public health concerns, scientific uncertainties, and data gaps in the sewage sludge standards. A response to the NRC review was published by the USEPA (2003c), and the plan to conform to the recommendations in the NRC report was presented. Beecher et al. (2004) reviewed the current understanding of risk perception, risk communication, and public participation with regard to biosolids management. They agree with the NRC (2002) that risk assessment is subjective and is a blending of science and judgment. Choosing models to address the issues is difficult, and data availability is limited.

For land application, sewage sludge or material derived from sewage sludge must as a minimum meet the pollutant ceiling concentrations, class B requirements for pathogens, and vector attraction reduction requirements. Cumulative pollutant loading rates are required for sewage sludges that meet the pollutant ceiling concentrations but do not satisfy the pollutant concentrations.

The concepts described in this section are generally limited to those operations designed for treatment or reuse of the sludge via land application or surface disposal. Landfills and other "high-quality" surface disposal practices are covered in other texts (USEPA, 1978, 1979, 1981b). Some degree of sludge stabilization is typically used prior to land application or surface disposal, and dewatering may be economically desirable; however, systems are designed so the receiving land surface provides the final sludge treatment as well as utilizing the sludge organic matter and nutrients. These natural sludge management systems can be grouped into two major types: land application and surface disposal.

Land application systems involve the vegetation, soils, and related ecosystem for final treatment and utilization of the sludge. The design sludge loadings are based on the nutrient and organic needs of the site as constrained by metals, toxics, vector control, and pathogen content of the sludge (USEPA, 1981b). Systems in this group include agricultural and forest operations where repetitive sludge applications are planned over a long term, as well as reclamation projects where the sludge is used to reclaim and revegetate disturbed land. The site is designed and then operated so no future restrictions are placed on the use of the land. The flowchart in Figure 9.3 presents a series of steps to follow that make it easy to determine if land application of sludge is appropriate (Sieger and Herman, 1993). A process design manual for land application of sewage sludge and domestic septage is available from the USEPA (1995c).

Surface disposal systems depend almost entirely on reactions in the upper soil profile for treatment. Vegetation is typically not an active treatment component, and no attempt is made to design for the beneficial utilization of sludge organic matter or nutrients. The site is often dedicated for this purpose, and restrictions may be placed on future use of the land, especially for crop production involving the human food chain. Systems receiving biodegradable sludges utilize acclimated soil organisms for that purpose and are designed for periodic loading and rest periods. Petroleum sludges and similar industrial wastes are often managed in this way. Figure 9.4 is a flowchart that makes it easy to determine the applicability of surface disposal of sludges (Sigmund and Sieger, 1993). A process

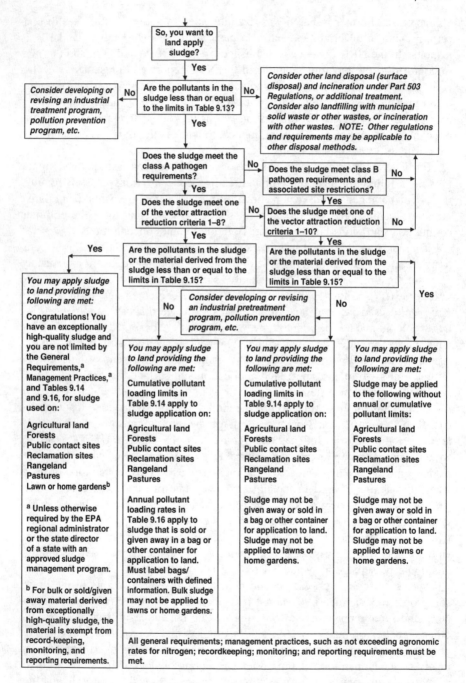

FIGURE 9.3 Flowchart to determine the applicability of land application of sludge. (From Sieger, R.B. and Herman, G.J., *Water Eng. Manage.*, 140(8), 30–31, 1993. With permission.)

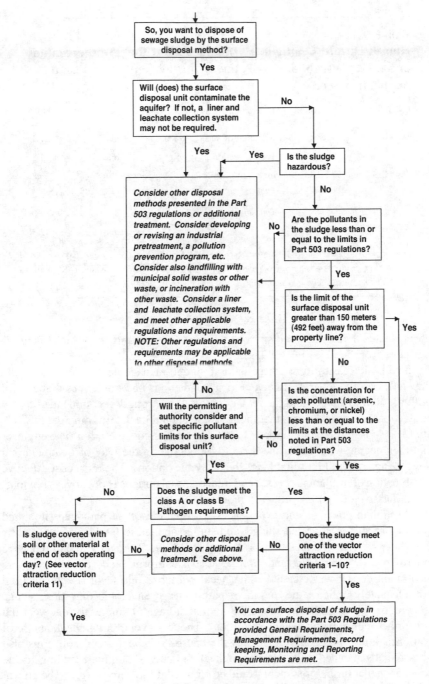

FIGURE 9.4 Flowchart to determine the applicability of surface disposal of sludge. (From Sigmund, T.W. and Sieger, R.B., *Water Eng. Manage.*, 140(9), 18–19, 1993. With permission.)

TABLE 9.12
Human Hazard Quotient (HQ) Values >1 at the 95th Percentile of the HQ Distribution by Pathway for the Agricultural Land Application Scenario

CASRN	Chemical	Pathway	Receptor	HQ
14797-65-0	Nitrite	Irrigation of surface water	Child	1.1
		Total ingestion	Child	1.3
7440-22-4	Silver	Ingestion of milk	Adult	3.8
			Child	12.0
		Total ingestion	Adult	4.0
			Child	12.3

Source: USEPA, *Technical Background Document for the Sewage Sludge Exposure and Hazard Screening Assessment*, Document No. 822-B-03-001, Office of Water, U.S. Environmental Protection Agency, Washington, D.C., 2003.

design manual for surface disposal of sewage sludge and domestic septage is available from the USEPA (1995a).

The basic feasibility of these natural sludge management options is totally dependent on the federal, state, and local regulations and guidelines that control both the sludge quality and the methodology. It is strongly recommended for all sludge management designs that the first step should be determining the possible sludge disposal/utilization options for the area under consideration. The engineer can then decide what has to be done to the sludge in the way of treatment and dewatering so it will be suitable for the available options. The most cost-effective combination of in-plant processes and final disposal options is not always obvious, so an iterative design procedure is required.

Australian practices and experiences with the die-off of pathogens in stored wastewater sludge, digested sludge, and sludge applied to land were reported by the Water Services Association of Australia (1995). Pathogens monitored were enteroviruses, *Salmonella*, and *Giardia*. Fecal coliforms and streptococci were also monitored. It was concluded that fecal coliforms and fecal streptococci did not adequately indicate the die-off of pathogens in anaerobic digestion, sludge storage, or soil amendment. It was found that storage of sludge for one year did not further reduce pathogen concentrations. The regrowth of *Salmonella* and fecal coliforms after a year of storage was the cause. *Salmonella* and coliforms also occurred in amended soils after rainfall at the end of summer, leading to the conclusion that further treatment would be required if food crops were to be grown.

Documents describing methods to assess biosolids risk and hazards screening are available (USEPA, 1995b, 2003b). Hazard quotients (HQs) for 40 pollutants were developed and the results are presented in Table 9.12, Table 9.13, and Table

TABLE 9.13
Hazard Quotient (HQ) Values ≥1 at the 95th Percentile of the HQ Distribution for Aquatic and Terrestrial Wildlife Via Direct-Contact Pathways[a]

CASRN	Chemical	Receptor[b]	HQ
67-64-1	Acetone	Sediment biota	356.2
120-12-7	Anthracene	Sediment biota	2.9
7440-39-3	Barium	Aquatic community	235.7
7440-41-7	Beryllium	Aquatic community	7.8
75-15-0	Carbon disulfide	Sediment biota	1.9
106-47-8	4-Chloroaniline	Aquatic invertebrates	1.3
333-41-5	Diazinon	Sediment biota	1.1
206-44-0	Fluoranthene	Aquatic community	10.7
		Sediment biota	4.2
7439-96-5	Manganese	Aquatic community	13.9
78-93-3	Methyl ethyl ketone	Sediment biota	5.8
108-95-2	Phenol	Sediment biota	102.4
129-00-0	Pyrene	Aquatic community	41.9
		Sediment biota	21.1
		Soil biota	4.5
7440-22-4	Silver	Aquatic community	246.6
		Aquatic invertebrates	28.2
		Fish	4.8

[a] No pollutant resulted in an HQ ≥ 1 for any wildlife species based on ingestion pathways.
[b] Sediment biota organisms include sediment invertebrates; aquatic community organisms include fish, aquatic invertebrates, aquatic plants, and amphibians; soil biota organisms include soil invertebrates.

Source: USEPA, *Technical Background Document for the Sewage Sludge Exposure and Hazard Screening Assessment*, Document No. 822-B-03-001, Office of Water, U.S. Environmental Protection Agency, Washington, D.C., 2003.

9.14. HQ values greater than 1 are considered to have failed the human health screen and the ecological screen.

Several attributes necessary to characterize and manage the potential risks for organic chemicals in biosolids are toxicity and dose response, transport potential, chemical structure, environmental stability, analytical capability in the matrix of interest, concentrations and persistence in waste streams, plant uptake, availability from surface application vs. incorporation, solubility factors, and environmental

TABLE 9.14

Human Hazard Quotient (HQ) Values >1 at the 95th Percentile of the HQ Distribution by Pathway for the Sewage Sludge Lagoon Scenario

CASRN	Chemical	Pathway	Receptor	HQ
7440-39-3	Barium	Drinking water from groundwater	Adult	1.5
			Child	3.5
106-47-8	4-Chloroaniline	Drinking water from groundwater	Adult	2.7
			Child	6.4
7439-96-5	Manganese	Drinking water from groundwater	Adult	32.3
			Child	76.3
14797-65-0	Nitrite	Drinking water from groundwater	Adult	13.6
			Child	33.8
14797-55-8	Nitrate	Drinking water from groundwater	Adult	9.2
			Child	23.0

Source: USEPA, *Technical Background Document for the Sewage Sludge Exposure and Hazard Screening Assessment*, Document No. 822-B-03-001, Office of Water, U.S. Environmental Protection Agency, Washington, D.C., 2003.

fate. Kester et al. (2004) present examples of deterministic and probabilistic models for quantitative risk assessment for polychlorinated biphenyls (PCBs) and dioxin, but they pointed out that, unfortunately, this information is available for only a small number of chemicals. As more information becomes available, better assessments of risk and management techniques will become available.

9.8.1 CONCEPT AND SITE SELECTION

A preliminary evaluation should identify the available options as well as the expected physical, chemical, and biological characteristics for the sludge. The chemical characteristics will control the following:

- Can the sludge be applied in a cost-effective manner?
- Which options are technically feasible?
- What is the amount of sludge permitted per unit area on an annual and a design-life basis?
- What types and frequencies of site monitoring and other regulatory controls are imposed on the operation?

The biological characteristics of greatest concern are the presence of toxic organics and pathogens and the potential for odors during transport, storage, and application. The most important physical characteristic is the sludge moisture

TABLE 9.15
Preliminary Sludge Loadings for Site Identification

Option	Application Schedule	Typical Rate (mt/ha)
Agricultural	Annual	10
Forest	One time, or at 3- to 5-year intervals	45
Reclamation	One time	100
Surface	Annual	340

Source: USEPA, *Process Design Manual: Land Application of Municipal Sludge*, EPA 625/1-83-016, CERI, U.S. Environmental Protection Agency, Cincinnati, OH, 1983.

content. When the amount of sludge to be managed has been estimated, it is necessary to conduct a map survey, as described in Chapter 2, to identify sites with potential feasibility for agriculture, forests, reclamation, or surface treatment. Table 9.15 presents preliminary loading rates for the four application options. These values should be used only for this preliminary screening and not for design. A guide to field storage of biosolids and organic by-products to be used in agriculture and soil management has been prepared by the USEPA and the USDA (2000). Water quality and control of pathogens are discussed.

The land area estimates produced with the values in Table 9.15 are the treatment area only, with no allowance for sludge storage, buffer zones, and other requirements. The preliminary screening to identify suitable sites can be a desktop analysis using commonly available information. Numerical rating procedures based on soil and groundwater conditions, slopes, existing land use, flood potential, and economic factors were described in Chapter 2 and in Reed and Crites (1984) and USEPA (1978, 1983). These procedures should be used to identify the most desirable sites if a choice exists. This preliminary screening is advised because it is very costly to conduct detailed field investigations on every potential site. The final site selection is based on the technical data obtained by the site investigation, on a cost-effectiveness evaluation of capital and operating costs, and on the social acceptability of both the site and the intended sludge management option. The requirements for pathogen reduction were discussed in Chapter 3, and details can be obtained by consulting 40 CFR Part 503 (*Federal Register*, February 19, 1993) (Bastian, 1993; Crites et al., 2000).

9.8.2 PROCESS DESIGN, LAND APPLICATION

The basic design approach is based on the underlying assumption that, if sludge is applied at rates that are equal to the requirements of the design vegetation, over the time period of concern there should not be any greater impact on the groundwater than from normal agricultural operations. The design loading, based

TABLE 9.16
Ceiling Concentrations

Pollutant	Ceiling Concentration (mg/kg, dry weight basis)
Arsenic	75
Cadmium	85
Copper	4300
Lead	840
Mercury	57
Molybdenum	75
Nickel	420
Selenium	100
Zinc	7500

Source: Data from Bastian (1993) and Crites et al. (2000).

TABLE 9.17
Cumulative Pollutant Loading Rates

Pollutant	Cumulative Pollutant Loading Rate (kg/ha)
Arsenic	41
Cadmium	39
Copper	1500
Lead	300
Mercury	17
Molybdenum[a]	—
Nickel	420
Selenium	100
Zinc	2800

[a] Molybdenum was dropped in 1994 and a new value has not been set. Check the USEPA website for current values.

Source: Bastian, R.K., *Summary of 40CFR Part 503, Standards for the Use or Disposal of Sewage Sludge*, U.S. Environmental Protection Agency, Washington, D.C., 1993.

initially on nutrient requirements, is modified as required to satisfy limits on metals and toxic organics. As a result of this design approach, extensive monitoring should not be required, and the use of sludge by private farmers is made possible. As the loading increases, as it may in forests and on dedicated sites, the potential for nitrate contamination of the groundwater increases, and it is then usually necessary to design a municipally owned and operated site to ensure proper management and monitoring.

9.8.2.1 Metals

The following is extracted from the 40 CFR Part 503.13 pollutant limits (Bastian, 1993). Bulk sewage sludge or sewage sludge sold or given away in a bag or other container must not be applied to the land if the concentration of any pollutant in the sewage sludge exceeds the ceiling concentration shown in Table 9.16 (chromium was removed from Table 9.16 in 1994). If bulk sewage sludge is applied to agricultural land, forest, a public contact site, or a reclamation site, either the cumulative loading rate for each pollutant must not exceed the cumulative loading rate for the pollutants shown in Table 9.17 or the concentration of any pollutant in the sewage sludge must not exceed the ceiling concentration shown in Table 9.16. If bulk sludge meets the "high-quality" pollutant concentrations shown in Table 9.18, the cumulative pollutant loading rates (Table 9.17) do not apply because these materials can be applied at agronomic rates for 100 years without concerns about limiting cumulative loading rates.

If bulk sewage sludge is applied to a lawn or a home garden, the concentration of each pollutant in the sewage sludge must not exceed the concentrations shown in Table 9.18. If sewage sludge products are sold or given away in a bag or other container for application to the land, either the concentration of each pollutant in the sewage sludge must meet the pollutant concentrations in Table 9.18 or the product label will provide product use directions to limit the annual pollutant loading rates shown in Table 9.19. Equation 9.6 shows the relationship between the annual pollutant loading rate (APLR) and the annual whole sludge application rate (AWSAR):

$$APLR = C \times AWSAR \times 0.001 \tag{9.6}$$

or

$$AWSAR = \frac{APLR}{C \times 0.001}$$

where

APLR = Annual pollutant loading rate (kg/ha per 365-d period).

C = Pollutant concentration (mg/kg of total solids on a dry weight basis).

AWSAR = Annual whole sludge application rate (mt/ha per 365-d period on a dry weight basis).

0.001 = A conversion factor.

TABLE 9.18
Pollutant Concentrations (High Quality)

Pollutant	Monthly Average Concentration (mg/kg, dry weight basis)
Arsenic	41
Cadmium	39
Copper	1500
Lead	300
Mercury	17
Molybdenum[a]	—
Nickel	420
Selenium	100
Zinc	2800

[a] Molybdenum was dropped in 1994 and a new value has not been set. Check the USEPA website for current values.

Source: Bastian, R.K., *Summary of 40CFR Part 503, Standards for the Use or Disposal of Sewage Sludge*, U.S. Environmental Protection Agency, Washington, D.C., 1993.

Equation 9.6 can be modified to calculate the lifetime loading of the heavy metals:

$$\text{LWSAR} = \frac{\text{CPLR}}{C \times 0.001} \qquad (9.7)$$

where LWSAR is the lifetime whole sludge application rate (mt/ha), CPLR is the cumulative pollutant loading rate (kg/ha), and the other terms are as defined previously.

To determine the AWSAR or LWSAR for a sewage sludge, analyze a sample of the sewage sludge to determine the concentration of each pollutant listed in Table 9.19. Insert the proper APLRs from Table 9.19 or the proper CPLRs from Table 9.17 and the milligrams of pollutants per kilogram of dry solids into Equation 9.6 or 9.7 to determine the annual or cumulative whole sludge application rate. The AWSAR or LWSAR for the sewage sludge is the lowest value calculated for the various metals. For example, a measured concentration of copper of 2000 mg/L and an APLR of 75 kg/ha per 365-d period would yield an AWSAR of $2000/(75 \times 0.001) = 26{,}667$ mt/ha per 365 d. Calculate values for the other metals and select the lowest AWSAR for design.

TABLE 9.19
Annual Pollutant Loading Rate

Pollutant	Annual Pollutant Loading Rate (kg/ha per 365-d Period)
Arsenic	2
Cadmium	1.9
Copper	75
Lead	15
Mercury	0.85
Molybdenum[a]	—
Nickel	21
Selenium	5
Zinc	140

[a] Molybdenum was dropped in 1994 and a new value has not been set. Check the USEPA website for current values.

Source: Bastian, R.K., *Summary of 40CFR Part 503, Standards for the Use or Disposal of Sewage Sludge*, U.S. Environmental Protection Agency, Washington, D.C., 1993.

Some states may have more stringent metal limits than those presented here; therefore, it is essential to consult local regulations prior to design of a specific system.

9.8.2.2 Phosphorus

Some states require that the nutrient-limited sludge loading be based on the phosphorus needs of the design vegetation to ensure even more positive protection. This also provides a safety factor against nitrate contamination, as most sludges contain far less phosphorus than nitrogen, but most crops require far more nitrogen than phosphorus, as shown in Table 8.8. If optimum crop production is a project goal, this approach will require supplemental nitrogen fertilization. Equation 9.8 can be used to determine the phosphorus-limiting sludge loading; it is based on the common assumption (USEPA, 1978) that only 50% of the total phosphorus in the sludge is available:

$$R_P = \left(K_P\right)\left(\frac{U_P}{C_P}\right) \tag{9.8}$$

where

R_P = Phosphorus-limited annual sludge application rate, assuming 50% availability in the sludge (mt/ha; ton/ac).

K_P = 0.001 (metric units) or 0.002 (U.S. units).

U_P = Annual crop uptake of phosphorus (kg/ha; lb/ac); see Chapter 3 of this text for further discussion and USEPA (1983) for more exact data for midwestern crops.

C_P = Total phosphorus in sludge, as a decimal fraction (equation has already been adjusted for 50% availability, but this could be adjusted if data are available to indicate a higher or lower percent available total phosphorus)

9.8.2.3 Nitrogen

Calculation of the nitrogen-limited sludge loading rate is the most complicated of the calculations involved because of the various forms of nitrogen available in the sludge, the various application techniques, and the pathways nitrogen can take following land application. Most of the nitrogen in municipal sludges is in organic form, tied up as protein in the solid matter. The balance of the nitrogen is in ammonia form (NH_3). When liquid sludges are applied to the soil surface and allowed to dry before incorporation, about 50% of the ammonia content is lost to the atmosphere through volatilization (Sommers et al., 1981). As a result, only 50% of the ammonia is assumed to be available for plant uptake if the sludge is surface applied. If the liquid sludge is injected or immediately incorporated, 100% of the ammonia is considered to be available.

The availability of the organic nitrogen is dependent on the "mineralization" of the organic content of the sludge. Only a portion of the organic nitrogen is available in the year the sludge is applied, and a decreasing amount continues to be available for many years thereafter. The rate will be higher for sludges with higher initial organic nitrogen content. The rate drops rapidly with time, so for almost all sludges after the third year it is down to about 3% per year of the remaining organic nitrogen.

For the first few years of a sludge application, the nitrogen contribution from mineralization can still be significant. It is essential to include this factor when the design is based on annual applications and nitrogen is the potential limiting parameter. The nitrogen available (to plants) during the application year is given by Equation 9.9, and the available nitrogen from that same sludge in subsequent years is given by Equation 9.10. When annual applications are planned, it is necessary to repeat the calculations using Equation 9.10 and then add the results to those of Equation 9.9 to determine the total available nitrogen in a given year. These results will converge on a relatively constant value after 5 to 6 yr if sludge characteristics and application rates remain about the same.

Available nitrogen in the application year is given by:

$$N_a = K_N[NO_3 + k_v(NH_4) + f_n(N_o)] \tag{9.9}$$

TABLE 9.20
Typically Assumed Mineralization Rates for Organic Nitrogen in Wastewater Sludges

Time after Sludge Application	Mineralization Rate (%)		
	Raw Sludge	Anaerobic Digested	Composted
1	40	20	10
2	20	10	5
3	10	5	3
4	5	3	3
5	3	3	3
6	3	3	3
7	3	3	3
8	3	3	3
9	3	3	3
10	3	3	3

Source: Data from Sommers et al. (1981) and USEPA (1983).

where

N_a = Plant-available nitrogen in the sludge during the application year (kg/mt dry solids; lb/ton dry solids).

K_N = 1000 (metric units) or 2000 (U.S. units).

NO_3 = Percent nitrate in the sludge (% as a decimal).

k_v = Volatilization factor = 0.5 for surface-applied liquid sludge, 1.0 for incorporated liquid sludge and dewatered digested sludge applied in any manner.

NH_4 = Fraction of ammonia nitrogen in sludge (as a decimal).

f_n = Mineralization factor for organic nitrogen in first year $n = 1$ (see Table 9.20 for values).

N_o = Fraction of organic nitrogen in sludge (as a decimal).

Nitrogen available in subsequent years is

$$N_{pn} = K_N[f_2(N_o)_2 + f_3(N_o)_3 + \dots + f_n(N_n)] \qquad (9.10)$$

where N_{pn} is the plant-available nitrogen available in year n from mineralization of sludge applied in a previous year (kg/mt or lb/ton dry solids), $(N_o)_n$ is the decimal fraction of organic nitrogen remaining in the sludge in year n, and the other terms are as defined previously.

The nitrogen-limiting annual sludge loading is then calculated using Equation 9.11:

$$R_N = \frac{U_N}{N_a + N_{pn}}$$ (9.11)

where

R_N = Annual sludge loading in year of concern (mt/ha; ton/ac).

U_N = Annual crop uptake of nitrogen (kg/ha; lb/ac) (see Tables 8.8).

N_a = Plant-available nitrogen from current year's sludge, from Equation 9.9 (kg/mt or lb/ton dry solids).

N_{pn} = Plant-available nitrogen from mineralization of all previous applications (kg/mt or lb/ton dry solids).

In addition to the available nitrogen calculated above, it is also necessary that nitrogen from any other source be included when calculating agronomic rates.

9.8.2.4 Calculation of Land Area

Equation 9.6, Equation 9.8, and Equation 9.11 should be solved to determine the parameter limiting the sludge loading. Some regulatory authorities require limits on constituents other than nitrogen, phosphorus, or metals. The limiting parameter for design will then be the constituent that results in the lowest calculated sludge loading. The application area can then be determined using Equation 9.12. The area calculated using this equation is only the actual application area; it does not include any allowances for roads, buffer zones, and seasonal storage:

$$A = \frac{Q_s}{R_L}$$ (9.12)

where

A = Application area required (ha; ac).

Q_s = Total sludge production for the time period of concern (mt or ton dry solids).

R_L = Limiting sludge loading rate as defined by previous equations (mt/ha·yr; ton/ac·yr) for annual systems or for the time period of concern.

It is not likely that the design procedure described above will result in the ideal balance of nitrogen, phosphorus, and potassium for optimum crop production in an agricultural operation. The amounts of these nutrients in the sludge to be applied should be compared with the fertilizer recommendations for the desired crop yield, and supplemental fertilizer applied if necessary. USEPA (1983) gives typical nutrient requirements for crops in the midwestern states; agricultural agents and extension services can provide similar data for most other locations.

Annual applications are a common practice on agricultural operations. Forested systems typically apply sludge on a 3- to 5-yr interval due to the more difficult site access and distribution. The total sludge loading is designed using the equations presented above; however, because of mineralization of the larger single application, there may be a brief period of nitrate loss during the year of sludge application.

Reclamation and revegetation of disturbed land generally require a large quantity of organic matter and nutrients at the start of the effort to be effective. As a result, the sludge application is typically designed as a one-time operation, and the lifetime metal limits given in Table 9.17 are controlling on the assumption that the site might someday be used for agriculture. A single large application of sludge may result in a temporary nitrate impact on the site groundwater. That impact should be brief and preferable to the long-term environmental impact from the unreclaimed area. When cumulative metal loading limits control the sludge loading, the same total application area will be necessary for either agricultural or reclamation projects.

Forest systems may require the largest total land area of the three concepts because of access and application difficulties. Application of liquid sludge has been limited to tank trucks with sprinklers or spray guns. The maximum range of these devices is about 120 ft (37 m). To ensure uniform coverage, the site will require a road grid on about 250-ft (76-m) centers or limit applications to 120 ft (37 m) on each side of the existing road and firebreak network.

Experience has shown that tree seedlings do poorly in fresh anaerobically digested sludge (Cole et al., 1983). It may be necessary to wait for 6 months before planting to allow for aging of the sludge. Weeds and other undergrowth will crowd out new seedlings, so herbicides and cultivation may be necessary for at least 3 years (Sopper and Kerr, 1979). Sludge spraying on young deciduous trees should be limited to their dormant period to avoid heavy sludge deposits on the leaves.

Example 9.6

Find the area required for sludge application in an agricultural operation. Assume the following characteristics and conditions: anaerobically digested sludge production, 3 mt/d dry solids; sludge solids content, 7%; total nitrogen, 3%; ammonia nitrogen, 2%; nitrate, 0; arsenic, 50 ppm; cadmium, 18 ppm; copper, 400 ppm; lead, 430 ppm; mercury, 20 ppm; nickel, 80 ppm; selenium, 50 ppm; zinc, 900 ppm (ppm = mg/kg). A marketable crop is not intended, but the site will be planted with a grass mixture. It is expected that the orchard grass will eventually dominate. The local regulatory authorities accept the USEPA metal limitations and allow a design based on nitrogen fertilization requirements. A parcel of land is available within 6 km of the treatment plant.

Solution

1. A preliminary cost analysis indicates that transport of the liquid sludge to the nearby site will be cost effective, so further dewatering will not be required, and the application technique will be surface application.

2. Metal limits (from Table 9.17) are As, 41 kg/ha; Cd, 39 kg/ha; Cu, 1500 kg/ha; Pb, 300 kg/ha; Hg, 17 kg/ha; Ni, 420 kg/ha; Se, 100 kg/ha; and Zn, 2800 kg/ha. The annual nitrogen uptake of the grass will be 224 kg/ha·yr (from Table 8.5). The mineralization rates for anaerobically digested sludge will be 20, 10, 5, and 3%, etc. (from Table 9.20).

3. The lifetime metal loadings are calculated using Equation 9.7:

$$\text{LWSAR} = \frac{\text{CPLR}}{C \times 0.001}$$

For arsenic,

$$\text{LWSAR} = \frac{41 \text{ kg/ha}}{50 \times 0.001} = 820 \text{ mt dry sludge per ha}$$

Similarly,

    Cd LWSAR = 2167 mt dry sludge per ha.
    Cu LWSAR = 3750 mt dry sludge per ha.
    Pb LWSAR = 698 mt dry sludge per ha.
    Hg LWSAR = 850 mt dry sludge per ha.
    Ni LWSAR = 5250 mt dry sludge per ha.
    Se LWSAR = 2000 mt dry sludge per ha.
    Zn LWSAR = 3111 mt dry sludge per ha.

Lead results in the lowest sludge loading and is therefore the limiting metal parameter. As a result, 698 mt/ha of sludge can be applied during the useful life of the site if sludge conditions remain the same. If all of the metal concentrations had been below the pollutant concentration limits shown in Table 9.18, heavy metal constraints would not affect the sizing of the facility.

4. Use Equation 9.9 and Equation 9.10 to calculate the available nitrogen in the sludge. Because the liquid sludge will be surface applied, volatilization losses will occur, and k_v will equal 0.5. Assume that organic nitrogen equals total nitrogen less ammonium nitrogen:

$$N = (K_N)\left[(NO_3) + k_v(NH_4) + f_n(N_o)\right]$$

$$= (1000)\left[(0) + (0.5)(0.02) + (0.20)(0.01)\right]$$

$$= (1000)(0.012)$$

$$= 12 \text{ kg / mt dry solids}$$

The residual nitrogen in this sludge in the second year is:

$$(N_o)_1 - (f_1)(N_o) = (0.01) - (0.20)(0.01$$

$$= 0.008 \text{ (as a decimal fraction)}$$

The second year mineralization is:

$$(f_2)(N_o)_2 = (0.10)(0.008) = 0.0008$$

Residual nitrogen in the third year is

$$(N_o)_3 = (0.008) - (0.0008) = 0.00072$$

Similarly, mineralization in the third year is 0.0004; in the fourth year, 0.0002; in the fifth year, 0.0002; etc. The total available nitrogen in the second year is the second-year contribution plus the residual from the first year:

$$(N_a)_2 = (N_a)_1 + K_N f_2 (N_a)_2$$

$$= 12 + (1000)(0.0008)$$

$$= 12.8 \text{ kg / mt dry sludge}$$

Similarly,

$$(N_a)_3 = (N_a)_1 + K_N \left[f_2(N_o)_2 + f_3(N_o)_3 \right]$$

$$= 12 + (1000)(0.0008 + 0.0004) = 13.3 \text{ kg / mt dry sludge}$$

$(N_a)_4 = 13.4$ kg/mt dry sludge; $(N_a)_5 = 13.6$ kg/mt, etc. Assuming that the sludge characteristics stay the same, the available nitrogen will remain at about 13.6 kg/mt dry sludge from the fifth year on.

5. Use Equation 9.11 to calculate the annual nitrogen-limited sludge loading. Use 13.6 kg/mt as the steady-state value from step 4:

$$R_N = \frac{U_N}{N_a + N_{pn}} = \frac{224}{13.6} = 16.5 \text{ mt / ha / yr of dry sludge}$$

Higher loadings may be applied during the first 2 years if desired, as the full cumulative effects of mineralization will not be realized until the third year.

6. Use Equation 9.12 to find the required application area. Because food chain crops are not involved, the annual loading is based on the nitrogen limits:

$$A = \frac{Q_s}{R_L} = (3 \text{ mt / d})(365 \text{ d / yr}) / (16.5 \text{ mt / ha} \cdot \text{yr}) = 66 \text{ ha}$$

7. Determine the useful life of the site for sludge application. This will eliminate restrictions on potential future land uses, including production of human food crops. The lead-limited sludge loading calculated in step 3 will control.

Useful life = (698 mt/ha)/(16.5 mt/yr) = 42.3 yr

A system design for a reclamation site would typically use a single sludge application. The total annual sludge production is 1095 mt/yr (3 mt/d × 365 d/yr). At a single loading of 698 mt/ha, 1.6 ha of land would require reclamation each year. Reclamation project designs must ensure that sufficient land will be available for each year of the intended operational life.

9.8.3 DESIGN OF SURFACE DISPOSAL SYSTEMS

The design of surface disposal systems typically includes all of the factors discussed for land application systems, as metals and nutrients may still control the sludge loading and useful life of the site. In addition, sludges intended for surface disposal systems may contain a larger fraction of biodegradable material than typical municipal sludges and have significant concentrations of toxic or hazardous substances. These materials, more common in petroleum and many industrial sludges, are quite often organic compounds. Their presence, if degradable, may control the frequency as well as the size of the design unit loading on the system. If the pollutants are nondegradable, the application site should more properly be considered as a disposal or containment operation; information on such systems may be found elsewhere (Sittig, 1979; USACE, 1984). The primary mechanism for degradation of organic chemicals in soil is due to the activity of the soil microorganisms. Volatilization may be significant for some compounds (Brown, 1983; Jenkins and Palazzo, 1981) and plant uptake may be a factor if vegetation is a system component, but biological reactions are the major treatment mechanism.

9.8.3.1 Design Approach

The design approach for these organic materials is based on their half-life in the soil system. This is analogous in some respects to the mineralization rate approach for nitrogen management. If, for example, a substance in the sludge has a 1-yr half-life and the sludge is applied on an annual basis, half of the mass of the substance will still be left in the soil at the end of the first year. At the end of the second year, three quarters of the annual mass applied will still be in the soil, and so forth, until at the seventh year the mass remaining in the soil will be very close to the amount of the annual application. It is suggested that, for compounds with a half-life of up to 1 year, the amount allowed to accumulate in the soil should not exceed twice the annual application of the substance (Brown, 1983; Burnside, 1974). This can be achieved by adopting an application schedule that is equal to one half-life for the substance of concern.

Soil texture and structure, moisture content, temperature, oxygen level, nutrient status, pH, and the type and number of microorganisms present influence the biological reactions in the soil. The optimum conditions for all of these factors are essentially the same as those required for successful operation of an agricultural land application system. An aerobic soil with a pH of 6 to 7, a temperature of at least 50°F (10°C), and soil moisture at field capacity would represent near-optimum conditions for most situations. An additional special concern with toxic organics is their impact on soil microbes. A unit loading that is too high may actually sterilize the soil. Mixing of the soil and the sludge reduces this risk and promotes aeration and contact between the microbes and the waste. As a result of this need for mixing, surface vegetation is not typically a treatment component in systems designed for short-half-life sludges.

9.8.3.2 Data Requirements

Characterization of the sludge constituents is a critical first step in design, especially if potentially toxic or hazardous organic compounds are present. Essential data include inorganic chemicals, electrical conductivity, pH, titratable acids and bases, moisture (water) content, total organic matter, volatile organic compounds, extractable organic compounds, residual solids, and a biological assessment to determine acute and genetic toxicity. The inorganic chemicals might include the same metals, nutrients, and halides and other salts that would be included in an analysis for land application designs.

9.8.3.3 Half-Life Determination

The degradation and half-life of complex organic compounds are typically determined in the laboratory by a series of soil respirometer tests. Representative samples of soil and sludge are mixed in a proportional range and placed in sealed flasks, which in turn are placed in an incubation chamber. Humidified, carbon-dioxide-free air is passed through each flask. The carbon dioxide evolved from microbial activity in the flask is picked up by the air and then collected in columns containing $0.1\text{-}N$ sodium hydroxide. The sodium hydroxide solutions are changed about three times a week and then titrated with hydrochloric acid. Detailed procedures can be found in Brown (1983) and Stotzky (1965). The typical incubation period is up to 6 months. The control tests are run at 68°F (20°C), but, if field temperatures are expected to vary by more than 18°F (10°C), the half-life at these other temperatures should also be determined. In some cases it is desirable to verify laboratory results with pilot studies in the field. Soil samples are taken on a routine basis after application and mixing of the sludge and soil. The analysis should include total organics as well as compounds of specific concern. In addition to measurements of carbon dioxide evolution by the respirometer tests, it is recommended that the organic fractions of the original sample and that of the final soil-sludge mixture be determined. The degradation rates are then determined using Equation 9.13 and Equation 9.14. For total carbon degradation,

$$D_t = \frac{(0.27)\left([CO_2]_w - [CO_2]_s\right)}{C} \tag{9.13}$$

where

D_t = Fraction of total carbon degraded over time t.
$[CO_2]_w$ = Cumulative CO_2 evolved by soil-waste mixture.
$[CO_2]_s$ = Cumulative CO_2 evolved by unamended soil.
C = Carbon applied with the sludge.

For organic carbon degradation,

$$D_{t,0} = \frac{1 - \left(C_{r,0} - C_S\right)}{C_{a,0}} \tag{9.14}$$

where

$D_{t,0}$ = Fraction of organic carbon degraded over time t.
$C_{r,0}$ = Amount of residual carbon in the organic fraction of the final sludge–soil mixture.
C_S = Amount of organic carbon extracted from the unamended soil.
$C_{a,0}$ = Amount of carbon in the organic fraction of the applied sludge.

The degradation rates of individual organic subfractions are also determined by Equation 9.14. The half-life for the total organics or for a specific waste is determined using Equation 9.15:

$$t_{1/2} = \frac{(0.5)(t)}{D_t} \tag{9.15}$$

where

$t_{1/2}$ = Half-life of the organics of concern (d).
t = Time period used to produce the data for Equation 9.13 or Equation 9.14 (d).
D_t = Fraction of carbon degraded over time t.

If vegetation is to be a routine treatment component in the operational system, greenhouse and pilot field studies are necessary to evaluate toxicity and develop optimum loading rates. Greenhouse studies are easier and less costly to run, but field studies are more reliable. Systems designed only for soil treatment need not be tested unless vegetation is planned as a post closure activity.

Because a range of sludge–soil mixtures is tested in the respirometers, it is also possible to determine the concentration at which acceptable microbial activity occurs. It is then possible to determine the annual loading from this value and the previously determined half-life:

$$C_{yr} = \frac{(0.5)(C_c)}{t_{1/2}} \qquad (9.16)$$

where

C_{yr} = Annual application rate for the organic of concern (kg/ha/yr; lb/ac/yr).

C_c = Critical concentration at which acceptable microbial toxicity occurs (kg/ha; lb/ac).

$t_{1/2}$ = Half-life of the organic of concern (yr).

The loading rate is then calculated using a variation of Equation 9.6:

$$R_{0,C} = \frac{C_{yr}}{C_w} \qquad (9.17)$$

where

$R_{0,C}$ = Loading rate limited by organics (kg/ha/yr; lb/ac·yr).

C_{yr} = Annual application rate for organic of concern (Equation 9.16) (kg/ha·yr; lb/ac·yr).

C_w = Fraction of the organic of concern in the sludge (as a decimal).

If the half-life of the organic of concern is less than 1 year, the $R_{0,C}$ calculated from Equation 9.16 may be applied on a more frequent schedule. In this case, the number of applications becomes:

$$N = \frac{1}{t_{1/2}} \qquad (9.18)$$

where N is the number of applications per year, and $t_{1/2}$ is the half-life (yr).

The land area required is then determined using Equation 9.12. As with land application systems, the calculations are performed for nutrients, metals, and other potentially limiting factors. The limiting parameter for design is then the constituent requiring the largest land area as calculated by Equation 9.12.

9.8.3.4 Loading Nomenclature

Depending on industrial conventions and practices, the loading rates and application rates used in the design calculations may be expressed in a variety of units; for example, in the petroleum industry, it is common to express the loading in terms of barrels per hectare. In most cases, the sludge is mixed with the surface soil. This surface zone, termed the *incorporation zone*, is typically 6 in. (15 cm) thick. As a result, the loading is also often expressed as kilograms per meter of incorporation zone or as a percentage of a contaminant (on a mass basis) in the incorporation zone. The calculations below illustrate the various possibilities.

One barrel (bbl) of oil contains 42 gal (159 L), which is about 315 lb (143 kg) of oil. One cubic foot of "typical" soil contains about 80 lb of soil (1270 kg of soil). One acre of treatment area with a 6-in. (15-cm) incorporation zone contains $(0.5)(43,560) = 21,780$ ft^3/ac (1520 m^3/ha). At 41 bbl oil per acre (100 bbl/ha), the mass loading will be $(41)(316)/21,780$ ft^3/ac $= 0.595$ lb/ft^3 (9.53 kg/m^3) of incorporation zone. At 203 bbl/ac (500 bbl/ha), the mass loading (on a percentage basis) will be $(203)(316)/(21,780)(79) = 3.75\%$ oil in the incorporation zone.

Example 9.7

Find the land area required for treatment of a petroleum sludge produced at a rate of 5 mt/d, containing 15% critical organics. The following data were obtained with respirometer tests:

- Applied carbon (C) = 3000 mg.
- CO_2 produced (90 d) = 1500 mg (waste + soil) = 100 mg (soil only).
- A field test indicated that the critical application (C_c) for maintenance of the soil microbes was 71,500 kg/ha·yr (3.75%).

Solution

1. Use Equation 9.13 to determine evolved CO_2 on a total carbon basis:

$$D_t = \frac{(0.27)\left([CO_2]_w - [CO_2]_s\right)}{C}$$

$$D_{90} = \frac{(0.27)(1500 - 100)}{3000} = 0.13$$

2. Determine the half-life for the organic compounds using Equation 9.15:

$$t_{1/2} = \frac{(0.5)(t)}{D_t} = \frac{(0.5)(90)}{0.13}$$

$$= 346 \text{ d} = 0.95 \text{ yr}$$

3. Determine the application rate for the critical compounds using Equation 9.16:

$$C_{yr} = \frac{(0.5)(C_c)}{t_{1/2}} = \frac{(0.5)(71,500)}{0.95}$$

$$= 37,632 \text{ kg / ha} \cdot \text{yr}$$

4. Determine the organic-controlled loading rate using Equation 9.17:

$$R_{0,C} = \frac{C_{yr}}{C_w}$$

$$R_{0,C} = \frac{37,632}{0.15}$$

$$= 250,880 \text{ kg / ha} \cdot \text{yr} = 251 \text{ mt / ha} \cdot \text{yr}$$

5. Determine the required land area using Equation 9.12:

$$A = \frac{Q_s}{R_L} = \frac{(5)(365)}{251} = 7.3 \text{ ha}$$

6. To complete the design calculations, the area required for nutrients, metals, and any other limiting substances should be determined. The largest of these calculated areas will then be the design treatment area.

9.8.3.5 Site Details for Surface Disposal Systems

The site selection procedure and design will depend on whether the site is to be permanently dedicated for a treatment/disposal operation or if it is to be restored and made available for unrestricted use following the operational life. A system of the former type may be operated as a treatment system, but ultimately one or more of the waste constituents will exceed the specified cumulative limits, so the site must be planned as a disposal operation. Criteria for these disposal operations can be found in Sittig (1979) and USACE (1984). The general site characteristics are similar for both land application and surface disposal systems. A major difference is often the control of runoff. Off-site runoff is not generally permitted for either type of operation; however, in the case of agricultural sludge operations, runoff is contained but then may be allowed to infiltrate on the application site. Runoff is a more serious concern for surface disposal operations, as the sludge may contain mobile toxic or hazardous constituents.

The site is typically selected, constructed on a gentle slope (1 to 3%), and subdivided into diked plots. The purpose is to induce controlled runoff and ensure minimum infiltration and percolation. A complete hydrographic analysis is required to determine the criteria for design of collection channels, retention basins, and structures to prevent off-site runoff from entering the site. Such designs should be based on the peak discharge from a 25-year storm, and the retention basins for a 25-year, 24-hour-return-period storm. The discharge pathway from the retention basin will depend on the composition of the water. In many cases, it may be land applied using one or more of the techniques described in Chapter 8. Special treatments may be required for critical materials; sprinklers or aeration in the retention basin are often used to reduce the concentration of volatile organics.

If clay or other liners are a site requirement, then underdrains will be necessary. Under drains may also be required to control groundwater levels in an unlined site and to ensure maintenance of aerobic conditions in the incorporation zone. Any water collected with these drains must also be retained and possibly treated.

The site design must also consider the application method to be used, and appropriate access for vehicles must be provided. Sprinklers and portable spray guns have been used with liquid sludges. In this case, the civil engineering aspects of site design are quite similar to those for the overland-flow concept described in Chapter 8. Dry sludges can be spread and mixed with the same type of equipment that would be used for land application operations.

On-site temporary storage may also be a requirement, particularly in colder climates. Optimal soil temperatures for microbial activity are 68°F (20°C) or higher. If lower temperatures are expected, the interval between applications can be extended (as determined by field or respirometer tests), or the sludge can be stored during the cold periods.

The soil temperatures for bare soil surfaces are commonly greater than the ambient air temperature by 5 to 9°F (3 to 5°C) during daylight hours. Surface soils at many land treatment sites may exceed ambient temperatures by 9 to 18°F (5 to 10°C) because of microbial activity and increased radiation absorption when dark, oily wastes are incorporated (Loehr and Ryan, 1983). In the general case, it can be assumed that active degradation is possible when the ambient air temperatures are 50°F (10°C) or higher and no frost remains in the soil profile. On this basis, the operational season for a surface disposal system may be slightly longer than for a land application system in the same location.

REFERENCES

Alam, M.Z., Fakhru'l-Razi, A., and Molla, A.H. (2004). Evaluation of fungal potential for bioconversion of domestic wastewater sludge, *J. Environ. Sci. (China)*, 16(1), 132–137.

Banks, L. and Davis, S.F. (1983a). Desiccation and treatment of sewage sludge and chemical slimes with the aid of higher plants, in *Proceedings of Symposium on Municipal and Industrial Sludge Utilization and Disposal*, Rutgers University, Atlantic City, NJ, April 6–8, 1983, 172–173.

Banks, L. and Davis, S.F. (1983b). Wastewater and sludge treatment by rooted aquatic plants in sand and gravel basins, in *Proceedings of Workshop on Low Cost Wastewater Treatment*, Clemson University, Clemson, SC, April 1983, 205–218.

Bastian, R.K. (1993). *Summary of 40CFR Part 503, Standards for the Use or Disposal of Sewage Sludge*, U.S. Environmental Protection Agency, Washington, D.C.

Beeceher, N., Harrison, E., Goldstein, N., Field, P., and Susskind, L. (2004). Risk perception, communication, and stakeholder involvement, in *Sustainable Land Application Conference Abstracts*, Soil and Water Science Department, University of Florida, Gainesville.

Brown, K. W. (1983). *Hazardous Waste Land Treatment*, EPA Report SW-874, Office of Solid Waste and Emergency Response, U.S. Environmental Protection Agency, Washington, D.C.

Burnside, O.V. (1974). Prevention and detoxification of pesticide residues in the soil, in *Pesticides in Soil and Water*, Guenzi, W.D., Ed., Soil Scientists of America, Madison, WI, 387–412.

Cole, D.W., Henry, C.L., Schiess, P., and Zasoski, R.J. (1983). The role of forests in sludge and wastewater utilization programs, in *Proceedings of 1983 Workshop on Utilization of Municipal Wastewater and Sludge on Land*, University of California, Riverside, 125–143.

Costic & Associates (1983). *Engineers Report: Washington Township Utilities Authority Sludge Treatment Facility*, Costic & Associates, Long Valley, NJ.

Crites, R.W. and Tchobanoglous, G. (1998). *Small and Decentralized Wastewater Management Systems*, McGraw-Hill, New York.

Crites, R.W., Reed, S.C., and Bastian, R.K. (2000). *Land Treatment Systems for Municipal and Industrial Wastes*, McGraw-Hill, New York.

Donovan, J. (1981). *Engineering Assessment of Vermicomposting Municipal Wastewater Sludges*, EPA-600/2-81-075, U.S. Environmental Protection Agency, Washington, D.C.; available as PB 81-196933 from National Technical Information Service, Springfield, VA.

Eastman, B.R., Kane, P.N., Edwards, C.A., Trytec, L., Gunadi, B., Stermer, A.L., and Mobley, J.R. (2001). The effectiveness of vermiculture in human pathogen reduction for USEPA biosolids stabilization, *Compost Sci. Utilization*, 9(1), 38–49.

Farrell, J.B., Smith, Jr., J.E., Dean, R.B., Grossman, E., and Grant, O.L. (1970). Natural freezing for dewatering of aluminum hydroxide sludges, *J. Am. Water Works Assoc.*, 62(12), 787–794.

Grady, C.P.L., Daigger, G.T., and Lim, H.C. (1999). *Biological Wastewater Treatment*, 2nd ed., Marcel Dekker, New York.

ICE. (2002). *Wastewater Sludge Disposal Strategy*, Water Board Position Statement, Institution of Civil Engineers, Great Britain.

Jenkins, T.F. and Palazzo, A.J. (1981). *Wastewater Treatment by a Prototype Slow Rate Land Treatment System*, CRREL Report 81-14, Cold Regions Research and Engineering Laboratory (CRREL), Hanover, NH.

Kester, G.B., Brobst, R.B., Carpenter, A., Chaney, R.L., Rubin, A.B., Schoof, R.A., and Taylor, D.S. (2004). Risk characterization, assessment, and management of organic pollutants in beneficially used residual products, in *Sustainable Land Application Conference Abstracts*, Soil and Water Science Department, University of Florida, Gainesville.

Kuter, G.A., Hoitink, H.A.J., and Rossman, L.A. (1985). Effects of aeration and temperature on composting of municipal sludge in a full-scale vessel system, *J. Water Pollut. Control Fed.*, 57(4), 309–315.

Lang, L.E., Bandy, J.T., and Smith, E.D. (1985). *Procedures for Evaluating and Improving Water Treatment Plant Processes at Fixed Army Facilities*, Report of the U.S. Army Construction Engineering Research Laboratory, Champaign, IL.

Loehr, R.C. and Ryan, J. (1983). *Land Treatment Practices in the Petroleum Industry*, American Petroleum Institute, Washington, D.C.

Loehr, R.C., Martin, J.H., Neuthauser, E.F., and Malecki, M.R. (1984). *Waste Management Using Earthworms, Engineering and Scientific Relationships*, National Science Foundation ISP-8016764, Cornell University, Ithaca, NY.

Metcalf & Eddy. (1991). *Wastewater Engineering: Treatment, Disposal, and Reuse*, 3rd ed., McGraw-Hill, New York.

Metcalf & Eddy. (2003). *Wastewater Engineering: Treatment and Reuse*, 4th ed., McGraw-Hill, New York.

Miller, F.C. and Finstein, M.S. (1985). Materials balance in the composting of wastewater sludge as affected by process control strategy, *J. Water Pollut. Control Fed.*, 57(2), 122–127.

NRC. (2002). *Biosolids Applied to Land: Advancing Standards, and Practices*, National Academy Press, Washington, D.C.

Penner, E. (1962). Ground freezing and frost heave, CBD-26, *Canadian Building Dig.* (http://irc.nrc-cnrc.gc.ca/cbd/cbd026e.html).

Pietz, R.I., Peterson, J.R., Prater, J.E., and Zenz, D.R. (1984). Metal concentrations in earthworms from sewage sludge amended soils at a strip mine reclamation site, *J. Environ. Qual.*, 13(4), 651–654.

Reed, S.C. and Crites, R.W. (1984). *Handbook of Land Treatment Systems for Industrial and Municipal Wastes*, Noyes Publications, Park Ridge, NJ.

Reed, S.C., Bouzoun, J., and Medding, W.S. (1984). A rational method for sludge dewatering via freezing, *Comptes Rendus, 7e Symposium sur le traitment des euax us lees*, Montreal, November 20–21, 1984, 109–117.

Reynolds, T.D. and Richards, P.A. (1996). *Unit Operations and Processes in Environmental Engineering*, PWS Publishing, New York.

Rush, R.J. and Stickney, A.R. (1979). *Natural Freeze–Thaw Sludge Conditioning and Dewatering*, EPS 4-WP-79-1, Environment Canada, Ottawa.

Schleppenbach, F.X. (1983). *Water Filtration at Duluth, Minnesota*, EPA 600/2-84-083, U.S. Environmental Protection Agency, Washington, D.C.; available as PB 84-177 807 from National Technical Information Service, Springfield, VA.

Schneiter, R.W., Middlebrooks, E.J., Sletten, R.S., and Reed, S.C. (1984). *Accumulation, Characterization and Stabilization of Sludges from Cold Regions Lagoons*, CRREL Special Report 84-8, U.S. Army Cold Regions Research and Engineering Laboratory (CRREL), Hanover, NH.

Sieger, R.B. and Herman, G.J. (1993). Land application requirements of the new sludge rules, *Water Eng. Manage.*, 140(8), 30–31.

Sigmund, T.W. and Sieger, R.B. (1993). The new surface disposal requirements, *Water Eng. Manage.*, 140(9), 18–19.

Sittig, M. (1979). *Landfill Disposal of Hazardous Wastes and Sludges*, Noyes Data Corp., Park Ridge, NJ.

Sommers, L.E., Parker, C.F., and Meyers, G.J. (1981). *Volatilization, Plant Uptake and Mineralization of Nitrogen in Soils Treated with Sewage Sludge*, Technical Report 133, Water Resources Research Center, Purdue University, West Lafayette, IN.

Sopper, W.E. and Kerr, S.N. (1979). *Utilization of Municipal Sewage Effluent and Sludge on Forest and Disturbed Land*, Pennsylvania State University Press, University Park.

Stotzky, G. (1965). Microbial respiration, in *Methods of Soil Analysis*. Part 2. *Chemical and Microbial Properties*, American Society of Agronomy, Madison, WI, 1550–1572.

USACE. (1984). *Technical Manual Hazardous Waste Land Disposal and Land Treatment Facilities*, TM 5-814-7, Huntsville Division, U.S. Army Corps of Engineers, Huntsville, AL.

USDA/USEPA. (1980). *Manual for Composting Sewage Sludge by the Beltsville Aerated Pile Method*, EPA 600/8-80-022, EPA Municipal Environmental Research Laboratory, Cincinnati, OH.

USEPA. (1978). *Process Design Manual: Municipal Sludge Landfills*, EPA 625/1-78-010, U.S. Environmental Protection Agency, Washington, D.C.; available as PB-279 675 from National Technical Information Service, Springfield, VA.

USEPA. (1979). *Process Design Manual: Sludge Treatment and Disposal*, EPA 625/1-79-011, U.S. Environmental Protection Agency, Cincinnati, OH.

USEPA. (1981a). *Composting Processes To Stabilize and Disinfect Municipal Sewage Sludge*, EPA 430/9-81-011, Office of Water Program Operations, U.S. Environmental Protection Agency, Washington, D.C.

USEPA. (1981b). *Process Design Manual: Land Treatment of Municipal Wastewater*, EPA 625/1-81-013, Center for Environmental Research Information, U.S. Environmental Protection Agency, Cincinnati, OH.

USEPA. (1982). *Process Design Manual: Dewatering Municipal Wastewater Sludges*, EPA 625/1-82-014, Center for Environmental Research Information, U.S. Environmental Protection Agency, Cincinnati, OH.

USEPA. (1983). *Process Design Manual: Land Application of Municipal Sludge*, EPA 625/1-83-016, Center for Environmental Research Information, U.S. Environmental Protection Agency, Cincinnati, OH.

USEPA. (1984). *Sludge Composting and Improved Incinerator Performance*, Technology Transfer Seminar Report, Center for Environmental Research Information, U.S. Environmental Protection Agency, Cincinnati, OH.

USEPA. (1990). National sewage sludge survey, *Fed. Reg.*, 55(218), 47210–47283.

USEPA. (1994a). *Biosolids Recycling: Beneficial Technology for a Better Environment*, EPA/832-R-94-009, Office of Water, U.S. Environmental Protection Agency, Washington, D.C.

USEPA. (1994b). *A Plain English Guide to the EPA Part 503 Biosolids Rule*, EPA/832/R-93/003, Office of Wastewater Management, U.S. Environmental Protection Agency, Washington, D.C.

USEPA. (1995a). *Process Design Manual: Surface Disposal of Sewage Sludge and Domestic Septage*, EPA/625/R-95/002, Office of Research and Development, U.S. Environmental Protection Agency, Cincinnati, OH.

USEPA. (1995b). *A Guide to the Biosolids Risk Assessments for the EPA Part 503 Rule*, EPA/832-B-93-005, Office of Wastewater Management, U.S. Environmental Protection Agency, Washington, D.C.

USEPA. (1995c). *Process Design Manual: Land Application of Sewage Sludge and Domestic Septage*, EPA/625/R-95/001, Office of Research and Development, U.S. Environmental Protection Agency, Washington, D.C.

USEPA. (1999). *Biosolids Generation, Use, and Disposal in the United States*, EPA/530-R-99-009, Office of Solid Waste, U.S. Environmental Protection Agency, Washington, D.C.

USEPA. (2003a). *Environmental Regulations and Technology: Control of Pathogens and Vector Attraction in Sewage Sludge*, EPA/625/R-92/013, Office of Research and Development, U.S. Environmental Protection Agency, Cincinnati, OH.

USEPA. (2003b). *Technical Background Document for the Sewage Sludge Exposure and Hazard Screening Assessment*, Document No. 822-B-03-001, Office of Water, U.S. Environmental Protection Agency, Washington, D.C.

USEPA. (2003c). *Use and Disposal of Biosolids (Sewage Sludge): Agency Final Response to the National Research Council Report on Biosolids Applied to Land and the Results of the Review of Existing Sewage Sludge Regulations*, EPA/822-F-03-010, Office of Water, U.S. Environmental Protection Agency, Washington, D.C.

USEPA/USDA. (2000). *Guide to Field Storage of Biosolids and Other Organic By-Products Used in Agriculture and for Soil Resource Management*, EPA/832-B-00-007, Office of Wastewater Management, U.S. Environmental Protection Agency, Washington, D.C., and Agricultural Research Service, Beltsville, MD.

USEPA/USDA/FDA. (1981). *Land Application of Municipal Sewage Sludge for the Production of Fruits and Vegetables: A Statement of Federal Policy and Guidance*, Office of Municipal Pollution Control, U.S. Environmental Protection Agency, Washington, D.C.

Water Services Association of Australia. (1995). *Die-Off of Human Pathogens in Stored Wastewater Sludge and Sludge Applied to Land*, Report No. WSAA 92; available through www.wsaa.aun.au or via e-mail (info@wsaa,asn.au).

Whiting, D.M. (1975). *Use of Climatic Data in Design of Soil Treatment Systems*, EPA 660/2-75-018, Corvallis Environmental Research Laboratory, U.S. Environmental Protection Agency, Corvallis, OR.

10 On-Site Wastewater Systems

Effluent disposal options for on-site systems range from soil absorption in conventional gravity leachfields to water reuse after high-tech membrane treatment. Individual on-site systems are the most prevalent wastewater management systems in the country. This chapter describes the various types of on-site wastewater systems, wastewater disposal options, site evaluation and assessment procedures, cumulative areal nitrogen loadings, nutrient removal alternatives, disposal of variously treated effluents in soils, design criteria for on-site disposal alternatives, design criteria for on-site reuse alternatives, correction of failed systems, and role of on-site management systems.

10.1 TYPES OF ON-SITE SYSTEMS

While many types of on-site systems exist, most involve some variation of subsurface disposal of septic tank effluent. The four major categories of on-site systems are:

- Conventional on-site systems
- Modified conventional on-site systems
- Alternative on-site systems
- On-site systems with additional treatment

The most common on-site system is the conventional on-site system that consists of a septic tank and a soil absorption system (see Figure 10.1). The septic tank is the wastewater pretreatment unit used prior to on-site treatment and disposal. Modified conventional on-site systems include shallow trenches and pressure-dosed systems. Alternative on-site disposal systems include mounds, evapotranspiration systems, and constructed wetlands. Additional treatment of septic tank effluent is sometimes needed, and intermittent and recirculating granular-medium filters are often the economical choice. Where further nitrogen removal is required, one or more of the alternatives for nitrogen removal (see Section 10.4) may be considered. The types of disposal and reuse systems used for individual on-site systems are presented in Table 10.1.

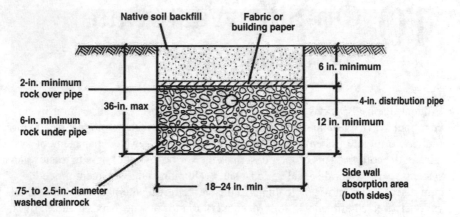

FIGURE 10.1 Typical cross-section through conventional soil absorption system.

10.2 EFFLUENT DISPOSAL AND REUSE OPTIONS

Alternative infiltration systems (presented in Table 10.2) have been developed to overcome restrictive conditions such as:

- Very rapidly permeable soils
- Very slowly permeable soils
- Shallow soil over bedrock
- Shallow groundwater
- Steep slopes
- Groundwater quality restrictions
- Limited space

The alternatives for reuse of on-site system effluent include drip irrigation, spray irrigation, groundwater recharge, and toilet flushing. Drip irrigation is becoming more popular for water reuse and is described in this chapter. Spray irrigation is more suited to larger flows (commercial, industrial, and small community flows) and is described in detail in Chapter 8. Groundwater recharge, which is used in areas of deep permeable soils, is also described in Chapter 8.

10.3 SITE EVALUATION AND ASSESSMENT

The process of selecting a suitable on-site location for on-site disposal involves multiple steps of identification, reconnaissance, and assessment. The process begins with a thorough examination of the soil characteristics, which include permeability, depth, texture, structure, and pore sizes. The nature of the soil profile and the soil permeability are of critical concern in the evaluation and assessment of the site. Other important aspects of the site are the depth to groundwater, site

TABLE 10.1
Types of On-Site Wastewater Disposal/Reuse Systems

Disposal/Reuse System	Remarks
Conventional Systems	
Gravity leachfields/conventional trench	Most common system
Gravity absorption beds	—
Modified Conventional Systems	
Gravity leachfields:	
Deep trench	To get below restrictive layers
Shallow trench	Enhanced soil treatment
Pressure-dosed:	
Conventional trench	To reach uphill fields
Shallow trench	Uphill and shallow sites
Drip application	Following additional treatment of septic tank effluent; to optimize use of available land area
Alternative Systems	
Sand-filled trenches	Added treatment
At-grade systems	Less expensive than mounds
Fill systems	Import soil
Mound Systems	
Evapotranspiration systems	Zero discharge
Evaporation ponds	See Chapter 4
Constructed wetlands	Requires a discharge or subsequent infiltration (see Chapter 7)
Reuse Systems	
Drip irrigation	Usually follows added treatment
Spray irrigation	Requires disinfection
Graywater reuse	—
Other Systems	
Holding tanks	Seasonal use alternative
Surface water discharge	Allowed in some states following added treatment

slope, existing landscape and vegetation, and surface drainage features. After a potential site has been located, the site evaluation and assessment proceeds, generally in two phases: preliminary site evaluation and detailed site assessment.

TABLE 10.2
Appropriate On-Site Disposal Methods To Overcome Site Constraints

Method	Soil Permeability			Bedrock		Groundwater		Slope		Small Lot Size
	Very Rapid	Moderately Rapid	Very Slow	Shallow	Deep	Shallow	Deep	0–5%	>5%	
Trenches		•			•		•	•	•	•
Beds		•			•		•	•	•	•
Pits		•			•		•	•	•	•
Mounds	•	•			•		•	•	•	•
Fill systems	•	•	•	•	•	•	•	•	•	
Sand-lined trenches and beds	•	•	•	•	•	•	•	•	•	
Drained systems		•			•		•	•	•	
Evaporation ponds	•	•	•	•	•	•	•	•	•	
ET beds	•	•	•	•	•	•	•	•	•	
ETA beds	•	•	•	•	•	•	•	•	•	
Spray irrigation	•	•	•		•		•	•	•	•
Drip irrigation	•	•	•		•		•	•	•	

Note: The symbol • indicates appropriate system; ET, evapotranspiration; ETA, evapotranspiration–absorption.

TABLE 10.3
Typical Regulatory Factors in On-Site Systems

Factor	Unit	Typical Value
Setback distances (horizontal, separation from wells, springs, surface waters, escarpments, site boundaries, buildings)	ft	(See Table 10.12)
Maximum slope for on-site disposal field	%	25-30
Soil characteristics:		
Depth	ft	2
Percolation rate	min/in.	>1 to <120
Minimum depth to groundwater	ft	3
Septic tank (minimum size)	gal	750
Maximum hydraulic loading rates for leachfields	gal/ft²·d	1.5
Maximum loading rates for sand filters	gal/ft²·d	1.2

10.3.1 PRELIMINARY SITE EVALUATION

The initial step in conducting a preliminary site evaluation is to determine the current and proposed land use, the expected flow and characteristics of the wastewater, and to observe the site characteristics. The next step is to gather information on the following characteristics:

- Soil depth
- Soil permeability (general or qualitative)
- Site slope
- Site drainage
- Existence of streams, drainage courses, or wetlands
- Existing and proposed structures
- Water wells
- Zoning
- Vegetation and landscape

10.3.2 APPLICABLE REGULATIONS

When the pertinent data have been collected, the local regulatory agency should be contacted to determine the regulatory requirements. The tests required for the phase 2 investigation, which can include identifying depth to groundwater during the wettest period of the year and permeability tests to determine water absorption rates, can also be determined at this time. A list of typical regulatory factors for on-site disposal is presented in Table 10.3.

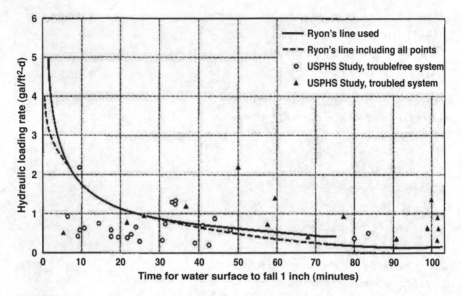

FIGURE 10.2 Percolation rate vs. hydraulic loading rate for soil absorption systems. (From Winneberger, J.H.T., *Septic-Tank Systems: A Consultant's Toolkit*. Vol. 1. *Subsurface Disposal of Septic-Tank Effluents*, Butterworth, Boston, MA, 1984. With permission.)

10.3.3 DETAILED SITE ASSESSMENT

The important parameters that require field investigation are soil type, structure, permeability, and depth, as well as depth to groundwater. The use of backhoe pits, soil augers, piezometers, and percolation tests may be required to characterize the soil. Backhoe pits are useful to allow a detailed examination of the soil profile for soil texture, color, degree of saturation, horizons, discontinuities, and restrictions to water movement. Soil augers are useful in determining the soil depth, soil type, and soil moisture, and many hand borings can be made across a site prior to the siting of a backhoe pit location. Piezometers are occasionally required by regulatory agencies to determine the level and fluctuation of groundwater.

In most parts of the country, the results of percolation tests are used to determine the required size of the soil absorption area. The allowable hydraulic loading rate for the soil absorption system is determined from a curve or table that relates allowable loading rates to the measured percolation rate. A typical curve relating percolation rate to hydraulic loading rate for subsurface soil absorption systems is shown in Figure 10.2.

In the percolation test, test holes that vary in diameter from 4 to 12 in. (100 to 300 mm) are bored in the location of the proposed soil absorption area. The bottom of the test hole is placed at the same depth as the proposed bottom of the absorption area. Prior to measuring the percolation rate, the hole should be soaked for a period of 24 hr. Tests and acceptable procedures used by local regulatory agencies should be checked prior to site investigations.

Although used commonly, the percolation test results, because of the nature of the test, are not related to the performance of the actual leachfields. Many agencies and states are abandoning the test in favor of detailed soil profile evaluations. The percolation test is only useful in identifying soil permeabilities that are very rapid or very slow. Percolation tests should not be used as the sole basis for design of soil absorption systems because of the inherent inaccuracies.

10.3.4 HYDRAULIC ASSIMILATIVE CAPACITY

For facilities that are designed for larger flows than those generated by individual households or for sites where the hydraulic capacity is borderline within the local regulations, a shallow trench pump-in test or a basin infiltration test can be used. The absorption test has been developed for wastewater disposal (Wert, 1997). This procedure allows an experienced person to determine the site absorption capacity. In the shallow trench pump-in test, a trench 6 to 10 ft (2 to 3 m) long is excavated to the depth of the proposed disposal trenches. Gravel is placed in a wooden box in the trench to simulate a leachfield condition. A constant head is maintained using a pump, water meter, and float. The soil acceptance rate is then calculated by measuring the amount of water that is pumped into the soil over a period of 2 to 8 d.

10.4 CUMULATIVE AREAL NITROGEN LOADINGS

As described in Chapter 3, nitrogen forms can be transformed when released to the environment. Because the oxidized form of nitrogen, nitrate nitrogen, is a public health concern in drinking water supplies, the areal loading of nitrogen is important.

10.4.1 NITROGEN LOADING FROM CONVENTIONAL
EFFLUENT LEACHFIELDS

The nitrogen loading from conventional leachfields depends on the density of housing and the nitrogen in the applied effluent. The impact of the nitrate nitrogen on groundwater quality depends on the nitrogen loading, the water balance, and the background concentration of nitrate nitrogen. To determine the nitrogen loading, the following procedure is suggested:

1. Determine the wastewater loading rate. The unit generation factor is multiplied by the density of the units per acre; for example, 150-gal/household $\times$ 4 houses per acre yields 600 gal/d·ac.
2. Determine the nitrogen concentration in the applied effluent (use 60 mg/L).
3. Calculate the nitrogen loading. Multiply the nitrogen concentration by the wastewater loading:

$$\text{Nitrogen loading (lb/ac·d)} = L \times N_c \times C \times 10^{-6} \qquad (10.1)$$

where

L =   Wastewater loading (gal/ac·d).

N_c =   Nitrogen concentration (mg/L).

C   =   8.34 lb/gal.

10^{-6} =   Parts per million = mg/L.

4. In this example,

$$\text{Nitrogen loading} = (600 \text{ gal/ac·d})(60 \text{ mg/L})(8.34)(10^{-6})$$
$$= 0.30 \text{ lb/ac·d } (135 \text{ gal/ac·d})$$

10.4.2 CUMULATIVE NITROGEN LOADINGS

The loadings of nitrate nitrogen to the groundwater are reduced by denitrification in the soil column. As indicated in Chapter 8, denitrification depends on the carbon available in the soil or the percolating wastewater and on the soil percolation rate. For sandy, well-drained soils, the denitrification fraction is 15%. For heavier soils or where high groundwater or slowly permeable subsoils reduce the rate of percolation, the denitrification fraction can be estimated at 25%. The percolate nitrate concentration can be calculated from Equation 10.2:

$$N_p = N_c(1 - f) \qquad\qquad (10.2)$$

where

N_p =   Nitrate nitrogen in the leachfield percolate (mg/L).

N_c =   Nitrogen concentration in the applied effluent (mg/L).

f =   Denitrification decimal fraction (0.15 to 0.25).

Example 10.1. Nitrogen Loading Rate in On-Site Systems

A local environmental health ordinance limits the application of septic tank effluent on an areal basis to 45 g/ac·d. Determine the housing density with conventional septic tank effluent–soil absorption systems that will comply with the ordinance. Assume a total nitrogen content in the septic tank effluent of 60 mg/L and a household wastewater generation of 175 gal/d.

Solution

1. Determine the acceptable loading rate in lb/ac·d:

$N_L = 45 \text{ g/ac·d} \times 1/454 \text{ g/lb} = 0.099 \text{ lb/ac·d}$

2. Calculate the corresponding wastewater application rate using Equation 10.1:

$L = \text{Nitrogen loading/(nitrogen concentration} \times 8.34)(10^{-6})$

$L = 0.099 \text{ lb/ac·d}/(60 \text{ mg/L} \times 8.34 \text{ lb/gal})(10^{-6})$

$L = 197.8 \text{ gal/ac·d}$

3. Determine the number of households per acre:

$$\text{Households per acre} = L/175 \text{ gal/d} = 1.13$$

4. Calculate the minimum lot size for compliance:

$$\text{Lot size} = 1/1.13 = 0.88 \text{ ac}$$

Comment

This would be a very conservative ordinance. If a 25% denitrification fraction were recognized in the ordinance, the nitrogen loading rate would be increased to 60 g/ac·d.

10.5 ALTERNATIVE NUTRIENT REMOVAL PROCESSES

Alternative nutrient removal processes have been and continue to be developed for the cost-effective control of nutrients from on-site systems. Nitrogen removal is the most critical of the nutrients because nitrogen can have public health effects as well as eutrophication and toxicological impacts. A large group of attached growth and suspended growth biological systems are available for pretreatment (Tchobanoglous et al., 2003). A listing of attached growth bioreactors used with on-site systems is presented in Table 10.4.

10.5.1 NITROGEN REMOVAL

Removal of nitrogen is a critical issue in most on-site disposal systems. On-site nitrogen removal processes include intermittent sand filters and recirculating granular medium filters, as well as septic tanks with attached growth reactors (internal trickling filters in septic tanks).

10.5.1.1 Intermittent Sand Filters

As described in Chapter 5, intermittent sand filters are shallow beds (2 ft thick) of fine to medium sand with a surface distribution system and an underdrain system. In the late 1880s, many Massachusetts communities used the intermittent sand filter (ISF) to treat septic tanks effluent (Mancl and Peeples, 1991). The ISFs were the forerunners of rapid infiltration and vertical flow wetlands, with hydraulic loading rates of 0.48 to 2.77 gal/d·ft² (19 to 113 mm/d).

A typical ISF is shown in Figure 10.3. Septic tank effluent is applied intermittently to the surface of the sand bed. The treated water is collected an underdrain system that is located at the bottom of the filter. Intermittent filters are either open or buried, but the majority of on-site ISFs have buried distribution systems. The treatment performance of ISF systems is presented in Table 10.5. Suspended solids and bacteria are removed by filtration and sedimentation. BOD and ammonia are removed by bacterial oxidation. Intermittent application and venting of

TABLE 10.4
Types of Trickling Biofilter Media for Pretreatment of On-Site System Wastewater

Granular Media Biofilters	Organic Media Biofilters	Synthetic Media Biofilters
Activated carbon	Ecoflow®	Advantex
AIRR (alternating intermittent recirculating reactor)	ECO-PURE Peat	Aerocell
	Peat moss	Bioclere
Ashco-A RSF III™	Puraflo® peat	Rubber (shredded tires)
Crushed brick	Woodchip trickling	SCAT™
Envirofilter™ modular recirculating media filter		Septi Tech
		Waterloo
Eparco		
Expanded aggregate		
Glass (crushed)		
Glass (sintered)		
Gravel (recirculating gravel filter [RGF])		
Phosphex™ system		
RIGHT®		
Sand		
Stratified sand		
Slag		
Zeolite		

Source: Leverenz, H. et al., *Review of Technologies for the Onsite Treatment of Wastewater in California*, Report No. 02-2, prepared for the California State Water Resources Control Board, Sacramento, CA, Department of Civil and Environmental Engineering, University of California, Davis, 2002.

the underdrains help to maintain aerobic conditions within the filter. Denitrification can be enhanced by flooding the underdrains.

The key design factors for ISFs are sand size, sand depth, hydraulic loading rate, and dosing frequency. The smaller sand sizes (0.25 mm) generally cause eventual failure due to clogging and therefore require periodic raking to remove solids. With buried systems the medium sands (0.35 to 0.5 mm) can result in long-term operation without raking or solids removal, providing the hydraulic loading rate is kept around 1.2 gal/d·ft^2 or less (<50 mm/d). The sand must be washed and free of fines (Crites and Tchobanoglous, 1998). Typical design criteria for ISFs are presented in Table 10.6.

10.5.1.2 Recirculating Gravel Filters

The recirculating sand filter was developed by Michael Hines (Hines and Favreau, 1974). The modern recirculating filter uses fine gravel, as shown in Figure 10.4.

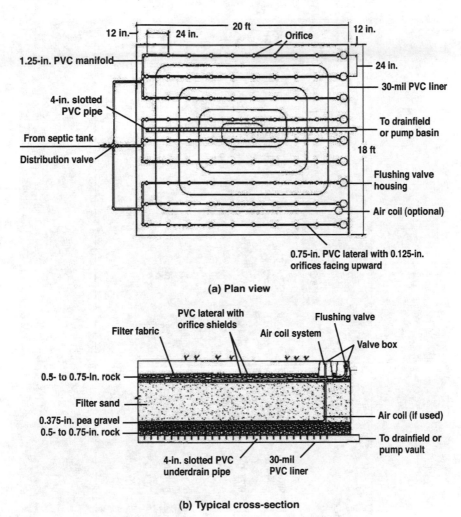

FIGURE 10.3 Schematic of an intermittent sand filter: (a) plan view, and (b) profile of a 2-ft-deep sand filter. (Courtesy of Orenco Systems, Inc., Sutherlin, OR.)

A recirculation tank is used to allow multiple passes of wastewater over the bed. A valve in the recirculation tank allows filtered effluent to be discharged. Recirculating fine gravel filters (RFGFs) use coarser media and higher hydraulic loading rates than ISFs. The performance of RFGFs is presented in Table 10.7. Recirculating gravel filters can nitrify effectively (over 90%). One consideration in nitrification, particularly with ammonia levels that can exceed 60 mg/L, is adequate alkalinity in the applied wastewater. As ammonia is nitrified, 7 mg of alkalinity is destroyed for every 1 mg of ammonia oxidized to nitrate. Denitrification will recover a portion of the alkalinity, but lack of alkalinity in a soft, low-alkalinity wastewater may cause the pH to drop, which will impact the ability to

TABLE 10.5
Performance of Intermittent Sand Filters

Location (Ref.)	Effective Sand Size (mm)	Loading Rate (gal/ft²-d)	BOD₅			Total Nitrogen		
			Influent (mg/L)	Effluent (mg/L)	Percent Removal (%)	Influent (mg/L)	Effluent (mg/L)	Percent Removal (%)
Florida (Grantham et al., 1949)	0.25–0.46	1.7–4.0	148	14	90	37	32	14
Florida (Furman et al., 1955)	0.25–1.04	2.0–13.0	57	4.8	92	30	16	47
Oregon (Ronayne et al., 1984)	0.14–0.3	0.33–0.88	217	3.2	98	58	30	48
Stinson Beach, California (Nolte Associates, 1992a)	0.25–0.3	1.23	203	11	94	57	41	28
University of California, Davis (Nor, 1991)	0.29–0.93	1.0–4.0	82	0.5	99	14	7.2	47
Paradise, California (Nolte Associates, 1992a)	0.3–0.5	0.5	148	6	96	38	19	50
Placer County, California (Cagle and Johnson, 1994)	0.25–0.65	1.23	—	2	98	—	37	40
Gloucester, Maine (Jantrania et al., 1998)	0.8	86	—	15	—	—	61.3	—

TABLE 10.6
Design Criteria for Intermittent Sand Filters Treating Septic Tank Effluent

Design Factor	Unit	Range	Typical
Filter Medium			
Material			Medium sand
Effective size	mm	0.25–0.75	0.35
Uniformity coefficient	U.C.	<4	3.5
Depth	in.	18–36	24
Underdrain Bedding			
Type		Gravel or stone	Gravel
Size	in.	0.375–0.75	0.5
Underdrain Piping			
Type		Slotted	Perforated
Size	in.	3–4	4
Slope	%	0–1	0
Pressure Distribution			
Pipe size	in.	1–2	1.5
Orifice size	in.	0.125–0.25	0.125
Head on orifice	ft	3–6	5
Lateral spacing	ft	1.5–4	2
Orifice spacing	ft	1.5–4	2
Design Parameters			
Hydraulic loading[a]	gal/ft²·d	0.6–1.5	1.25
BOD loading	lb/ft²·d	0.0005–0.002	<0.001
Dosing frequency	times/d	4–24	16
Dosing tank volume	days flow	0.5–1.5	1.0
Filter medium temperature	°F	—	<41

[a] Based on peak flow.

completely nitrify the wastewater. The design criteria for recirculating gravel filters are presented Table 10.8.

10.5.1.3 Septic Tank with Attached Growth Reactor

This system involves a small trickling filter unit placed above the septic tank. Septic tank effluent, which is pumped over the filter, is nitrified as it passes

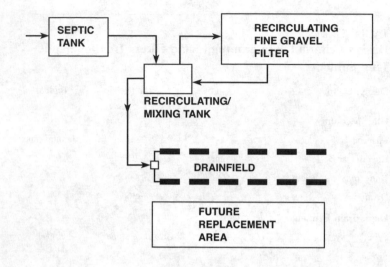

System Schematic

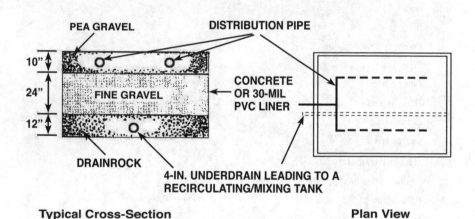

Typical Cross-Section   **Plan View**

FIGURE 10.4 Recirculating gravel filter.

through and over the plastic medium. The system is shown schematically in Figure 10.5. A number of experimental units have been installed in septic tanks. The best performance with a plastic trickling filter medium has been achieved with a hydraulic loading rate of 2.5 gal/min (9.5 L/min) over a unit 3 ft (0.9 m) deep containing hexagonally corrugated plastic with a surface area of 67 ft^2/ft^3 (226 m^2/m^3). A total nitrogen removal of 78% has been reported with an effluent nitrogen concentration of less than 15 mg/L (Ball, 1995). The performance of these systems is summarized in Table 10.9. Recent studies have shown the variability of performance (Loomis et al., 2004). Alternative filter media that have

TABLE 10.7
Analysis of Volume per Dose for Various Hydraulic Loading Rates and Dosing Frequencies for Intermittent Sand Filters[a]

Hydraulic Loading Rate (gal/ft²·d)	Dosing Frequency (times/d)	Hydraulic Application Rate		Field Capacity Filled (%)[b]
		(mm/dose)	(gal/ft²·dose)	
1	1	40	1	217
	2	20	0.5	107
	4	10	0.25	53
	8	5	0.12	26
	12	3.3	0.083	18
	24	1.67	0.042	9.0
2	1	81	2	427
	2	40	1	217
	4	20	0.5	107
	8	10	0.25	53
	12	6.75	0.12	26
	24	3.38	0.083	18
4	1	163	4	855
	2	82	2	427
	4	41	1	217
	8	20	0.5	107
	12	14	0.33	71
	24	6.79	0.17	36

[a] For 1 ft² of surface area and depth of 1.25 ft.
[b] Five% as volumetric water content (water volume/total volume) (Bouwer, 1978).

Source: Crites, R.W. and Tchobanoglous, G., *Small and Decentralized Wastewater Management Systems*, McGraw-Hill, New York, 1998. With permission.

been tested include the foam medium used in the Waterloo filter and the textile chips used in the textile bioreactor.

10.5.1.4 RSF2 Systems

In the RSF2 system, a recirculating sand filter is used for nitrification and is combined with an anaerobic filter for denitrification (Sandy et al., 1988). A flow diagram for the RSF2 system is presented in Figure 10.6. Septic tank effluent is discharged to one end of a rock storage filter, which is directly below and in the same compartment as the RSF. Septic tank effluent flows horizontally through the

TABLE 10.8
Performance of Recirculating Gravel Filters

Location (Ref.)	Effective Medium Size (mm)	Loading Rate (gal/ft²·d)	BOD₅			Total Nitrogen		
			Influent (mg/L)	Effluent (mg/L)	Percent Removal (%)	Influent (mg/L)	Effluent (mg/L)	Percent Removal (%)
Michigan (Loudon et al., 1984)	0.3	3.0	240	25	90	92	34	60
Oregon (Ronayne et al., 1984)	1.2	1.45	217	2.7	99	58	32	45
Paradise, California (Nolte Associates, 1992)	3.0	4.4	134	12	91	63	35	44
Paradise, California (Nolte Associates, 1992)	3.0	2.5	60	8	87	57	26	54
Martinez, California (Crites et al., 1997)	3.0	3.0	—	<5	96	—	12.6	80
Minnesota (Christopherson et al., 2001)	—	5.0	—	18	93	—	43	47
Gloucester, Massachusetts (Jantrania et al., 1998)	—	3.0	—	7	96	—	60.8	36

Source: Adapted from Reed et al. (1995) and Leverenz et al. (2002).

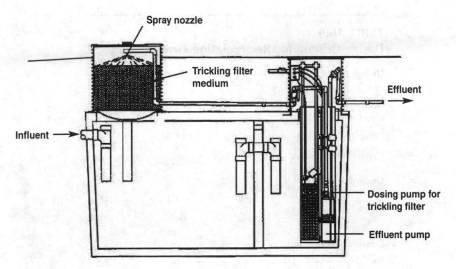

FIGURE 10.5 Septic tank with attached-growth reactor for the removal of nitrogen. (Courtesy of Orenco Systems, Inc., Sutherlin, OR.)

rock and enters a pump chamber at the other end. The septic tank effluent is pumped over the RSF, where it is nitrified. Filtrate is collected from near the top of the rock storage filter, directed into a second pump chamber, and returned to the anaerobic environment of the septic tank, where raw wastewater can serve as a carbon source for denitrification. A portion of effluent from the second pump chamber is discharged for disposal. Experiments with the RSF2 system produced nitrogen removals of 80 to 90%. Total nitrogen concentrations in the effluent ranged from 7.2 to 9.6 mg/L (Sandy et al., 1988). The rock storage zone, filled with 1.5-in. (38-mm) rock, was effective in promoting denitrification. An alternative modification is to add the fixed medium (plastic, textile sheets) for biomass growth into the recirculation tank. Nitrified effluent from the recirculating sand filter is mixed with the incoming septic tank effluent and flows past the attached biomass, where any residual dissolved oxygen is consumed rapidly and the nitrate is denitrified using the organic matter in the septic tank effluent as the carbon source.

10.5.1.5 Other Nitrogen Removal Methods

Other types of media have been used in bioreactors, including crushed glass, sintered glass, expanded aggregate, and crushed brick (Leverenz et al., 2002). The performance of three of these media filters is presented in Table 10.10. Other nitrogen methods that have been conceptualized include ammonia removal by ion exchange and nitrogen removal by denitrification in soil trenches. Attempts have been made to remove ammonia by ion exchange using zeolite at Los Osos, California, and other locations (Nolte Associates, 1994). The attempts have been generally unsuccessful to date because of inadequate volumes of zeolite used and the high cost of frequent regeneration or replacement of the ion exchange medium.

TABLE 10.9
Design Criteria for Recirculating Gravel Filters

Design Factor	Unit	Range	Typical
Filter Medium			
Effective size	in.	1–5	2.5
Depth	in.	18–36	24
Uniformity coefficient	U.C.	<2.5	2.0
Underdrains			
Size	in.	3-4	4
Slope	%	0–0.1	0
Pressure Distribution			
Pipe size	in.	1–2	1.5
Orifice size	in.	1/8–1/4	1/8
Head on orifice	ft	3–6	5
Lateral spacing	ft	1.5–4	2
Orifice spacing	ft	1.5–4	2
Design Parameters			
Hydraulic loading[a]	gal/ft^2·d	3–5	4
BOD loading	lb/ft^2·d	0.002–0.008	<0.005
Recirculation ratio	Unitless	3:1–5:1	4:1
Dosing Times			
Time on	min	<2–3	<2–2
Time off	min	15–25	20
Dosing			
Frequency	times/d	48–120	—
Dosing tank volume	flow/d	0.5–1.5	1

[a] Based on peak flow.

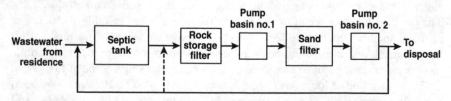

FIGURE 10.6 Flow diagram for RSF2 system for the removal of nitrogen.

TABLE 10.10
Performance Studies of Alternative Media

Parameter	Expanded Shale[a]	Advantex[b]	Crushed Glass[c]
Hydraulic loading rate[d]	1.35	—	1.8
Effluent BOD[e]	1 (99)	5 (98)	10.7 (94)
Effluent total suspended solids[e]	5 (95)	3 (90)	2.5 (95)
Effluent nitrogen[e]	29 (39)	7 (78)	19.7 (55)
Effluent phosphorus[e]	0.5 (94)	—	—

[a] 24 in. of LECA® (light expanded clay aggregate) (Anderson et al., 1998).
[b] Roseburg, Oregon (Bounds et al., 2000).
[c] Oswego, New York (Elliott, 2001).
[d] In gal/ft^2·d.
[e] In mg/L (% removal).

Source: Leverenz, H. et al., *Review of Technologies for the Onsite Treatment of Wastewater in California*, Report No. 02-2, prepared for the California State Water Resources Control Board, Sacramento, CA, Department of Civil and Environmental Engineering, University of California, Davis, 2002.

10.5.2 PHOSPHORUS REMOVAL

Phosphorus removal is seldom required for on-site systems; however, when it is required, the soil mantle is the most cost-effective place to remove and retain phosphorus (see Chapter 8). Attempts to remove phosphorus in peat beds have usually been unsuccessful unless iron or limestone is present or added to the bed. In Maryland, the use of iron filings plowed into the peat bed was successful in removing phosphorus.

10.6 DISPOSAL OF VARIOUSLY TREATED EFFLUENTS IN SOILS

The disposal of partially treated wastewater into soils involves two major considerations: (1) treatment of the effluent so it does not contaminate surface or groundwater, and (2) hydraulic flow of the effluent through the soil and away from the site. Pretreatment of the raw wastewater affects the degree of treatment that the soil–aquifer must achieve after the pretreated effluent is applied to the soil absorption system. Treatment of wastewater in soil has long been recognized (Crites et al., 2000). The soil is a combined biological, chemical, and physical filter. Wastewater flowing through soil is purified of organic and biological constituents, as described in Chapter 8. Septic tank effluent has sufficient solids and organic matter to form a biological mat ("biomat") in the subsurface,

TABLE 10.11
Allowable Hydraulic Loading Rates for Variously Treated Effluent

Type of Effluent	Allowable Hydraulic Loading Rates			Mass Loading Rate $(g/m^2 \cdot d)$		
	(in./d)	(gal/ft²·d)	(mm/d)	BOD_5	TSS	TKN
Restaurant septic tank[a]	0.12	0.07	3	2.4	0.9	0.24
Domestic septic tank	0.4	0.25	10	1.5	0.8	0.55
Graywater septic tank	0.6	0.37	15	1.8	0.6	0.22
Domestic aerobic unit	0.8	0.50	20	0.7	0.8	0.30
Domestic sand filter	3.0	1.87	76	0.3	0.75	0.75

[a] Increased from Siegrist's values for BOD (800 mg/L), TSS (300 mg/L), and TKN (80 mg/L)
    and lowered hydraulic loading rate from 4 mm/d to 3 mm/d.

Note: BOD_5, biochemical oxygen demand; TSS, total suspended solids; TKN, total Kjeldahl
nitrogen.

Source: Adapted from Siegrist, R.L., in *Proceedings of the Fifth National Symposium on Individual and Small Community Sewage Systems*, American Society of Agricultural Engineers,
Chicago, IL, December 14–15, 1987.

particularly if gravity flow application is used. More highly treated effluent and
pressure-dosed application results in little, if any, biomat formation, and the
flow through the soil is only inhibited by the hydraulic conductivity of the soil.
Allowable hydraulic loading rates for variously treated effluents are presented
in Table 10.11.

10.7 DESIGN CRITERIA FOR ON-SITE DISPOSAL ALTERNATIVES

Gravity-flow leachfields are the most common type of on-site wastewater disposal. This type of on-site disposal functions well for sites with deep, relatively
permeable soils, where groundwater is deep and the site is relatively level.

10.7.1 GRAVITY LEACHFIELDS

Septic tank effluent flows by gravity into a series of trenches or beds for subsurface
disposal. Trenches are usually shallow, level excavations that range in depth from
1 to 5 ft (0.3 to 1.5 m) and in width from 1 to 3 ft (0.3 to 0.9 m). The bottom
of the trench is filled with 6 in. (150 mm) of washed drain rock. The 4-in. (100-
mm) perforated distribution pipe is next placed in the center of the trench.

Additional drain rock is placed over the top of the distribution pipe, followed by a layer of barrier material, typically building paper or fabric. The purpose of the barrier material is to prevent migration of fines from the backfill into the drain rock and avoid clogging of the drain rock by the clay or silt particles. The infiltrative surfaces in a leachfield trench are the bottom and the sidewalls; however, as a clogging layer of biological solids or "biomat" develops, the infiltration through the bottom of the trench decreases and the sidewalls become effective and become the long-term route for water passage.

Bed systems consist of an excavated area or bed with perforated distribution pipes that are 3 to 6 ft (0.9 to 1.8 m) apart. The route for water passage out of the bed is through the bottom. Bed systems can also use infiltration chambers, which create underground caverns over the soil's infiltrative surface and therefore do not need the gravel or barrier material.

Leaching chambers constructed out of concrete are open-bottomed shells that replace perforated pipe and gravel for distribution and storage of the wastewater. The chambers interlock to form an underground cavern over the soil. Wastewater is discharged into the cavern through a central weir, trough, or splash plate and allowed to flow over the infiltrative surface in any direction. Access holes in the top of the chambers allow the surface to be inspected and maintained as necessary. Many leaching chamber systems have been installed in the northeastern United States.

Typical criteria for siting of leachfield systems are presented in Table 10.12. Loading rates for trench and bed systems can be based on percolation test results and regulatory tables, on soil characteristics, or a combination of both. Disposal field loading rates recommended by the USEPA for design, based on bottom area, for various types of soils and observed percolation rates are shown in Table 10.13.

The loading rate based on the most conservative criterion is to assume that the percolation rate through the soil will eventually be reduced to coincide with the percolation rate through the biomat. On this basis, the hydraulic loading rate is 0.125 gal/ft^2·d (5 L/m^2·d) based on trench sidewall area only (Winneberger, 1984).

Where the site soils contain significant amounts of clay, it is suggested that the disposal field be divided into two fields and that the two fields be used alternately every 6 months. When two fields are used, the actual hydraulic loading rate for the field in operation is 0.25 gal/ft^2·d (10 L/m^2·d).

10.7.2 SHALLOW GRAVITY DISTRIBUTION

Shallow leachfields offer the benefits of lower cost and higher biological treatment potential because the upper soil layers have the most bacteria and fungi for wastewater renovation (Reed and Crites, 1984). The State of Oregon recently allowed the use of leachfield trenches without gravel that are 10 in. (250 mm) deep and 12 in. (300 mm) wide (Ball, 1994).

TABLE 10.12
Design Considerations in Siting Leachfields

Item	Criteria
Landscape Form[a]	Level, well-drained areas; crests of slopes; convex slopes are most desirable. Avoid depressions, bases of slopes, and concave slopes unless suitable surface drainage is provided.
Slope[a]	0–25%; slopes in excess of 25% can be used, but construction equipment selection is limited.

Typical Horizontal Setbacks[b]

Water supply sells	50–100 ft
Surface waters, springs	50–100 ft
Escarpments, man-made cuts	10–20 ft
Boundary of property	5–10 ft
Building foundations	10–20 ft

Soil

Unsaturated depth	2–4 ft (0.6–1.2 m) of unsaturated soil should exist between the bottom of the disposal field and the seasonally high water table or bedrock.
Texture	Soils with sandy or loamy textures are best suited; gravelly and cobbley soils with open pores and slowly permeable clay soils are less desirable.
Structure	Strong granular, blocky, or prismatic structures are desirable; platey or unstructured massive soils should be avoided.
Color	Bright, uniform colors indicate well-drained, well-aerated soils; dull, gray, or mottled soils indicate continuous or seasonal saturation and are unsuitable.
Layering	Soils exhibiting layers with distinct textural or structural changes should be evaluated carefully to ensure that water movement will not be severely restricted.
Swelling clays	Presence of swelling clays requires special consideration in construction; location may be unsuitable if extensive.

[a] Landscape position and slope are more restrictive for seepage beds because of the depth of cut on the upslope side.

[b] Intended only as a guide. Safe distance varies from site to site, based on local codes, topography, soil permeability, groundwater gradients, geology, etc.

Source: Adapted from USEPA, *Design Manual: Onsite Wastewater Treatment and Disposal Systems*, Municipal Environmental Research Laboratory, U.S. Environmental Protection Agency, Cincinnati, OH, 1980.

TABLE 10.13
Recommended Rates of Wastewater Application for Trench and Bed Bottom Areas

Soil Texture	Percolation Rate (min/in.)	Application Rate (gal/ft²·d)[a,b]
Gravel, coarse sand	<1	Not suitable[c]
Coarse to medium sand	1–5	1.2
Fine sand, loamy sand	6–15	0.8
Sand loam, loam	16–30	0.6
Loam, porous silt loam	31–60	0.45
Silty clay loam, clay loam[d,e]	61–120	0.2
Clays, colloidal clays	>120	Not suitable[f]

[a] Rates based on septic tank effluent from a domestic waste source. A safety factor may be desirable for wastewaters of significantly different strength or character.

[b] May be suitable for sidewall infiltration rates.

[c] Soils with percolation rates <1 min/in. may be suitable for septic tank effluent if a 2-ft layer of loamy sand or other suitable soil is placed above or in place of the native topsoil.

[d] These soils are suitable if they are without significant amounts of expandable clays.

[e] Soil is easily damaged during construction.

[f] Alternative pretreatment may be required, as well as alternative disposal (wetlands or evapotranspiration systems).

Source: Adapted from USEPA, *Design Manual: Onsite Wastewater Treatment and Disposal Systems*, Municipal Environmental Research Laboratory, U.S. Environmental Protection Agency, Cincinnati, OH, 1980.

10.7.3 PRESSURE-DOSED DISTRIBUTION

Pressure dosing can be achieved using either a dosing siphon or a pump. A pressure distribution system has the advantages over gravity distribution of providing a uniform dose to the entire absorption area, promoting unsaturated flow, and providing a consistent drying and reaeration period between doses. Pressure-dosed distribution can allow the absorption site to be at a higher elevation from the septic tank and will also allow a shallow (6- to 12-in.) distribution network. With screened septic tank effluent or sand filter effluent, the distribution system can use 0.125-in. (3-mm) orifices, typically spaced 2 to 4 ft (0.6 to 1.2 m) apart. For septic tank effluent, the orifice size is typically 0.25 in. (6 mm). The spacing and sizing of orifices should be uniform because the objective of pressure dosing is to provide uniform distribution with unsaturated flow beneath the pipe. In heavier soils, the spacing can be increased to 4 to 6 ft (1.2 to 1.8 m).

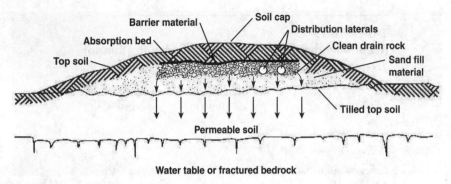

FIGURE 10.7 Schematic of a typical mound system.

10.7.4 IMPORTED FILL SYSTEMS

Fill systems involve importing suitable off-site soils and placing them over the soil absorption area to overcome limited depth of soil or limited depth to groundwater. Care must be taken when selecting suitable soil to use in a fill system and in the timing and conditions of importing the soil. Several conditions must be satisfied to construct a successful fill system:

- Native soil should be scarified prior to import of fill.
- The fill should be placed when the soil is dry.
- The fill material should also be dry to prevent compaction.
- The first 6 in. (150 mm) of fill should be mixed thoroughly with the native soil.

10.7.5 AT-GRADE SYSTEMS

The concept of the at-grade system was developed in Wisconsin as an intermediate system between conventional in-ground distribution and the mound system. The aggregate or drain rock is placed on the soil surface (at-grade) and a soil cap is added over the top. Typically, the area for the at-grade system is tilled, the drain rock is placed on the tilled area, the distribution pipe is positioned within the drain rock, synthetic fabric is spread over the drain rock, and final soil cover (12 in. or 300 mm) is placed over the system. At-grade systems do not require the 24 in. (600 mm) of sand that mounds have and, therefore, are less expensive.

10.7.6 MOUND SYSTEMS

Mound systems are, in effect, bottomless intermittent sand filters. Components of a typical mound, as shown in Figure 10.7, include a 24-in. (600-mm) layer of sand, clean drain rock, distribution laterals, barrier material, and the soil cap. Mounds are pressure dosed, usually 4 to 6 times per day. Mounds were first developed by the North Dakota Agricultural College in the late 1940s. They were known as NODAK systems and were designed to overcome problems with slowly

TABLE 10.14
Infiltration Rates for Determining Base Area of Mound

Native On-Site Soil	Percolation Rate (min/in.)	Infiltration Rate (gal/ft²·d)
Sand, sandy loam	0–30	1.2
Loam, silt loams	31–45	0.75
Silt loams, silty clay loams	46–60	0.50
Clay loams, clay	61–120	0.25

Source: Adapted from USEPA, *Design Manual: Onsite Wastewater Treatment and Disposal Systems*, Municipal Environmental Research Laboratory, U.S. Environmental Protection Agency, Cincinnati, OH, 1980.

permeable soils and areas that had high groundwater tables (Ingham, 1980; WPCF, 1990). Mounds may be used on sites that have slopes up to 12%, provided the soils are permeable. If the native soils are slowly permeable, the use of mounds should be restricted to slopes of less than 6%. The design of mound systems is a two-step process. Percolation tests are conducted on the native soils on the site at the depth at which the mound base will exist. The values of the measured percolation rate are correlated to the design infiltration rate in Table 10.14, and the infiltration rate is then used to calculate the base area of the mound. The second step is to design the mound section. On the basis of the type of material used to construct the mound, the area of the application bed in the mound is determined. Mound fill materials are listed in Table 10.15 along with the corresponding design infiltration rate for determining the bed area (Otis, 1982).

10.7.7 ARTIFICIALLY DRAINED SYSTEMS

Sometimes a high-groundwater condition can be overcome by draining the groundwater away from the site. High groundwater tables in the area of the soil absorption fields may be artificially lowered by vertical drains or underdrains. Underdrains can be perimeter drains, used for level sites and sites up to 12% in slope, or curtain drains (upslope side only), for sites with slopes greater than 12% (Nolte Associates, 1992b).

10.7.8 CONSTRUCTED WETLANDS

Constructed wetlands can be used for on-site treatment as well as on-site disposal and reuse. As described in Chapter 6, constructed wetlands can be either the free water surface type or the subsurface flow type. For on-site systems in close proximity to children, the subsurface flow wetlands are most appropriate. A large number of subsurface wetlands have been constructed and placed in operation in Louisiana, Arkansas, Kentucky, Mississippi, Tennessee, Colorado, and New

TABLE 10.15
Mound Fill Materials and Infiltration Rates

Material	Characteristics (% by weight)	Infiltration Rate (gal/ft²·d)
Medium sand	>25%, 0.25–0.2 mm	1.2
	<30–35%, 0.05–0.25 mm	
	<5–10%, 0.002–0.05 mm	
Sandy loam	5–15% clay	0.6
Sand/sandy loam	88–93% sand	1.2

Source: Adapted from USEPA, *Design Manual: Onsite Wastewater Treatment and Disposal Systems*, Municipal Environmental Research Laboratory, U.S. Environmental Protection Agency, Cincinnati, OH, 1980.

Mexico. These systems serve single-family dwellings, public facilities and parks, apartments, and commercial developments (Reed, 1993). On-site wetlands are SSF wetlands and are described in Chapter 7.

10.7.9 EVAPOTRANSPIRATION SYSTEMS

In arid climates, evapotranspiration (ET) systems can be used for effluent disposal. Effluent from the septic tank is applied through perforated pipes to a sand bed underlain by a liner. The sand depth is typically 24 to 30 in. (0.6 to 0.75 m). Bernhart (1973) recommended a sand depth of 18 in. (0.45 m). The surface of the sand bed is covered with a shallow layer of topsoil, which can be planted to water-tolerant vegetation. Treated wastewater is drawn up through the sand by capillary forces and by the plant roots, and it is evaporated or transpired to the atmosphere. A fine sand (0.1 mm) is recommended to maximize the capillary rise. Observation wells are used to monitor the depth of water in the sand beds.

The ET system can also be designed without a liner, and the resultant system is referred to as an evapotranspiration–absorption (ETA) system. The ETA approach can be used where percolation is acceptable and possible. An ETA system is similar to an at-grade system, except for the addition of surface vegetation. Both ET and ETA systems are designed using the hydraulic loading rate. For ET systems, the hydraulic loading rate is the minimum monthly net evapotranspiration rate for at least 10 years of record. For ETA systems, the minimum monthly percolation rate is added to the minimum ET rate to determine the design hydraulic loading rate. The bed area for ET and ETA systems can be determined using Equation 10.3:

$$A = Q \div (ET - Pr + P) \tag{10.3}$$

where:

A = Bed area (ft^2).

Q = Annual flow (ft^3/yr).

ET = Annual potential evapotranspiration rate (ft/yr).

Pr = Annual precipitation rate (ft/yr).

P = Annual percolation rate (ft/yr).

For ET systems, the percolation rate is zero; for ETA systems, the percolation rate should be determined based on long-term saturated flow conditions.

Example 10.2. Design of an Evapotranspiration System

Design an evapotranspiration system for a cluster of homes with a design flow of 1800 gal/d. The annual lake evaporation rate is 50 in./yr, and the precipitation rate for the wettest year in 10 is 20 in./yr.

Solution

1. Convert the daily flow to an annual flow:

Q = 1800 gal/d $\times$ 365 d/yr = 657,000 gal/yr $\times$ 1/7.48 gal/ft^3 = 87,834 ft^3/yr

2. Calculate the hydraulic loading rate:

HLR = $ET - Pr$ = 50 − 20 = 30 in./yr = 2.5 ft/yr

3. Calculate the bed area:

$A = Q \div$ HLR

A = 87,834 ft^3/yr $\div$ 2.5 ft/yr

A = 35,133 ft^2

Comment

A factor of safety, typically 15 to 20%, should be added to the bed area to account for variations in precipitation and flow rate.

10.8 DESIGN CRITERIA FOR ON-SITE REUSE ALTERNATIVES

Reuse alternatives for on-site systems include drip irrigation and spray irrigation.

10.8.1 DRIP IRRIGATION

Drip irrigation technology has advanced over the years to where non-clog emitters are available for both surface and subsurface uses. Sand filter and other high-quality effluent can be used in drip irrigation of landscape and other crops. Periodic chlorination of the drip tubing has been found to be necessary to avoid clogging growths in the distribution lines and emitters. Modern drip emitters have been designed not to be clogged by roots. For example, the Geoflow™ emitter has been treated with a herbicide to protect it from root intrusion. The emitters are designed with a turbulent flow path to minimize clogging from suspended solids. These emitters operate at a flow rate of 1 to 2 gal/hr with 0.06- to 0.07-in. (1.5- to 1.8-mm)-diameter openings. The drip irrigation system usually requires 15 to 25 lb/in^2 pressure. It may be necessary to flush the lines and to apply periodic doses of chlorine for control of clogging from bacterial growth. A typical on-site drip irrigation system consists of emitter lines placed on 2-ft (0.6-m) centers with a 2-ft (0.6-m) emitter spacing. This spacing is typical for sandy and loamy soils. Closer spacings of 15 to 18 in. (0.4 to 0.45 m) are used on clay soils where lateral movement of water is restricted. The emitter lines are placed at depths of 6 to 10 in. (150 to 250 mm). Drip systems can be optimized to minimize nitrate movement through the soil. Nitrification of septic tank effluent without denitrification can increase nitrate movement. Short daily pulses increase nitrogen removal compared to continuous applications (Beggs et al., 2004).

Example 10.3. Design of a Drip Irrigation System

Design a drip irrigation system for the reuse of 300 gal/d of treated effluent. Use a design infiltration rate of 0.25 gal/ft^2·d.

Solution

1. Determine the area needed for irrigation.

    $$A = 300 \text{ gal/d} \div 0.25 \text{ gal/ft}^2\text{·d} = 1200 \text{ ft}^2$$

2. Lay out the 1200 ft^2 as a 40-ft by 30-ft rectangle.
3. Select a spacing of the drip emitter lines of 2 ft. Use 20 emitter lines that are 30 ft long.
4. Use 1-gal/hr emitters, spaced at 2-ft intervals. Calculate the number of emitters.

    $$30 \text{ ft per line} \div 2\text{-ft spacing} = 15 \text{ emitters per line}$$

    $$20 \text{ lines} \times 15 \text{ emitters per line} = 300 \text{ emitters}$$

5. Calculate the flow discharged from 300 emitters.

    $$\text{Flow} = 300 \text{ emitters} \times 1 \text{ gal/hr} = 300 \text{ gal/hr}$$

6. Calculate the time of operation per day.

    $$300 \text{ gal/d} \div 300 \text{ gal/hr} = 1.0 \text{ hr/d}$$

7. Select a pump for the application. The pump must be able to supply 300 gal/d ÷ (1.0 hr × 60 min/hr) = 5.0 gal/min at a pressure of 20 lb/ft^2.

Comment

The emitters should be buried at a depth of 10 in.

10.8.2 SPRAY IRRIGATION

The use of spray irrigation for on-site disposal is relatively limited except in areas where housing density is low and other less expensive alternatives are not appropriate. Flows need to exceed 3 to 5 gal/min (11 to 19 L/min) to operate most single sprinklers. This relatively high flow generally means that spray irrigation is better suited to flows from an industrial, commercial, or institutional facility. In addition, for residential on-site systems, the additional treatment may need to include sand filtration and disinfection. The details of spray irrigation site assessment and design are presented in Chapter 8 in the discussion of slow-rate land treatment.

10.8.3 GRAYWATER SYSTEMS

In older homes and in areas where water conservation or reuse is practiced because of water shortages or lack of wastewater disposal capacity, the laundry water and other non-toilet wastewater is often reused or disposed of separately from the "black" water that goes into the septic tank. The graywater includes organics, nutrients, and pathogens; however, it is perceived as being a benign source of wastewater that can be reused directly for landscape irrigation. Local health departments have allowed graywater reuse in rural areas but have often denied graywater reuse in urban areas. In California, regulations specify safe and acceptable methods of on-site reuse of graywater (California Resources Agency, 1994). California's graywater standards are now part of the state plumbing code, making it legal to use graywater everywhere in California.

10.9 CORRECTION OF FAILED SYSTEMS

The failure of subsurface on-site disposal systems is defined as the inability of the system to accept and absorb the design flow of effluent at the expected rate. When failure occurs soon after the system is put into operation, the failure may be the result of poor construction (Winneberger, 1987), poor design, or unanticipated high groundwater, or a combination of the three. If high groundwater is the problem, a curtain drain or other drainage improvements may be necessary.

10.9.1 USE OF EFFLUENT SCREENS

When failure occurs after several years of successful operation, the reasons may be unanticipated flow increases, solids carryover from the septic tank, or biological clogging at the infiltration surfaces. A comparison of current flows to design flows should be made to ensure that hydraulic overloading is not the problem. The septic tank should be checked for scum and sludge layers and pumped, if necessary. The use of Orenco or Zabel (or equivalent) effluent screens will minimize the discharge of suspended solids to the disposal system. If biological clogging is occurring, then drying for a period of months, rehabilitation using oxidants such as hydrogen peroxide, or upgrading the septic tank effluent using a sand filter or equivalent treatment may be tried.

10.9.2 Use of Hydrogen Peroxide

For rehabilitation, a procedure developed at the University of Wisconsin uses hydrogen peroxide, a very strong oxidizing agent to destroy the organic deposition and restore the infiltration capacity (Harkin and Jawson, 1977). Lysimeter work at the University of New Hampshire was successful in rehabilitating sandy and loamy sand soil infiltration systems that had clogged because of the buildup of organic material. A 30% solution of hydrogen peroxide and water was successful in all cases (Bishop and Logsdon, 1981). A weaker solution of 7.5% hydrogen peroxide was successful only for sandy soils. The loading rates for hydrogen peroxide were 0.25 lb/ft^2 of surface for sands and at least 0.5 lb/ft^2 for silty soils. In subsequent research, it was found that one or two applications of hydrogen peroxide may be required to renovate clean sands (Mickelson et al., 1989) and that the infiltration rates may be reduced significantly by peroxide treatment of some soils (Hargett et al., 1985).

10.9.3 Use of Upgraded Pretreatment

A study was conducted in Wisconsin in which failing soils absorption units were rehabilitated by reducing the organic and solids loadings by upgrading the pretreatment. In 1994, 15 failing systems were upgraded using aerobically treated effluent. Of the 15 systems, 12 were able to resume successful operation accepting the higher quality effluent. Of the three systems that continued to have problems, two continued to require frequent pumping to operate, and one system could not be rehabilitated (Converse and Tyler, 1995).

10.9.4 Retrofitting Failed Systems

Other methods of retrofitting failed systems include upgrading the treatment of the septic tank effluent with intermittent sand filters, plastic medium, or textile bioreactors, followed by disposal in shallow trenches or drip irrigation. In some cases, the new shallow trenches can be above or in between the existing failed leachfields.

10.9.5 Long-Term Effects of Sodium on Clay Soils

If clay soil slaking is occurring (see Chapter 8 for effects of sodium on soil permeability), calcium can be added to the system to reverse the effects of the sodium (Patterson, 1997). Changing to a low-sodium laundry detergent may also reduce the sodium adsorption ratio and alleviate the problem.

10.10 ROLE OF ON-SITE MANAGEMENT

On-site management can play a number of roles with regard to individual and decentralized on-site disposal or reuse. The functions of on-site management districts include monitoring for system failures in addition to monitoring for

TABLE 10.16
On-Site Wastewater Management Districts

Location	Year Formed	Ref.
Georgetown, California	1971/1985	Prince and Davis (1988)
Allen County, Ohio	1972	Washington State DOH (1996)
Stinson Beach, California	1978	Crites and Tchobanoglous (1998)
Santa Cruz County, California	1985	Crites and Tchobanoglous (1998)
Jefferson County, Washington	1987	Washington State DOH (1996)
Sea Ranch, California	1987	Crites and Tchobanoglous (1998)
Island County, Washington	1989	Washington State DOH (1996)
Thurston County, Washington	1990	Washington State DOH (1996)
Clark County, Washington	1992	Washington State DOH (1996)
Paradise, California	1992	Crites and Tchobanoglous (1998)
Cayuga County, New York	1993	Washington State DOH (1996)
Kitsap County, Washington	1995	Washington State DOH (1996)
Mason County, Washington	1995	Washington State DOH (1996)

environmental and public health protection. On-site management can reduce the risk of using innovative, cost-effective technologies and can increase the opportunities for local water reuse. A list of on-site management districts and the years in which they were formed initially is presented in Table 10.16. On-site systems without any management must be designed very conservatively because failure could mean abandonment of a residence or business. With management and oversight, innovative technologies can be tried, with the assurance that, should the technology fail, corrective measures or replacement technology can be used. As an example, the Town of Paradise, California, has had an on-site management district since 1992 and has encouraged the use of sand filters, aerobic pretreatment of restaurant wastewater, and pressure-dosed distribution. Paradise and Gloucester, Massachusetts, participated in the USEPA-sponsored National On-Site Demonstration Program (NDOP), which paid for installation and monitoring of on-site treatment and disposal and reuse technologies.

REFERENCES

Anderson, D.L., Tyl, M.B., Otis, R.J., Mayer, T.G., and Sherman, K.M. (1998). Onsite wastewater nutrient reduction systems (OWNRS) for nutrient sensitive environments, in *Proceedings of the Eighth International Symposium on Individual and Small Community Sewage Systems*, American Society of Agricultural Engineering, Orlando, FL, March 8–10, 1998.

Ball, H.L. (1994). Nitrogen reduction in an onsite trickling filter/upflow filter wastewater treatment system, in *Proceedings of the Seventh International Symposium on Individual and Small Community Sewage Systems*, American Society of Agricultural Engineers, Atlanta, GA, December 11–13, 1994, 499–503.

Ball, H.L. (1995). Nitrogen reduction in an onsite trickling filter/upflow filter system, in *Proceedings of the 8th Northwest Onsite Wastewater Treatment Short Course and Equipment Exhibition*, University of Washington, Seattle.

Ball, H.L. (pers. comm., 1996). Roseburg, OR.

Beggs, R.A., Tchobanoglous, G., Hills, D., and Crites, R.W. (2004). Modeling subsurface drip application of onsite wastewater treatment system effluent, in *Proceedings of the Tenth International Symposium on Individual and Small Community Sewage Systems*, American Society of Agricultural Engineering, Sacramento, CA, March 21–24, 2004.

Bernhart, A.P. (1973). *Treatment and Disposal of Waste Water from Homes by Soil Infiltration and Evapotranspiration*, University of Toronto Press, Toronto, Canada.

Bishop, P.L. and Logsdon, H.S. (1981). Rejuvenation of failed soil absorption systems, *J. Am. Soc. Chem. Eng. Environ. Eng. Div.*, 107(EE1), 47.

Bounds, T., Ball, E.S., and Ball, H.L. (2000). Performance of packed bed filters, in *Proceedings of the National Onsite Wastewater Recycling Association Conference*, Jekyll Island, GA, November 3–6, 1999.

Bouwer, H. (1978). *Groundwater Hydrology*, McGraw-Hill, New York.

Burks, B.D. and Minnis, M.M. (1994). *Onsite Wastewater Treatment Systems*, Hogarth House, Madison, WI.

Cagle, W.A. and Johnson, L.A. (1994). Onsite intermittent sand filter systems: a regulatory/scientific approach to their study in Placer County, California, in *Proceedings of the Seventh International Symposium of Individual and Small Community Sewage Systems*, American Society of Agricultural Engineers, Atlanta, GA, December 11–13, 1994.

California State Resources Agency (1994). *Graywater Guide*, The Resources Agency, Sacramento, CA.

Christopherson, S.H., Anderson, J.L., and Gustafson, D.M. (2001). Evaluation of recirculating sand filters in Minnesota, in *Proceedings of the Ninth International Symposium on Individual and Small Community Sewage Systems*, American Society of Agricultural Engineering, Ft. Worth, TX, March 11–14, 2001.

Converse, J.C. and Tyler, E.J. (1995). Aerobically treated domestic wastewater to renovate failing septic tank-soil absorption fields, in *Proceedings of the Eighth Northwest Onsite Wastewater Treatment Short Course and Equipment Exhibition*, University of Washington, Seattle.

Crites, R.W. and Tchobanoglous, G. (1998). *Small and Decentralized Wastewater Management Systems*, McGraw-Hill, New York.

Crites, R.W., Lekven, C.C., Wert, S., and Tchobanoglous, G. (1997). A decentralized wastewater system for a small residential development in California, *Small Flows J.*, 3(1), 3–11.

Crites, R.W., Reed, S.C., and Bastian, R.K. (2000). *Land Treatment Systems for Municipal and Industrial Wastes*, McGraw-Hill, New York.

Darby, J., Tchobanoglous, G., Nor, M.A., and Maciolek, D. (1996). Shallow intermittent sand filtration: performance evaluation, *Small Flows J.*, 2(1), 3–15.

Elliott, R. (2001). Evaluation of the use of crushed recycled glass as a filter medium, *Water Eng. Manage.*, July/August.

Furman, T.S., Calaway, W.T., and Gratham, G.R. (1955). Intermittent sand filters: multiple loadings, *Sewage Indust. Wastes*, 27(3), 261–276.

Grantham, G.R., Emerson, D.L., and Henry, A.K. (1949). Intermittent sand filter studies, *Sewage Indust. Wastes*, 21(6), 1002–1015.

Hargett, D.L., Tyler, E.J., Converse, J.C., and Apfel, R.A. (1985). Effects of hydrogen peroxide as a chemical treatment for clogged wastewater absorption systems, in *Onsite Wastewater Treatment*, ASAE no. 701P0101, American Society of Agricultural Engineers, St. Joseph, MI.

Harkin, J.N. and Jawson, M.D. (1977). Clogging and unclogging of septic system seepage beds, in *Proceedings of the Second Illinois Symposium on Private Sewage Disposal Systems*, Illinois Department of Public Health, Springfield.

Hines, M.J. and Favreau, R.F. (1974). Recirculating sand filter: an alternative to traditional sewage absorption system, in *Proceedings of the National Home Sewage Disposal Symposium*, Chicago, IL, December 9–10, 1974, 130–136.

Ingham, A.T. (1980). *Guidelines for Mound Systems*, California State Water Resources Control Board, Sacramento, CA.

Jantrania, A.R., Sheu, K.C., Cooperman, A.N., and Pancorbo, O.C. (1998). Performance evaluation of alternative systems: Gloucester, MA, demonstration project, in *Proceedings of the Eighth International Symposium on Individual and Small Community Sewage Systems*, American Society of Agricultural Engineering, Orlando, FL, March 8–10, 1998.

Jowett, E.C. and McMaster, M.M. (1994). A new single-pass aerobic biofilter for onsite wastewater treatment, *Environ. Sci. Eng.*, 6(6), 28–29.

Lesikar, B.J., Garza, O.A., Persyn, R.A., Anderson, M.T., and Kenimer A.L. (2004). Food service establishments wastewater characterization, in *Proceedings of the Tenth International Symposium on Individual and Small Community Sewage Systems*, American Society of Agricultural Engineering, Sacramento, CA, March 21–24, 2004.

Leverenz, H., Tchobanoglous, G., and Darby, J.L. (2002). *Review of Technologies for the Onsite Treatment of Wastewater in California*, Report No. 02-2, prepared for the California State Water Resources Control Board, Sacramento, CA, Department of Civil and Environmental Engineering, University of California, Davis.

Loomis, G., Dow, D., Jobin, J., Green, L., Herron, E., Gold, A., Stolt, M., and Blezejewski, G. (2004). Long-term treatment performance of innovative systems, in *Proceedings of the Tenth International Symposium on Individual and Small Community Sewage Systems*, American Society of Agricultural Engineering, Sacramento, CA, March 21–24, 2004.

Loudon, T.L. (1995). Design of recirculating sand filters, in *Proceedings of the Eighth Northwest Onsite Wastewater Treatment Short Course and Equipment Exhibition*, University of Washington, Seattle.

Mancl, K.M. and Peeples, J.A. (1991). One hundred years later: reviewing the work of the Massachusetts State Board of Health on the intermittent sand filtration of wastewater from small communities, in *Proceedings of the Sixth National Symposium Individual and Small Community Sewage Systems*, ASAE Publ. No. 10-91, American Society of Agricultural Engineers, St. Joseph, MI.

Mickelson, M.J., Converse, C., and Tyler, E.J. (1989). Hydrogen peroxide renovation of clogged wastewater soil absorption systems in sands, *Trans. Am. Soc. Agric. Eng.*, 32(5), 1662–1668.

Nolte Associates (1992a). *Literature Review of Recirculating and Intermittent Sand Filters: Operation and Performance, Town of Paradise, California*, prepared for the California Regional Water Quality Control Board, Sacramento.

Nolte Associates (1992b). *Manual for the Onsite Treatment of Wastewater, Town of Paradise, California*, Sacramento.

Nolte Associates (1994). *Fatal Flaw Analysis of Onsite Alternatives for Los Osos, California*, prepared for Metcalf & Eddy and San Luis Obispo County, Sacramento, CA.

Nor, M.A. (1991). Performance of Intermittent Sand Filters: Effects of Hydraulic Loading Rate, Dosing Frequency, Media Effective Size, and Uniformity Coefficient, Ph.D. thesis, Department of Civil Engineering, University of California, Davis.

Otis, R.J. (1982). Pressure distribution design for septic tank systems, *J. Am. Soc. Chem. Eng. Environ. Eng. Div.*, 108(EE1), 123.

Patterson, R.A. (1997). Domestic wastewater and sodium factor, in *Site Characterization and Design of Onsite Septic Systems*, Bedinger, M.S. et al., Eds., ASTM STP 1324, American Society for Testing and Materials, Philadelphia, PA.

Prince, R.N. and Davis, M.E. (1988). Onsite System Management, paper presented at the Third Annual Midyear Conference of the National Environmental Health Association, Mobile, AL.

Reed, S.C. (1993). *Subsurface Flow Constructed Wetlands for Wastewater Treatment: A Technology Assessment*, EPA 832-R-93-001, U.S. Environmental Protection Agency, Washington, D.C.

Reed, S.C., Crites, R.W., and Middlebrooks, E.J. (1995). *Natural Systems for Waste Management and Treatment*, 2nd ed., McGraw-Hill, New York.

Ronayne, M.A., Paeth, R.A., and Wilson, S.A. (1984). *Oregon Onsite Experimental Systems Program, Oregon Department of Environmental Quality*, EPA/600/14, U.S. Environmental Protection Agency, Cincinnati, OH.

Sandy, A.T., Sack, W.A., and Dix, S.P. (1988). Enhanced nitrogen removal using a modified recirculating sand filter (RSF2), in *Proceedings of the Fifth National Symposium on Individual and Small Community Sewage Systems*, American Society of Agricultural Engineers, Chicago, IL, December 14–15, 1987.

Siegrist, R.L. (1988). Hydraulic loading rates for soil absorption systems based on wastewater quality, in *Proceedings of the Fifth National Symposium on Individual and Small Community Sewage Systems*, American Society of Agricultural Engineers, Chicago, IL, December 14–15, 1987.

Stuth, W. and Garrison, C. (1995). Crushed glass as a filter media for the onsite treatment of wastewater, in *Proceedings of the Eighth Northwest Onsite Wastewater Treatment Short Course and Equipment Exhibition*, University of Washington, Seattle.

Tchobanoglous, G. and Burton, F.L. (1991). *Wastewater Engineering: Treatment, Disposal and Reuse*, 3rd ed., McGraw-Hill, New York.

Tchobanoglous, G., Burton, F.L., and Stensel, H.D. (2003). *Wastewater Engineering, Treatment, and Reuse*, 4th ed., McGraw-Hill, New York.

USEPA. (1980). *Design Manual: Onsite Wastewater Treatment and Disposal Systems*, Municipal Environmental Research Laboratory, U.S. Environmental Protection Agency, Cincinnati, OH.

USEPA. (2002). *Onsite Wastewater Treatment Systems Manual*, EPA/625/R-00/008, Office of Water, Office of Research and Development, U.S. Environmental Protection Agency, Cincinnati, OH.

University of Wisconsin, Madison (1978). *Management of Small Wastewater Flows: Small Scale Waste Management Project*, EPA-600/2-78-173, U.S. Environmental Protection Agency, Washington, D.C.

Washington State DOH (1996). *Onsite Sewage System Monitoring Programs in Washington State*, Community Environmental Health Programs, Washington State Department of Health, Olympia, WA.

Wert, S. (pers. comm., 1997). Roseburg, OR.

Winneberger, J.H.T. (1984). *Septic-Tank Systems: A Consultant's Toolkit*. Vol. 1. *Subsurface Disposal of Septic-Tank Effluents*, Butterworth, Boston, MA.

WPCF. (1990). *Natural Systems for Wastewater Treatment*, Manual of Practice FD-16, Water Pollution Control Federation, Alexandria, VA.

Appendix 1
Metric Conversion Factors
(SI to U.S. Customary Units)

Multiply the SI Unit			To Obtain the U.S. Unit	
Name	Symbol	by	Symbol	Name
Area				
Hectare (10,000 m^2)	ha	2.4711	ac	Acre
Square centimeter	cm^2	0.1550	in.^{2}	Square inch
Square kilometer	km^2	0.3861	mi^2	Square mile
Square kilometer	km^2	247.1054	ac	Acre
Square meter	m^2	10.7639	ft^2	Square foot
Square meter	m^2	1.1960	yd^2	Square yard
Energy				
Kilojoule	kJ	0.9478	Btu	British thermal unit
Joule	J	2.7778×10^{-7}	kWh	Kilowatt-hour
Megajoule	MJ	0.3725	hp·hr	Horsepower-hour
Conductance, thermal	W/m^2·°C	0.1761	Btu/hr·ft^2·°F	Conductance
Conductivity, thermal	W/m·°C	0.5778	Btu/hr·ft·°F	Conductivity
Heat-transfer coefficient	W/m^2·°C	0.1761	Btu/hr·ft^2·°F	Heat-transfer coefficient
Latent heat of water	344,944 J/kg	—	144 Btu/lb	Latent heat of water
Specific heat, water	4215 J/kg·°C	—	1.007 Btu/lb·°F	Specific heat of water
Flow Rate				
Cubic meters per day	m^3/d	264.1720	gal/d	Gallons per day
Cubic meters per day	m^3/d	2.6417×10^{-4}	mgd	Million gallons per day
Cubic meters per second	m^3/d	35.3157	ft^3/s	Cubic feet per second
Cubic meters per second	m^3/s	22.8245	mgd	Million gallons per day
Cubic meters per second	m^3/s	15.8503	gal/min	Gallons per minute
Liters per second	L/s	22.8245	gal/d	Gallons per day
Length				
Centimeter	cm	0.3937	in.	Inch
Kilometer	km	0.6214	mi	Mile
Meter	m	39.3701	in.	Inch
Meter	m	3.2808	ft	Foot

| Multiply the SI Unit | | | To Obtain the U.S. Unit | |
Name	Symbol	by	Symbol	Name
Length (cont.)				
Meter	m	1.0936	yd	Yard
Millimeter	mm	0.03937	in.	Inch
Mass				
Gram	g	0.0353	oz.	Ounce
Gram	g	0.0022	lb	Pound
Kilogram	kg	2.2046	lb	Pound
Megagram (10^3 kg) (metric ton)	Mg (mt)	1.1023	ton (t)	Ton (short: 2000 lb)
Megagram	Mg	0.9842	ton	Ton (long: 2240 lb)
Power				
Kilowatt	kW	0.9478	Btu/s	British thermal units per second
Kilowatt	kW	1.3410	hp	Horsepower
Pressure				
Pascal	Pa (N/m^2)	1.4505×10^{-4}	lb/in.^{2}	Pounds per square inch
Temperature				
Degree Celsius	°C	1.8 (°C) + 32	°F	Degree Fahrenheit
Kelvin	K	1.8 (K) − 459.67	°F	Degree Fahrenheit
Velocity				
Kilometers per second	km/s	2.2369	mi/hr	Miles per hour
Meters per second	m/s	3.2808	ft/s	Feet per second
Volume				
Cubic centimeter	cm^3	0.0610	in.^{3}	Cubic inch
Cubic meter	m^3	35.3147	ft^3	Cubic foot
Cubic meter	m^3	1.3079	yd^3	Cubic yard
Cubic meter	m^3	264.1720	gal	Gallon
Cubic meter	m^3	8.1071×10^{-4}	ac-ft	Acre-foot
Liter	L	0.2642	gal	Gallon
Liter	L	0.0353	ft^2	Square foot
Liter	L	33.8150	oz.	Ounce
Megaliter (L × 10^6)	ML	0.2642	MG	Million gallons

Appendix 2
Conversion Factors
for Commonly Used
Design Parameters

Multiply the SI Unit			To Obtain the U.S. Customary Unit	
Parameter	Symbol	by	Symbol	Parameter
Cubic meters per second	m^3/s	22.727	mgd	Million gallons per day
Cubic meters per day	m^3/d	264.1720	gal/d	Gallons per day
Kilogram per hectare	kg/ha	0.8922	lb/ac	Pounds per acre
Metric ton per hectare	Mg/ha	0.4461	ton/ac	Tons (short) per acre
Cubic meter per hectare per day	$m^3/ha·d$	106.9064	gal/ac·d	Gallons per acre per day
Kilograms per square meter per day	$kg/m^2·d$	0.2048	$lb/ft^2·d$	Pounds per square foot per day
Cubic meter (solids) per 10^3 cubic meters (liquid)	$m^3/10^3\ m^3$	133.681	ft^3/MG	Cubic feet per million gallons
Cubic meters (liquid) per square meter (area)	m^3/m^2	24.5424	gal/ft^2	Gallons per square foot
Grams (solids) per cubic meter	g/m^3	8.3454	lb/MG	Pounds per million gallons
Cubic meters (air) per cubic meter (liquid) per minute	m^3/m^2	1000.0	$ft^3/10^3·min$	Cubic feet of air per minute per 1000 ft^3
Kilowatts per 10^3 cubic meters (tank volume)	$kW/10^3\ m^3$	0.0380	$hp/10^3\ ft^3$	Horsepower per 1000 ft^3
Kilograms per cubic meter	kg/m^3	62.4280	$lb/10^3\ ft^3$	Pounds per 1000 ft^3
Cubic meter per capita	$m^3/capita$	35.3147	$ft^3/capita$	Cubic feet per capita
Bushels per hectare	bu/ha	0.4047	bu/ac	Bushels per acre

Appendix 3
Physical Properties
of Water

Temperature (°C)	Density (kg/m³)	Dynamic Viscosity × 10³ (N·s/m²)	Kinematic Viscosity (g) × 10⁶ (m²/s)
0	999.8	1.781	1.785
5	1000.0	1.518	1.519
10	999.7	1.307	1.306
15	999.1	1.139	1.139
20	998.2	1.002	1.003
25	997.0	0.890	0.893
30	995.7	0.798	0.800
40	992.2	0.653	0.658
50	988.0	0.547	0.553
60	983.2	0.466	0.474
70	977.8	0.404	0.413
80	971.8	0.354	0.364
90	965.3	0.315	0.326
100	958.4	0.282	0.294

Appendix 4
Dissolved Oxygen Solubility in Freshwater

Temperature (°C)	Dissolved Oxygen Solubility (mg/L)	Temperature (°C)	Dissolved Oxygen Solubility (mg/L)
0	14.62	16	9.95
1	14.23	17	9.74
2	13.84	18	9.54
3	13.48	19	9.35
4	13.13	20	9.17
5	12.80	21	8.99
6	12.48	22	8.83
7	12.17	23	8.68
8	11.87	24	8.53
9	11.59	25	8.38
10	11.33	26	8.22
11	11.08	27	8.07
12	10.83	28	7.92
13	10.60	29	7.77
14	10.37	30	7.63
15	10.15		

Note: Saturation values of dissolved oxygen when exposed to dry air containing 20.90% oxygen under a total pressure of 760 mmHg.

Index

A

Acid mine drainage, 291–296
Acidity, total, 429–430
Activated sludge, characteristics of, 438
Adsorption, 43, 46, 62, 65, 73, 89
   in aquatic systems, 71, 82
   in land treatment systems, 75, 84
   in pond systems, 87, 173
   in soil systems, 75
   of boron, 91
   of metals, 84
   of phosphorus, 89
   of sodium, 91, 92
Advanced integrated wastewater ponds,
   144–149
   in wetland systems, 84
Aerated pond design considerations
   BOD removal, 96, 107, 110, 111,
     141, 143
   cell configuration, 126–127
   floating baffles, 120
   kinetics, 110, 111, 123–125,
     141–144
   mixing and aeration, 127, 133,
     141–144
   nitrogen removal, 3, 172–199
   number of cells, 125–126
   organic loading, 113–120
   pathogen removal, 3
   phosphorus removal, 202–203
   solids removal, 83
   temperature effects, 126
Aerated ponds, 83, 109
   area estimates for planning, 15, 19
   description of, 110, 141, 147, 182
   design model for, 110–111, 124,
     125
   performance expectations for, 2, 3,
     5, 110, 143, 144, 172
   site selection for, 12
   sludge accumulation in, 83,
     440–441
Aeration
   in aquatic systems, 60, 64, 88, 91
   in compost systems, 459–464
   in pond systems, 60, 127–133
Aerators, 71
   diffused air, 116, 118
   energy requirements for, 127–133
   in ponds, 113–133
   surface units, 113–115, 117–119
Aerobic conditions
   in aquatic systems, 60
   in land treatment systems, 427–429
   in pond systems, 60, 113, 133
   in wetland systems, 60, 348–350
Aerobic ponds, 4
Aerosols, 71, 77, 78, 79, 80, 94
   bacterial, 77, 78, 79, 80
   compost, 81
   sludge, 71
   viral, 77, 78, 79, 80
Agricultural crops
   function of, in overland flow, 412
   in slow-rate systems, 388–390
   management of, 402
   selection of, 388–390
Agricultural runoff, 384, 401
Air entry permeameter, 37, 39, 391
Air stripping, of ammonia, 87; *see also*
   Volatilization
Airport deicing fluids, 357
Alfalfa, 70, 91, 94, 390, 400
Algae, 2, 3, 43, 60, 62
   effect on oxygen supply, 3

effect on pH, 3
removal in wetland systems, 62,
    269, 271
suppression of, in aquatic systems,
    3, 200, 249, 251
Alkalinity, 91
importance of, for ammonia
    removal, 361–362
Alum
for phosphorus removal, 203–204
sludge with, 441–442
Ammonia form of nitrogen, 87
Anaerobic conditions
in aquatic systems, 62
in pond systems, 62, 134–135
in wetland systems, 62, 261, 265,
    269, 274, 280
Anaerobic digestion, for sludge
    stabilization, 438
Anaerobic ponds, 133–140
Animals, 2
control of, in pond systems, 248–249
grazing, in land treatment, 71, 75,
    402
Application methods
in land treatment, 8, 88, 400, 412,
    424
in wetlands systems, 316–318
Application period, 55, 89, 409
Application rates
for land treatment, 379–380
for overland flow, 408–409
for sludge systems, 8, 472–482
for soil aquifer treatment systems,
    422
slow rate, *see* Slow-rate land
    treatment
Application scheduling, 401–402
Aquatic systems, 2–4, 9, 43, 47
trace organics, removal of, 47
Aquifer, 5, 48, 49
flow characteristics of, 33–35, 39
properties of, 35–37
protection for, 388, 424
Areal loading on wetlands, 309,
    312–313
Artificially drained onsite system, 517

Ascaris, 71, 76
Aspect ratio (L, W)
of FWS wetlands, 300, 314–315
of SSF wetlands, 339, 352–353
Aspergillus fumigatus, 81
At-grade onsite systems, 516
Autoflocculation, 247

B

Bacillus thuringiensis israelensis (BTI),
    323–324
Backfill for rapid infiltration
    construction, 425–426
Background concentrations in wetlands,
    277–278
Bacteria, 43, 71
Bacteria removal
in land treatment, 75, 76, 78, 407,
    420
in pond systems, 71
in sludge systems, 76, 442, 450,
    458, 463, 465, 468, 470, 471
in wetland systems, 73, 275–277,
    345
Baffles and attached growth in ponds,
    247–248
Barley, 91, 390
straw, for algae control, 249, 251
Basins, 10
in SAT systems, 57, 62, 69, 422
infiltration testing, 37–39, 391, 422
Batch chemical treatment for
    phosphorus removal in ponds,
    203–204
Batch loading for wetlands, 355, 359,
    369
Bedrock location, 25, 494–496
Benzene, 47, 64, 67, 68
Bermuda grass, 390, 412
Biochemical oxygen demand (BOD)
    removal
in land treatment, 403–404,
    413–414
in pond systems, 86–106
in wetland systems, 269–270, 344,
    346, 360, 368–369, 371

BIOLAC® processes, 95, 154–164
Bioreactor, textile, 505–507, 511
Bogs, 259
Borings, 29–30
Boron, 84, 90, 91, 387
Brewery wastes, 394
Brome grass, 390
Bromoform, 64, 67, 68
Buffer zones, 40
Bulking agents, 8, 460
Bulrushes (*Scirpus*), 261–263
    aquatic systems and, 3
    characteristics of, 262
    land treatment and, 6
    pond systems and, 3, 60
    wetland systems and, 5, 60, 279–281,
        318–323, 327, 356, 358, 370

C

Cadmium, 81, 82, 84, 86
Cadmium removal
    in aquatic systems, 84
    in land treatment, 84, 85
Carbon sources for denitrification, 2, 88,
        269–271, 392–393, 414–415, 509
Carbon tetrachloride, 65, 70, 427
Cation exchange capacity (CEC), 31–32
Cattails (*Typha*), 70, 71, 74, 88,
        261–262, 348–349
    characteristics of, 262
    performance expectations, 74, 88,
        279, 283, 285, 318–323, 327
Chambers, leaching, 513
Cheese processing wastes, 357, 394
Chemistry, soil, 31–32, 402
Chlorides, 46, 387
Chlorobenzene, 47, 64, 67, 68
Chloroform, 47, 64, 67, 68, 69
Citrus processing wastewater, 394
Cle Elum, WA, wetlands, 270, 279–280
Climatic influences
    in land treatment, 27, 390–392,
        396–399, 411–412, 426
    in pond systems, 96–104
    in sludge systems, 27, 443–450,
        453–454, 461–464, 488

in wetland systems, 5, 19, 264–271,
        274, 276, 280, 298–299,
        302–308, 310–314, 339–343,
        275, 345
Clogging in SSF wetlands, 336–337
Clostridium perfringes, 77
Coagulation–flocculation, 238–239
Coastal Bermuda grass, 390
Combined pond systems, 144–149
Combined sewer overflows, 283–286
Complete mix model for facultative
        ponds, 101–102
Complete retention ponds, 246–247
Composting, 8, 81
    aeration, 463
    area required for, 462
    design example, 463–464
    odors, 462
    performance expectations, 460
    types, 459
Concept evaluation, 8, 9, 11–19
Concept selection, 9, 41, 418
Conductivity
    electrical (EC), 31–32
    hydraulic, 33–35
Configuration, 58, 90; *see also* Aspect
        ratio
    of pond systems, 112–113
    of wetland systems, 299–301, 315,
        339, 352
Constructed wetland applications for
    free water surface wetlands,
        278–296; *see also* Constructed
        wetlands
    agricultural runoff, 286–288
    combined sewer overflow,
        283–286
    commercial and industrial
        wastewaters, 281
    design procedures, 296–314
    food processing wastewater, 289
    landfill leachates, 289–291
    livestock wastewaters, 288–289,
        290
    mine drainage, 291–296
    municipal wastewaters, 278–281
    stormwater runoff, 282–283

Constructed wetland applications for
　subsurface flow wetlands,
　356–357; *see also* Constructed
　wetlands
　airport deicing fluids treatment, 357
　cheese processing wastewater, 357
　design procedures, 346–355
　domestic wastewater , 356
　landfill leachate, 357
　onsite systems, 364–366
Constructed wetland design
　considerations, 296–320; *see also*
　Constructed wetlands
　BOD removal, 3, 269–270,
　　310–313, 344, 346
　cell configuration, 299–301, 315
　evapotranspiration, 264–265
　example, 74, 314, 357–360
　kinetics, 308–314
　metals removal, 273–275, 345
　nitrogen removal, 3, 269–272, 311,
　　313, 344, 347–352
　organic loading, 153, 315
　oxygen requirements, 265–266
　pathogen removal, 73, 275–277,
　　345
　phosphorus removal, 272–273, 345
　solids removal, 5, 269, 271, 344,
　　347
　temperature effects, 280–281,
　　298–299, 302–308, 310–314,
　　339–343, 346, 350, 352
　temperature reduction, 274, 276
Constructed wetlands, 4, 5, 8, 70, 74, 84
　area estimate for planning, 15–16
　components of, 261–268
　costs, 324–328, 373–374
　description of, 259–261
　hydraulics, 299–301, 335–339
　operation and maintenance,
　　320–324, 373
　performance expectations, 5,
　　268–278, 343–345
　site selection, 12, 15–16, 19, 297
　troubleshooting, 328–329, 374
　vegetation selection, 318–320,
　　353–354, 372–373

Construction
　of land treatment systems, 401–402,
　　412, 425
　of wetland systems, 314–320,
　　370–373
Controlled discharge pond, 2, 3, 10,
　243, 246
Coontail, 264
Copper, 81, 82, 84, 85, 90, 345
Corn, 91, 390
Costs and energy, 8, 9, 59, 87, 250–252,
　324–328
Cotton, 91, 390
Crop selection, 388–392
Crop uptake, 390
Crushed glass media, 511

D

Darcy's law, 34, 335–336
DDT (dichlorodiphenyl-
　trichloroethane), 46
Denitrification, 87, 88
　in land treatment systems, 392–393,
　　405–406, 413–415
　in onsite systems, 501–511
　in pond systems, 88, 172–181
　in wetland systems, 269, 271,
　　347–349, 351–352
Depth, 44, 45, 51, 55, 56, 57, 58, 59, 60,
　62, 63, 66, 70, 89
　of water and gravel in SSF wetlands,
　　336–343, 352–354, 358
　to bedrock and groundwater, 21–23,
　　25, 51, 495–497
Design models
　aerated ponds, 110–111, 124–125
　facultative ponds, 97–109
　heat transfer in wetland systems,
　　302–308, 339–343
　overland flow, 407–408
　soil-aquifer treatment, 413–415,
　　420–423
　slow-rate systems, 389–395
　sludge freezing, 443–450
　sludge land application, 470–482
　wetlands, 308–314, 346–352

Dewatering, 2
  by composting, 459–464
  by reed beds, 450–455, 459
  by sludge freezing, 76, 443–450, 459
  by vermistabilization, 456–459
Diffused aeration, 116, 118
Disinfection, 75, 76, 141, 165, 166–167,
    169, 189, 192
Dispersion, 45, 46, 48, 49, 78
Disposal, 5
  of harvested plants, 8
  of sludge, 2, 8, 71
Dissolved air flotation, 239–243
Dissolved oxygen
  in ponds, 60, 124, 128, 134, 157,
    182, 184–185, 187–188
  in wetlands, 60, 349–350
Distribution techniques
  in land treatment, 400–401, 412, 424
  in wetlands, 277–278, 354–355
Diversity in wetlands, 266
Domestic wastewater treatment in
    wetlands, 356, 364–366
Drain spacing, 59, 401
Drainage, subsurface/surface, 401
Drinking water requirements, 5, 50, 65,
    82, 86, 87, 91
Drip irrigation, 400, 519–520
Drying beds; *see* Dewatering
Dual power multicellular ponds, 141–144
Duckweed (*Lemna*), 164–165, 248, 264,
    280, 324
Dyes for algae control, 249, 264

E

Earthworms for sludge treatment
    (vermistabilization), 76, 456–459
Economic factors for site selection, 11,
    22
Effluent characteristics
  land treatment, 7, 403–407,
    413–420
  ponds, 3, 110–111, 141–144
  wetlands, 5, 344–348, 360, 369, 371
Eichhornia crassipes (water hyacinth),
    83, 89

Electrical conductivity (EC), 31–32
Elodea, 264
Emergent plants, 262–263
Endocrine disrupting compounds,
    419–420
Estrogenic compounds, 417, 419–420
Evapotranspiration (ET), 264–265,
    380–381, 391–392
  beds, 518–519
Expanded shale media, 511

F

Fabric shade for algae control, 249–250
Facultative pond design considerations
  BOD removal, 3, 101–104
  kinetics, 96–109
  nitrogen removal, 172–181
  organic loading, 97
  pathogen removal, 3
  phosphorus removal, 202–203
  sludge accumulation, 83
  solids removal, 83
  temperature effects, 97, 104
Facultative ponds, 70, 71, 72, 73
  area estimate for planning, 14–15
  description of, 96
  models for, 97–109
Fecal coliform removal, 7
  in aerosols, 77, 78, 79
  in land treatment, 75, 420
  in ponds, 71, 72
  in wetlands, 2, 3, 74
Fecal streptococci, 77, 78
Federal Register, 40 CFR Part 503,
    464–482
Federal regulations, 442, 463–482
Fertilizer value, 70
  of sludge, 458, 478
  of wastewater, 388–390
Fescue, 390
Field area, 394–396, 410–411, 423
Field crops, 390
  regulations, 1, 2, 8, 9
Field investigations, 28–40, 44, 46, 81,
    89
Fill for onsite systems, 516

Filters
  intermittent sand, 501–503
  rapid sand, 230, 231, 237
  recirculating gravel, 502–505
  rock, 227–236
Fish
  effect of ammonia on, 86, 87, 248
  for mosquito control, 323–324
Floating plants, 264
Flood plain, 26–27
Flooding basin test, 37–39, 422
Food chain, 81, 84
Forage grasses, 390
Forested systems
  for sludges, 24–25
  for wastewater, 25, 389–390
  site selection for, 24–25
Free water surface wetlands, 70
  advantages of, 5, 259–261
  area estimate for planning, 15–16
  aspect ratio, 299–301, 315
  construction of, 5, 314–318
  costs, 324–328
  description of, 259–261
  design procedures for, 296–320
  management of, 320, 323–324
  monitoring, 324
  mosquito control, 323–324
  operation and maintenance of,
    320–324
  performance expectations, 5,
    268–278
  preapplication treatment, 297
  site selection for, 12–13, 15–16, 19,
    297, 318–320
  troubleshooting, 328
  vegetation, establishment of,
    318–323
Free water surface wetlands design
    considerations
  BOD removal, 5, 269–270
  cell configuration, 299–301, 315
  heat transfer, 302–308
  hydraulics, 299–301
  kinetics, 308–312
  metals removal, 273–276
  nitrogen removal, 5, 269–272

organic loading, 315
organisms, 268
oxygen transfer, 265
pathogen removal, 275–277
phosphorus removal, 272–273
soils in, 267–268
solids removal, 5, 269, 271
temperature effects, 280–281,
    298–299, 302–308, 310–314
temperature reduction, 274, 276
trace organics removal, 274–277
Freezing sludges
  description of, 8, 443
  design examples of, 444, 449–450
  operation and maintenance of,
    449–450
  process requirements, 443–450
  system construction of, 448–449
Fungus in composting, 81

G

Gambusia affinis (Gambusia fish), 323
Gas transfer
  in emergent plants, 265
  in ponds, 16, 113–119
  in soils, 427–429
Giardia, 468
Gloyna model, 99, 101, 107–109,
    427–429
Glycol, 357
Grade
  land applied sludge systems and,
    23–24, 487
  land treatment and, 403, 412
  wetlands and, 299–301, 314–318,
    336–339, 358
Grass
  management of, 402
  selection of, 390
Graywater onsite systems, 495, 512,
    521
Groundwater, 43, 45, 51, 52, 53, 58
  depths for onsite systems, 494–497
  mounding, *see* Mounding,
    groundwater
  pathogen contamination of, 71, 76

pollutant travel, 46, 47, 48, 75
    protection of, 87, 89, 388
Guide to project development, 8–99
Gypsum, 92, 402

H

Habitat values in wetlands, 259–261,
    280–281
Half-life for organic compounds, 62,
    481–484
Harvest
    of aquatic plants, 87
    of reed bed plants, 8
    of terrestrial plants, 89, 402, 412
    of wetland plants, 88, 322
Heat transfer in wetland systems,
    302–308
Heavy metals, see Metals
Helminths, 71
Hormones, 419–420
Human hazard quotient, 468–470
Hyacinth ponds, 2, 3, 61, 73, 83, 84, 87,
    88, 89, 91, 248
Hydraulic conductivity, 90; see also
    Permeability
Hydraulic control in ponds, 200–201
Hydraulic design
    of FWS wetlands, 5, 299–301
    of SSF wetlands, 5, 335–339
Hydraulic gradient, 5, 52, 59, 70, 90,
    344
    Darcy's law, and, 5, 344
    wetlands, and, 34, 299–301, 336–338
Hydraulic loading
    land treatment, and, 6, 69, 76, 86,
        380, 390–392, 409, 422–423
    nitrogen limits, and, 5, 347, 393,
        422
    type 2 slow-rate land treatment, and,
        391–392
    wetlands, and, 309, 312, 347
Hydraulic residence time (HRT)
    in ponds, 3, 100–102, 104–106,
        118–124
    in wetlands, 5, 84, 298–299,
        310–315, 343–352, 358

Hydrogen peroxide, use of in onsite
    systems, 522
Hydrograph-controlled release (HCR)
    pond, 245–246

I

Ice cover
    on soil aquifer treatment, 426
    on wetlands, 302–307
Imhoff tanks, 297
Industrial wastewaters, 9, 91, 357, 369,
    394, 427–431
Infiltration, 33–39, 44, 46, 51, 54, 55, 60
    rapid, in soils, 47, 48, 49, 61, 68, 86,
        88, 89; see also Rapid infiltration
Infiltration rates
    slow-rate systems, 390–392
    soil–aquifer treatment systems,
        422–424
Infiltration testing
    with air entry permeameter, 39 , 422
    with flooding basin, 36–39, 422
Infiltrometers, 38–39, 422
Inlet structures, 316–318
Insects, 323–324; see also Mosquito
    control
Intermittent sand filtration, 501–503
Ion exchange, 84, 509
Iron deficiency, in hyacinth systems, 91
Irrigation
    efficiency, 83, 84, 392
    in sludges, 83
    requirement, 392

K

Kentucky bluegrass, 390
Kinetics
    of facultative ponds, 96–109
    of nitrogen removal, 176–180,
        360–363
    of pathogen removal, 72–73
    of phosphorus removal, 65, 89–90
    of toxic organics removal, 62–69
    of wetlands, 346–352, 360–363,
        365, 368

partial mix pond model for,
109–113
plug-flow pond model for, 102–103

L

Lagoons, *see* Aerated ponds; Facultative
ponds; Hyacinth ponds
Land application of sludge, 2, 8, 81; *see
also* Sludge
concept selection, 464
nitrogen limits, 476–482
site selection, 13, 20–26
sludge characteristics, 437–442
Land application of sludge design
procedures
metal limits, 473–475
municipal sludges, 464–482
toxic sludges, 486, 488
Land requirements
estimates for planning, 13–19
for pond systems, 14–15
for sludge systems, 20–26
for wetland systems, 15–16, 346,
350–351, 361–362, 365
overland flow, 16–17, 410–411
slow rate, 17–18, 394–396
soil-aquifer treatment, 18, 423
Land treatment of wastewater
area estimates for planning, 16–18
climate and storage, 17–18, 391,
396–399, 411, 426
crop management, 402, 412
crop selection, 388–392
design examples, 395–399, 411, 425
design objectives, 379–380,
382–384
distribution techniques, 400, 412,
424
hydraulic loading, 390–392, 409,
422
land requirements, 16–18, 380,
394–395, 410, 423
nitrogen loading, 392–394, 422–423
organic loading, 394, 403, 423,
427–429
pathogen concern, 386, 407, 420

phosphorus removal, 406, 415, 417
potassium requirements, 90
preapplication treatment, 384–386,
407–408, 420
site selection, 19–26, 28–41, 403,
413
slow-rate, *see* Slow-rate land
treatment
surface runoff control, 401
Landfill leachate, 289–294, 357
Leachfields, 512–515
Leaching requirements, 390–392,
430–431
Lead, 81, 82, 83, 84, 85, 86, 91
Legumes, 390
Lemna (duckweed), 164–165, 248, 264,
280, 324
Lemna systems, 164–172
Limiting design factor (LDF), 43, 60,
61, 84, 88, 384, 409, 420–421
Liners
for sludges, 449, 451
for wetlands, 316
Livestock wastewaters, 288–290
Loading rates, 61
for land treatment systems, 61, 76,
390–394, 408–410, 420–423
for onsite systems, 499–501,
511–515
for ponds, 61, 97
for sludge systems, 464–482
for wetlands, 315, 347, 368–369

M

Macrophytes
emergent plants, 261–263
floating plants, 248, 264
Maintenance
of land treatment systems, 402, 412,
426
of wetland systems, 322–324, 373
Management of onsite systems,
522–523
Manning's equation, 300
Marais and Shaw design model,
107–109

Marshes; *see* Constructed wetlands; Free water surface wetlands; Subsurface flow wetlands

Media filtration for onsite treatment, 501–511

Membrane liners, *see* Liners

Membrane use in ponds, 198

Mercury, 83, 275

Metals, 46, 81
  content in sludges, 70
  limits for land application of sludges, 8, 61, 70, 472–475
  sulfide precipitation, 274

Metals removal
  from ponds, 82
  land treatment and, 84, 86, 415
  sludge systems and, 475
  wetland systems and, 84, 273–276, 345

Metric conversion factors, 529–532

Micronutrients, 43, 90, 291, 295

Microstrainers, 252

Mine drainage, 291–296

Mineralization of sludge nitrogen
  example of, 479–482
  rates of, 476–477

Minoa, New York wetlands, 357–360

Mixing in ponds, 110, 113, 117, 119

Monitoring
  of land treatment systems, 402, 412, 426
  of sludge systems, 455, 463, 466–467, 470, 473
  of wetland systems, 324, 373

Mosquito control, in wetlands, 323–324

Mound systems, 516–517

Mounding, groundwater, 51, 424–425
  description of, 51
  design example of, 56, 57
  design procedure for, 52

Municipal wastewater treatment in wetlands, 278–281, 356–360

N

Natural wetlands, 70, 259

Nickel, 81, 82, 84, 85, 275, 292, 294

Nitrates
  in groundwater, 392–393
  in wetlands, 269, 271, 278, 351–352, 369, 371
  removal of, 269, 271

Nitrification, 87
  hyacinth ponds and, 87, 88
  land treatment and, 88, 392–393, 405–406, 413–416
  of filter bed, 360–363
  ponds and, 87, 172–199
  wetlands and, 88, 271–272, 347–351

Nitrogen, 43, 86, 90
  land treatment systems, and, 61, 390, 392–395, 405–406, 413–415
  slow-rate land treatment, 392–395
  sludge systems, 476–482

Nitrogen removal, 87
  aquatic systems and, 3, 87
  hyacinth systems and, 88
  land treatment and, 7, 88, 392–393, 405–406, 413–415, 422–423
  onsite systems and, 499–511
  ponds and, 2, 3, 87, 172–181
  sludge systems and, 476–482
  wetlands and, 5, 88, 269, 271–272, 344, 347–352

O

Octanol–water partition coefficient, 67

Odor control, 41

Onsite disposal systems, 493–523

Onsite treatment systems, 501–522

Onsite wastewater management districts, 522–523

Operation
  of composting systems, 81
  of land treatment, 57, 59, 76, 79, 86, 402, 412, 426
  of sludge systems, 76
  of wetlands, 320–324, 373

Organic loading
  on aquatic systems, 5, 81
  on intermittent sand filters, 504

on land treatment, 394, 403, 413–415, 427–429
on ponds, 97, 102
on recirculating gravel filters, 504–508
on wetlands, 315
Organics, 4, 43, 47, 81
biodegradable, 60, 61, 62
on ponds, 3, 81
on sludge systems, 4, 81
on wetlands, 4, 81
removal, 65, 66
travel time in soils, 70–71
Outlet structures, 316–318, 355
Overland flow (OF), 5, 61, 62
application methods, 379–391
application period, 409
application rate of, 407–409
area estimate for planning, 16–17
description of, 379–391
design example, 411
design objectives, 402–403
design procedure for, 409–410
distribution system for, 412
land requirements for, 410
loading, 61, 409–410
management, 412
monitoring, 412
nitrogen removal in, 7, 405–406
performance expectations for, 7, 403–407
planning, 12, 16–17, 21
preapplication treatment for, 407–408
recycle of, 409–410
site selection for, 403
storage, 17, 410–411
volatilization, 62, 64, 65, 66
winter operation and storage, 411
Overland flow removal of
BOD, 7, 404
nitrogen, 7, 405–406
pathogens, 407
phosphorus, 7, 406
runoff collection, 412
suspended solids, 7, 403, 405
toxic organics, 406

Overland flow slope
construction of, 412
length and grade of, 408–409, 412
maintenance of, 412
Overland flow vegetation
function of, 412
management of, 412
selection of, 412
Oxygen balance, 427–429
Oxygen demand, 60, 87, 348–349, 427–429
Oxygen diffusion in soil, 427–429
Oxygen solubility in water, 537–538
Oxygen transfer, 61, 427–429

P

Parasites, 75–76, 468
Partial-mix aerated ponds, 72, 83, 109–123; see also Aerated ponds
Pathogens, removal of, 43, 71, 78
in land treatment, 71, 407, 420
in sludges and ponds, 71, 73, 76, 81
in wetlands, 71, 73, 275–278, 345
PCBs (polychlorinated biphenyls), 47, 69
Percolation, 44, 47, 55, 89, 391–393, 422, 497–498
Percolation tests for onsite systems, 498–499
Permeability, 33–35, 43
Darcy's law, and, 34, 336–338
subsurface, 33–37, 43, 44, 45, 51, 52, 53, 59, 70
test procedures for, 36–39
pH, 29, 31–32, 46, 87, 89, 90, 99, 134, 148, 151, 174, 176, 177, 178, 179, 180, 218, 223, 232, 234, 237, 238, 240, 248, 261, 262, 263, 264, 281, 282, 291, 292, 293, 294, 296, 297, 402, 429, 430, 439, 441, 451, 454, 456, 460, 463, 483, 503
Pharmaceutical compounds, 419
Phase isolation in ponds, 247
Phosphorus, 43, 86, 88

Phosphorus removal
 in land treatment, 7, 89, 406, 415,
  417
 in onsite systems, 511
 in ponds, 3, 202–203
 in wetlands, 89, 272–273, 345
Phragmites (common reed), 261, 262,
  263, 266–267, 279, 283–285, 345,
  348, 349, 354, 356, 358, 365, 367,
  451, 452, 454,
Physical properties of water, 535–536
Phytoremediation, 93, 426–427
Planning, 2, 9, 11–27, 93, 94
Plug-flow model in facultative ponds,
  102–103
Plugging of SSF wetlands, 374
Pollutants, movement of, 47
Polychlorinated biphenyls (PCBs), 47,
  69
Ponds, 43, 47, 60, 61, 62, 71, 72, 73, 83,
  84, 87, 88, 95–209; *see also*
  Aerated ponds; Facultative ponds;
  Hyacinth ponds
Pondweed, 264
Porosity, 35, 45, 46, 48, 49, 52, 70
Potassium, 43, 86, 90
Potato processing wastes, 289
Praxair system, 172–173
Preapplication treatment
 for land treatment, 384–388,
  407–408, 420
 for wetlands, 297, 353
Precipitation, 46, 59, 75, 82, 84, 89,
  391–392, 395–399
Pressure-dosed disposal, 515
Project development, 9, 10
Pseudomonas, 77, 80
Pure oxygen in ponds, 172–173

Q

Quality of effluent
 from land treatment, 7, 388,
  404–406, 413–420
 from ponds, 3
 from wetlands, 268–278, 315,
  344–345, 348, 360, 369, 371

R

Rapid infiltration (RI), 46, 48, 49, 61,
  86, 88, 89; *see also* Soil aquifer
  treatment
Reciprocating dosing wetlands, 360
Recirculating gravel filters, 502–508
Reclamation with sludge, 8, 10
Reed bed systems, sludge drying and, 8,
  76; *see also* Subsurface flow
  wetlands
Reed canary grass, 390
Reeds, 263, 266–267, 283–284, 346,
  349, 356, 358, 365, 367; *see also*
  Phragmites
Regulations for onsite systems,
  497–499
Retardation of organics, 46, 47, 48, 49,
  50
Retrofitting failed onsite systems, 522
Rock filters, 227–236
Root zone, 49, 88
Runoff, 69, 75, 76
 control of, 401
 in overland flow, 412
Rushes, *see* Bulrushes

S

Sacramento County wetlands, 270–276,
  279, 324, 327–329
Salinity, 386–388, 430–431
Salmonella, 76, 468
Sand filtration, 230–231, 238
Scirpus, *see* Bulrushes
Sealing, of wetland systems, 316
Sedges, 263
Seepage velocity, 46, 48, 52
Septic tank effluent screens, 515, 522
Septic tanks, 512–513
Site assessment for onsite systems,
  494–499
Site evaluation, 12–13, 28–40
Site identification, 19–25
Site selection, 41
Slope, 52, 59, 63, 64, 66, 69, 75; *see
also* Hydraulic gradient

of wetland systems, 299–301,
    314–318, 335–339
overland flow, 66, 379–380, 408–410
site investigation of, 20–25
Slow-rate (SR) land treatment
    application methods, 380–381, 386,
        400
    application rate of, 390–392
    application scheduling, 401
    area estimate for planning, 17–18
    climate influences in, 187
    crop selection, 388–390
    description of, 379, 384
    design example of, 395–399
    design objectives, 384
    design procedure for, 388–392
    distribution system for, 400
    land requirements for, 394–396
    management, 402
    monitoring, 402
    nitrogen removal in, 7, 390–393
    performance expectations for, 7
    planning, 17–18, 20–22
    preapplication treatment for,
        384–388
    runoff control, 401
    site selection for, 12, 17–18, 41
    underdrainage in, 401
    winter operation and storage,
        396–400
Slow-rate land treatment loading
    hydraulic, 390–392
    nitrogen, 392–393
    organic, 7, 61, 394, 427–429
Slow-rate land treatment removal of
    BOD, 7
    nitrogen, 7, 88
    phosphorus, 7
    suspended solids, 7
Slow-rate land treatment vegetation
    function of, 388–390
    management of, 402
    selection of, 388–390
Sludge, 69, 71, 76, 77, 78, 81, 82, 83,
        84, 86, 88
    characteristics of, 43, 437–442
    composting of, 459–464

design examples for, 444, 449–450,
    455, 457–458, 463–464,
    479–482, 486–487
dewatering and, 8, 442–459
for reed drying beds, 450–455,
    458–459
loading rates and, 472–482
metal limitations for, 473–475
nitrogen limitations for, 476–482
pathogens and, 76
performance expectations for,
    437–487
petroleum sludges, 486, 488
phosphorus limits for, 475–476
site selection for, 13, 19–26
toxic, 488
vermistabilization and, 456–459
Sludge design procedures
    for composting, 459–464
    for freezing, 443–450
    for land application of municipal
        sludges, 8, 464–487
Sodium, 90, 91, 92, 386–387, 522
Sodium adsorption ratio (SAR), 91, 92,
    387
Soil, 43, 44, 45, 46, 51, 53, 55, 59, 61,
    62, 68, 70, 75, 82, 84, 86, 88, 90, 91
    absorption systems, 65, 511–522
    borings, 29–30
    chemical characteristics of, 31–32
    infiltration capacity, 33–36
    physical characteristics of, 29–30,
        33–40
    permeability, 33–40
    porosity, 29–40
Soil aquifer treatment
    ANAMMOX process, 414
    application methods, 424
    application rate, 422
    area estimate for planning, 18
    basin construction, 425
    description of, 382–383, 413
    design example of, 425
    design procedure for, 420–421
    distribution system for, 424
    groundwater mounding, 51,
        424–425

land requirements for, 423
management, 426
nitrogen removal in, 413–416
performance expectations for,
    413–420
phosphorus removal, 415, 417
planning, 12, 18–22, 26–41
preapplication treatment for, 420
site selection for, 413
winter operation, 426
Soil aquifer treatment and removal of
    BOD, 414
nitrogen, 413–415
pathogens, 420
phosphorus, 415, 417
suspended solids, 413
toxic organics, 415, 418–420
Soil aquifer treatment loading
hydraulic, 422
nitrogen, 422
organic, 423
solids, 413
Solar radiation, 341, 343, 399
Specific retention, 35–36
Specific surface area, 359–361
Specific yield, 35–36, 53, 56
Spray irrigation, 381, 400, 521
Sprinkler application for irrigation, 381
Sprinkler infiltrometers, 38
Stabilization ponds, 72; see also
    Aerated ponds; Facultative ponds;
    Hyacinth ponds
Stable organics, see also Priority
    pollutants
in sludges, 469–470, 483–488
removal of, 406, 415–420
travel time in soils, 70–71
Storage ponds
for land treatment, 17–18, 396–399
sizing, 396–399
Stormwater wetlands, 282–283
Submerged plants, 264
Subsurface flow (SSF) wetlands
advantages of, 260, 335
batch flow, 355
construction of, 370–373
costs, 373–374

depth of gravel and water, 353
description of, 335–336
design of, 353–364
hydraulic design, 335–339
management, 373–374
nitrogen removal in, 347–352
onsite systems, 364–366, 517–518
oxygen transfer, 349
performance expectations for,
    343–353
preapplication treatment for, 353
reciprocating dosing, 356
site selection for, 12, 16
slope, construction of, 337, 353
thermal aspects, 339–343
winter operations, 339–343,
    354–355
Subsurface flow wetlands and removal
    of
BOD, 344, 346
metals, 345
nitrogen, 5, 347–352
pathogens, 5, 345
phosphorus, 345
suspended solids, 5, 344, 347
Subsurface flow wetlands loading
hydraulic, 347, 355, 358
organic, 346, 348
solids, 347–348
Subsurface flow wetlands vegetation in
function of, 348–349
management of, 372–373
selection of, 348, 353–354
Sulfides, 274, 358–360
Suspended solids removal
in land treatment, 7, 403–405, 413
in ponds, 2, 3, 143–153, 156,
    162–165, 171
in wetlands, 5, 269, 271, 344, 347,
    360
Swine waste, 289–290

T

Temperature, 96, 97, 107, 126, 136,
    138, 139, 146, 147, 151, 155,
    174–180, 182, 187–189, 201

land treatment and, 396–399, 411, 426

ponds and, 99, 101–105, 111–122, 128–131, 133, 140, 150–155, 175, 177, 179–180, 190–191

reduction in wetlands, 274–276

wetland systems and, 275–276, 280–281, 298–299, 302–308, 310–314, 339–343, 346–347, 350–352, 362, 368

Test pits, 29–30

Thermal conductance, 302–307, 341–342

Thermal conductivity, 306, 341

Thermal models, 302–308

Thirumurthi design model, 103–109

Tidal vertical wetlands, 369, 371

Toluene, 47, 64, 65, 66, 68, 69

Tomato processing wastes, 394

Toxic wastes, *see* Priority pollutants

Tracer studies, 46, 329

Transmissivity, 45

Transpiration, *see* Evapotranspiration

Travel time in soil, 70–71

Trickling filter, 505–511

Troubleshooting in wetlands, 328, 372

Typha, 61; *see also* Cattails

U

Ultrafiltration membrane use in ponds, 198

Underdrains, spacing design of, 58, 401

V

Vegetation, function of, 43, 60
in land treatment systems, 390–392, 412
in sludge systems, 450–452, 475–482
in wetland systems, 261–267, 348–349, 353–354, 372–373

Vegetation, management of
in land treatment, 75, 402, 412

in sludge systems, 75, 452–454, 475–478
in wetlands, 75, 320, 323–324, 373

Vegetation, selection of
in land treatment systems, 24, 388–390, 412
in sludge systems, 450–451, 475–478
in wetlands, 318–320, 353–354

Vermistabilization, 76, 456–459

Vertical flow wetlands beds, 367–370

Virus removal
in land treatment, 71, 73
in ponds, 71, 73
in sludge systems, 71
in wetlands, 71, 277

Volatile organics, removal of, 62, 63, 65, 66, 274–277

Volatilization, 46, 47, 62, 63, 65, 87, 88
of ammonia, 87
of organics, 66, 70

W

Water hyacinths, 83, 89, 248

Water rights, 27

Wehner–Wilhelm model, 103–106

Wetlands, 43, 60, 62, 70, 71, 73, 74, 84, 88; *see also* Constructed wetlands; Free water surface wetlands; Subsurface flow wetlands

Winery wastes, 369

Winter operation
of FWS wetlands, 302–308, 317, 320–321
of land treatment, 396–399, 411–412, 426
of ponds, 97–115
of SSF wetlands, 339–343, 357–360
on sludge systems, 443–450, 453–454, 461–464, 488

Z

Zeolite, 509

Zinc, 81, 82, 84, 85, 90, 275–276, 345